D0007717

Avec les portes isothermes

POLYKO

vous réduirez vos échanges
de chaleur

★ *Portes en Polyester armé*
    (moulé)
★ *Automatiques*
★ *Semi-automatiques*
★ *Modèle coulissant*

POLYKO S. A.   87, Eymoutiers — Tél. 82

Les Hydrocarbures Chlorofluorés
sont les plus utilisés des

FLUIDES
FRIGORIGÈNES

parce qu'ils sont
très stables,
ininflammables et inexplosibles
non toxiques.

Ugine Kuhlmann fabrique
la Gamme la plus complète
d'Hydrocarbures Chlorofluorés

La marque

est une garantie de
très haute pureté

C.P.D.I. de Swarte

Froid :

Conditionnement d'air	R 11	**R 22**	R 113	R 114
Froid domestique et industriel	**R 12**	R 114		
Conservation à basse température	**R 12**	R 22		
Congélation	**R 22**	R 115	**R 502**	
Basses températures	R 22	R 13B1		
Très basses températures	R 13	R 23		

Isolation :
Les mousses rigides de polyuréthane expansées au
FORANE 11 sont actuellement le matériau isolant
industriel le plus efficace :
Indice de conductibilité thermique
en kcal/mh °C 0,016 à 0,021.

Société des Usines Chimiques
UGINE KUHLMANN
DIRECTION COMMERCIALE :
25, BD DE L'AMIRAL BRUIX ● 75 PARIS 16 ● TEL. : 525.52.00
SERVICE TECHNIQUE DES VENTES ET APPLICATIONS (S.T.V.A.) :
3, R. COLLANGE ● 92 LEVALLOIS-PERRET ● TEL. : 270.03.10

F comme Froid

Froid
COMEF ❄

Compresseurs ouverts
type CB
Faut-il le rappeler,
des centaines de milliers
de compresseurs sont
en service dans le monde entier

COMEF 29, Rue des Tuilliers - 69 LYON 8ᵉ FRANCE - Télex 30412 - Tél. (78) 69.44.56 +

Avec le gaz, les moteurs équipant les installations de production de froid fonctionnent dans des conditions avantageuses et permettent de réaliser d'importantes économies d'exploitation. Grâce à la récupération de chaleur le rendement global des équipements Gaz est très élevé : ils produisent 1 kWh pour le prix d'1 thermie. Il existe en France un matériel sûr et robuste de longue durée de vie et de haute performance. Les moteurs à gaz entraînent les alternateurs ou les compresseurs fournissant l'énergie de l'établissement. Des échangeurs, placés sur le circuit de refroidissement des moteurs et sur les gaz d'échappement, produisent de l'eau chaude ou de la vapeur alimentant par exemple un groupe à absorption.

Renseignez-vous auprès des Centres de Distribution du GAZ de FRANCE et demandez notre documentation à la

Direction des Services Economiques et Commerciaux du GAZ de FRANCE 23, r. Philibert-Delorme, PARIS 17ᵉ

DES
HOMMES
ET
DES
TECHNIQUES
A VOTRE
SERVICE

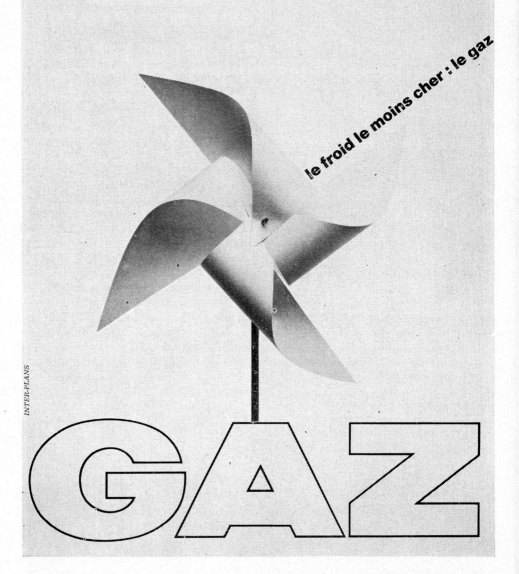

le froid le moins cher : le gaz

INTER-PLANS

POUR TOUS VOS PROBLÈMES

D'ISOLATION

THERMIQUE — PHONIQUE — ACOUSTIQUE

•

+ 1010 °C

Machines tous navires

Transports de bitume

Paquebots

Polythermes

Congélateurs

Méthaniers

- 260 °C

•

Fibre de verre ISOVER(D) - Verre cellulaire FOAMGLAS(R)
Polystyrène extrudé STYROFOAM(R)

SAINT-GOBAIN

DIVISION MATÉRIAUX ISOLANTS — SERVICE TRANSPORTS :
62, boulevard Victor-Hugo
92 - NEUILLY - S/SEINE — Tél. 637 10 00 — Poste 4080

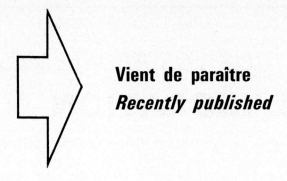

Vient de paraître
Recently published

Cryoélectrotechnique
Low temperature and electric power

Annexe 1969-1 au Bulletin de l'Institut International du Froid, 450 pages (60 F - US $ 11.00)

Gaz naturels liquéfiés
Liquefied natural gas

Annexe 1969-2 au Bulletin de l'Institut International du Froid, 720 pages (42 F - US $ 8.00)

Développements récents en lyophilisation

Tome 1 : Aspects thermodynamiques (transfert de chaleur et de masse)

Recent developments in freeze drying

Tome 1 : Thermodynamic aspects (heat and mass transfer)

Annexe 1969-4 au Bulletin de l'Institut International du Froid, 209 pages (35 F - US $ 6.50)

REFRIGERATION SCIENCE AND TECHNOLOGY
SCIENCE ET TECHNIQUE DU FROID

Heat transfer

Non-stationary heat transfer through walls
Measurement of thermal conductivity
Heat transfer with two phase refrigerants

Current applications of air conditioning

Testing air treatment equipment
Topical problems and special applications
of air conditioning

Échanges thermiques

Transmission de chaleur en régime variable
dans les parois
Mesures de conductivité thermique
Échanges thermiques avec les fluides frigorigènes biphasés

Actualités en conditionnement d'air

Essais des équipements de traitement d'air
Problèmes d'actualité et applications spéciales
du conditionnement d'air

PERGAMON PRESS

OXFORD · NEW YORK · TORONTO
SYDNEY · BRAUNSCHWEIG

399363

Mitchell Memorial Library
Mississippi State University

Pergamon Press Ltd., Headington Hill Hall, Oxford

Pergamon Press Inc., Maxwell House, Fairview Park, Elmsford,
New York 10523

Pergamon of Canada Ltd., 207 Queen's Quay West, Toronto 1

Pergamon Press (Aust.) Pty. Ltd., 19a Boundary Street,
Rushcutters Bay, N.S.W. 2011, Australia

Vieweg & Sohn GmbH, Burgplatz 1, Braunschweig

Copyright © 1971 Pergamon Press Ltd and
The International Institute of Refrigeration

*All Rights Reserved. No part of this publication may be
reproduced, stored in a retrieval system, or transmitted, in any
form or by any means, electronic, mechanical, photocopying,
recording or otherwise, without the prior permission of
Pergamon Press Ltd.*

First edition 1971

Library of Congress Catalog Card No. 72–157655

Printed in Belgium by Ceutrick

08 016597 4

TABLE DES MATIÈRES — CONTENTS

COMMUNICATIONS

Section 1

Généralités	Generalities

Section 2

Transmission de chaleur en régime variable dans les parois Étude des enceintes confinées Détermination des charges calorifiques	Non-stationary heat transfer through walls Study of confined enclosures Calculation of cooling loads

3

Section 3

Mesures de conductivité thermique   **Measurement of thermal conductivity**

Section 4

| Échanges thermiques avec les fluides frigorigènes biphasés | Heat transfer with two-phase refrigerants |

Section 5

Essais des équipements de traitement d'air   Testing air treatment equipment

7

Section 6

AVANT-PROPOS

FOREWORD

MEETING OF COMMISSIONS II AND VI OF THE I.I.R.
LIÈGE (BELGIUM)

SEPTEMBER 9-11, 1969

Commissions II and VI of the I.I.R. met in Liege (Belgium) between September 9 and 11, 1969. Commission II held its meeting under the presidency of Mr. MATTAROLO (Italy), assisted by Miss GRIFFITH (United Kingdom) and Mr. NESSELMANN (Germany), Vice-Presidents, and Messrs HALASZ (Hungary) and MYNCKE (Belgium), Secretaries; Commissions VI met under the presidency of Mr. BURNAY (Belgium), assisted by Messrs BOEKE (the Netherlands) and TIREL (France), Vice-Presidents and Mr. LASSERRE, Secretary. Also participating in these meetings were Messrs LORENTZEN, President of the Scientific Council, KAYAN, past President of Commission II, YATE-PITTS, past Vice-President of this Commission, EIDSVIK, Vice-President of the Executive Committee, and ANQUEZ, Deputy-Director of the I.I.R.

About 125 participants were at these meetings (plus about 20 accompanying members), from 19 countries. 60 papers were read during the three-day meeting. The following items were studied in joint sessions:

— non-stationary heat transfer through walls;

— study of confined rooms or enclosures;

— calculation of cooling loads.

In individual meetings the two Commissions studied the following items:

Commission II:

— heat transfer with two-phase refrigerants;

— measurement of thermal conductivity and water vapour permeability of insulating materials.

Commission VI:

— tests on air handling equipment (room air-conditioners, induction or fan coil air-conditioners);

— present problems and special application of air-conditioning.

Also, a panel discussion on methods of measuring thermal conductivity was presided over by Miss GRIFFITH in order to further the studies by a special working party which published its conclusions in 1967.

Participants visited the laboratory of CEDRIC at Liège University, directed by Prof. BURNAY. A reception was offered by the organisers in the restaurant of the new Campus of the University of Sart Tilman, a few kilometres from Liege.

This very successful meeting was organised by Prof. BURNAY and his assistants at CEDRIC and we would convey our heart thanks and congratulations for the excellent work accomplished.

RÉUNIONS DES COMMISSIONS II ET VI DE L'I.I.F.
LIÈGE (BELGIQUE)

9-11 SEPTEMBRE 1969

Les Commissions II et VI de l'I.I.F. se sont réunies à Liège (Belgique) du 9 au 11 septembre 1969. La Commission II a tenu sa réunion sous la présidence de M. MATTAROLO (Italie), assisté de Miss GRIFFITH (Royaume-Uni), et de M. NESSELMANN (Allemagne), Vice-Présidents, et de MM. HALASZ (Hongrie) et MYNCKE (Belgique), Secrétaires; la Commission VI sous la présidence de M. BURNAY (Belgique), assisté de MM. BOEKE (Pays-Bas) et TIREL (France), Vice-Présidents et de M. LASSERRE, Secrétaire. Participaient en outre à ces réunions MM. LORENTZEN, Président du Conseil Scientifique, KAYAN, ancien Président de la Commission II, YATE-PITTS, ancien Vice-Président de cette même Commission, EIDSVIK, Vice-Président du Comité Exécutif, ANQUEZ, Directeur adjoint de l'I.I.F.

Environ 125 personnes ont participé à ces réunions (auxquelles il faut ajouter une vingtaine de membres accompagnateurs), venant de 19 pays. 60 rapports ont été présentés au cours de ces trois journées de travail. Les thèmes suivants ont été étudiés en séance commune :

— transmission de chaleur en régime variable dans les parois;
— études des enceintes confinées;
— détermination des charges frigorifiques.

En séances séparées, les deux Commissions ont étudié les sujets suivants :

Commission II :

— échanges thermiques avec les fluides frigorigènes biphasés;
— mesures de conductivité thermique et de perméabilité à la vapeur d'eau des matériaux isolants.

Commission VI :

— essais des équipements de traitement d'air (conditionneurs autonomes, éjecto-convecteurs, ventilo-convecteurs);
— problèmes d'actualité et applications spéciales du conditionnement d'air.

En outre, une discussion orientée a été présidée par Miss GRIFFITH sur le thème des méthodes de mesure de conductivité thermique, afin de poursuivre les études d'un groupe de travail spécialisé qui a déposé ses conclusions en 1967.

Les participants ont visité le laboratoire du CEDRIC, de l'Université de Liège, dirigé par le Professeur BURNAY. Une réception a été offerte par les organisateurs de la réunion au restaurant du nouveau Campus universitaire du Sart Tilman, à quelques kilomètres de Liège.

L'organisation de ces réunions très réussies était assurée par le Professeur BURNAY et ses assistants du CEDRIC. Ils doivent être chaleureusement remerciés et félicités pour l'excellence du travail qu'ils ont accompli.

LISTE DES PARTICIPANTS   LIST OF PARTICIPANTS

ALLEMAGNE (Ouest-Rép. Féd.) — GERMANY (West-Fed. Rep.)

AURACHER H., *Stuttgart*
CHAWLA J. M., *Karlsruhe*
GENTHNER K., *Berlin*
GERTIS K., *Holzkirchen*
GLASER H., *Stuttgart*

GUTTINGER M., *Stuttgart*
NESSELMANN K., *Bad Dürrheim*
RAKOCZY T., *Köln*
SCHNEIDER H. W., *Stuttgart*
WALLNER R., *Stuttgart*

ALLEMAGNE (Est) — GERMANY (East)

DÖLZ H., *Dresde*
JUNGNICKEL H., *Dresde*

SCHEUNEMANN K. H., *Dresde*

BELGIQUE — BELGIUM

VAN BEVEREN, *Bruxelles*
BILLEN R., *Alleur*
BURNAY G., *Liège*
DANS P., *Bruxelles*
DECOSTER M., *Bruxelles*
DELVAUX L., *Liège*
DUMONT A., *Bruxelles*
FAFCHAMPS J. N., *Embourg*
FLAMAND Ch., *Bruxelles*
HANNAY J., *Liège*
JACQUEMIN A., *Waret l'Evêque*
LAUREYS E., *Bruxelles*
LEBRUN J., *Liège*
LENDERS A. Jr., *Schoten*

LEPELEIRE G. de, *Heverlee*
LETIST, *Bruxelles*
LOOZE de, *Anvers*
MARCQ H., *Bruxelles*
MARÉCHAL G., *Bruxelles*
MEUNIER D., *Heverlee*
MYNCKE H., *Louvain*
NUSGENS P., *Liège*
OPHOFF P., *Schoten*
PORTIER J. J., *Liège*
SNEESSENS H., *Liège*
UYTTENBROEK J., *Bruxelles*
VANDENBULCKE G., *Bruxelles*
WHITLEY G. H., *Schoten*

CANADA

TAMBLYN R. T., *Toronto*

DANEMARK — DENMARK

HUELLE Z. R., *Nordborg*
KORSGAARD V., *Lyngby*

LUND MADSEN Th., *Lyngby*

ÉTATS-UNIS — U.S.A.

KAYAN C. F., *New York*

TYE R. P., *Cambridge*

FRANCE

BOURGERIE Ph., *Paris*
BOURRIN L., *Paris*

CAMIA F. M., *Marseille*
CLAIN F., *St. Rémy-les-Chevreuse*

DOUSSAIN R., *Paris*
FOURNIER D., *Rantigny*
FRUY G., *Chatou*
JUDET DE LA COMBE A., *Lille*
KLARSFELD S., *Rantigny*
LASSERRE P., *Paris*
LAVIGNE P., *Grenoble*

LEFAURE L., *Paris*
LOTZ R., *St. Étienne du Rouvray*
MARÉCHAL J.C., *Paris*
MICHEL H., *Écuelles*
OLIVET J., *Le Raincy*
TIREL J.J., *Carquefou*
VETZEL P., *Paris*

HONGRIE — HUNGARY

GAÁL F., *Budapest*
HALÁSZ L., *Budapest*
KESZY-HARMATH, *Budapest*

KOVÁCS G., *Budapest*
VÁRJON D., *Budapest*

ISRAEL

LANDSBERG R., *Haïfa*
ROTHSCHILD J., *Tel-Aviv*

SCHAUDINISCHKY L., *Haïfa*
SCHWARTZ A., *Haïfa*

ITALIE — ITALY

BERTOLINI G., *Milan*
BETTANINI E., *Padoue*
BONAUGURI E., *Milan*
CAVALLINI A.C., *Padoue*
DI FILIPPO P., *Padoue*
MATTAROLO F., *Padoue*
MAGRINI U., *Gênes*

DE PONTE F., *Padoue*
REALE G., *Gênes*
SACCHI A., *Turin*
SAGGESE G., *Turin*
TUBERGA, *Turin*
ZECCHIN R. F., *Padoue*

JAPON — JAPAN

WATANABE T., *Tokyo*

NORVÈGE — NORWAY

BRENDENG E., *Trondheim*
EIDSVIK O., *Bergen*
FRIVIK P.E., *Trondheim*

LORENTZEN G., *Trondheim*
TVEIT A., *Trondheim*

PAYS-BAS — THE NETHERLANDS

BOEKE A.W., *Delft*
BOUWMAN H.B., *Amersfoort*
CAUBERG J.J.M., *Valkenburg*
ENGELS M.A., *Nijmegen*
EUSER P., *Delft*
LEIJENDECKERS P.H.H., *Milsbeek*

LIEM K.G., *Amersfoort*
MULDER J., *Waalre*
POSTHUMA S.A., *Amsterdam*
VAN VELDHOVEN L.J.A., s'*Hertogen-bosch*
VOS B.H., *Delft*

POLOGNE — POLAND

STANISZEWSKI B.E., *Varsovie*

SUSKI J.F., *Varsovie*

13

ROYAUME-UNI — UNITED KINGDOM

BAKER G.J., *Bracknell*
COPCUTT D.P.T., *Egham*
EMERSON W.H., *Glasgow*
GRIFFITH M.V., *Leatherhead*

HURD R., *Brackley*
MOLNAR W., *Edmonton*
SCOTT D.R., *London*
YATE PITTS G., *Stockport*

SUÈDE — SWEDEN

PALJAK I., *Stockholm*

SUISSE — SWITZERLAND

EMBLIK E., *Winterthur*
SAINT GIRONS G., *Blonay*

SCHUCHARDT H., *Berne*

TCHÉCOSLOVAQUIE — CZECHOSLOVAKIA

CHYSKY J., *Prague*
PROUSEK Z., *Prague*

SEDLÁK P., *Prague*

YOUGOSLAVIE — YUGOSLAVIA

JUHART A., *Skofja Loka*

TODOROVIĆ B., *Belgrade*

I.I.F. — I.I.R.

ANQUEZ Michel, *Directeur Adjoint*

14

SECTION I

GÉNÉRALITÉS

GENERALITIES

THE QUEST FOR ULTIMATE RESOLUTION
OF THE UNIT-SYSTEM CONTROVERSY IN U.S.A.

C. F. KAYAN

Columbia University, New York (U.S.A.)

Recherche d'une solution finale à la controverse sur les systèmes d'unités aux U.S.A.

RÉSUMÉ : *Depuis longtemps, le problème de l'adoption du système métrique aux États-Unis a été étudié sur une base réaliste dans des milieux officiels. En août 1968, le Congrès a voté un texte autorisant le « National Bureau of Standards » à entreprendre une étude approfondie sur le pour et le contre de cette question. Les avantages paraissent attrayants : le vieux problème du coût entraîné par le changement est apparu à beaucoup comme effrayant. On attend avec beaucoup d'intérêt la solution du problème en Angleterre et l'importance de l'arithmétique décimale a trouvé un regain d'intérêt. On passe rapidement en revue les attitudes et les circonstances actuelles.*

Imagine a conversation regarding possible changeover to metric between an engineer and a business man in the USA: The business man: "I see that England is finally changing over to decimal coinage, after long usage of the Pound Sterling, with its 20 shillings to the pound and 12 pence to the shilling, 240 pence all told". The engineer: "Oh yes, it will be fine; they are going to decimalize with 100 pennies to the pound... same arrangement as in most other countries. Then the plan is to metricate... to introduce a metric system like so many other countries. My friends on the Continent seem most enthusiastic about it, although it will be something of a problem to make the change from the English system of units, the pound avoirdupois, you know, and the foot, etc."

"Oh yes", replies the business man, "This is intriguing, using the meter, the kilogram and so forth, along with the decimal system. Great! The kilogram has 1000 grams and the meter 1000 millimeters, and then of course we could have the metric clock — 24 hours to the day, just like they have in the foreign timetables. But wait, why not decimal time? We would have a complete system: the ton of 1000 kilograms—1 000 000 grams, the kilometer with 1000 meters—1 000 000 millimeters, and then the day 1 000 000 seconds or something. That really would be A System!" "Yes", says the engineer, "That has been suggested, only we couldn't really call it the 'Second'. The 'bing' has been suggested, with 100 'bings' to the 'bang', 100 'bangs' to the 'boom' and 100 'booms' to the day. Then we would have one million bings to the day, or the bing would be one micro-day". "Well", says the business man, "Why not?" I am going over to Switzerland next week on business, they are the great clock people, I'll ask them about it. If we are going to drastically change our own system, why not make it complete! I'll tell you all about it when I get back."

A few weeks later the business man returned, and lunched with his friend. "They tell me over there I'm crazy. The present system of time, no matter how awkward, must stay... They couldn't afford to make such a monumental change; everything they have depends on the present unit of time—the cost of change would be astronomical! Why wasn't it done when the metric system was first instituted? This might then have been possible, instead of the compromise, but now... IMPOSSIBLE! Forget it."

Yes, indeed, it would be COLOSSAL ! Changes would start with re-evaluation of acceleration of gravity, on through a monstrous array of practically all technical data, as well as all time-dependent instruments, etc. Yes, Tremendous: but the decision to retain old-fashioned non-decimal time was an understandable compromise with practicality and tradition, plus Costs. Perhaps such changeover wasn't even considered.

In 1866, Congress of the United States enacted:"It shall be lawful throughout the United States of America to employ the weights and measures of the metric system."... Then, in 1893 an Executive Order directed the "Office of Weights and Measures (now the National Bureau of Standards)... will in the future regard the international prototype meter and kilogram as fundamental standards, and the customary units, the yard and the pound, will be derived therefrom in accordance with the Act of July 28, 1866." Thus the legality of the use of metric values has been long established; it is now the practical question of reality of international usage that has kindled vigorous fires of unit-system reform.

Perhaps an important "milestone" was a symposium staged at the 125th Annual Meeting of the American Association for the Advancement of Science (AAAS) in 1958 at Washington. 28 different papers on "National and International Aspects of Systems of Units" were contributed by as many worldwide authorities. This Symposium punctuated growing interest in fuller international use of a metric system of units. The papers covered many kaleidoscopic facets, and have been published by AAAS [1].

The Washington program was continued in a second AAAS Symposium in Chicago in 1959. "Operation SCUDS—Simplification, Clarification, Unification, Decimalization, Standardization" served further to focus attention on the problems and to establish some guidelines for action.

Since then there has been progress at an increasing rate. Many technical, business and government people (exlusive of the general public) are concerned about the unit-system confusion and chaos. Some call for mitigation of troublesome problems through modification of the existing American system. Others advocate a complete changeover to a common metric system. Many feel that progress has been needlessly thwarted by lack of recognition of the profound problems confronting us, and that a step-wise progressive program is in order: Evolution rather than Revolution. The critical question underlying all thoughts of ultimate conversion continues to be the awesome COSTS of a complete changeover, roughly estimated at many BILLIONS of dollars. Yes, economic and practical questions dictate many of our decisions in the World of Reality!

Nevertheless, there has been an impressive array of moves to further explore the problem; among these are:

In the July 1962 American Society of Mechanical Engineers (ASME) publication "Mechanical Engineering", the changeover problem was explored under the heading, "The Metric System—Should We Convert?" In a group of four papers [2, 3, 4, 5], ASME explored the possibility of conversion to an international system of units, noting that world-wide use of the inch-system was decreasing, and that competitive world- trade made this a matter of critical importance to all engineers.

In August 1962 the Executive Committee of the American Society for Testing and Materials (ASTM) Board of Directors authorized the appointment of an Ad Hoc committee "to study and advise the Board on (1) the use of the metric system in ASTM standards and publications, and (2) whether or not the Society should take a position on the adoption of the metric system of mensuration by the United States, and report to the Board of Directors, recommending a course of action for the Society."

In 1963 note was taken of the publication by the British Standards Institution (BSI) of the results of wide consultation with British industry, indicating that they were in favor of starting the change without delay and without waiting for change in the rest of the Commonwealth and the United States. The spirit of possible change had been fired up soon after the War, about 1951. (The ultimate action in undertaking decimalization of the Pound Sterling, and initiating national Metrication with completion by 1975 − roughly a 25-year span from hope to reality, is being watched in USA with great interest.)

The first edition of the "ASTM Metric Practice Guide" [6] was published in January 1964 to provide the technical committees of ASTM with conversion procedures and factors to implement the Society's policy of 1962. This publication was revised in December 1966, and represents a most-important present-day document. Noting that in some industries in USA and abroad, U.S. customary (and British) units are gradually being replaced by those of a modernized metric system known as Système International d'Unités (SI System), the Society recognized this trend and therefore considered it important to prepare for broader use of the modernized metric system through coexistence of these two major systems of units.

At the December 1964 Winter Annual Meeting, citing a spirit of "Decision on the Inch, and Favoring Evolution rather than Disastrous Revolution", ASME adopted the following resolution: "The American Society of Mechanical Engineers, in the interest of national economy and industrial efficiency, advocates the continued use of the existing American, British and Canadian sizes, modules, designs and ratings. Further, the Society is of the opinion that legislative action directed to the alternate system of dimensional standards, such as the metric, will be at this time confusing and disturbing to the productive capacity of the United States, and is not, therefore, in the best of public interest."

During this general period, a special committee of the American Standards Association (ASA), now United States of America Standards Institute (USASI) continued its activity to preserve and promote the use of the decimalized inch, presenting tables and preferred systems for the expression and use of decimalized systems. In effect, their document staunchly reaffirms the preferential use of the present US unit of length, the inch.

A special Society of Automotive Engineers (SAE) Report was issued in 1965: "Metric Equivalents of U.S. Conventional Units of Measure—SAE J916, SAE Recommended Practice", the report being designed to assist users of the SAE Handbook in conversions of dimensions and units of measure, between the U.S. conventional system and the metric system.

Since then, there has been further action in other spheres of interest:

In 1968 at the annual meeting of the American Society of Heating, Refrigerating and Air Conditioning Engineers (ASHRAE), a special committee on metric system parallel use in Society publications was established.

USASI established a Metric Advisory Committee to provide U.S. industry with a forum for consideration of metric problems. And at the 50th anniversary meeting of the Institute in Washington in December 1968, a full session was devoted to the topic "Why Adopt the Metric System?" [7]. Controversial views were presented by a distinguished group of international panelistics, including the director of BSI and a member of Congress.

ASME, also, in 1968 formed an Ad Hoc Metric Study Committee to consider and evaluate types of problems and the magnitudes of each, which might be encountered in the mechanical engineering field in the conversion of US engineering-oriented industry to the metric system, and to provide suggestions for solving or minimizing the problems.

And now, as of June 1969, the Engineers Joint Council (EJC)—an overall organization composed of representatives of some 15 different professional engineering societies of the Nation—has set up a special committee to broadly explore our metric problem.

So it may be seen, the metric study in USA represents real action.

But perhaps the most important action of all is that of the Congress of the United States: On August 9, 1968, the 90th Congress enacted Public Law 90-472—"An Act, to authorize the Secretary of Commerce to make a study to determine the advantages and disadvantages of increased use of the metric system in the United States." This study will be conducted by the National Bureau of Standards, "with a program of investigation, research and survey to determine the impact of increasing world-wide use of the metric system on the United States; to appraise the desirability of increasing the use of metric weights and measures in the United States; to study the feasibility of retaining and promoting by international use of dimensional and other engineering standards based on the customary units of the United States; and to evaluate the costs and benefits of alternative course of action which may be feasible for the United States."

Thus it may be seen, there is spirited activity in store for us in USA.

Many questions are being asked, such as:

Why not re-organize our teaching of grade-school arithmetic to further the use of decimal and decade values, to de-emphasize common-(vulgar) fraction arithmetic?

In preparation for possible future metric changeover, why not vigorously push decimalization of the inch, the pound ("bounce-the-ounce") and other units in popular use? The move in this direction, as noted, is already strong in industrial circles, and well under way. The decimal system is also the cornerstone of the metric system.

Why not decimalize the stock-market (bourse) quotations and abandon the common fractions of the dollar (eighths, sixteenths, etc.), and thus promote decimal arithmetic?

What will be the impact ultimately of metric use in USA on our electric power picture; 50-cycle alternating current is in wide use in metric countries; will we be asked to abandon our 60 cycles to strengthen our foreign equipment position?

How truly is our international trade dependent on the adoption of the international metric system? A recent Department of Commerce survey [8] on our trade abroad does not mention any disadvantage due to our unit-system.

With teacher re-education important, can this new look in unit-systems be successfully promulgated?

Can public re-education be made successful? It has been estimated that probably three-quarters of the adult population in the USA do not know what the metric system is, much less how to use it.

Why not develop the international operations in the metric system as the first step, leaving the domestic arrangements as now, i.e., encourage the idea of co-existence of the two systems for immediate use?

The spirit of co-existence on units seems strong (as with languages), inspired by the horror of tremendous costs in codes, laws, standards, tooling, machines, habits and publications.

The Torch of Evolution now seems lighted, and the prospect of correction of our chaos and confusion appears now much brighter!

REFERENCES

[1] C. F. KAYAN, Editor, "Systems of Units", *AAAS Symposium*, Volume 57.
[2] J. E. KJEMTRUP, "All Out—Now", *ASME Mechanical Engineering* (July 1962).
[3] R. P. TROWBRIDGE, "Let Industry Decide". *ASME Mechanical Engineering* (July 1962).
[4] W. G. WALTERMIRE, "Decimalize the Inch". *ASME Mechanical Engineering* (July 1962).
[5] C. F. KAYAN, "The Transition". *ASME Mechanical Engineering* (July 1962).
[6] Ad Hoc Committee on Metric Practice, "ASTM Metric Practice Guide", *ASTM*, January 1964, First Edition; Second Edition (December 1966).
[7] USASI Metric Panel, "Metrication for the United States". *ASME Mechanical Engineering* (May 1969); Panelists as follows:
   CLAIBORNE PELL, "Congress Needs Answers", p. 12;
   H. E. CHESEBROUGH, "Is It Really Necessary?", p. 14;
   H. A. R. BINNEY, "How Britain is Doing It", p. 16;
   J. H. JENKINS, "The Canadian View", p. 19.
      Addendum to above:
   Louis POLK, "What are the Options", p. 22, *Mechanical Engineering* (May 1969).
[8] L. A. FOX, Bureau of International Commerce, "U.S. Foreign Trade : A Five-Year Outlook, with Recommendations for Action", *U.S. Dept. of Commerce* (April 1969); U.S. Government Printing Office.

DISCUSSION

G. LORENTZEN (Norway) — The question of the SI unit system is a rather serious one and deserves really to be taken seriously. Its introduction as a common system for all the world will simplify greatly the exchange of data and scientific publications. Every international institution should do its best to push the development in this direction. The difficulties are certainly great but should not be impossible to overcome. It is obviously necessary to start with university books, with the education of the teachers. Much of the problem consists in overcoming the natural human laziness; in reality the change in a system is a fairly simple operation from a technical viewpoint. The real obstacle is in the enormous costs involved in the change of measuring equipment, data collections, etc. But if we don't start, we will never arrive.

E. EMBLIK (Switzerland) — The switch-over to the SI system is a delicate problem. We should begin with children in elementary schools. But what about the teachers? How would you resolve this problem?

C. F. KAYAN — I agree with Prof. EMBLIK that propagation of the SI metric system must encompass the reeducation of schoolteachers. This is a real problem since teachers now earn very good salaries (often by striking) and I would hope that affluence will not dull their will to learn new techniques, such as the teaching of the SI metric system. Also to be noted is that the ladies may not be charmed by the substitution of "Joules" for "calories" in the diet-conscious world. Joules will be hard for them to appreciate.

TIME-WISE PERSISTENCE OF ERROR
IN PRESENTATION OF THERMODYNAMIC DATA,
SUCH AS ON "ENTROPY"

C. F. KAYAN

Columbia University, New York (U.S.A.)

Persistance dans le temps de l'erreur de présentation de données thermodynamiques telles que l'« entropie »

RÉSUMÉ : *Pendant de nombreuses années les unités fondamentales définissant la valeur de l'entropie, par exemple, ont été publiées sur une base erronée. On a peine à croire que c'est dû à une mauvaise compréhension des concepts fondamentaux, mais c'est sans doute plutôt dû à une négligence dans la répétition d'erreurs faites au départ dans des tables de valeurs thermodynamiques largement répandues, telles que les tables de vapeur et de fluide frigorigène. L'aspect du point de vue physique opposé au point de vue mathématique entre aussi en jeu. Cette situation semble internationale et continue à se manifester même dans des publications très sérieuses même encore en 1969.*

For many years the basic units defining the value of entropy, for example, have appeared in print on an erroneous basis. Associated with this problem is also the tendency to express property values in terms of reduced dimensions. Since this has a bearing on the entropy question, this latter phase will be discussed first.

In our enthusiasm for mathematically-inspired simplification in the presentation of thermodynamic data, we have all-too-often been tempted to use "reduced" dimensions with equivalent units. Thereby meanings have been lost or distorted. For example, as every experimenter knows, numerically, thermal conductivity for a slab of material is arrived at as the heat-flow per unit of time per unit of heat-flow area per unit of thickness temperature-gradient. The temperature-gradient in American practice is often based on the inch, with the heat-flow area in square feet. Thus, thermal conductivity k is expressed as Btu per hour per square foot per (Δt-degree Fahrenheit per inch, or per foot, thickness) — admittedly a unique situation:

$$k_{\text{in.}} = \frac{\text{Btu}}{\text{hr} - \text{ft}^2 - (\text{deg}_\text{F}/\text{in.})} \quad \text{or} \quad k_{\text{ft}} = \frac{\text{Btu}}{\text{hr} - \text{ft}^2 - (\text{deg}_\text{F}/\text{ft})}$$

but not, for the latter, $k_{ft} = \text{Btu}/(\text{hr} - \text{ft} - {}^\circ\text{F})$. The reduced forms involve physical values and units, but lose the real physical significance of the property value. (This, in retrospect, may prove important in the present trend to increased international usage of a metric system of units.) Moreover, there is false implication in the use of $^\circ$F (the practical temperature value), rather than deg$_\text{F}$ which properly implies temperature difference (on the Fahrenheit scale). This is of great importance in the following discussion of entropy.

Specific entropy values, based on dQ/T, and which obviously involve the Absolute Temperature, have in wide-spread fashion been shown similar to specific heat capacity (derived from dQ/dt, dQ/dT) and obviously involving temperature-gradient. They are dimensionally the same, leading to similar expressions:

Group A

Specific entropy $s = \text{Btu}/(1\text{b}_m - {}^\circ\text{F})$, or $= \text{kcal}/(\text{kg}_m - {}^\circ\text{C})$
Specific heat capacity $c = \text{Btu}/(1\text{b}_m - {}^\circ\text{F})$, or $= \text{kcal}/(\text{kg}_m - {}^\circ\text{C})$

Fortunately, in informed circles this is now recognized as incorrect and misleading, and is preferably shown:

Group B

Specific entropy $s = \text{Btu}/(1\text{b}_m - °\text{R})$, or $= \text{kcal}/(\text{kg}_m - °\text{K})$

Specific heat capacity $c = \text{Btu}/(1\text{b}_m - \deg_F)$ or $= \text{kcal}/(\text{kg}_m - \deg_C)$

To avoid ambiguity here, the subscript m has been added to clearly indicate that mass is intended; F and C have been added as subscript to deg to represent Fahrenheit and Celsius degree, respectively (unnecessary if only one unit-system is involved); in the Anglo-American system of units °R represents absolute temperature in degrees Rankine. [Modern practice leaves off the degree symbol (°), shows temperatures as F and R, C and K.] It is of course understood that similar presentations may be made in SI units, for example, $s = \text{kJ}/(\text{kg} - \text{K})$.

But most unfortunately, though Group B is known by many to be the correct and preferable way to identify entropy and specific heat, the old system of Group A continues to persist right up to the present year 1969, even in highly respected circles throughout the world. This is particularly vicious in connection with entropy; it seems the basic concepts have been forgotten, and the errors of more than fifty years (at least) are continued in many tables of thermodynamic properties, presumably by "oversight", or through misguided and misleading application of the dimensional technique. I join others in deploring the situation.

It is particularly inappropriate to draw Ts diagrams, not with the full vertical scale of absolute temperature, Rankine or Kelvin, but rather with negative and positive values of the practical Fahrenheit or Celsius temperature scales. This turns out to be especially bad when the negative values, below 0°, are merely continued as far as convenient for graphical presentation. Thus the area below the temperature line between two entropy values no longer has the fundamental physical significance as being equivalent to the involved energy.

Needless to say, wherever temperature difference Δt is involved, deg (grd) should be employed rather than °F or °C (or F,C). Examples, in addition to thermal conductivity (k or λ), are local heat-transfer conductance (h or α), overall coefficient of heat transfer U, etc.

From the classic work of Lewis and Randall: "Thermodynamics" (1923) we quote: "Clausius summed up the findings of thermodynamics in the statement, "Die Energie der Welt ist konstant; die Entropie der Welt strebt einem Maximum zu", and it was this quotation which headed the great memoir of Gibbs on "The Equilibrium of Heterogeneous Substances". What is this entropy, which masters have placed in a position of coordinate importance with energy, but which has proved a bugbear to so many a student of thermodynamics?"

The above statement apparently is still applicable, right down to the present day!

SECTION 2

**TRANSMISSION DE CHALEUR EN RÉGIME VARIABLE
DANS LES PAROIS
ÉTUDE DES ENCEINTES CONFINÉES
DÉTERMINATION DES CHARGES CALORIFIQUES**

*NON-STATIONARY HEAT TRANSFER THROUGH WALLS
STUDY OF CONFINED ENCLOSURES
CALCULATION OF COOLING LOADS*

RÉSOLUTION DES PROBLÈMES DE TRANSFERT CONDUCTIF DE CHALEUR PAR LA MÉTHODE DES ÉLÉMENTS FINIS*

H. SNEESSENS

Institut de Thermodynamique et CEDRIC
Université de Liège (Belgique)

The resolution of conductive heat transfer problems using the finite element method

SUMMARY: *The search of a solution for Fourier's partial derivative equation relative to the phenomena of conductive heat transfer may be advantageously replaced, in using a variational principle, by the problem of the minimalisation of a functional which takes into account the imposed boundary conditions. The method of finite elements consists of breaking down the domain under study into small elements within which one introduces a law for temperature distribution or for flux: linear variation of temperature for 3-node triangles, 4-node tetrahedrons,.. parabolic variation for 6-node triangles, 8-node tetrahedrons...*

All these characteristics are expressed in relation to their values in nodes (peaks in the case of linear variation, peaks and the region of slopes with a parabolic variation). These characteristics values are thus determined by a numerical minimalisation process from the built-up functional. The calculations being made in a purely numerical manner, all the physical characteristics of the body could be variable, in very broad terms. This method, very simple to use, provides an easy study of thermal fields under both the steady and variable states.

INTRODUCTION

La méthode des éléments finis, développée tout d'abord pour la résolution des problèmes d'élasticité, est basée sur l'utilisation d'un principe variationnel remplaçant l'intégration d'une équation aux dérivées partielles par l'extrêmalisation d'une fonctionnelle correspondant à l'énergie potentielle totale en élasticité et à la variation du taux de production d'entropie pour l'étude de la conduction.

De plus, le corps continu qu'on étudie est remplacé par une série d'éléments accolés les uns aux autres et interconnectés en un nombre fini de points, les nœuds. Toutes les grandeurs physiques cherchées sont exprimées en fonction de leurs valeurs en ces nœuds par des lois arbitraires dont dépend la forme des éléments à utiliser.

La théorie de la méthode montrera que l'extrêmalisation de la fonctionnelle entraîne la résolution d'un système algébrique d'équations linéaires, problème particulièrement plus simple que celui de l'intégration d'équations aux dérivées partielles.

Dans une première partie, nous établirons la fonctionnelle à extrêmaliser.

La seconde partie sera consacrée à l'étude assez détaillée de la méthode des éléments finis pour l'étude des régimes permanent et dynamique du transfert conductif de chaleur.

PREMIÈRE PARTIE

CONSTRUCTION DE LA FONCTIONNELLE À EXTRÊMALISER

Deux moyens différents permettent de construire cette fonctionnelle :

— une approche purement mathématique : l'application du principe vari ationnel de Ritz à l'équation classique de la chaleur;

— une approche purement physique : l'utilisation du principe du minimum de la variation de taux de production d'entropie mis au point notamment par Glansdorff et Prigogine.

(*) Cette communication est extraite d'une étude de synthèse en préparation sous la direction du Professeur G. BURNAY

Notre but n'étant pas ici la démonstration de ces principes, nous montrerons uniquement notre façon de les utiliser.

1. *Approche mathématique*

L'équation de la chaleur s'écrit :

$$\frac{\partial}{\partial x}\left(K_x\,\frac{\partial T}{\partial x}\right) + \frac{\partial}{\partial y}\left(K_y\,\frac{\partial T}{\partial y}\right) + \frac{\partial}{\partial z}\left(K_z\,\frac{\partial T}{\partial z}\right) + Q - c\rho\,\frac{\partial T}{\partial t} = 0$$

avec

T	la température;
K_x, K_y, K_z	les conductivités thermiques dans les trois directions x, y et z;
Q	les sources de chaleur éventuelles;
c	la capacité calorifique;
ρ	la masse volumique;
t	le temps.

Si nous nous plaçons à un instant particulier, la dérivée temporelle de la température ne dépend que de la position. Le principe de Ritz peut alors être utilisé et l'équation de la chaleur se met sous la forme

$$\frac{\partial F}{\partial T} - \frac{\partial}{\partial x}\left(\frac{\partial F}{\partial \frac{\partial T}{\partial x}}\right) - \frac{\partial}{\partial y}\left(\frac{\partial F}{\partial \frac{\partial T}{\partial y}}\right) - \frac{\partial}{\partial z}\left(\frac{\partial F}{\partial \frac{\partial T}{\partial z}}\right) = 0$$

en prenant

$$F = \frac{1}{2}\left\{K_x\left(\frac{\partial T}{\partial x}\right)^2 + K_y\left(\frac{\partial T}{\partial y}\right)^2 + K_z\left(\frac{\partial T}{\partial z}\right)^2\right\} - \left\{Q - c\rho\,\frac{\partial T}{\partial t}\right\}T$$

D'après Ritz, la recherche de la solution de l'équation de la chaleur dans le volume V et la recherche de l'extrêmalisation de la fonctionnelle

$$\chi = \int_V F\,dV$$

sont deux problèmes identiques : l'un peut donc être remplacé par l'autre.

Nous devons encore tenir compte des conditions aux limites. Un des grands avantages de l'utilisation de principes variationnels réside dans le fait que ces conditions sont introduites directement au niveau de la fonctionnelle à extrêmaliser (en fait, il s'agit d'une minimalisation comme une étude de la variation seconde peut le montrer). Les conditions aux limites peuvent être de deux types :

— la température est imposée sur la surface S entourant V;

— une condition du type de Fourier est imposée sur cette surface

$$K_x\,\frac{\partial T}{\partial x}\,1_x + K_y\,\frac{\partial T}{\partial y}\,1_y + K_z\,\frac{\partial T}{\partial z}\,1_z + q + \alpha T = 0$$

où 1_x, 1_y et 1_z sont les cosinus directeurs de la normale extérieure à cette surface, αT, la perte par convection et rayonnement et q, le flux de chaleur imposé par unité de longueur.

28

La fonctionnelle à minimaliser s'écrit dans le premier cas

$$\chi = \int_V F \, dV$$

et dans le second

$$\chi = \int_V F \, dV + \int_S \left(qT + \frac{\alpha}{2} T^2 \right) dS$$

2. *Approche physique.*

Pour cette approche, nous nous plaçons aussi à un instant particulier. La dérivée temporelle de la température ne dépendant plus, à ce moment, du temps, nous l'introduirons dans le terme sources de chaleur et noterons

$$Q_1 = Q - c\rho \frac{\partial T}{\partial t}$$

Le troisième principe variationnel de Glansdorff-Prigogine s'exprime : « La fonction ϕ, potentiel local ou production d'entropie généralisée, ne peut que décroître au cours du temps et est minimale à l'état stationnaire ». La variation du potentiel local est, dans le cas du transfert de chaleur par conduction

$$\frac{\partial \phi}{\partial t} = \int_V \psi \, dV = \int_V \frac{\partial}{\partial t} \left(\frac{1}{T} \right) \frac{\partial \rho u}{\partial t} \, dV$$

où u est l'énergie interne spécifique donnée par l'équation

$$\frac{\partial \rho u}{\partial t} = - \frac{\partial q_j}{\partial x_j} + Q_1$$

En permutant les dérivées par rapport au temps et par rapport à l'espace, et en appliquant le théorème de la divergence, la variation du potentiel local peut se mettre sous la forme

$$\frac{\partial \phi}{\partial t} = - \int_S \frac{\partial}{\partial t} \left(\frac{1}{T} \right) q_j \, n_j \, dS + \int_V \frac{\partial}{\partial x_j} \left(\frac{\partial}{\partial t} \left(\frac{1}{T} \right) \right) . \, q_j \, dV +$$

$$+ \int_V \frac{\partial}{\partial t} \left(\frac{1}{T} \right) . \, Q_1 \, dV$$

où n_j est la direction de la normale extérieure au volume.

En tenant compte de la relation de Fourier

$$q_j = - K_j \frac{\partial T}{\partial x_j}$$

à l'intérieur du volume et de

$$q_j \, n_j = q + \alpha T$$

sur la surface de pourtour, cette variation s'écrit encore

$$\frac{\partial \phi}{\partial t} = \int_s \frac{1}{T^2}\left(q + \alpha T\right)\frac{\partial T}{\partial t}\,dS + \int_v \frac{K_j}{T_2}\frac{\partial T}{\partial x_j}\frac{\partial}{\partial t}\left(\frac{\partial T}{\partial x_j}\right)dV - \int_v \frac{Q_1}{T_2}\frac{\partial T}{\partial t}\,dV$$

Pour intégrer cette variation de potentiel local, il faut linéariser T et K_j autour de l'état stationnaire

$$T(x_j, t) \simeq T_o\,(x_j)$$

$$K(T) \simeq K(T_o) = K_o$$

d'où la fonction potentiel local

$$\phi = \int_v \frac{1}{T_o{}^2}\left[\frac{K_j}{2}\left(\frac{\partial T}{\partial x_j}\right)^2 - Q_1\,T\right]dV + \int_s \frac{1}{T_o{}^2}\left(q\,T + \frac{\alpha}{2}\,T^2\right)dS$$

Comme la variation de ce potentiel doit être nulle pour toute variation de la température, nous trouvons comme fonctionnelle à minimaliser

$$\chi = \int_v \left[\frac{1}{2}K_j\left(\frac{\partial T}{\partial x_j}\right)^2 - \left(Q - c\rho\,\frac{\partial T}{\partial t}\right)T\right]dV + \int_s \left(q\,T + \frac{\alpha}{2}\,T^2\right)dS.$$

Dans le cas où la température est imposée sur S, l'intégrale sur cette surface disparaît.

Nous avons montré au cours de la première partie, comment remplacer le problème de l'intégration de l'équation aux dérivées partielles en tenant compte de conditions aux limites par la minimalisation d'une fonctionnelle. Voyons comment cette minimalisation peut être réalisée assez simplement.

SECONDE PARTIE

THÉORIE DE LA MÉTHODE DES ÉLÉMENTS FINIS

Afin de ne pas trop alourdir les notations, nous supposerons le problème à étudier à deux dimensions, l'extension de la théorie à trois dimensions se faisant sans aucune difficulté.

La fonctionnelle à minimaliser s'écrit

$$\chi = \iint_D \left\{\frac{1}{2}\left[K_x\left(\frac{\partial T}{\partial x}\right)^2 + K_y\left(\frac{\partial T}{\partial y}\right)^2\right] - \left[Q - c\rho\,\frac{\partial T}{\partial t}\right]T\right\}dx\,dy$$

$$+ \int_C \left(q\,T + \frac{\alpha}{2}\,T^2\right)ds$$

D étant le domaine à étudier et C, le contour de celui-ci.

Occupons-nous du terme de surface χ_1

$$\chi = \chi_1 + \int_C \left(q\,T + \frac{\alpha}{2}\,T^2\right)ds$$

Découpons le domaine à étudier en petits éléments *e* caractérisés par un certain nombre de nœuds (par exemple les sommets) et laissons la température à l'intérieur de chaque élément définie par rapport aux températures des nœuds de cet élément. Afin uniquement de ne pas allonger l'exposé de la méthode par des calculs inutiles à sa compréhension, nous supposerons D découpé en petits triangles à trois nœuds, les trois sommets *i, j, m*.

Pour chaque élément, nous avons

$$T = [N_i \, N_j \, N_m] \, \{T\}^e$$

avec

$$\{T\}^e = \{T_i \, T_j \, T_m\}^T$$

Comme nous avons choisi un découpage en triangles à trois nœuds, nous devons prendre une loi de variation de température linéaire en fonction de la position (pour des éléments rectangulaires à quatre nœuds—les quatre sommets—, loi de variation linéaire avec un terme en xy; pour des éléments triangulaires à six nœuds—les trois sommets et les milieux des trois côtés—, loi de variation parabolique;...)

$$T = \alpha_1 + \alpha_2 x + \alpha_3 y$$

Cette loi de variation devant naturellement être vérifiée aux trois nœuds, nous trouvons les valeurs de N_i, N_j, N_m

$$N_i = \frac{1}{2\Delta} \, (a_i + b_i x + c_i y) \qquad\qquad (i, j, m)$$

où Δ est la surface du triangle *ijm* et les a_i, b_i, c_i étant des combinaisons des coordonnées des nœuds.

Les N_i ne dépendant que de la position, nous avons aussi

$$\frac{\partial T}{\partial t} = [N_i \, N_j \, N_m] \left\{ \frac{\partial T}{\partial t} \right\}^e$$

avec

$$\left\{ \frac{\partial T}{\partial t} \right\}^e = \left\{ \frac{\partial T}{\partial t}\bigg|_i \quad \frac{\partial T}{\partial t}\bigg|_j \quad \frac{\partial T}{\partial t}\bigg|_m \right\}^T$$

Soit {T}, le vecteur des températures aux N nœuds du domaine

$$\{T\} = \{T_1, ..., T_i, ..., T_n\}^T$$

Les températures à l'intérieur d'un élément étant définies en fonction des températures aux nœuds de cet élément, les températures à l'intérieur du domaine D sont définies en fonction de {T}. La minimalisation de la fonctionnelle par rapport à la température est donc équivalente à sa minimalisation par rapport à {T}

$$\frac{\partial \chi_1}{\partial T} = \frac{\partial \chi_1}{\partial \{T\}} = 0$$

avec

$$\frac{\partial \chi_1}{\partial \{T\}} = \left\{ \frac{\partial \chi_1}{\partial T_1}, ..., \frac{\partial \chi_1}{\partial T_i}, ..., \frac{\partial \chi_1}{\partial T_n} \right\}^T$$

31

La valeur de la fonctionnelle pour chaque élément e est liée à χ par les relations

$$\chi_1 = \sum_e \chi_1^e$$

et

$$\frac{\partial \chi_1}{\partial T_i} = \sum_e \frac{\partial \chi_1^e}{\partial T_i}$$

les sommations s'étendant sur tous les éléments.

En faisant l'hypothèse, non restrictive d'ailleurs, de la constance des caractéristiques physiques à l'intérieur d'un élément, nous trouvons

$$\frac{\partial \chi_1}{\partial T_i} = \sum_e \sum_l h_{il} \, T_l + \sum_e \sum_l p_{il} \left(\frac{\partial T}{\partial t}\right)_l + \sum_e F_i^e = 0$$

les sommations Σ_e se faisant sur tous les éléments et les sommations Σ_l sur tous les nœuds d'un élément. Cette relation s'écrit encore sous forme matricielle

$$[H] \, \{T\} + [P] \left\{\frac{\partial T}{\partial t}\right\} + \{F\} = 0$$

avec [H] et [P], deux matrices carrées (nombre de nœuds $\times$ nombre de nœuds) d'éléments

$$h_{ij} = \frac{1}{4\Delta} (K_x \, b_i \, b_j + K_y \, c_i \, c_j)$$

$$p_{ij} = c\rho \left\{ \frac{\Delta}{9} + \frac{1}{48\Delta} \left[b_i \, b_j (x_i^2 + x_j^2 + x_m^2) + c_i \, c_j (y_i^2 + y_j^2 + y_m^2) + \right. \right.$$

$$\left. \left. + (b_i \, c_j + b_j \, c_i)(x_i \, y_i + x_j \, y_j + x_m \, y_m) \right] \right\}$$

les x_k, y_k étant les coordonnées des nœuds d'un élément pris dans le système du centre de masse de cet élément, et $\{F\}$, une matrice-colonne (nombre de nœuds) d'éléments

$$f_i = - \frac{Q\Delta}{3}$$

Si les conditions aux limites sont du type de Fourier, il reste à ajouter le résultat de la minimalisation de

$$\int_c \left(qT + \frac{\alpha}{2} T^2 \right) ds$$

à celle de χ_1. Cela conduit à un système d'équations semblable à celui obtenu dans le cas où les températures sont imposées aux limites

$$[H'] \, \{T\} + [P] \left\{\frac{\partial T}{\partial t}\right\} + \{F'\} = 0$$

avec

$$[H'] = [H] + [D]; \quad \{F'\} = \{F\} + \{QQ\}$$

où [D] est une matrice carrée (nombre de nœuds × nombre de nœuds) d'éléments

$$d_{ii} = \frac{1}{3} (\alpha_h L_h + \alpha_i L_i)$$

$$d_{hi} = \frac{1}{6} \alpha_h L_h$$

$$d_{ij} = \frac{1}{6} \alpha_i L_i$$

h, i, j étant trois nœuds consécutifs se trouvant sur le contour (L_i est la distance de i à j); les autres éléments sont nuls et $\{QQ\}$, une matrice-colonne (nombre de nœuds) d'éléments

$$qq_i = \frac{1}{2} (q_h L_h + q_i L_i)$$

h et i étant deux nœuds consécutifs sur le contour; les autres éléments sont nuls.

Quel que soit le type de conditions aux limites, le système d'équations obtenu est

$$[H]\{T\} + [P]\left\{\frac{\partial T}{\partial t}\right\} + \{F\} = 0$$

Dans le cas du régime permanent, ce système est un système algébrique d'équations linéaires. La résolution de ce système ne pose aucun problème.

Dans le cas du régime dynamique, il n'en est plus de même : nous sommes en présence d'un système différentiel d'équations linéaires. À l'intégration directe d'un tel système, nous avons préféré la méthode suivante.

Considérons que, dans l'intervalle de temps Δt, les valeurs de la dérivée temporelle de la température varient linéairement avec le temps. Cette hypothèse peut toujours être faite, l'intervalle de temps pouvant être pris aussi petit que l'on veut. Dans l'intervalle de temps $t - \Delta t$, nous avons

$$\left\{\frac{\partial T}{\partial t}\right\} = \left\{\frac{\partial T}{\partial t}\right\}_{t-\Delta t} + \left[\left\{\frac{\partial T}{\partial t}\right\}_t - \left\{\frac{\partial T}{\partial t}\right\}_{t-\Delta t}\right] \frac{\tau - (t-\Delta t)}{\Delta t}$$

et en intégrant cette relation entre $t - \Delta t$ et t, nous trouvons

$$\left\{\frac{\partial T}{\partial t}\right\}_t = \frac{2}{\Delta t}\left[\left\{T\right\}_t - \left\{T\right\}_{t-\Delta t}\right] - \left\{\frac{\partial T}{\partial t}\right\}_{t-\Delta t}$$

Remplaçons $\left\{\dfrac{\partial T}{\partial t}\right\}_t$ par sa valeur dans l'équation matricielle, solution du problème au temps t.

Nous obtenons, pour chaque instant, un système algébrique d'équations linéaires

$$[C]\{T\}_t = \{E\}_{t-\Delta t} - \{F\}_t$$

en posant

$$[C] = [H] + \frac{2}{\Delta t} [P]$$

$$\{E\}_{t-\Delta t} = [P]\left\{\left\{\frac{\partial T}{\partial t}\right\}_{t-\Delta t} + \frac{2}{\Delta t} \{T\}_{t-\Delta t}\right\}$$

33

La température en tout point, au temps t, peut donc être calculée par la résolution du système algébrique dont le second membre ne contient, en dehors des termes de sources, que des valeurs calculées au temps précédent. A chaque instant, nous faisons le contrôle de l'hypothèse de la linéarité de la dérivée temporelle dans l'intervalle de temps utilisé : un processus à pas variable s'impose.

CONCLUSIONS

La méthode des éléments finis remplaçant l'intégration d'une équation aux dérivées partielles en tenant compte de conditions aux limites par la résolution d'un système algébrique d'équations linéaires pour chaque temps, est d'une grande simplicité d'utilisation. Elle permet d'étudier aisément le transfert de chaleur par conduction dans des corps hétérogènes de géométrie quelconque soumis à n'importe quel type de conditions aux limites.

RÉFÉRENCES

[1] O. C. ZIENKEWICZ. The finite element method in structural and continuum mechanics. McGraw-Hill (1967).
2] W. VISSER. A finite method for the determination of non-stationary temperature distribution and thermal deformation. Proceedings of the conference hold at Wright-Patterson Air Force Base Ohio, 26-28 Oct. 1965 (Matrix methods in structural mechanics, Nov. 1966).
[3] E. L. WILSON and R. W. CLOUGH. Dynamic response by step by step matrix analysis. Symposium on the use of computers in civil engineering. Lisbon (October 1962).
[4] H. SNEESSENS. Application de la méthode des éléments finis au problème de la conduction. Institut de Thermodynamique de l'Université de Liège (1968).

DISCUSSION

F. M. CAMIA (France) — Nous sommes tout à fait d'accord avec le Prof. BURNAY; jusqu'ici nous avons travaillé de façon similaire sans nous connaître. La distinction de plusieurs domaines dans une paroi peut s'étendre à deux ou trois dimensions par réticulation.

Alors que la solution analytique précise, qui n'est d'ailleurs que l'expression d'une réticulation à domaines arbitrairement petits, donne pour la température à l'intérieur d'un solide la formule :

$$\theta = \alpha U . e$$

α vecteur ligne de fonctions des productions de chaleur dans la masse;
U matrice diagonale de fonctions de Sturm-Liouville des variables d'espace;
e vecteur colonne d'exponentielles du temps.

La division du solide en domaines suffisamment petits donne la même expression mais les fonctions U sont des fonctions trigonométriques. Pour un solide discontinu ou pour un chemin thermique de section variable, la température s'exprime par la somme :

$$q = \alpha_{0m}\mu_{pm} \cos(w_{pm}\xi + \varphi_{pm}) \, e^{-\omega_{pm}^2 h_p t} \qquad m = 1, 2, ..., \alpha$$

(ξ = abscisse dans le domaine considéré, t = temps, φ_p, μ_p, ω_p, et h_p sont respectivement les valeurs, dans le domaine d'ordre p, un angle de phase, une constante, un coefficient angulaire et la diffusivité thermique). α_0 décrit le programme imposé par la considération de conditions aux limites ou d'adiabaticité ou d'isotherme, on peut introduire de façon simple les problèmes usuels. Quant aux cas les plus complexes : convection, rayonnement, paramètres thermiques variables ou fonction de la température, l'introduction à l'ordinateur du programme de solution et des données permet l'utilisation de la méthode « pas à pas ». Nous avons ainsi comparé les résultats du calcul à l'I. B. M. avec ceux que nous avions publiés dans « Flamme et thermique » (en 1958 je crois) par calcul différent, relativement à la température d'un mur soumis aux influences diurnes et aux échanges de chaleur avec l'intérieur du local (ces calculs sont valables autant pour la chaleur que pour le froid).

L'accord est excellent. L'avantage de l'ordinateur, c'est que, avec des conditions complexes et variables, nous avons eu plus de 50 000 résultats de température et de flux en moins de 20 minutes, et le même programme peut servir indéfiniment.

E. EMBLIK (Suisse) — Est-ce que votre système permet de calculer le champ de température non stationnaire quand les constantes (λ, ρ, c) varient ? Cela arrive par exemple dans un paquet de denrées surgelées pendant leur transport dans une ambiance avec température élevée.

H. SNEESSENS — 1. La méthode proposée permet d'étudier des corps extrêmement hétérogènes. Il suffit de découper assez finement le domaine que l'on étudie pour satisfaire l'hypothèse de la constance des caractéristiques physiques à l'intérieur de cet élément.

2. Je suis entièrement d'accord que la conduction n'est qu'une partie d'une étude totale d'une salle de réfrigération. Je compte aussi étudier la convection par la méthode des éléments finis ou par une extension de celle-ci puisque en me basant sur le troisième principe de Glansdorff-Prigogine cette méthode semble le permettre.

THE SOLUTION OF UNSTEADY HEAT CONDUCTION PROBLEMS IN COMPOSITE WALLS VIA THE MONTE CARLO METHOD

B. STANISZEWSKI and J. SUSKI

Polytechnical School, Warsaw (Poland)

Solution des problèmes de conduction de chaleur variable dans les parois composées suivant les méthodes de Monte Carlo

RÉSUMÉ : *Ce rapport donne les résultats de la solution des problèmes de conduction de chaleur variable dans les parois à couches multiples suivant les méthodes de Monte Carlo. On a considéré le cas de conduction de chaleur variable à deux dimensions dans une paroi à couches multiples, de section transversale rectangulaire, dans des conditions instables limites.*

Les conditions instables étaient les suivantes :
a) La température d'une surface de paroi $T_w = f(t)$ *dépendait du temps ;*
b) À la surface de la paroi parallèle, on connaissait le coefficient de transfert de chaleur $\alpha = f(x)$ *et la température du fluide ambiant* $T_f = f(t)$. *Le coefficient* $\alpha = f(x)$ *dépendait de la coordonnée et la température* T_f *dépendait du temps ;*
c) Les deux surfaces restantes étaient isolées. Dans le rapport on présente la méthode de calcul de la température en un point arbitraire de l'intérieur de la paroi composée de 3 couches différentes.

On effectue aussi l'analyse du temps de calcul, la précision des calculs et la comparaison des méthodes de Monte Carlo avec des méthodes numériques.
On considère les possibilités d'application des méthodes de Monte Carlo dans le cas de conditions géométriques plus complexes et des conditions limites.

1. INTRODUCTION

Monte Carlo methods are more and more employed in solving partial differential equations in physics and engineering.

It has long been recognized that probability methods (Monte Carlo) constitute a potential means of solving differential and difference equations encountered in heat conduction.

A basic ingredient of the Monte Carlo solution of differential equations is the random displacement which can be realized in different ways [1, 2, 3].

The most important advantages of Monte Carlo methods are:

a) the possibility of solving cases with non-elementary geometrical configurations;

b) the possibility of solving non-linear problems;

c) the possibility of computating the temperature at any isolated point in a solid without solving the whole problem.

Its most serious disadvantages are long computation times, frequent low accuracies and a strong influence upon the particular random-number generator used. Computer programs based upon the Monte Carlo method do not appear to achieve equal success with different computer applications.

This paper discusses the solution of unsteady heat conduction problems in multi-layer walls with complicated boundary conditions via the Monte Carlo method. Such problems are met frequently in refrigeration techniques.

The analysis uses the application of fixed random displacement, i.e., when the step and the pathways are prefixed.

2. THE PRINCIPLES OF THE MONTE CARLO METHOD

The solution of heat conduction problems by probability methods is based on random displacement [1, 4, 5], which is very closely related to the phenomena of diffusion and heat conduction. In order to show this connection let us consider unsteady heat conduction in a two-dimensional body shown in figure 1. Heat balance

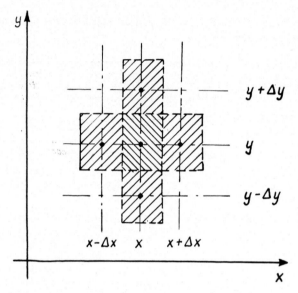

Fig. 1

for an element of the body can be expressed in a finite difference as follows:

$$\Delta x \Delta y \rho c \left[T(x, y, t+\Delta t) - T(x, y, t) \right] = \lambda'_x \frac{T(x-\Delta x, y, t) - T(x, y, t)}{\Delta x} \Delta y \Delta t -$$

$$- \lambda''_x \frac{T(x, y, t) - T(x+\Delta x, y, t)}{\Delta x} \Delta y \Delta t + \lambda'_y \frac{T(x, y-\Delta y, t) - T(x, y, t)}{\Delta y} .$$

$$\cdot \Delta x \Delta t - \lambda''_y \frac{T(x, y, t) - T(x, y+\Delta y) t}{\Delta y} \cdot \Delta x \Delta t . \qquad (1)$$

Setting

$$M'_x = \frac{\lambda'_x \Delta t}{c\rho \Delta x^2} ; \quad M''_x = \frac{\lambda''_x \Delta t}{c\rho \Delta x^2} ; \quad M'_y = \frac{\lambda'_y \Delta t}{c\rho \Delta y^2} ; \quad M''_y = \frac{\lambda''_y \Delta t}{c\rho \Delta y^2}$$

the difference equation (1) can be written

$$T(x, y, t+\Delta t) = T(x, y, t) (1 - M'_x - M''_x - M'_y - M''_y) +$$

$$+ M'_x T(x-\Delta x, y, t) + M''_x T(x+\Delta x, y, t) +$$

$$+ M'_y T(x, y-\Delta y, t) + M''_y T(x, y+\Delta y, t) . \qquad (2)$$

38

Assuming that $\Delta x = \Delta y$ and $M'_x = M''_x = M'_y = M''_y = M$ one obtains

$$T(x, y, t + \Delta t) = T(x, y, t)(1 - 4M) + M[T(x - \Delta x, y, t) +$$

$$+ T(x + \Delta x, y, t) + T(x, y - \Delta y, t) + T(x, y + \Delta y, t)] \qquad (3)$$

The stability condition is $M \geqslant 4$. If $M = 4$, then

$$T(x, y, t + \Delta t)$$

$$= \frac{T(x - \Delta x, y, t) + T(x + \Delta x, y, t) + T(x, y - \Delta y, t) + T(x, y + \Delta y, t)}{4}. \qquad (4)$$

Replacing the expressions M'_x, M''_x, M'_y, M''_y by p'_x, p''_x, p'_y, p''_y and puting $p_0 = 1 - M'_x - M''_x - M'_y - M''_y$ one can write equation (2) in the form,

$$T(x, y, t + \Delta t) = p_0 T(x, y, t) + p'_x T(x - \Delta x, y, t) + p''_x T(x + \Delta x, y, t) +$$

$$+ p'_y T(x, y - \Delta y, t) + p''_y T(x, y + \Delta y, t). \qquad (5)$$

This equation may be given a probabilistic interpretation: if the point (x, y) is the momentary position of a random displacing particle, then p'_x, p''_x, p'_y, p''_y represent the probabilities of stepping respectively to points $x - \Delta x$, $x + \Delta x$, $y - \Delta y$, $y + \Delta y$. The probability of remaining at the point (x, y) is represented by p_0.

The individual probabilities must be non-negative and their sum must be unity

$$p_0 + p'_x + p''_x + p'_y + p''_y = 1. \qquad (6)$$

If the conditions of validity of equation (4) are selected, one obtains $p_0 = 0$ and

$$p'_x = p''_x = p'_y = p''_y = p = 1/4.$$

The choice of the direction in which the particle will move can be found by chance and may be realized by the use of a digital computer. Suppose that this computer is programmed to supply, upon request, a number having random magnitudes between 0 and 1. Then the interval $(0,1)$ can be divided among the probabilities $p_0, p'_x, p''_x, p'_y, p''_y$. For instance in the case when $p_0 = 0$ and $p'_x = p''_x = p'_y = p''_y = 0.25$ if the random number lies between 0 and 0.25, the particle is instructed to displace from (x, y) to $(x - \Delta x, y)$; on the other hand, if the random number lies between 0.25 and 0.5, the particle displaces from (x, y) to $(x + \Delta x, y)$, and so forth. Moreover equation (2) states that at each step the time associated with the moving particle decreases by an increment Δt.

Boundary conditions may be represented in the following way.

a) *Boundary conditions of the second order*

The suitable boundary element is shown in figure 2. In the considered case, the value of heat flux at the boundary is given. Heat balance equation for this element can be written as:

$$\Delta x \frac{\Delta y}{2} \rho c [T(x, y, t + \Delta t) - T(x, y, t)] = \lambda'_y \frac{T(x, y - \Delta y, t) - T(x, y, t)}{\Delta y} \cdot$$

$$\cdot \Delta x \Delta t - q \Delta x \Delta t + \lambda'_x \frac{T(x - \Delta x, y, t) - T(x, y, t)}{\Delta x} \frac{\Delta y}{2} \Delta t -$$

$$- \lambda''_x \frac{T(x, y, t) - T(x + \Delta x, y, t)}{\Delta x} \frac{\Delta y}{2} \Delta t. \qquad (7)$$

39

Using these same notations as before and putting

$$P = \frac{2q\,\Delta t}{c\rho\,\Delta y}$$

one obtains

$$T(x, y, t+\Delta t) = T(x, y, t)(1-2M'_y-M'_x-M''_x)+2M'_y\,T(x, y-\Delta y, t)+$$
$$+ M'_x(x-\Delta x, y, t)+M''_x\,T(x+\Delta x, y, t)-P. \qquad (8)$$

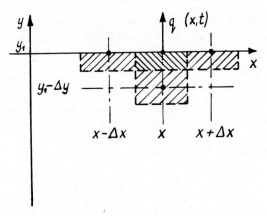

Fig. 2

Such a boundary is called a reflecting barrier. The probabilities of remaining and stepping to the adjacent elements are equal, respectively

$$p_0 = 1-2M'_y-M'_x-M''_x;\quad p_1 = 2M'_y;\quad p_2 = M'_x;\quad p_3 = M''_x.$$

Moreover if a particle is at a boundary, the quantity P tallies. By the suitable choice of constants M, probability p_0 can be equal to zero.

b) *Boundary conditions of the third order*

A suitable boundary element is shown in figure 3. In the case considered, the ambient temperature $T_0(x, t)$ and coefficient of heat transfer $\alpha(y, t)$ are known.
Heat balance equation for this element is:

$$\frac{\Delta x}{2}\,\Delta y\,\rho c\,[T(x, y, t+\Delta t)-T(x, y, t)] = -\alpha\Delta y\,[T(x, y, t)-T_0(y, t)]\,\Delta t+$$

$$+\lambda'_x\,\Delta y\,\frac{T(x-\Delta x, y, t)-T(x, y, t)}{\Delta x}\,\Delta t+$$

$$+\lambda'_y\,\frac{T(x, y-\Delta y, t)-T(x, y, t)}{\Delta y}\,\frac{\Delta x}{2}\,\Delta t-$$

$$-\lambda''_y\,\frac{T(x, y, t)-T(x, y+\Delta y, t)}{\Delta y}\,\frac{\Delta x}{2}\,\Delta t. \qquad (9)$$

40

Introducing the notations M'_s, M'_y, M''_y and

$$R = \frac{2\alpha \Delta t}{c\rho \Delta x},$$

one obtains

$$T(x, y, t+\Delta t) = T(x, y, t)(1-2M'_x-M'_y-M''_y-R)+2M'_x\, T(x-\Delta x, y, t)+$$

$$+M'_y\, T(x, y-\Delta y, t)+M''_y\, T(x, y+\Delta y, t)-R\, T_0(y, t). \tag{10}$$

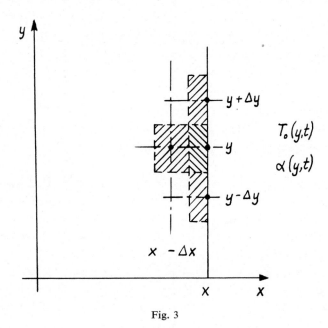

Fig. 3

Such a boundary condition is equivalent to a partially absorbing, partially reflecting barrier.

The probabilities of immobility, transfer to adjacent elements and absorbing are equal respectively:

$$p_0 = 1-2M'_x-M'_y-M''_y-R\,; \quad p_1 = 2M'_x\,; \quad p_2 = M'_y\,; \quad p_3 = M''_y\,; \quad p_4 = R.$$

If a particle is absorbed at a boundary, the temperature T_0 is recorded.

c) *Boundary condition of the first order*

A boundary now has a prescribed temperature and is called an absorbing barrier. This boundary condition can be treated as a particular case for a boundary condition of the third order for $\alpha \rightarrow \infty$. When a random displaced particle arrives at a boundary, its movement is terminated and boundary temperature is recorded.

d) Boundary condition of the fourth order

This case concerns two different walls in contact, as shown in figure 4. Heat balance equation for the considered element can be written as:

$$(\rho_1 c_1 + \rho_2 c_2) \frac{\Delta x}{2} \Delta y [T(x, y, t+\Delta t) - T(x, y, t)] =$$

$$= \lambda'_{1x} \frac{T(x-\Delta x, y, t) - T(x, y, t)}{\Delta x} \Delta y \Delta t +$$

$$+ \lambda'_{1y} \frac{T(x, y-\Delta y, t) - T(x, y, t)}{\Delta y} \cdot \frac{\Delta x}{2} \Delta t -$$

$$- \lambda''_{1y} \frac{T(x, y, t) - T(x, y+\Delta y, t)}{\Delta y} \frac{\Delta x}{2} \Delta t -$$

$$- \lambda''_{2x} \frac{T(x, y, t) - T(x+\Delta x, y, t)}{\Delta x} \Delta y \Delta t +$$

$$+ \lambda'_{2y} \frac{T(x, y-\Delta y, t) - T(x, y, t)}{\Delta y} \cdot \frac{\Delta x}{2} \Delta t -$$

$$- \lambda''_{2y} \frac{T(x, y, t) - T(x, y+\Delta y, t)}{\Delta y} \frac{\Delta x}{2} \Delta t. \qquad (11)$$

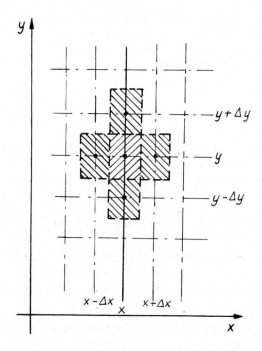

Fig. 4

Using the following notations:

$$M'_{1x} = \frac{\lambda'_{1x}\Delta t}{(\rho_1 c_1 + \rho_2 c_2)\Delta x^2} \; ; \; M''_{2x} = \frac{\lambda''_{2x}\Delta t}{(\rho_1 c_1 + \rho_2 c_2)\Delta x^2} \; ; \; M'_{1y} = \frac{\lambda'_{1y}\Delta t}{(\rho_1 c_1 + \rho_2 c_2)\Delta y^2}$$

$$M''_{1y} = \frac{\lambda''_{1y}\Delta t}{(\rho_1 c_1 + \rho_2 c_2)\Delta y^2} \; ; \; M'_{2y} = \frac{\lambda'_{2y}\Delta t}{(\rho_1 c_1 + \rho_2 c_2)\Delta y^2} \; ; \; M''_{2y} = \frac{\lambda''_{2y}\Delta t}{(\rho_1 c_1 + \rho_2 c_2)\Delta y^2}$$

one obtains:

$$T(x, y, t+\Delta t) = T(x, y, t)\,(1 - 2M'_{1x} - 2M''_{2x} - M'_{1y} - M''_{1y} - M'_{2y} - M''_{2y}) +$$

$$+ 2M'_{1x}\,T(x-\Delta x, y, t) + 2M''_{2x}\,T(x+\Delta x, y, t) +$$

$$+ (M'_{1y} + M'_{2y})\,T(x, y-\Delta y, t) + (M''_{1y} + M''_{2y})\,T(x, y+\Delta y, t). \qquad (12)$$

The probabilities of immobility and transfer in one of four possible directions are now equal:

$$p_0 = 1 - 2M'_{1x} - 2M''_{2x} - M'_{1y} - M''_{1y} - M'_{2y} - M''_{2y}$$

$$p_1 = 2M'_{1x}; \; p_2 = 2M''_{2x}; \; p_3 = M'_{1y} + M'_{2y}; \; p_4 = M''_{1y} + M''_{2y}.$$

If $\quad M'_{1x} = M'_{1y} = M''_{1y} = M_1 \quad$ and $\quad M''_{2x} = M'_{2y} = M''_{2y} = M_2$, $\quad$ then

$$p_0 = 1 - 4(M_1 + M_2); \; p_1 = 2M_1; \; p_2 = 2M_2; \; p_3 = p_4 = M_1 + M_2.$$

In the case of a composite wall with all forms of boundary conditions the Monte Carlo solution for the temperature $T(x_i, y_i, t_c)$ in an element (x_i, y_i) at a moment t_c can be expressed as:

$$T(x_i, y_i, t_c) = \frac{1}{N}\sum_{j=1}^{N} [T_b(x, y, t_c - \Sigma\Delta t) + T_0(x, y, t_c - \Sigma\Delta t) +$$

$$+ T_p(x, y) + P(x, y, t_c - \Sigma\Delta t)]. \qquad (13)$$

In equation (13) N is the total number of random displaced particles. Some of them arrive at the boundary in a time less than $n\Delta t = t_c$ and then, according to boundary conditions, one of the following quantities are recorded: boundary temperature T_b, ambient temperature T_0, or parameter P; the values of these quantities are taken when the particle reaches the boundary.

If a particle is still within a body when $n\Delta t = t_c$ the initial temperature T_p at this point is recorded.

It is worth mentioning that, apart from fixed random displacement, some other techniques of the Monte Carlo application have been developed [1, 2].

3. ACCURACY OF MONTE CARLO METHODS

The errors from Monte Carlo techniques are caused by space and time separation and by variation in sampling.

The former error is the same as would be found in a finite difference method and will not be considered here.

43

The second error can only be calculated with some probability. If this probability is 0.95, the error may be expressed as

$$B = \frac{2\sigma}{\sqrt{N}} \tag{14}$$

assuming the error probability of 0.9973, one obtains

$$B = \frac{3\sigma}{\sqrt{N}} \tag{15}$$

where σ is the mean standard deviation.

The theoretical calculation of the variance σ^2 before the solution of a considered problem is difficult and the obtained results are usually too high. The accuracy of a Monte Carlo method can be evaluated correctly in the process of solving only. It is analogous to the fact that the correct evaluation of the accuracy of any physical experiment is possible only in this experiment itself.

From the equations (14) and (15) two conclusions may be drawn:

a) the increase in accuracy requires a significant increase in the number of random displaced particles and computing time;

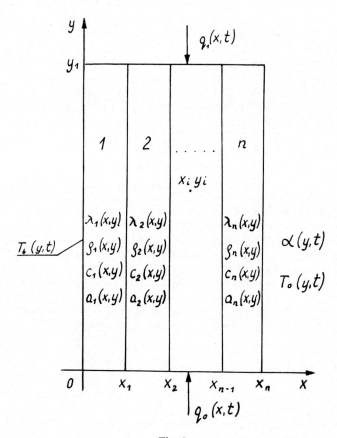

Fig. 5

b) in many engineering problems when extreme accuracy is not necessary, it is possible to obtain satisfactory results by relatively low N. In practice the value of $N = 10^3 \div 10^5$ is normally used.

4. A COMPUTED EXAMPLE

In order to check the usefulness of Monte Carlo methods, several computed experiments were made. They showed the practical possibility of applications to some technical problems when the temperature value at an isolated point is required.

One such experiment is briefly described. In this experiment the problem of unsteady heat conduction in a two-dimensional multi-layer wall has been considered.

The wall shown in figure 5 is composed of 3 layers bounded by the following coordinates: $x = 0$, $x = 0.27$ m, $y = 0$ and $y = 6$ m. Thermal characteristics of individual layers are as follows:

a) I. wall $\quad 0 \leqslant x \leqslant 0.02$ m $\qquad \lambda_1 = 1.3$ W/m deg
$\quad c_1 = 1{,}050$ J/kg deg; $\rho_1 = 2{,}300$ kg/m^3; $\qquad a_1 = 5.38 \cdot 10^{-7}$ m^2/s

b) II. wall 0.02 m $\leqslant x \leqslant 0.22$ m; $\qquad \lambda_2 = 0.03$ W/m deg
$\quad c_2 = 1{,}380$ J/kg deg; $\qquad \rho_2 = 20$ kg/m^3; $\qquad a_2 = 1.09 \cdot 10^{-6}$ m^2/s

c) III. wall $\quad 0.22$ m $\leqslant x \leqslant 0.27$ m $\qquad \lambda_3 = 1.5$ W/m deg
$\quad c_3 = 1{,}130$ J/kg deg; $\qquad \rho_3 = 2{,}300$ kg/m^3; $\qquad a_3 = 5.77 \cdot 10^{-7}$ m^2/s

Initial and boundary conditions:
at the moment $t = 0$ $\quad$ the temperature of the wall

$$T_p = 20°C$$

the temperature of the surface $x = 0$

$$T_b^{'} = 20 - 1.54 \cdot 10^{-4} t_j^{'}$$

the surfaces $y = 0$ and $y = 6$ are insulated ($q = 0$) the temperature of the fluid in contact with surface $x = 0.27$

$$T_0^{'} = 25 - 1.3 \cdot 10^{-4} t$$

and heat transfer coefficient $\alpha = 2y$.

The temperature in the middle of the wall at the point $x = 0.14$ m; $y = 3$ m at the moment $t_c = 8.63 \cdot 10^4$ s is equal to

$$T = 14.3°C \ .$$

This result was obtained from an analysis of $N = 110$ random displacements, the error $B = \pm 0.588°C$ (from an error probability of 0.95). The number of computed operations was 11,700,000 meaning about 100,000 per random displaced particle.

NOMENCLATURE

a $\qquad$ thermal diffusivity;

B $\qquad$ error;

c $\qquad$ specific heat;

M $\qquad \dfrac{\lambda \Delta t}{c\rho \Delta x^2}$ or $\dfrac{\lambda \Delta t}{c\rho \Delta y^2}$

N $\qquad$ number of random walks;

p_0, p_1, p_2, p_3, p_4 probabilities of immobility, transfer to adjacent elements and eventual absorbing;

P $\quad \dfrac{2q\,\Delta t}{c\rho\Delta y}$;

q $\quad$ heat flux;

R $\quad \dfrac{2\alpha\,\Delta t}{c\rho\Delta x}$;

t $\quad$ time;

Δt $\quad$ time increment;

T $\quad$ temperature;

T_0 $\quad$ ambient temperature;

T_b $\quad$ temperature at the boundary;

T_p $\quad$ initial temperature;

x $\quad$ coordinate;

Δx $\quad$ step size in x direction;

y $\quad$ coordinate;

Δy $\quad$ step size in y direction;

α $\quad$ heat transfer coefficient;

λ $\quad$ thermal conductivity;

ρ $\quad$ density;

σ $\quad$ mean standard deviation.

SUBSCRIPTS

$'x$ $\quad$ refers to the element $x - \Delta x$;

$''x$ $\quad$ refers to the element $x + \Delta x$;

$'y$ $\quad$ refers to the element $y - \Delta y$;

$''y$ $\quad$ refers to the element $y + \Delta y$;

1,2 $\quad$ refers to the 2 different bodies in contact.

REFERENCES

[1] A. HAJI-SHEIKH and E. M. SPARROW, The Solution of Heat Conduction Problems by Probability Methods. *ASME Paper*, No. 66-WA/HT-1.
[2] A.F. EMERY and W.W. CARSON, A Modification to the Monte Carlo Method — The Exodus Method. *ASME Paper*, No. 66-WA/HT-61.
[3] J. ROZEWICZ, Zastosowanie metody Monte Carlo do zagadnień przewodzenia ciepła. *Politechnika Śląska, Zeszyt Naukowy*, Nr 180, Gliwice (1967).
[4] G.W. KING, Monte Carlo Method for Solving Diffusion Problems *Ind. Eng. Chemistry*, vol. 43 (1951).
[5] E. F. BECKENBACH, editor, *Modern Mathematics for the Engineer*, McGraw Hill Book Co., Inc., New York, N.Y. (1956).

THERMAL BEHAVIOUR OF COMPOSITE WALLS UNDER TRANSIENT CONDITIONS. THEIR CHARACTERIZATION BY TWO PARAMETERS. SIMPLIFIED CALCULATION METHOD

P. DI FILIPPO, M. SOVRANO and G. ZORZINI

Istituto di Fisica Tecnica e Laboratorio per la Tecnica
del Freddo dell'Università di Padova (Italy)

Comportement thermique des parois composées en régime variable. Leur caractérisation au moyen de deux paramètres. Calcul simplifié

RÉSUMÉ : *On caractérise les parois composées en régime variable au moyen du coefficient global de transmission thermique H et du temps de réponse τ_0, dont la définition est rappelée.*
   Les paramètres proposés permettent de simplifier le calcul de transmission de la chaleur en régime variable dans les parois composées qui enveloppent un espace.
   On simplifie le problème, chaque paroi composée pouvant être remplacée par une paro i homogène équivalente, définie par H et τ_0.
   Les AA. illustrent l'installation pour l'étude expérimentale des parois en régime variable.

1. INTRODUCTION

The thermal behaviour of composite walls in transient conditions has been studied by several authors [1, 2, 3, 4, 5].

We can find an exact, analytical solution of the problem only in the case of composite walls, made up of parallel, homogeneous layers: it, however, requires calculations far more complex than those we need to solve the problem in the case of a homogeneous wall. Hence the necessity of finding a simpler, but obviously, less precise calculation method.

To this end some authors based themselves on the concept of a homogeneous, equivalent wall, defined in different ways.

Particularly, according to Mackey and Wright a composite wall, made up of n homogeneous layers of specific resistance R_j and of specific capacity C_j, is equivalent to a homogeneous wall, whose specific resistance and whose specific capacity are given by the following expressions:

$$R_{eq} = \sum_{j=1}^{n} R_j \tag{1}$$

$$C_{eq} = 1.1 \sum_{j=1}^{n-1} C_j + \frac{C_n}{R_n}\left[R_n - 0.1 \sum_{j=1}^{n-1} R_j\right] \tag{2}$$

where the value n is referred to the layer in contact with the external environment.

It is to be noted that by expression (2), composite walls made up of n different layers [keeping the nth layer fixed and changing the others $(n-1)$ as desired] cannot be distinguished from each other.

Generally, such walls can behave differently. Besides, the relation of Mackey and Wright can be used only in the case of sinusoidal periodic steady condition corresponding to a period of 24 hours.

According to Nessi's and Nissolle's work one could introduce another principle of equivalence, based on the identity of the influence functions [6]. It is very difficult, however, to find two different walls having the same influence functions, since this happens under special conditions which are rarely found in the materials usually employed.

As the method of Mackey and Wright and the considerations drawn by Nessi's and Nissole's definitions do not always give satisfactory results, we have suggested a principle of equivalence between a composite and a homogeneous wall based on the equality of their total coefficients of heat transfer H and their response times τ_0, under standard conditions of the surface resistances.

2. CRITERION OF EQUIVALENCE H, τ_0

If one cannot find a homogeneous wall having the same influence function as a given composite wall, we can, however, univocally define a homogeneous wall whose influence function has an average shifting equal to zero in relation to the corresponding curve of the given wall, i.e., after the surface coefficients of heat transfer have been fixed, the two walls have an equal heat transfer coefficient and an equal response time τ_0 [7, 8].

We have checked the validity of this principle of equivalence by experimentation on electrical analogic patterns in relation to transient heating and to the periodic heat flow produced by temperature fluctuations of a sinusoidal form or corresponding to the curve of the fictitious sol-air temperature [8].

In the case of a sinusoidal period T = 24 hrs such criterion gives percentage differences between the response of the real wall and that of the equivalent homogeneous wall with the same amplitude and time lag. They don't exceed ten percent.

3. DEFINITION OF THE RESPONSE TIME OF A WALL

Consider a composite wall in thermal steady state between two fluids, one at temperature $(t_f')_0$ and the other at temperature $(t_f'')_0$. Suppose that at $\tau = 0$ the temperature t_f' of the first fluid changes from the value $(t_f')_0$ to $(t_f')_1$ following any law, while the temperature of the second fluid is kept constant at the starting value $(t_f'')_0$.

The surface temperature t_s'' at the second fluid side will gradually change from the value $(t_s'')_0$ to a new value $(t_s'')_1$ following a law which is a function of the characteristics of the system and of the form of temperature variation (fig. 1).

If we call $\Delta t_f'(\tau) = t_f'(\tau) - (t_f')_0$ and $\Delta t_s''(\tau) = t_s''(\tau) - (t_s'')_0$ the values at any time respectively of the changes of t_f' and t_s'' referred to the starting values $(t_f')_0$ and $(t_s'')_0$, the response time τ_o of the wall will be given by the following integral:

$$\tau_0 = \int_0^\infty \left[\frac{\Delta t_f'(\tau)}{\overline{\Delta t_f'}} - \frac{\Delta t_s''(\tau)}{\overline{\Delta t_s''}} \right] d\tau \tag{3}$$

where:

$$\overline{\Delta t_f'} = (t_f')_1 - (t_f')_0 ,$$

$$\overline{\Delta t_s''} = (t_s'')_1 - (t_s'')_0 .$$

The τ_0 value is represented by the dotted area in figure 2.

According to the analytical definition of τ_0 and of its geometric interpretation, it is clear that the equality of the response times of two walls takes place when the average shifting between the respective influence functions is equal to zero.

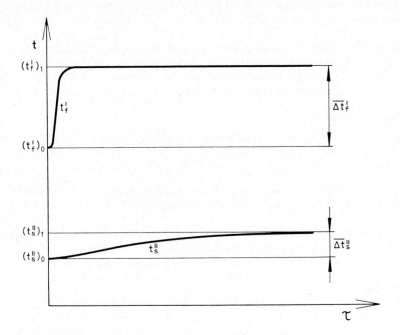

Fig. 1

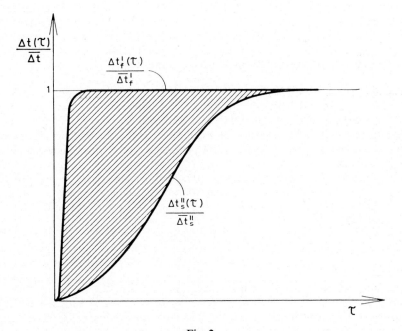

Fig. 2

4. COMPUTING SIMPLIFICATIONS OBTAINED FROM THE USE OF THE EQUIVALENCE PRINCIPLES

Via the principles of equivalence we can substitute a flat multilayer wall by an equivalent homogeneous wall and then we can determine the thermal behaviour of the composite wall by the expressions one can use for a homogeneous wall. Referring to the principle of equivalence (H, τ_0) the response time of a homogeneous wall depends on its resistance **R**, on its capacity C and on the surface resistance according to the following equation:

$$\tau_0^{\neg} = \frac{C}{6} \frac{R^2 + 3R(R_e + R_i) + 6R_e R_i}{R + R_e + R_i} \tag{4}$$

Expression (4) permits us to determine the homogeneous wall equivalent to a given one whose thermal resistance R and response time τ_0 have been found experimentally.

In fact, for fixed values of the surface thermal resistances, the equation (4) gives the value of the thermal capacity of the equivalent wall.

It is to be noted, at this point, that the definition of response time given in the previous paragraph refers to the connection of a wall with an inner environment at steady temperature, equivalent to an infinite thermal capacity.

In addition to the outside wall, the capacity of internal environment and some of the structures assisting it contribute to an unsteady thermal phenomenon.

Thus the relation wall-environment is different from that we assume in the definition for τ_0.

However some tests on electrical analogic patterns showed that for usual values of the thermal resistances and capacities, two walls, equivalent according to the principle (H, τ_0), still practically behave in the same way in the two cases [9].

In conclusion, a homogeneous wall equivalent to a given wall under the conditions fixed by the principle of equivalence still represents it acceptably even under the usual conditions of employment.

5. EXPERIMENTAL DETERMINATION OF THE PARAMETERS H, τ_0

The apparatus, previously described in note [10], consists of two chambers at controlled temperature separated by the wall in question; this has been used to determine the parameters H and τ_0 experimentally (fig. 3).

The installation, an ASTM type with a calorimetric chamber and a guard chamber, has been modified in order to determine the parameter τ_0 and other quantities peculiar to thermal transient heat flow.

We measured the coefficient of heat transfer H under steady conditions, integrating the heat flow rate produced in the calorimetric chamber as well as the difference of the average air temperatures of the two parts of the wall for a suitable time T'.

We assumed:

$$H = \frac{\int_0^{T'} \Delta q \, d\tau}{\int_0^{T'} \Delta t \, d\tau} . \tag{5}$$

The integrations, obtained with suitable instruments, are necessary since the calorimetric chamber is supplied with a steady heat flux as well as with an intermittent flux to regulate its temperature and bring it to the same average value of the temperature fixed in the guard chamber. In order to determine τ_0 the system air-wall-air is,

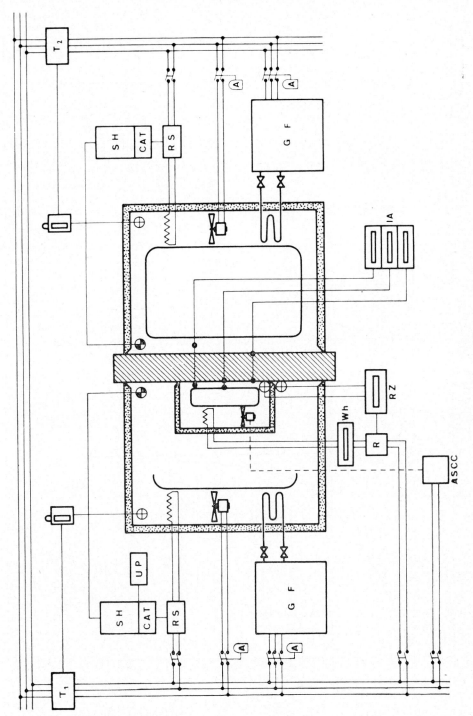

Fig. 3 — (see legend pag. 56)

51

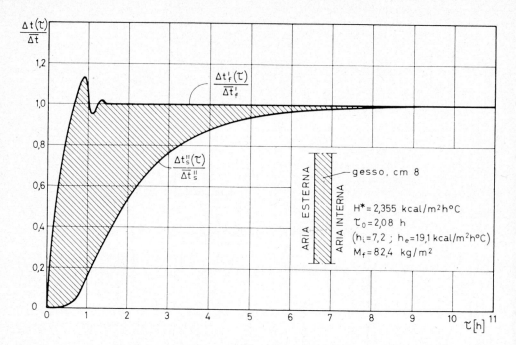

Fig. 4

Aria esterna – Outside air   Aria interna – Inside air   Gesso – Plaster

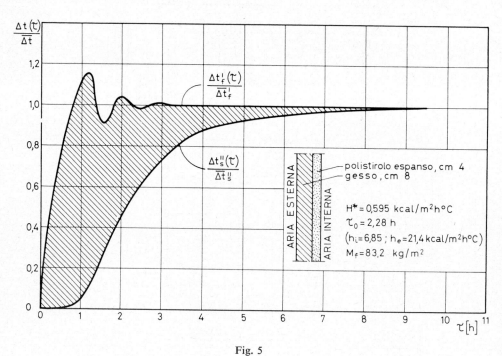

Fig. 5

Aria esterna   – Outside air
Polistirolo espanso – Expanded polystyrene

Aria interna – Inside air
Gesso   – Plaster

at the beginning, under steady conditions since the difference between the warm air temperature and the cold air temperature is also steady. Usually the warm air is kept at $+20°C$ and the cold air at $+5°C$. At the instant $\tau = 0$, the set-point of the temperature regulator of the cold chamber is brought to $-5\,°C$. The temperature changes of the cold air and the warm surface of the wall are recorded.

The temperatures are taken through two thermopiles made up of five differential copper-constantan elements with the reference junctions at fixed steady temperature.

One can obtain the response time from the experimental curves, according to what has already been said in paragraph 3. Figures 4 and 5 give the results of the tests carried out on two walls.

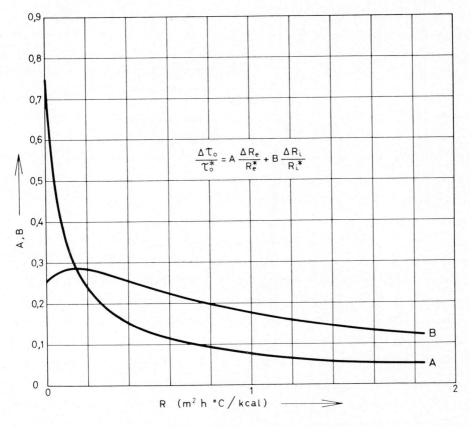

Fig. 6

The τ_0 values we have previously given represent the average of at least five determinations, each being different by $\pm5\%$ from the corresponding average value. It is to be noted that, as the response time depends on the surface resistances R_e and R_i, in order to compare different walls, we must give standard values to these resistances. It is difficult to obtain these standard values during the experimental determination of τ_0.

Therefore, one must know the connection between the changes in τ_0 and those of R_e and R_i in relation to their standard values. This is simple for a homogeneous wall,

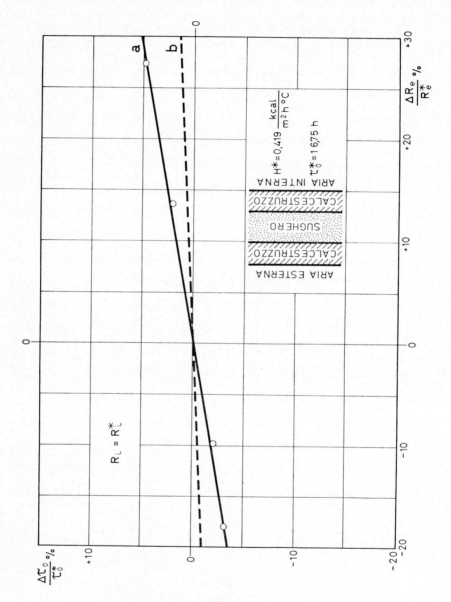

Fig. 7

Aria esterna	–	Outside air
Calcestruzzo	–	Concrete
Sughero	–	Cork
Aria interna	–	Inside air

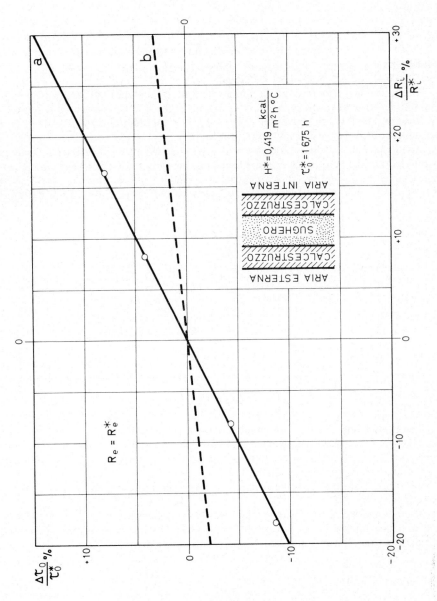

Fig. 8

Aria esterna – Outside air
Calcestruzzo – Concrete
Sughero – Cork
Aria interna – Inside air

as from the relation (4) we get:

$$\frac{\Delta\tau_0}{\tau_0^*} = A\,\frac{\Delta R_e}{R_e^*} + B\,\frac{\Delta R_i}{R_i^*} \tag{6}$$

where R_e^* and R_i^* represent the standard values of the surface specific resistances, τ_0^* is the value of the response time corresponding to them. A and B take over the values quoted in the diagrams of figure 6 [11], against R, being $R_e^* = 0.05$ m^2 h °C/kcal and $R_i^* = 0.15$ m^2 h °C/kcal.

One must remember that for the composite walls, the relation (4) can be used when we consider the equivalent capacity as a function of the surface resistances R_e and R_i.

Usually however, the function C (R_e, R_i) and the diagrams of A and B type are unknown in the case of any composite wall.

The behaviour of some multilayer walls has been examined by electrical analogy with regard to the percentage changes in the response time τ_0 from τ_0^* against percentage changes of R_e and R_i from the respective standard values. Figures 7 and 8 show the behaviour of a wall having the characteristics indicated in the following table

Layer	Thermal specific resistance (m^2 h °C/kcal)	Thermal specific capacity (kcal/m^2 °C)
Inside	0.0685	57.11
Central	2.048	4.878
Outside	0.0685	57.11

The dotted line *b* shows the result obtained from relation (4), considering the homogeneous wall equivalent under standard conditions, i.e., neglecting the dependence of C_{eq} on R_e and R_i while the continuous line *a* interpolates the experimental points obtained from the analogic pattern.

Diagrams like these have been obtained for other walls and show results of the same type.

Therefore it is clear that one cannot neglect the dependence of C_{eq} on R_e and R_i and that one must take measurements under conditions as near as possible to the standard ones and, later, to correct the results by interpolation.

LEGEND OF FIGURE 3

T_1, T_2	safety remote control switches;
SH + CAT + SR	temperature regulation units;
UP	program unit (optionally connected on the cold side or warm side temperature regulation unit);
GF	Refrigerating units;
ASCC	Stabilized direct current supply;
I.A.	Integration units;
Wh	Watt-hour meter;
RZ + R	Zero regulator unit.

LIST OF SYMBOLS

C	thermal specific capacity of wall	$(kcal/m^2 \, ^\circ C)$
H	thermal transmittance	$(kcal/m^2 \, h \, ^\circ C)$
q	rate of heat flow	$(kcal/m^2 h)$
R	thermal resistance	$(m^2 \, h \, ^\circ C/kcal)$
t	temperature	$(^\circ C)$
τ, T	time	(h)

REFERENCES

[1] C.O. Mackey and L.T. Wright, Periodic heat flow: composite walls or roofs. *A.S.H.V.E. Transactions* (1946), vol. 52, pp. 283-296.
[2] Z.U.A. Warsi and N.K.D. Choudhury, Weighting function and transient thermal response of buildings. *Int. J. Heat Mass Transfer* (1964), vol. 7, pp. 1309-1334.
[3] U. Magrini, Aspetto tecnico della trasmissione di un onda termica attraverso una parete. *La Termotecnica* (1965), vol. 19, pp. 235-251.
[4] J.S. Alford, J.E. Ryan and F.O. Urban, Effect of heat storage and variation in outdoor temperature and solar intensity on heat transfer through walls. *A.S.H.V.E. Transactions* (1939), vol. 45, pp. 369-396.
[5] D. Le Febve de Vivy, Dynamique des températures d'une paroi. Fonctions de transfert et courbes de référence. *Revue Générale de Thermique*, nᵒ 49 (1966), pp. 21-31, et nᵒ 78 (1968), pp. 619-628.
[6] A. Nessi et L. Nisolle, Fonctions d'influence de flux de chaleur des parois de construction. *Rapport Comité Tech. Indust. Chauffage*, Paris (1947).
[7] M. Sovrano, Le temps de réponse dans la détermination de la diffusivité thermique des matériaux. *Annexe 1964-2 au Bulletin de l'I.I.F.*, pp. 23-43.
[8] M. Sovrano and G. Zorzini, Scelta e misura delle grandezze atte a caratterizzare il comportamento termico delle pareti in regime variabile. *La Termotecnica*, nᵒ 7 (1966), pp. 409-415.
[9] M. Sovrano and G. Zorzini, Il coefficiente di trasmissione K ed il tempo di risposta τ_0 nello studio del comportamento termico delle pareti. *Ricerche di Termotecnica*, nᵒ 16, (1966), pp. 70-81.
[10] P. Di Filippo, Nuovo impianto per le prove di trasmissione termica sulle pareti prefabbricate. *Ricerche di Termotecnica*, nᵒ 16 (1966), pp. 63-69.
[11] P. Di Filippo and G. Zorzini, Misura del coefficiente di trasmissione termica H e del tempo di risposta τ_0 delle pareti. *La Termotecnica*, nᵒ 5 (1969), pp. 249-254.

DISCUSSION

J.N. FAFCHAMPS (Belgique) — La méthode de calcul simplifiée de la réponse d'une paroi composite à une sollicitation variable, proposée par les auteurs consiste à exploiter les éléments de la réponse d'une paroi homogène équivalente. Cette dernière (la paroi homogène) est définie sur la base de deux critères d'équivalence: H (coefficient total de transfert) et τ_0 (temps de réponse) déterminés expérimentalement, sur la paroi composite, dans des conditions (Re, Ri) aussi voisines que possible des conditions standards, le critère τ_0 résultant d'une intégration dans le temps de la courbe de réponse à une sollicitation échelonnée.

Or les conditions expérimentales sont conformes aux hypothèses : 1. de linéarité du système; 2. de repos ou d'équilibre au temps initial; 3. de valeur finale (c'est-à-dire aux grandes valeurs du temps) invariable, de telle sorte que la courbe expérimentale (différence des ordonnées des deux courbes enregistrées) fournit la réponse indirecte de la paroi réelle, à partir de laquelle il est possible de déduire, par différentes méthodes classiques, les caractéristiques de la réponse harmonique : atténuation et déphasage, pour toute fréquence souhaitée de même d'ailleurs que la réponse à une sollicitation quelconque non périodique à partir de l'application du théorème de Duhamel mais alors il faut traduire analytiquement la courbe expérimentale.

Il nous paraît dès lors que la courbe expérimentale relevée constitue, comme telle, une source d'informations plus riche que son intégrale τ_0. Je voudrais alors terminer par une question : les auteurs ont-ils eu l'occasion de comparer, par le biais de l'analogie électrique, les caractéristiques de la réponse de la paroi homogène équivalente avec celles déduites, comme je viens de le rappeler, de la réponse indirecte de la paroi réelle ? et si oui, quelle est l'importance des écarts ?

P. DI FILIPPO — On a effectué beaucoup d'essais comparatifs au moyen du modèle analogique entre la paroi réelle et la paroi homogène équivalente déterminée par H et τ_0.

Nous avons trouvé des erreurs au maximum de 5 % pour l'amplitude et de 10 % jusqu'à 15 % pour le déphasage respectivement d'une paroi indéfinie et d'une paroi connectée avec d'autres structures.

Les résultats se réfèrent à des variations sinusoïdales de la température le long de la journée et à des variations du type « sol-air » selon Mackey et Wright.

Les erreurs dépendent avant tout de la nature et de la disposition des matériaux constituant la paroi.

Les résultats obtenus soit pour les mesures d'amplitude soit pour celles du déphasage de réponse, concordent mieux selon la méthode proposée par les AA. que selon celle de Mackey et Wright [8, 9].

J. J. TIREL (France) — M. DI FILIPPO nous a exposé une méthode simplifiée de calcul de transmission de chaleur à travers des parois composées en les caractérisant uniquement par deux paramètres.

Quel est l'intérêt et le domaine d'application de cette méthode simplifiée à un moment où l'ordinateur devrait permettre l'utilisation de méthodes de plus en plus complexes et rigoureuses ?

F. MATTAROLO (Italie) — Le but du travail de M. DI FILIPPO a été de définir une grandeur qui soit mesurable.

Pour les murs minces l'indication du coefficient de transmission K ne suffit pas ; il faut ajouter quelque chose qui serve à caractériser leur comportement en régime variable. On pourrait introduire le rapport d'amplitude et le déphasage pour une variation sinusoïdale. M. DI FILIPPO a introduit le temps de réponse τ_0 qu'on obtient en réalisant sur un côté du mur une variation quelconque de la température de l'air qui arrive à une condition de régime et en mesurant la réponse sur la surface opposée, pendant que la température de l'air de ce côté est maintenue constante. Les valeurs de K et τ_0 caractérisent le mur.

ÉTUDE ANALYTIQUE DU RÉGIME VARIABLE DE STRUCTURES THERMIQUEMENT ACTIVES*

J. J. PORTIER

Institut de Thermodynamique et CEDRIC, Université de Liège (Belgique)

Analytical study of thermally active structures in unsteady state conditions

SUMMARY: *This report gives an analytical method for the study of the propagation of heat by unsteady state conduction which is based on the possibility that one can develop a function satisfying Dirichlet's conditions in an interval bearing, in a series of orthogonal functions in this same domain. Assuming the linearity of the partial derivatives equation and that of boundary conditions, the process is then to define, by a series of first order differential equations whose second members are directly expressed from non-homogeneous data, coefficients of development of the investigated value in the infinite system of space harmonics of the homogeneous problem. The systematic nature of its application is demonstrated by a two-dimensional example of the "heating ceiling" from which one can outline a parametric analysis.*

INTRODUCTION

L'importance pratique de la connaissance de la distribution des températures dans les structures de nombreuses applications de la technique, et la difficulté de leur détermination par les méthodes analytiques usuelles suffisent à justifier l'intérêt de la présentation d'une méthode générale de résolution systématique des problèmes non homogènes de propagation de la chaleur par conduction.

Extension de celui de Fourier, le formalisme proposé postule, comme lui, la linéarité de l'équation aux dérivées partielles et celle des conditions aux limites, ce qui restreint son application à la considération d'écarts de température permettant d'admettre la constance des caractéristiques thermophysiques des matériaux envisagés. Sa seule exigence théorique supplémentaire est la possibilité de définir le problème dans un domaine physique d'étendue limitée où il résout l'équation

$$\text{div} \left[\lambda(\vec{r}) \, \text{grad} \, T(\vec{r}, t) \right] = \rho(\vec{r}) \, c(\vec{r}) \frac{\partial T}{\partial t}(\vec{r}, t) - Q_v(\vec{r}, t)$$

avec des conditions aux limites des trois types classiques imposées, pour la généralité de son exposé, sur des portions adjacentes de la frontière

$$T(\vec{r}, t) = T_a(\vec{r}, t) \qquad \text{sur } A_1$$

$$\lambda \frac{\partial T}{\partial n}(\vec{r}, t) = Q_a(\vec{r}, t) \qquad \text{sur } A_2$$

$$\lambda \frac{\partial T}{\partial n}(\vec{r}, t) + h T(\vec{r}, t) = {}_t h T_e(\vec{r}, t) \qquad \text{sur } A_3$$

et la répartition initiale

$$T(\vec{r}, t) = T_0(\vec{r}).$$

Son principe revient à définir, par une suite d'équations différentielles du premier ordre, les coefficients du développement de la grandeur recherchée dans le système infini

---

* Cette communication est extraite d'une étude de synthèse en préparation sous la direction du Professeur G. BURNAY.

des fonctions orthogonales que constitue, dans le volume étudié, l'ensemble des harmoniques spatiaux de la solution du problème homogénéisé.

Dans ces relations, outre les symboles λ, ρ et c qui traduisent la conductivité, la masse volumique et la chaleur spécifique du corps considéré, h désigne son coefficient de transfert global par convection et rayonnement avec le milieu continu à la température locale T_e qui l'entoure, Q_v représente le débit des sources volumiques qui s'y trouvent localisées, tandis que T_a et Q_a précisent respectivement la température et l'intensité du flux calorifique qui caractérisent les conditions de Dirichlet et de Neumann imposées sur sa surface.

LA MÉTHODE DE RÉSOLUTION

La solution du problème homogénéisé exprime le processus d'évolution de la distribution initiale dans la structure finie considérée, supposée passive et limitée par une paroi, soit athermane, soit maintenue à une température nulle ou en contact avec un milieu de même définition. Sa recherche, par séparation de la variable temporelle

$$T_p(\vec{r}, t) = U(\vec{r}).u(t)$$

définit un spectre discret de constantes

$$p \equiv p_n \qquad (n = 1, 2,...)$$

qui déterminent la décroissance exponentielle, dans le temps, de chacun des modes spatiaux de relaxation de la température

$$U(\vec{r}) \equiv U_n(\vec{r})$$

lesquels vérifient les équations

$$\text{div}\,(\lambda\,\text{grad}\,U_n) + p_n^2 \rho c U_n = 0$$

avec les conditions aux limites

$$U_n = 0 \qquad\qquad \text{sur } A_1$$

$$\frac{\partial U_n}{\partial n} = 0 \qquad\qquad \text{sur } A_2$$

$$\lambda \frac{\partial U_n}{\partial n} + h U_n = 0 \qquad\qquad \text{sur } A_3$$

et satisfont les relations

$$\int_V \rho c U_n U_m dV = 0 \qquad\qquad (m \neq n)$$

comme le montre l'intégration, sur le domaine envisagé, de l'égalité vectorielle

$$\text{div}\,[\lambda(U_n\,\text{grad}\,U_m - U_m\,\text{grad}\,U_n)] = U_n\text{div}(\lambda\,\text{grad}\,U_m) - U_m\,\text{div}(\lambda\,\text{grad}\,U_n).$$

Avec la suite infinie de ces fonctions orthogonales, la température du problème non homogène peut, si elle remplit les conditions de Dirichlet, s'exprimer par le développement en série

$$T(\vec{r}, t) = \sum_n u_n(t) \cdot U_n(\vec{r})$$

dont les coefficients, temporellement dépendants, sont définis par les intégrales

$$u_n(t) = \frac{1}{q_n^2} \int_V \rho c T U_n dV$$

où l'on a introduit la notation

$$q_n^2 = \int_V \rho c U_n^2 dV$$

et apparaissent comme les solutions d'équations différentielles du premier ordre dont le second membre s'explicite entièrement à partir du débit des sources de volume et des conditions aux limites par l'intégration de l'égalité

$$\text{div} \left[\lambda (T \, \text{grad} \, U_n - U_n \, \text{grad} \, T) \right] = T \, \text{div}(\lambda \, \text{grad} \, U_n) - U_n \, \text{div}(\lambda \, \text{grad} \, T)$$

qui livre, eu égard aux équations satisfaites par les fonctions qui y figurent, l'expression

$$\int_V \rho c \left(\frac{\partial T}{\partial t} + p_n^2 T \right) U_n dV = \int_V Q_v U_n dV + \int_{A_1 \, \text{U} A_2 \, \text{U} A_3} \lambda \left(\frac{\partial T}{\partial n} U_n - \frac{\partial U_n}{\partial n} T \right) dA$$

laquelle conduit, avec les conditions imposées sur la frontière, aux relations

$$\frac{du_n}{dt} + p_n^2 u_n = W_n(t)$$

écrites en posant

$$q_n^2 W_n(t) = \int_V Q_v U_n dV - \int_{A_1} \lambda T_a \frac{\partial U_n}{\partial n} dA + \int_{A_2} Q_a U_n dA + \int_{A_3} h T_e U_n dA$$

Par variation des constantes, il vient alors

$$u_n(t) = u_n(0) e^{-p_n^2 t} + \int_0^t W_n(\tau) e^{-p_n^2(t-\tau)} d\tau$$

où les inconnues sont précisées par la situation initiale

$$u_n(0) = \frac{1}{q_n^2} \int_V \rho c T_0 U_n dV$$

et la distribution des températures dans la structure considérée affecte enfin la forme générale

$$T(\vec{r}, t) = \sum_n \frac{1}{q_n^2} \left\{ \int_V \rho c T_0 U_n dV + \int_0^t \left[\int_V Q_v U_n dV - \int_{A_1} \lambda T_a \frac{\partial U_n}{\partial n} dA + \right. \right.$$

$$\left. \left. + \int_{A_2} Q_a U_n dA + \int_{A_3} h T_e U_n dA \right] e^{p_n^2 \tau} d\tau \right\} U_n e^{-p_n^2 t}$$

Quant à la densité de flux dans la direction du vecteur unitaire $\vec{1}_l$, elle se calcule simplement par dérivation, terme à terme, de cette série puisque la formule du gradient permet d'exprimer les coefficients de son propre développement en les mêmes fonctions orthogonales comme

$$v_n(t) = -\frac{1}{q_n^2}\sum_m u_m(t)\int_V \rho c\,\lambda\,\text{grad}\,U_m\,\vec{1}_l U_n\,\mathrm{d}V$$

et en conséquence d'écrire

$$\text{grad}\,T\,\vec{1}_l = \sum_n u_n(t)\,\text{grad}\,U_n\,\vec{1}_l.$$

Dans le cas particulier où la répartition initiale est uniforme, le théorème d'Ostrogradsky se prête encore à la transformation de la première intégrale apparaissant dans l'expression de la température qui devient alors

$$T(\vec{r},t) = -T_0\sum_n \frac{1}{p_n^2 q_n^2}\left[\int_{A_1}\lambda\frac{\partial U_n}{\partial n}\mathrm{d}A - \int_{A_3}hU_n\,\mathrm{d}A\right]U_n\mathrm{e}^{-p_n^2 t} +$$

$$+ \sum_n\left[\int_0^t W_n(\tau)\mathrm{e}^{p_n^2\tau}\mathrm{d}\tau\right]U_n\mathrm{e}^{-p_n^2 t}.$$

Si enfin une condition de Neumann est imposée à l'entièreté de la frontière, la solution évidente

$$U_1(\vec{r}) = 1$$

qui résulte de la compatibilité de la condition

$$\frac{\partial U_1}{\partial n} = 0 \qquad \text{sur A}$$

avec l'équation

$$\text{div}(\lambda\,\text{grad}\,U_1) = 0$$

pour le choix $p_1^2 = 0$ peut être traitée séparément pour définir la distribution recherchée, en omettant la présence de sources de volume, par

$$T(\vec{r},t) = \frac{1}{\int_V \rho c\mathrm{d}V}\left[\int_V\rho cT_0\,\mathrm{d}V + \int_0^t\int_A Q_a\mathrm{d}A\mathrm{d}\tau\right] +$$

$$+ \sum_n\frac{1}{q_n^2}\left[\int_V\rho cT_0\,U_n\,\mathrm{d}V + \int_0^t\mathrm{e}^{p_n^2\tau}\int_A Q_a U_n\,\mathrm{d}A\mathrm{d}\tau\right]U_n\mathrm{e}^{-p_n^2 t}.$$

UN EXEMPLE BIDIMENSIONNEL

À titre d'exemple, on développera le calcul de l'évolution de la répartition des températures dans un plafond chauffant, idéalisé par une plaque homogène où sont noyées des conduites minces, équidistantes et parcourues par un fluide actif. Ces sources, que l'on supposera filiformes, sont situées dans un plan parallèle aux faces de la structure que schématise la figure 1 et dont l'échange avec l'ambiance, externe et interne, est caractérisé par des coefficients de transfert constants. La température de ces milieux, stationnaire à l'intérieur, traduit par contre les variations extérieures quotidiennes en oscillant autour d'une valeur moyenne qui définit également la distribution uniforme imposée à l'origine des temps.

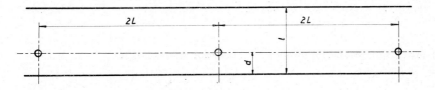

Fig. 1

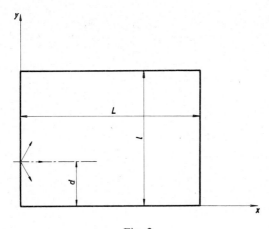

Fig. 2

Si l'on peut admettre que la surface des conduits est isotherme, le problème peut être traité comme bidimensionnel pour une section transversale; si de plus, le débit de chacune des sources suit la même loi de variation temporelle, on peut se borner, avec le choix des coordonnées de la seconde figure, à la considération du rectangle

$$0 \leqslant x \leqslant L \qquad\qquad 0 \leqslant y \leqslant l$$

athermane sur ses faces $x = 0$ et $x = L$, et qui contient une demi-source en $x = 0$, $y = d$. L'équation aux dérivées partielles

$$\text{div}(\text{grad } U) + \frac{p^2}{a^2} U = 0$$

où a^2 désigne la diffusivité du matériau constitutif, et les conditions aux limites

$$\frac{\partial U}{\partial x} = 0 \qquad\qquad \text{pour } x = 0 \text{ et } x = L$$

$$\lambda \frac{\partial U}{\partial y} - h_i U = 0 \qquad \text{pour } y = 0$$

$$\lambda \frac{\partial U}{\partial y} + h_e U = 0 \qquad \text{pour } y = l$$

63

que doivent vérifier les harmoniques spatiaux du problème homogénéisé, sont satisfaites par la double famille des fonctions

$$U_{nm}(x, y) = \cos \frac{n\pi}{L} x \cdot \left(\delta_m \cos \delta_m \frac{y}{l} + \frac{h_i l}{\lambda} \sin \delta_m \frac{y}{l} \right) \qquad (n = 0, 1, 2, \ldots)$$

où les δ_m sont les racines positives de l'équation transcendante

$$\delta \cotg \delta = \frac{\lambda^2 \delta^2 - h_i h_e \, l^2}{\lambda l (h_i + h_e)}$$

avec

$$p_{nm}^2 = a^2 \left(\frac{n^2 \pi^2}{L^2} + \frac{\delta_m^2}{l^2} \right).$$

De façon systématique, on trouve alors

$$q_{nm}^2 = \int_V \rho c U_{nm}^2 \, dV = \begin{cases} \rho c \, \dfrac{Ll}{2A_m} & \text{pour } n = 0 \\[2ex] \rho c \, \dfrac{Ll}{4A_m} & \text{pour } n \neq 0 \end{cases}$$

en introduisant la notation

$$A_m = \frac{\lambda^2 (\lambda^2 \delta_m^2 + h_e^2 \, l^2)}{(\lambda^2 \delta_m^2 + h_i^2 \, l^2)(\lambda^2 \delta_m^2 + h_e^2 \, l^2) + \lambda l (h_i + h_e)(\lambda^2 \delta_m^2 + h_i h_e \, l^2)}$$

puis successivement, pour l'effet des sources, si $Q(t)$ est le débit de l'une d'entre elles

$$\int_V Q_v \, U_{nm} \, dV = Ll U_{nm}(0, d) Q(t) = Ll \left(\delta_m \cos \delta_m \frac{d}{l} + \frac{h_i l}{\lambda} \sin \delta_m \frac{d}{l} \right) Q(t)$$

pour l'effet du transfert par la paroi $y = 0$, si T_i est la température constante du
milieu intérieur

$$\int_A h_i T_i U_{nm} \, dA = h_i T_i \int_0^L U_{nm}(x, 0) dx = \begin{cases} h_i T_i L \delta_m & \text{pour } n = 0 \\ 0 & \text{pour } n \neq 0 \end{cases}$$

et enfin, pour la contribution, par la face $y = l$, de l'ambiance à la température
$T_e + \Delta T \sin \omega t$

$$\int_A h_e (T_e + \Delta T \sin \omega t) U_{nm} \, dA = h_e (T_e + \Delta T \sin \omega t) \int_0^L U_{nm}(x, l) dx$$

$$= \begin{cases} h_e (T_e + \Delta T \sin \omega t) L \dfrac{\lambda^2 \delta_m^2 + h_i^2 \, l^2}{l\lambda (h_i + h_e)} \sin \delta_m & \text{pour } n = 0 \\[2ex] 0 & \text{pour } n \neq 0 \end{cases}$$

64

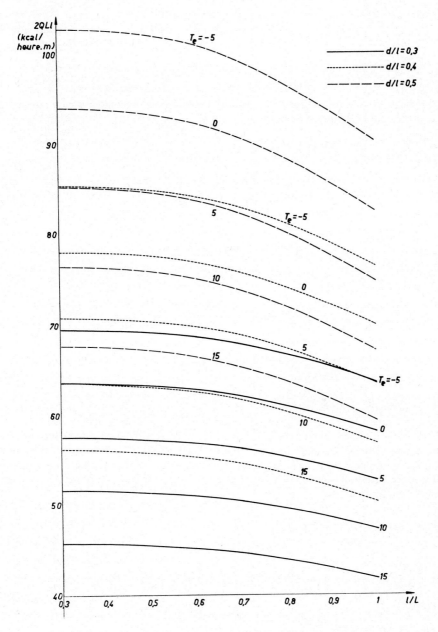

Fig. 3

Si l'on ajoute que la répartition initiale uniforme livre

$$\int_V \rho c T_e U_{nm} \, dV = \rho c T_e \int_0^L \int_0^l U_{nm}(x, y) dx \, dy$$

$$\rho c L l T_e \left[\frac{h_e}{\lambda^2 (h_i + h_e)} \frac{\lambda^2 \delta_m^2 + h_i^2 l^2}{\delta_m^2} \sin \delta_m + \frac{h_i l}{\lambda} \frac{1}{\delta_m} \right] \quad \text{pour } n = 0$$

$$= $$

$$0 \qquad\qquad\qquad\qquad\qquad\qquad\qquad\qquad\qquad \text{pour } n \neq 0$$

l'application de la formule générale définit l'évolution de la température en tout point de la structure considérée comme

$$T(x, y, t) = T_e + 2 \frac{h_i l}{\lambda} (T_i - T_e) \sum_m \frac{A_m}{\delta_m} (1 - e^{-p_{0m}^2 t}) U_{0m}(y)$$

$$+ 2 \frac{a^2}{l^2 \lambda^2} \frac{h_e}{(h_i + h_e)} \Delta T \sum_m A_m (\lambda^2 \delta_m^2 + h_i^2 l^2) \times$$

$$\times \sin \delta_m \frac{p_{0m}^2 \sin \omega t - \omega \cos \omega t + \omega \, e^{-p_{0m}^2 t}}{p_{0m}^4 + \omega^2} U_{0m}(y)$$

$$+ 2 \frac{l^2}{\lambda} Q \sum_m A_m U_{0m}(d) \left[\frac{1 - e^{-p_{0m}^2 t}}{\delta_m^2} U_{0m}(y) + \right.$$

$$\left. + 2 \sum_{n \neq 0} \frac{1 - e^{-p_{n,m}^2 t}}{\delta_m^2 + n^2 \pi^2 l^2 / L^2} U_{nm}(x, y) \right]$$

où le débit des sources volumiques a été supposé constant. Ce sont précisément les résultats du calcul de cette grandeur, rapportée à l'unité de longueur des conduits enrobés dans la paroi envisagée, pour que la température maximale de sa face interne ne dépasse pas 35°C, que montre la figure 3, en fonction de l/L, pour différentes valeurs du rapport d/l et de la température moyenne du milieu extérieur dont l'amplitude de la variation sinusoïdale a été fixée à 5°C sur une période de 24 heures. Définies par la relation

$$35 = T_e + 2 \frac{h_i l}{\lambda} (T_i - T_e) \sum_m A_m$$

$$+ 2 \frac{a^2}{l^2 \lambda^2} \frac{h_e}{(h_i + h_e)} \Delta T \, \max_{t \in [0, \, 2\pi/\omega]} \sum_m A_m (\lambda^2 \delta_m^2 + h_i^2 l^2) \delta_m \times$$

$$\times \sin \delta_m \frac{p_{0m}^2 \sin \omega t - \omega \cos \omega t}{p_{0m}^4 + \omega^2}$$

$$+ (2QLl) \frac{l}{\lambda L} \sum_m A_m U_{0m}(d) \delta_m \left(\frac{1}{\delta_m^2} + 2 \sum_{n \neq 0} \frac{1}{\delta_m^2 + n^2 \pi^2 l^2 / L^2} \right)$$

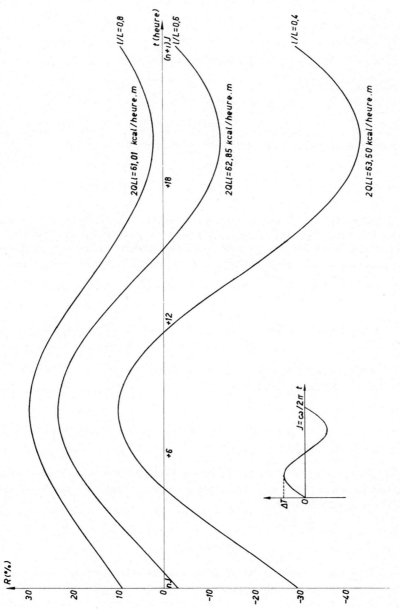

Fig. 4

les valeurs présentées correspondent à une paroi épaisse de 0,20 m, faite d'un matériau doué des caractéristiques

$$\lambda = 1 \text{ kcal/m.h. } °C$$

$$\rho = 2200 \text{ kg/m}^3$$

$$c = 0,2 \text{ kcal/kg. } °C$$

et dont l'échange avec le milieu intérieur à la température de 20°C est défini par le coefficient de transfert

$$h_i = 6 \text{ kcal/m}^2\text{.h. } °C$$

tandis que, sur l'autre face

$$h_e = 10 \text{ kcal/m}^2\text{.h. } °C.$$

Quant à la convergence de leur obtention, elle a été facilitée par la sommation, dans l'expression précédente, de la dernière série qui y apparaît et qui vaut

$$\sum_{n \neq 0} \frac{1}{\delta_m^2 + n^2 \pi^2 \, l^2/L^2} = \frac{1}{2}\left(\frac{L}{l}\frac{1}{\delta_m}\text{cotgh }\delta_m\frac{L}{l} - \frac{1}{\delta_m^2}\right).$$

La dernière figure traduit enfin l'influence considérable de l'écartement des sources sur le rendement de l'installation, par l'allure de variation, durant une période des oscillations extérieures et après qu'aient disparu les transitoires temporels de la mise en régime, du rapport des débits calorifiques émis dans le local à échauffer et injectés au sein de la structure elle-même :

$$R(t) = \frac{h_i \int_0^L [\overline{T(x, 0, t)} - T_i]dx}{QLl} = 2\frac{h_i l}{\lambda}\sum_m \frac{A_m}{\delta_m}U_{0m}(d)$$

$$+ \frac{2h_i}{Ql}\left[(T_i - T_e)\left(\frac{h_i l}{\lambda}\sum_m A_m - \frac{1}{2}\right)\right.$$

$$+ \frac{a^2}{l^2\lambda^2}\frac{h_e}{(h_i + h_e)}\Delta T\sum_m A_m(\lambda^2\delta_m^2 + h_i^2 l^2)\,\delta_m\sin\delta_m\frac{p_{0m}^2\sin\omega t - \omega\cos\omega t}{p_{0m}^4 + \omega^2}\right].$$

Son calcul a été conduit, pour une localisation des sources définie par $d/l = 0,3$ et une température extérieure moyenne de 0°C, avec les valeurs numériques des débits obtenus à l'application précédente.

REMARQUE

R. LANDSBERG (Israel) — There seems to be an exaggerated emphasis on problems of heat conduction, neglecting convection. In actual refrigeration, convection is a major factor; the cooling down of fruit (except when dipped in chilled water), the impact—together with radiation on the outside of buildings (strong winds may reduce radiation), etc. Several methods of conduction problems do not seem relevant to refrigeration or air conditioning, while the late distribution of preprints renders reasoned criticism difficult.

RESISTANCE NETWORK ANALOGUE METHOD FOR TWO-DIMENSIONAL UNSTEADY TEMPERATURE DISTRIBUTION UNDER REFRIGERATION INSTALLATIONS

E.S. KURYLEV and E.L. LICHTENSTEIN

Technological Institute of Refrigeration Industry, Leningrad (U.S.S.R)

Méthode d'analogie au moyen d'un réseau de résistances, pour la répartition de la température instable à deux dimensions dans les installations frigorifiques.

RÉSUMÉ : *Ce rapport indique les résultats de recherche d'un champ de température à deux dimensions dans une installation frigorifique. Les résultats ont été obtenus à l'aide de la méthode d'analogie pour réseau de résistances.*

Pour représenter les problèmes donnés, on a appliqué une méthode avantageuse de différence finie (dite méthode de division différentielle des opérateurs).

Au début, on a tiré par représentation la solution du problème à deux dimensions dans les mêmes conditions que celles obtenues par expérience physique. Le désaccord ne dépasse pas 7%.

D'après les méthodes décrites, on a représenté les problèmes de la répartition de température en régime instable dans des pistes de patinage et des anneaux de vitesse dans diverses conditions.

L'analyse des résultats de l'application de la méthode par analogie permet de donner des recommandations pratiques.

Resistance network analogue method is one of the most advantageous research methods of the temperature distribution under refrigeration installations, as it allows multidimensional problems to be solved quickly and accurately enough, as well as the effect of boundary conditions, variable thermal constants and the process duration to be ascertained. Accurate analytical solving of such problems is not available at present and receiving data by means of physical experiment is practically unfeasible.

To study a flat refrigeration installation temperature distribution with the negative temperature on the ground surface, a two-dimensional nonlinear differential equation for heat conductivity is to be modelled

$$\frac{\partial t}{\partial r} = a_i \left(\frac{\partial^2 t}{\partial x^2} + \frac{\partial^2 t}{\partial y^2} \right) \pm \frac{\omega}{(c\gamma)_i}, \tag{1}$$

where:

$a_i = K_i/(c\gamma)_i$ thermal diffusivity for thawed or frozen areas of the soil, m^2/s;

$(c\gamma)_i$ specific volume heat capacity, $kJ/m^3 \cdot deg$;

$\omega = Q_0/\delta r$ latent heat extracted or absorbed at the boundary of the areas per unit of time in a unit soil volume, $kJ/m^3 \cdot s$.

As the boundary conditions (fig. 1) temperatures at the ground surface ($y = 0$, $t = t_g$), that at some depth under the construction ($y = l$, $t = t_l$) as well as ambient temperature have been set. The temperature distribution in the soil under the construction and round it by the beginning of the process is also set and the temperature distribution in the soil at some distance from the construction is believed to be independent upon its temperature conditions, i.e., $dg_{x=\pm a} = 0$.

The methods of modelling such type of problems is presented in Ref. [1, 2]. However, direct analogue method for two-dimensional problems on the thermal state of soil under big constructions such as cold stores, artificial skating rinks, etc., are hardly feasible. Dividing the area under research even into big distance intervals requires that several hundred electrical resistances should be inserted into the network-analogue.

Therefore, for modelling the given problems, the finite difference method for calculating multidimensional heat conductivity equation (so called differential operator splitting method) offered by N. N. Yanenko [3] has been used.

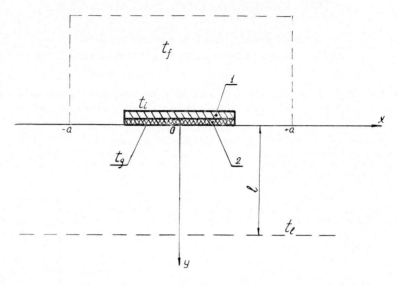

Fig. 1 — The scheme of the problem under solution: 1. ice; 2. cooling plate.

The following difference expression can be approximated to the heat conductivity equation in its simplest form for a two-dimensional problem

$$\frac{t^{k+1} - t^k}{r} = \Lambda \cdot t^{k+1} \tag{2}$$

where:

t^K temperature at the given point of the two-dimensional area during time interval K;
t^{k+1} temperature at the same point of the area during the next time interval $k+1$.

According to Ref. [3] this general expression may be substituted by non-obvious expression due to which the value of t^{k+1} can be determined from two one-dimensional difference equations:

$$\frac{t^{k+1/2} - t^k}{r} = \Lambda_1 \cdot t^{k+1/2} \tag{3}$$

$$\frac{t^{k+1} - t^{k+1/2}}{r} = \Lambda_2 \cdot t^{k+1} \tag{4}$$

Approximation and stability of such an expression has been studied in Ref. [3].

Application of the analogue method means obtaining a two-dimensional temperature distribution at the end of each time interval and dividing it into two half steps. Some fictitious temperature distribution at the end of the first half step is received by modelling a one-dimensional equation, taking into consideration the temperature variation only along the y-axis (netwerk "y"), the boundary conditions being with $y = 0$ and $y = l$. Real temperature distribution by the end of the whole time interval is obtained by modelling the second one-dimensional equation (network "x"), with the boundary conditions at the boundary $x = \pm a$.

70

The temperature distribution obtained at the end of the first half step is the initial datum for the second. Available freezing is taken into account by supplying additional current proportional to the spreading of the frozen area boundary into a corresponding unit [2]. "Freezing cell", however, is switched but to the one-dimensional network "y". When operating with network "x", the cell is switched off, the voltage in the given unit being kept in accordance with the freezing temperature.

Since the same network (after changing the appropriate resistances) can be applied for modelling both one-dimensional equations, the number of resistances necessary is decreased to 20-30. An original model was used to solve a two-dimensional problem with the same conditions as those obtained by physical experiment, in which the frozen silt loamy soil was $h = 1.64$ m deep [4]. When modelling with 240 hr time interval, $h = 1.53$ m was obtained. The discrepancy may be explained as the result of a possible lack of correspondence between real thermal constants of the soil and those assumed in the analogue as well as the use of average values k_i and $(c\gamma)_i$ within thawed and frozen areas. The major mistake is the assumption made in the case of modelling that freezing occurs instantly at the critical temperature point whereas it really takes place within a certain range of temperatures.

Ref. [5] proves that the application of average thermal constants in the process of loamy soil freezing shows the error of the order 1.5-4%, assuming an instant freezing error of 10%, which reduces the degree of freezing.

There are two ways for the analogue estimation:

1) coefficient k_i is considered in calculation of resistances R, modelling the distance intervals, and $(c\gamma)_i$—in other resistances R_k, modelling the time intervals;

2) thermal diffusivity $a_i = K_i/(c\gamma)_i$ is considered in calculation of resistances R_k. Due to the particular character of thermal constant variations during freezing, the frozen area depth obtained by means of the first way is larger.

According to the described methods, problems on unsteady state temperature distribution under artificial skating-rinks and speed skating-rinks have been modelled. As for the soil studied silt loamy soil capable of heaving with the following thermal constants was chosen:

solid volume weight	1,600 kg/m^3
soil weight moisture	23%
freezing temperature	-0.6°C
latent heat of freezing	91,641 kJ/m^3
thawed soil thermal diffusivity coefficient	$0.0061.10^{-4}$ m^2/s
frozen soil thermal diffusivity coefficient	$0.01238.10^{-4}$ m^2/s

Insulation heat transfer coefficient is assumed to be 0.4652 w/m^2.deg and taken into consideration in the analogue as the thermal resistance at the soil-cooling plate boundary. The coefficient of heat transfer from the surroundings of the construction to the air for the covered constructions and open-air ones was considered $\alpha = 8.72$ w/m^2 and $\alpha = 23.26$ w/m^2 respectively. For open-air constructions, meteorological conditions of Moscow were simulated, with the cooling plate temperature corresponding to the ambient temperature and maintaining constant the temperature of ice. In covered constructions the conditions were chosen as $t_f = +16$°C, $t_g = -15$°C.

The symmetry of the temperature distribution under the construction permitted only half of it to be modelled, setting the condition of zero heat flow on the axis. The area under research was divided into 121 units with various distance intervals,

71

smaller (to 0.2 m) in the greatest temperature variation range. The value of the time interval depended upon the duration of the studied conditions ranging from 120 hrs to 30 days. As the voltage divider, the integrator SEI-I was employed, permitting 0.1 °C accuracy of temperature distribution to be obtained.

The analysis of the results of the analogue method application allows to draw the following conclusions:

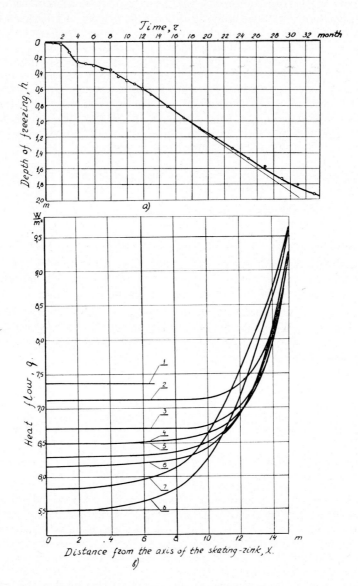

Fig. 2 — Freezing of the covered insulated practice skating-rink :

    a) spreading of the freezing boundary;
    b) heat flow from the soil to the cooling plate of the skating-rink: 1. in half a month time after starting the refrigeration installation; 2. in a month time; 3. in three months time; 4. in six months time; 5. in nine months time; 6. in twelve months time; 7. in twenty one months time; 8. in twenty nine months time.

1. *Covered practice skating-rink* (30 × 60 m)

The insulation layer and constant positive temperature surrounding slows down soil freezing to a large extent (fig. 2 a). During the first year, spreading of the frozen area boundary is affected by the heat previously accumulated by the soil under the skating-rink. By the end of the third year soil temperature variations become less than 0.1°C and the freezing process is slowed down to some extent, the maximum frozen layer in this case being 5-5,5 m deep [6]. The transient heat flow from the soil to the main

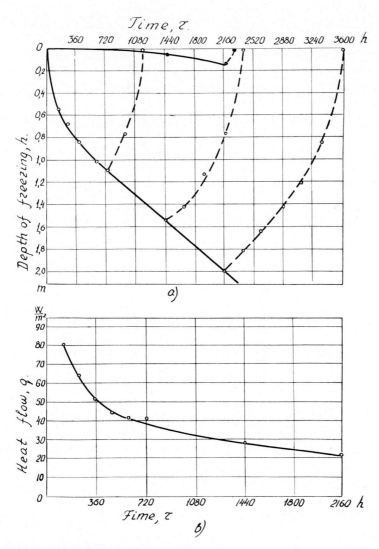

Fig. 3 — Freezing of the skating-rink inside sport arena:

   a) spreading of the freezing boundary:
   ● — skating-rink with insulation;
   ○ — skating-rink without insulation;
   ———— — freezing;
   .... — thawing.
   b) heat-flow from the soil to the central part of the skating-rink.

part of the plate does not exceed 7.56 w/m^2, the maximum heat flows to the boundaries of the skating-rink, however, being greater (fig. 2b). For maintaining uniform ice temperature, the cooling plate of the skating-rink should be 0.8-1 m in excess of the standard field dimensions.

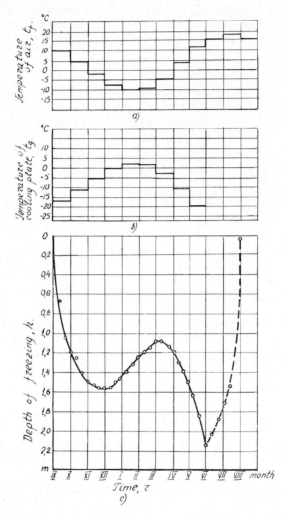

Fig. 4 — Freezing of the open-air uninsulated skating-rink (temperature of ice surface −4 °C):

    a) temperature of air;
    b) temperature of cooling plate;
    c) spreading of the freezing boundary.

2. *Sport arena* (30 × 60 m)

Freezing and subsequent thawing of the occasionally employed non-insulated skating-rink was considered. The maximum freezing layer 1.07 m was observed during the first month, during the next months the process slowed down. Due to the increasing frozen layer thermal resistance, the thawing period equal to one month commercial operation of the skating-rink is constant and involves 20 days (fig. 3a).

74

Heat flow from the soil to the central part of the skating-rink is rather considerable from 81.4 to 40.7 w/m² within the first month and 23.36 w/m² in three months' time (fig. 3b).

When the skating-rink is to be functioning continuously during three months, it should be insulated. The frozen layer in three months will not reach more than 0.12 m deep and the soil will completely thaw in 5 days after stopping the refrigeration installation.

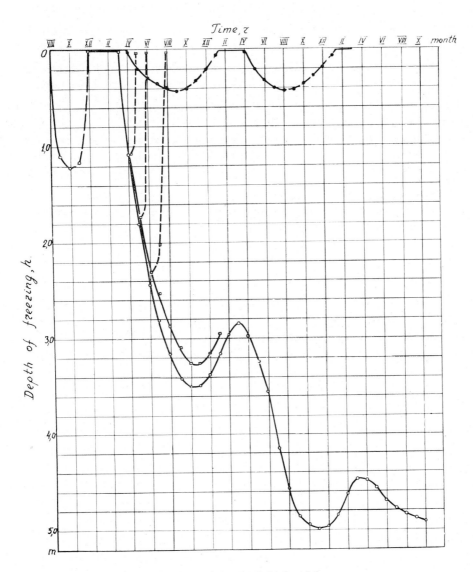

Fig. 5 — Freezing of the central part of the speed skating-rink:
    o — one-dimensional, non-insulated;
    □ — two-dimensional, non-insulated;
    • — insulated;
    ———— — freezing;
    - - - - - — thawing.

3. *Open-air uninsulated skating-rink* (30×60 m)

If a skating-rink is under seasonal commercial operation from September to May (9 months a year) with unfavourable meteorological and soil conditions available, the maximum soil freezing comes to 2.1 m. In this case the soil thaws before the new season (fig. 4). The transient heat flow from the soil to cooling plate of the skating-rink is 69.8 w/m^2 – 34.9 w/m^2. Thus, heat flow to uninsulated constructions such as artificial skating-rinks and speed skating-rinks may become rather a sensible value that can be explained not by the geothermal gradient but basically by the latent heat, which is particularly high for the quick soil freezing.

4. *Speed skating-rink* (400×14 m)

When modelling, it is assumed that even with a constant ice surface temperature of $t_i > t_f$, maximum soil freezing is supposed to occur within the spring and summer months (fig. 5). An insulated speed skating-rink, as can be seen in the picture, can be kept under commercial operation all the year round since the soil freezing does not exceed 0.5 m.

Though the speed-skating-rink is not wide, under variable air temperature conditions it does not induce slowing down the freezing process even if there is no insulation. If it is in seasonal operation (September–May), the soil under it thaws by the beginning of the new season.

Thus, applying the resistance network analogue method along with the splitting method, one can obtain the solution of various unsteady state heat exchanges under the refrigeration installations.

REFERENCES

[1] G. LIEBMANN, A new electrical analogue method for the solution of transient heat-conduction problems, *Transactions of ASME* (1956), No. 3.
[2] G. LIEBMANN, Solution of transient heat-transfer problems by the resistance-network analogue method, *Transactions of ASME* (1956), No. 6.
[3] N. N. YANENKO, Finite difference method for calculating of multidimensional heat conductivity equation. *Doklady AN SSSR*, 125 (1959), No. 6.
[4] V. S. LUKJANOV and M. D. GOLOVKO, Estimation of soil freezing depth. *Trudy VNIITC* (1957), issue 23.
[5] V. G. MELAMED, Solution of a general problem on temperature distribution with freezing available in the temperature range. Collected papers. *Heat and mass transfer* (1966), vol. 6.
[6] M.D. HASKIND, Heat transfer in soil under the cold store insulation, *Izvestia AN SSSR, OTN* (1958), No.10.

REPRÉSENTATION ÉLECTRIQUE DES CHAMPS THERMIQUES EN RÉGIME VARIABLE AU MOYEN D'UN MILIEU RÉSISTIF ET CAPACITIF CONTINU*

P. NUSGENS

Institut de Thermodynamique et CEDRIC
Université de Liège (Belgique)

Electrical representation of unsteady state thermal fields by means of a continuous resistive and capacitive medium

SUMMARY: *In the simulation of theoretical fields under dynamic conditions, one knows of, at the present moment, only those methods which introduce discontinuity into either the model or into the time notion. With the one, there are inherent mistakes in the finite difference method and with the other, physical facts are often hidden.*

The author has developed a method using continuous two-dimensional electrical models from both the resistive and capacitive aspects. This method, a variant of that of Benken's, uses a resistive model, a paper conductor being the first plate of a capacitor and the other plate, a perfect conductor, connected to earth; the originality of this method lies in the well-distributed capacity. Another feature is the low cost and easy manner of operation : one may consider models of very complicated geometry and there is no shortage in adequate measuring techniques.

Certain limitations of the frequencies being used, these models are capable of providing a very good Fourier's diffusion equation: using this formula, comparisons of measurements with theoretical results have given average errors below 2%.

INTRODUCTION

Parmi les méthodes utilisées en analogie thermoélectrique, celle qui utilise le papier conducteur est généralement réservée à l'étude de phénomènes stationnaires.

Cependant, les nombreux avantages qu'elle présente, à savoir : l'utilisation d'un modèle continu, la mise en œuvre aisée, le faible coût de réalisation, la précision satisfaisante ($\pm 5\%$) nous ont amené à mettre au point une méthode, mettant en œuvre un milieu résistif et capacitif continu. Le principe en est simple. Le modèle continu qui assure la fonction résistive (papier conducteur) sert également de premier plateau d'un condensateur dont le deuxième plateau, de même forme, parfaitement conducteur, est séparé du premier par une épaisseur constante de diélectrique et est mis à la masse.

On obtient ainsi un modèle bidimensionnel dont les caractéristiques électriques sont : d'une part la résistance d'un carré mesurée parallèlement au plan du modèle : $R_{c \cdot el} = r_{el}/e$, r_{el} étant la résistivité du papier conducteur et e son épaisseur, d'autre part la capacité par unité de surface : $C_{s \cdot el} = C_{el \cdot e}$, C_{el} étant exprimée en Farad/m^3. Notons ici qu'il est certain qu'aucun corps ne possède de capacité électrique par unité de volume; cependant cet artifice de notations se justifie par le souci de se rapprocher des caractéristiques thermiques, la capacité thermique d'un matériau s'exprimant par unité de volume.

On considère donc le modèle comme une tranche d'épaisseur e d'un modèle tridimensionnel de résistivité r_{el} et de "capacité volumique" C_{el}.
La caractéristique dynamique est :

$$K_{el} = \frac{1}{r_{el} \cdot C_{el}} = \frac{1}{r_{el}/e \cdot C_{el} \cdot e} = \frac{1}{R_{c \cdot el} \times C_{s \cdot el}}$$

(*) Cette communication est extraite d'une étude de synthèse en préparation sous la direction du Professeur G. BURNAY.

et l'équation de diffusion du potentiel dans un tel modèle peut s'écrire :

$$\frac{\partial^2 V}{\partial x^2} + \frac{\partial^2 V}{\partial y^2} = \frac{1}{K_{el}} \frac{\partial V}{\partial \tau} \tag{1}$$

que l'on doit rapprocher de l'équation du transfert de chaleur dans un plan

$$\frac{\partial^2 t}{\partial x^2} + \frac{\partial^2 t}{\partial y^2} = \frac{1}{K_{th}} \frac{\partial t}{\partial \tau} \tag{2}$$

avec

$$K_{th} = \frac{\lambda}{\rho . c} = \frac{1}{r_{th} . C_{th}}$$

en posant

$$r_{th} = \frac{1}{\lambda} \quad \text{résistivité thermique}$$

$C_{th} = \rho . c$ capacité thermique par unité de volume

en considérant que nous avons découpé mentalement une portion du modèle thermique dont la dimension perpendiculaire au plan où se déroulent les phénomènes est n. Dès lors, on simplifiera en ne parlant plus de résistivité ou de capacité volumique mais de résistance d'un carré et de capacité superficielle :

$$R_{c, th} = \frac{r_{th}}{n} \qquad\qquad C_{s,th} = C_{th \cdot n}$$

$$K_{th} = \frac{1}{R_{c\,th} \times C_{s\,th}}$$

On constate donc que le phénomène électrique dans le modèle envisagé est régi par une loi formellement semblable à celle du phénomène thermique : écoulement bidimensionnel, conduction simple. Si toutes les conditions qui suffisent à déterminer complètement ces phénomènes (en particulier la géométrie des domaines) sont elles aussi formellement semblables, il y aura correspondance biunivoque entre les deux phénomènes et toutes les constatations effectuées sur les modèles électriques pourront être transposées au phénomène thermique moyennant l'emploi d'échelles ou facteurs de conversion.

Les échelles

On ne peut établir de parallélisme entre les deux équations (1) et (2) que si on considère les grandeurs y intervenant comme analogues.
Ainsi on écrira $V = \mu.t$ où μ, exprimé en V/°C, donne en volt la tension qui dans le modèle électrique correspond à une différence de température de 1deg C.
Le tableau 1 reprend les différentes grandeurs analogiques. Notons que les grandeurs tant thermiques qu'électriques de ce tableau pouvant être définies à partir d'un certain nombre d'entre elles, il existe entre les échelles des relations de dépendance et seul un nombre limité d'entre elles peut être fixé arbitrairement, les autres s'en déduisant.

Tableau I

Les grandeurs thermiques et électriques analogues utilisées à deux dimensions

Grandeur	Symbole	Unités Thermiques Techniques	Unités Électriques MKSA	Échelle (él./th).
Différence de température (de potentiel).	Δt ΔV	°C	V	μ
Longueur	l	m	m	σ
Résistance d'un carré	R_c	h.°C/kcal	Ω	β
Capacité superficielle	C_s	kcal/°C.m²	F/m²	α
Constante dynamique	K	m²/h	m²/s	$\gamma = \alpha^{-1}\beta^{-1}$
Temps	τ	h	s	$n = \alpha.\beta.\sigma^2$
Épaisseur		$e(m) = \sigma' n(m)$		
Résistance de passage		$R_p = \beta\sigma\sigma' \dfrac{1}{e.\Delta l_{el}} \cdot \dfrac{1}{\alpha(\text{kcal/h.m}^2.°C)}$		
Densité superficielle de courant		$k(A/m) = e \dfrac{\mu}{\beta - \sigma\sigma'} \, i \,(\text{kcal/m}^2.\text{h}).$		

Les conditions aux limites

Outre la similitude géométrique des modèles, ce sont les conditions aux limites qui déterminent les phénomènes en jeu. Il convient donc d'imposer des conditions analogues sur les portions analogues des limites des modèles en tenant compte des échelles.

Analyse du modèle

Dans ce qui précède, nous avons considéré le modèle comme un condensateur parfait. Il convient cependant d'examiner les effets parasites et leur influence sur la précision des résultats.

Il s'agit notamment de savoir dans quelles limites et moyennant quelles restrictions le modèle, dont le plateau résistif possède une épaisseur e, vérifie l'approximation du modèle d'épaisseur nulle. Pour ce faire, on étudie analytiquement un cas soluble par séparation de variables, en fait un carré de longueur égale à $2\,l$ et d'épaisseur e, soumis à un échelon d'Heaviside. On a $V = \underset{m}{\Sigma}\, V_m$ (3) avec $V_m = C_m . U_m(x, y) . Z_m(z)$. $T_m(\tau)$, le plan xy étant parallèle au modèle, l'axe des z leur étant perpendiculaire, les deux faces du modèle résistif ayant les altitudes o et e on obtient la relation

$$\nabla^2_{xy} V_m = \frac{m.\,e}{th(me)} \cdot \frac{r.\,c_s}{e} \frac{\partial V_m}{\partial \tau} = \frac{1}{K'(me)} \frac{\partial V_m}{\partial \tau}$$

En tout plan $z =$ Cte, chaque produit V_m obéit à une équation de Fourier bidimensionnelle, mais la caractéristique dynamique $K'(me)$ varie d'un terme à l'autre, provoquant une distorsion de la réponse. Cependant, la série (3) convergente peut être limitée, pour une précision choisie, à un terme d'indice $m = m_{max}$.

Il suffit dès lors que, pour ce terme, $m_{max}e/th(m_{max}e)$ soit suffisamment voisin de l'unité pour que $K'(me)$ soit égal à K pour tous les termes non négligeables. Le calcul sur ordinateur numérique montre que l'épaisseur du papier e n'introduit aucune erreur perceptible tant qu'elle reste inférieure au deux millièmes des autres dimensions du modèle. Pratiquement, puisque l'on dispose de papier de 0,1 mm d'épaisseur, cela exige que la plus petite dimension du modèle soit de 20 cm.

Nous avons jusqu'ici fait l'hypothèse d'un modèle électrique parfait. L'étude de l'influence des capacités parasites et des conductances de fuites a conduit à restreindre la bande de fréquence f des sollicitations électriques suivant la relation

$$\frac{\sigma_d}{\varepsilon_d} \ll \omega \ll \frac{\sigma_c}{\varepsilon_c}$$

où $\omega = 2\pi f$, σ est la conductivité, ε la permitivité diélectrique, soit du plateau résistif (indice c), soit du diélectrique (indice d).

Cette relation montre simplement qu'il faut travailler avec des signaux de fréquence telle que l'on puisse considérer comme milieu conducteur pur et le diélectrique comme un diélectrique pur.

Enfin, l'anisotropie du papier due au laminage présente un effet identique à celui que l'on rencontre en régime stationnaire et qui peut être corrigé par une transformation géométrique.

MISE EN OEUVRE DE LA MÉTHODE.

Nous envisageons tout d'abord la réalisation du modèle. Le milieu résistif est constitué de papier rendu conducteur par l'adjonction d'un composé de graphite et vendu dans le commerce sous le nom de papier « Télédeltos ». Le diélectrique doit présenter une faible conductibilité, une bonne homogénéité, une faible épaisseur, une grande constante diélectrique. Diverses tentatives ont conduit à adopter une fine feuille de plastique adhésif collée sur le papier conducteur. Ce collage doit être réalisé avec soin car l'existence de bulles d'air entre le papier et le plastique créerait des hétérogénéités inacceptables. Pratiquement, ces bulles sont invisibles à l'oeil nu, et vu le caractère statistique de leur répartition, elles introduisent une inhomogénéité du même type que celle du papier. Le deuxième plateau du condensateur est obtenu en enduisant de peinture d'argent la surface libre du plastique. Ce procédé permet d'obtenir un matériau dont la capacité est voisine de 60 pf/cm².

Le modèle est alors découpé suivant une forme transformée du modèle thermique à étudier. Les conditions aux limites sont réalisées de la même façon que dans les méthodes continues.

La mesure du K_{el} peut difficilement se réaliser par les mesures séparées de r_{el}/e et C_s, la résistivité variant avec le degré hygrométrique de l'air ambiant et la mesure de C_s au capacimètre manquant d'exactitude. A défaut de pouvoir mesurer directement sur le modèle, on utilise un carré constitué du même matériau que le modèle et, en examinant sa réponse à un signal carré, on peut en déduire sa constante de temps et par là, le K_{el} du modèle.

Il est clair qu'un signal répété plusieurs fois par seconde est plus facile à mesurer qu'un signal non répétitif. C'est pourquoi, à toute sollicitation thermique, on fera correspondre une sollicitation électrique périodique qui sera appliquée aux frontières du modèle.

Les sollicitations thermiques peuvent se répartir en deux catégories : sollicitations périodiques et sollicitations non périodiques.

Dans la première catégorie, on trouve d'une part les sollicitations sinusoïdales et rectangulaires dont les analogues électriques sont faciles à réaliser et d'autre part des sollicitations de forme plus complexe. Pour ces dernières, on peut envisager, suivant la complexité de la forme, soit une analogie directe, soit une décomposition en série de Fourier, cette méthode s'alourdissant de calculs.

Parmi les sollicitations non périodiques, la montée brusque en température (fonction d'Heaviside) qui occupe une place importante sera représentée par un signal rectangulaire périodique dont la fréquence doit être suffisamment basse pour que la tension, en tout point du modèle, ait le temps d'atteindre la valeur de crête, et suffisamment élevée pour que les signaux-réponses du modèle soient nettement distincts de la sollicitation afin de permettre des mesures aisées (fig. 1). Les autres sollicitations non périodiques seront, soit décomposées en signaux périodiques, soit répétées un grand nombre de fois par seconde.

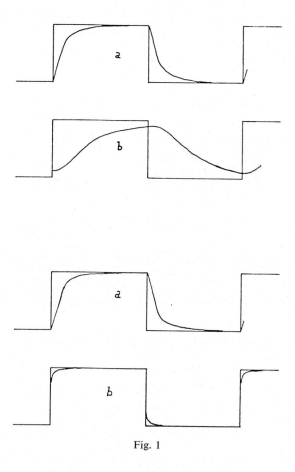

Fig. 1

En conclusion, toutes les sollicitations peuvent se ramener à quelques signaux simples, tous périodiques, qui sont réalisés suivant les cas par des générateurs habituels (sinusoïdaux, rectangulaires) ou des générateurs de fonctions à diodes.

Le principe de la mesure est d'effectuer une mesure relative consistant à faire le rapport de la valeur instantanée de la réponse à la valeur instantanée de la sollicitation afin de supprimer l'erreur due à des variations d'amplitude de cette dernière. Pour ce faire, on peut utiliser un oscilloscope bi-courbe. Cette technique, pas très précise, ($\pm 10\%$) offre les avantages de la rapidité et d'un faible prix quand on dispose déjà d'un oscilloscope à double trace; on peut la considérer comme un excellent moyen de contrôle.

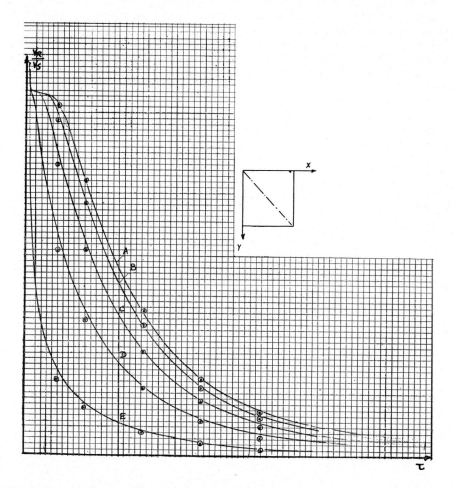

Fig. 2. Réponse d'un carré à un échelon d'Heaviside sans coefficient de passage. Les points sont des résultats théoriques.

Courbe	Point
A	(0.1, 0.1)
B	(0.3, 0.1)
C	(0.5, 0.1)
D	(0.7, 0.1)
E	(0.9, 0.1)

Pour obtenir plus de précision, il est indispensable de pouvoir tracer directement la réponse sur un diagramme. La fréquence des signaux utilisés n'est pas compatible avec la bande passante des enregistreurs *xy* habituels et exige l'emploi d'un appareil de transformation (oscilloscope à échantillonnage ou Waveform translator) qui convertit avec précision des signaux périodiques très rapides en un signal qui peut être reçu par un enregistreur *xy* à encre. Cette technique offre les avantages de posséder une bonne précision, et de présenter les résultats sous la forme de diagrammes directement utilisables.

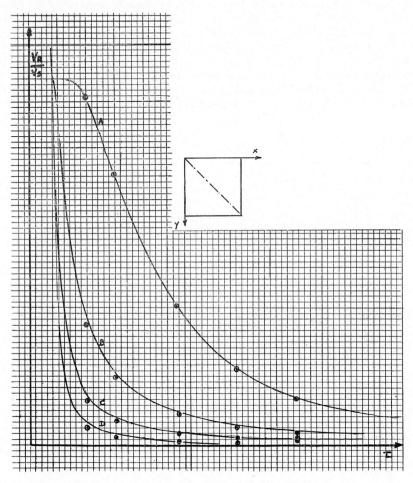

Fig. 3. Réponse d'un carré à un échelon d'Heaviside sans coefficient de passage. Les points sont des résultats théoriques.

Courbe	Point
A	(0.1, 0.1)
B	(0.7, 0.7)
C	(0.9, 0.7)
D	(0.9, 0.9)

Nous avons tout d'abord simulé le refroidissement d'une barre infinie de section carrée, sans coefficient de passage. Le modèle, réalisé suivant la méthode indiquée, est limité, étant donné la symétrie des conditions, au quart de la surface de la section de la barre. Afin de pouvoir négliger l'erreur introduite par l'épaisseur du papier, la longueur (1) du carré a été choisie égale à 50 cm. Les résistivités dans la direction du laminage et dans la direction perpendiculaire étant dans le rapport 25,8/29,3, le côté du modèle dans la direction perpendiculaire au laminage a été corrigé : $l' = 0,94$, $l = 47$ cm. La sollicitation est fournie par un générateur de signaux rectangulaires, et sa fréquence a été choisie à 1350 c/s. Les mesures, effectuées à l'aide d'un enregistreur xy à encre et d'un transformateur de signaux réalisé en nos laboratoires, ont été effectuées en quinze points uniformément répartis sur toute la surface du modèle. Les figures 2 et 3 et le tableau 2 donnent les résultats des mesures effectuées en cinq instants différents en comparaison avec les résultats théoriques afin d'en déduire les erreurs relatives à la valeur maximale de la sollicitation. On constate que l'erreur moyenne en module est de 1,2 % et l'erreur maximale de 4,4 %.

Tableau 2

RÉPONSE D'UN CARRÉ À UN ÉCHELON D'HEAVISIDE SANS COEFFICIENT DE PASSAGE: COMPARAISON DE RÉSULTATS EXPÉRIMENTAUX ET ANALYTIQUES

Courbe	15 µs	30 µs	60 µs	90 µs	120 µs
(0.1, 0.1)	959(957)+2	739(747)−8	374(377)−3	193(185)+8	103(90)+13
(0.1, 0.3)	931(913)+18	690(683)+2	338(340)−2	172(167)+5	91(82)+9
(0.1, 0.5)	829(793)+36	578(554)+22	277(271)+6	140(132)+8	80(65)+15
(0.1, 0.7)	600(556)+44	391(364)+27	180(174)+6	95(85)+10	52(42)+10
(0.1, 0.9)	245(202)+43	138(127)+11	65(60)+5	33(29)+4	21(14)+7
(0.3, 0.3)	906(871)+35	627(625)+2	301(308)−7	146(150)−4	80(74)+6
(0.3, 0.5)	792(757)+35	515(507)+8	242(244)−2	122(119)+3	71(58)+13
(0.3, 0.7)	575(531)+44	352(333)+19	161(157)+4	80(77)+3	43(37)+6
(0.3, 0.9)	231(193)+38	126(116)+10	59(54)+5	26(26)0	27(13)+14
(0.5, 0.5)	691(657)+34	421(412)+9	194(194)0	92(95)−3	50(46)+4
(0.5, 0.7)	502(461)+41	284(270)+14	123(125)−2	60 (61)−1	31(30)+1
(0.5, 0.9)	201(168)+33	109(94)+15	50(43)+7	29(21)+8	15(10)+5
(0.7, 0.7)	346(324)+22	188(177)+11	86(80)+6	44(39)+5	23(19)+4
(0.7, 0.9)	138(118)+20	76(61)+15	32(28)+4	20(13)+7	22(7)+15
(0.9, 0.9)	66(43)+23	35(21)+14	—(9)—	—(5)—	—(2)—

Les résultats expérimentaux sont suivis par les résultats analytiques correspondants, placés entre parenthèses. Le dernier chiffre est l'erreur. Tous ces résultats sont exprimés en 10^{-3} de la hauteur de l'échelon.

Le deuxième exemple illustre la mise en œuvre d'un oscilloscope bi-courbe. Il s'agit de l'étude dynamique des répartitions de températures dans un tuyau de longueur infinie soumis intérieurement à une montée linéaire de température sans coefficient de passage et dont la face externe est parfaitement isolée. La montée linéaire en température a été, ici, simulée par un générateur spécialement adapté fournissant un signal périodique dont la tension commence par monter linéairement, puis retombe brusquement à zéro, la période se terminant par une plage horizontale. Le rapport de la durée de la plage horizontale au temps de montée doit être suffisamment grand pour que la tension, en tout point du modèle, ait le temps de revenir à zéro, de manière à ce que la réponse à la rampe suivante puisse être considérée comme indépendante des sollicitations précédentes. Les figures 4 et 5 montrent la réponse du modèle d'une part au milieu de l'épaisseur du tuyau et d'autre part au 1/10 de l'épaisseur vers le centre.

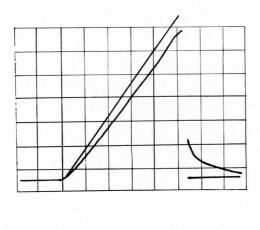

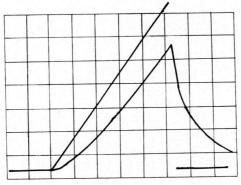

Fig. 4-5

CONCLUSION

Le modèle proposé possède des avantages certains : une grande facilité d'application, un faible coût, beaucoup de souplesse pour la représentation de modèles de géométrie compliquée et surtout l'absence de discrétisation. Les techniques de mesure sont d'un emploi facile et ne nécessitent pas la mise en œuvre d'appareillage très spécialisé.

Enfin, la précision obtenue par la méthode est bonne et est essentiellement limitée par la qualité du papier conducteur.

DISCUSSION p. 92.

THERMAL AND ELECTRICAL MODELS
FOR SOLVING PROBLEMS OF NON-STATIONARY HEAT
TRANSFER THROUGH WALLS

V. KORSGAARD

The Thermal Insulation Laboratory, The Technical University
Lyngby (Denmark)

Modèles thermiques et électriques pour la détermination du flux de chaleur à travers des murs en régime variable

RÉSUMÉ : *La connaissance du comportement d'une construction en régime variable est importante pour le calcul des températures intérieures et des charges frigorifiques.*

Le régime variable du flux de chaleur à travers un mur peut être représenté par un modèle électrique constitué d'un réseau de résistances et de capacités simples. De manière analogue on peut concevoir un modèle thermique du mur constitué de résistances et de capacités thermiques en série.

Dans l'article on décrit un tel modèle thermique où des plaques en aluminium représentent les capacités et des feuilles de mousse plastique les résistances.

L'amortissement de l'amplitude et le déphasage des variations sinusoïdales de la température ont été mesurés aussi bien sur un modèle électrique que sur un modèle thermique d'un mur défini. Les deux caractéristiques mesurées suivant les deux méthodes sont comparées avec un calcul utilisant la méthode de la matrice. On trouve une bonne concordance entre les trois méthodes.

SYMBOLS FOR PHYSICAL QUANTITIES

Symbol	Quantity	Unit
A	area	m^2
d	thickness	m
ρ	density	$kg\,m^{-3}$
c	specific heat	$kcal\,kg^{-1}\,C^{-1}$
λ	thermal conductivity	$kcal\,m^{-1}\,hr^{-1}\,C^{-1}$
t	time	hr
R	thermal resistance	$hr\,C\,kcal^{-1}$
ϑ	temperature	°C
Φ	rate of heat flow	$kcal\,hr^{-1}$
C	thermal capacity	$kcal\,C^{-1}$
a	thermal diffusivity	$m^2\,hr^{-1}$
ω	$2\pi \times$ frequency of the temperature oscillation	hr^{-1}.

THERMAL SYSTEM

Homogeneous wall exposed on one side to a room with constant temperature and free convection and on the other side to a sinusoidal temperature variation, see figure 1.

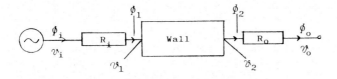

Fig. 1 — Block diagram of thermal system.

Problem: Find the amplitude ratio and phase lag between the two surface temperatures of the wall for various cycle periods.

Example:

Gas-concrete wall.

$d = 0.0455$ m $\lambda = 0.175$

$A = 0.04$ m^2 $c = 0.22$

$\rho = 700$

$R = \dfrac{d}{\lambda A} = 6.5$ $C = c\rho\, dA = 0.28$

$R_i = 0$ $R_0 = \dfrac{0.15}{0.04} = 3.75$

THE MATRIX METHOD

The sinusoidal heat flow through the surfaces and the temperature of the surfaces of a wall can be related by:

$$\begin{bmatrix} \vartheta_1 \\ \Phi_1 \end{bmatrix} = \begin{bmatrix} a_{11} & a_{12} \\ a_{21} & a_{22} \end{bmatrix} \cdot \begin{bmatrix} \vartheta_2 \\ \Phi_2 \end{bmatrix}.$$

For a homogeneous wall:

$$a_{11} = a_{22} = \cosh(\varphi d)$$

$$a_{12} = Z_0 \sinh(\varphi d) \qquad\qquad a_{21} = Z_0^{-1} \sinh(\varphi d)$$

$$\varphi = (1+i)\left[\frac{\omega}{2a}\right]^{1/2} \qquad Z_0 = \frac{R_\omega}{\varphi d}$$

For the thermal system shown in figure 1 we have:

$$\begin{bmatrix} \vartheta_0 \\ \Phi_0 \end{bmatrix} = \begin{bmatrix} 1 & R_i \\ 0 & 1 \end{bmatrix} \cdot \begin{bmatrix} a_{11} & a_{12} \\ a_{21} & a_{22} \end{bmatrix} \cdot \begin{bmatrix} 1 & R_0 \\ 0 & 1 \end{bmatrix} \cdot \begin{bmatrix} \vartheta_i \\ \Phi_i \end{bmatrix}.$$

For the gas-concrete wall given in the example, the amplitude ratio and the phase lag between the two surface temperatures ϑ_1 and ϑ_2 are calculated for 24, 8, 4 and 2 hr cycling period, and the results are shown in the table in the last lines of column A.

LUMPED THERMAL RC-MODEL

Scale values for model:
distance ratio $\Delta d/\Delta d_m = 1$, time ratio $\Delta t/\Delta t_m = 1$, resistance ratio $R/R_m = 1$, capacity ratio $C/C_m = 1$.

Subscript m denotes the corresponding symbols for the model.

The gas-concrete wall is divided into six equally thick layers. The thermal resistance of each section is 1.083, which corresponds to the thermal resistance of a 0.0013 m thick plate of Styropor cellular plastic with $A = 0.04$ and $\lambda = 0.03$.

The thermal capacity of each section is 0.0467, which corresponds to the thermal capacity of a 0.002 m thick plate of aluminium with A = 0.04, c = 0.214 and ρ = 2,700.

Omitting the thermal capacity of Styropor and the thermal resistance of aluminium, the lumped model of the wall will consist of 2 × 6 alternately plates of aluminium and Styropor, see figure 2.

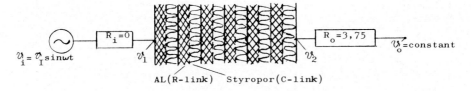

Fig. 2 — Lumped thermal RC-model of wall.

The temperature of the two surfaces and between the six RC-sections is measured using Cu-constantan thermocouples and recorded on a standard potentiometer electronic recorder, see figure 3. The amplitude ratio and the phase-lag of the recorded temperature waves in relation to ϑ_1 is calculated, and the results are shown in the table in column C.

As the thermal model used for the measurements was exposed on the output side to a room with constant temperature and free convection, the surface resistance will have a fixed value of app. 3.75, and this model can only simulate other walls for which the resistance ratio is unity.

Fig. 3 — Experimental set-up with lumped thermal RC-model of wall.

The model can therefore for example simulate a solid brick wall with A = 0.04, λ = 0.50, ρ = 1,800, c = 0.20. As thermal resistance ratio must be unity, the thickness of the brick wall is found to be 0.13.

The capacity ratio is given by

$$\frac{C}{C_m} = \frac{1{,}800 \cdot 0.40 \cdot 0.20 \cdot 0.13}{0.28} = 6.7 .$$

The time ratio is equal to the product of the resistance ratio, which was unity, and the capacity ratio and becomes the value of 6.7, which means that one hour in the original thermal system corresponds to 9 minutes in the model system.

DYNAMIC RESPONSE OF HOMOGENEOUS WALL

A. calculated, using the matrix method (6 layers with distributed parameters)
B. measured on lumped electrical model (6 RC-elements)
C. measured on lumped thermal model (6 RC-layers)

Cycle period 24 hrs.	Amplitude ratio			Phase lag radians			Time lag minutes
Layer	A	B	C	A	B	C	A
1	0.892	0.88	0.88	0.038	0.040	0.040	9
1 and 2	0.785	0.78	0.77	0.072	0.078	0.080	16
1 to 3	0.680	0.68	0.68	0.099	0.108	0.113	23
1 to 4	0.574	0.57	0.58	0.122	0.138	0.146	28
1 to 5	0.469	0.47	0.48	0.140	0.162	0.167	32
1 to 6	0.364	0.36	0.37	0.150	0.185	0.188	34
Wall considered as one layer	0.364			0.138			30
Cycle period 8 hrs.							
1	0.875	0.88	0.87	0.112	0.110	0.114	9
1 and 2	0.760	0.76	0.74	0.213	0.218	0.228	16
1 to 3	0.652	0.65	0.64	0.300	0.279	0.287	23
1 to 4	0.548	0.56	0.54	0.372	0.340	0.346	29
1 to 5	0.447	0.45	0.45	0.425	0.392	0.401	32
1 to 6	0.347	0.35	0.35	0.457	0.443	0.456	35
Wall considered as one layer	0.352			0.405			31
Cycle period 4 hrs.							
1	0.835	0.84	0.82	0.194	0.194	0.189	7
1 and 2	0.700	0.69	0.68	0.375	0.338	0.378	14
1 to 3	0.586	0.58	0.57	0.538	0.530	0.519	20
1 to 4	0.487	0.47	0.47	0.674	0.668	0.660	26
1 to 5	0.394	0.39	0.39	0.778	0.776	0.770	30
1 to 6	0.305	0.30	0.30	0.840	0.879	0.872	32
Wall considered as one layer	0.316			0.770			30
Cycle period 2 hrs.							
1	0.757	0.76	0.75	0.290	0.290	0.294	5
1 and 2	0.579	0.58	0.56	0.580	0.580	0.577	11
1 to 3	0.451	0.45	0.44	0.861	0.843	0.837	16
1 to 4	0.357	0.36	0.34	1.114	1.095	1.095	21
1 to 5	0.283	0.28	0.26	1.315	1.285	1.275	25
1 to 6	0.218	0.22	0.20	1.437	1.473	1.455	27
Wall considered as one layer	0.234			1.350			26

If instead of keeping the room temperature constant we had kept the room surface temperature constant by means of a waterbath, the model could simulate any homogeneous wall as in this case we can choose both the resistance and the capacity ratio freely.

Example: Foamglass wall, $d = 0.20$, $A = 10$, $\rho = 140$, $c = 0.20$, $\lambda = 0.045$, $R = 0.445$, $C = 56$.

$$\frac{R}{R_m} = \frac{0.445}{6.5} = 0.068, \quad \frac{C}{C_m} = \frac{56}{0.28} = 200, \quad \frac{\Delta t}{\Delta t_m} = 13.6.$$

LUMPED ELECTRICAL RC-MODEL

The lumped thermal model can be simulated by a lumped RC-electrical model, see figure 4.

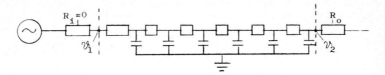

Fig. 4 — Lumped electrical RC-model of wall.

The voltage variations of the two "surfaces" and in the knots between the six RC-links are measured using a standard electronic recorder. For each cycle period a new model has been set up on our passive analog computer (fig. 5) with a time ratio so that the cycle period in the real system corresponds to 6 sec in the electrical system with the purpose of getting a good record on the strip chart recorder. The amplitude ratio and the phase lag of the recorded voltage variations in relation to ϑ_1 is calculated, and the results are shown in the table in column B.

Fig. 5 — The passive electrical analog computer at the Thermal Insulation Laboratory of the Technical University of Denmark in Copenhagen.

The matrix method has also been applied to the lumped thermal RC-model shown in figure 2. The results of the analysis are shown in the table in column A. The computations have been carried out on a GIER digital computer. It will be noticed that the agreement in amplitude ratio is good for the wall, when the latter is considered as one layer and when composed of six RC-layers, whereas there is a deviation in phase lag. This is probably due to insufficient accuracy in the procedure chosen for computing the phase lag.

ACKNOWLEDGEMENT is due to Mr. Poul Frederiksen for assistance with experimental and computational work.

REFERENCES

[1] D.G. STEPHENSON, Methods of determining non steady-state heat flow through walls and roofs of buildings. *J.I.H.V.E.* (May 1962).
[2] V. KORSGAARD and H. LUND, A passive-electrical analog computer for room-climate calculations. Copenhagen (1965).

DISCUSSION

J.J. TIREL (France) — Plusieurs méthodes de calcul de flux de chaleur à travers des murs composés sont exposées : deux méthodes analogiques et une méthode analytique, qui donnent des résultats assez concordants. Quelle conclusion pratique l'auteur tire-t-il, sur l'opportunité de l'utilisation de l'une ou l'autre des méthodes ?

V. KORSGAARD — In routine computation of cooling loads and indoor temperature calculations, the matrix method or other mathematical methods suitable for programming on at digital computer should be used. If you have a RE network analog computer available in your laboratory we find this method most suitable for special problems, and in courses for students on non-steady-state heat flow problems. The thermal models should be used in connection with investigation of scaled models of houses.

F.N. CAMIA (France) — Depuis l'invention de Beuken, un travail considérable a été fait sur les analogies électriques de la chaleur et leur utilisation. Il nous paraît souhaitable que l'I.I.F. puisse donner à ses collaborateurs une information systématique sur ce sujet; ce pourrait être un travail de Commission de coordonner et d'informer.
Nous nous permettons en outre de vous signaler les analogies hydrodynamiques, qui sont un moyen d'avoir un calculateur analogique bon marché et donnant de bons résultats. Lorsque la capacité thermique spécifique du milieu représenté est une fonction de la température, elles sont d'application plus simple que les analogies électriques. Elles permettent également de visualiser ne serait-ce que par un schéma, les problèmes de conduction complexes, avec apport d'énergie, imposition de température, etc., problèmes quelquefois difficiles à concevoir correctement. Le calcul de ces analogies introduit à un genre de calcul aux différences finies qu'on peut appeler mixte, ces différences finies étant relatives seulement à l'espace et non au temps, ce qui donne une formulation facile et efficace. Plusieurs laboratoires de la région parisienne emploient ces analogies. Le Dr. Beuken lui même a fait monter un appareil hydrodynamique à Maastricht.
En France, un travail de synthèse sur toutes les analogies de la chaleur a été fait par les soins de la S.F.T. et publié chez Dunod à Paris.

CALCULATION OF THE SOLAR HEAT GAIN THROUGH COMPOSITE CONSTRUCTIONS BY THE MATRIX METHOD

J.H. de BOER

Laboratory of Refrigeration of the Delft Technological University

and

P. EUSER

Institute of Applied Physics TNO-TH, Delft (The Netherlands)

Calcul des apports solaires à travers les constructions composites à l'aide d'une méthode matricielle

RÉSUMÉ : *Afin de calculer les apports de chaleur due au rayonnement solaire à travers les constructions composites, on a établi un programme d'ordinateur s'appuyant sur une méthode matricielle, appropriée à des conditions périodiques stables. Avec cette méthode on peut calculer la température et le flux de chaleur dans une construction, à l'aide d'un procédé de calcul numérique assez facile, sans discontinuité de temps ni de place. Le programme est initialement réalisé, et semble initialement convenir, pour les constructions simples, homogènes ou composites.*

1 — INTRODUCTION

In the computation of the heating and cooling load of buildings the usual methods applied result in a fairly rough evaluation of the non-steady heat transfer in the walls. Quite understandably, therefore, the tendency has become to take a somewhat large safety margin, which more often than not leads to the installation of an over-capacity system.

The exact, analytical computation of the non-steady or transient heat phenomena in a building, on which several authors have been engaged [1-5], is numerically also a time-cosuming affair. True, the analog methods have gained a certain degree of recognition, but these turned out to find a more ready application for research than for technical calculations.

However, at present some solutions for numerical processing by a computer have become known, which, in view of the growing use of the computer, may find increased technical application.

This article deals with the so-called matrix method, which, among other applications, is suited for the technical subject to be discussed here: the calculation of the solar heat gain through roofs and walls.

2 — PROBLEM DESCRIPTION, BOUNDARY CONDITIONS

The surface area of a multi-layer wall is exposed to solar radiation. Part of the incident solar radiation is absorbed by the surface area of the wall, the other part is reflected. On the surface area the heat is exchanged with the outside air (convection) and with the atmosphere (radiation). The ambient temperature at the inside of the wall is supposed to be constant.

The problem set is to find the temperature and the heat flow at a certain plane x in the wall, especially on the outer surface ($x = 0$) and on the inner surface ($x = d$).

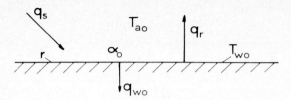

Fig. 1 — Heat fluxes and temperatures at a sun irradiated surface.

It is assumed that both solar radiation and external temperature go daily through a similar periodic cycle, and that this cycle is basically sinusoidal for both quantities. The net radiation into the atmosphere is assumed to be constant.

On basis of these assumptions, the boundary conditions can be formulated as follows:

incident solar radiation:

$$q_s = \bar{q}_s + \hat{q}_s \cos \omega(t - \varphi_s) \tag{1}$$

outside air temperature:

$$T_{ao} = \bar{T}_{ao} + \hat{T}_{ao} \cos \omega(t - \varphi_a) \tag{2}$$

net heat radiation flux:

$$q_r = \bar{q}_r . \tag{3}$$

The quantities q_s, q_r and T_{ao} can be combined to one single quantity T'_{ao}:

$$T'_{ao} = T_{ao} + \frac{(1-r)q_s - q_r}{\alpha_0} \tag{4}$$

in which q_s, T_{ao} and q_r result from (1), (2) and (3) respectively. For simplicity, further considerations will be based only on the exterior conditions according to (4).

A constant inside ambient temperature is assumed of T_{ai}. The heat transfer coefficient on the inner surface of the construction under consideration (α_i) is assumed to be the sum of the coefficient for the convection and of that for the radiation.

3 — METHOD OF SOLUTION

Based on the equation of Fourier

$$\frac{\partial T}{\partial t} = a \frac{\partial^2 T}{\partial x^2} \tag{5}$$

the equations valid for the temperature T_x and the heat flow per unit area q_x in a homogeneous wall with steady periodic sinusoidal fluctuations [6] are

$$T_x = [P \sin h\ kx(1+j) + Q \cos h\ kx(1+j)]\ e^{j\omega t} , \tag{6}$$

$$q_x = -\lambda k(1+j)\ [P \cos h\ kx(1+j) + Q \sin h\ kx(1+j)]\ e^{j\omega t} \tag{7}$$

94

with

$$k = \sqrt{\frac{\omega}{2a}}.$$

We mention

$$T_{x=0} = T_1 \qquad\qquad T_{x=d} = T_2$$

$$q_{x=0} = q_1 \qquad\qquad q_{x=d} = q_2$$

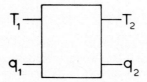

Fig. 2 — The surface quantities of a homogeneous slab.

If two of the four surface quantities are known, e.g. T_1 and q_1, it is possible to express P and Q, and also T_2 and q_2 in T_1 and q_1 [7-9]:

$$T_2 = AT_1 + Bq_1$$
$$q_2 = CT_1 + Dq_1$$

(8)

which with (6) and (7) gives:

$$A = D = \cos h\ kd(1+j) \tag{9}$$

$$B = -\frac{\sin h\ kd(1+j)}{\lambda k(1+j)} \tag{10}$$

$$C = -\lambda k(1+j)\sin h\ kd(1+j) \tag{11}$$

where

$$AD - BC = 1. \tag{12}$$

It follows further from (9)-(12) that

$$T_1 = DT_2 - Bq_2$$
$$q_1 = -CT_2 + Aq_2$$

(13)

Consequently, the relation between the four surface quantities can be written as a matrix equation:

$$\begin{pmatrix} T_2 \\ q_2 \end{pmatrix} = \begin{pmatrix} A & B \\ C & D \end{pmatrix} \begin{pmatrix} T_1 \\ q_1 \end{pmatrix}. \tag{14}$$

For a wall construction consisting of n layers the transform (14) can be applied to each successive layer. If the layers are in ideal contact, the values T and q on the boundary surfaces are equal for the respective layers, and the matrix equation shows as follows:

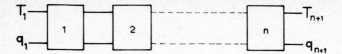

Fig. 3 — The surface quantities of a composite slab.

$$\begin{pmatrix} T_{n+1} \\ q_{n+1} \end{pmatrix} = \begin{pmatrix} A_n^{\eta} & B_n \\ C_n & D_n \end{pmatrix} \begin{pmatrix} A_{n-1} & B_{n-1} \\ C_{n-1} & D_{n-1} \end{pmatrix} - - - \begin{pmatrix} A_1 & B_1 \\ C_1 & D_1 \end{pmatrix} \begin{pmatrix} T_1 \\ q_1 \end{pmatrix}. \qquad (15)$$

By substituting (9)-(11) in (15) and by multiplying the n matrices explicit formulas for T_2 and q_2 could be obtained. This would involve a complicated process and render the method almost equally impracticable for technical calculations as other analytical methods. The multiplication of the *numerical* matrices, however, is quite well feasible, especially when making use of a computer.

If not two of the "surfaces quantities" are known, but, as in the present practical case, the temperatures of the adjoining media are known, another matrix is added for each side for the transfer through radiation and convection (fig. 2):

$$T_2 = T_1 - \frac{1}{\alpha} q_1$$
$$q_2 = q_1 \qquad (16)$$

so that

$$A = D = 1, \ B = -\frac{1}{\alpha}, \ C = 0$$

being the coefficients for these "surface matrices".

Then the complete formulation for the n-layer wall construction is as follows:
Notation:

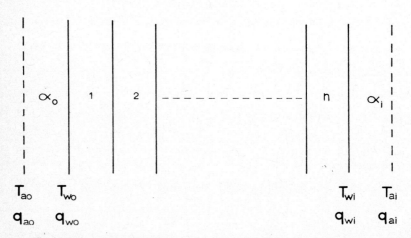

Fig. 4 — The boundary quantities in case of a composite slab with convection and radiation transfer at the surfaces.

Matrix equation:

$$\begin{pmatrix} T_{ai} \\ q_{ai} \end{pmatrix} = \begin{pmatrix} 1 & -1/\alpha_i \\ 0 & 1 \end{pmatrix} \begin{pmatrix} A_n & B_n \\ C_n & D_n \end{pmatrix} - - - \begin{pmatrix} A_1 & B \\ C_1 & D_1 \end{pmatrix} \begin{pmatrix} 1 & -1/\alpha_0 \\ 0 & 1 \end{pmatrix} \begin{pmatrix} T'_{ao} \\ q_{ao} \end{pmatrix} \quad (17)$$

where:

$$q_{ao} = q_{wo} \quad \text{and} \quad q_{ai} = q_{wi}.$$

The known quantities are T'_{ao} (sinusoidal) and T_{ai} (constant) as formulated in section 2. Quantities to be found are e.g. T_{wo}, T_{wi} and q_{wi}.

As to the programming, initially the numerical calculations were carried out as follows. First the matrix coefficients A, B, C and D were calculated for each separate layer with the aid of tables [Shirtliffe and Stephenson, 10], subsequently, the matrix equation (17) was solved with the aid of a computer. This procedure is described in [11].

Later on a computer program for the complete calculation was made, including both the calculation of the matrix coefficients and the numerical solution of the matrix equations. With this program, described in [12], the only input data are: λ, ρ, c and d for each layer, α_o and α_i, q_s, q_r, T_{ao} and T_{ai}, in the in section 2 mentioned form.

The program gives the results for one frequency. If the incident solar radiation cannot be approximated by a sinus function with sufficient accuracy, Fourier series simulation must be applied. The computation must then be repeated for some higher harmonics and the results be added according to the known procedure. For this purpose an additional program was made.

4 — APPLICATION TO ROOF CONSTRUCTIONS

For illustration the results are given for some roof constructions. The values which have been given to the incident solar radiation q_s, the net radiation q_r, the outside air temperature T_{ao} and the internal temperature T_{ai} are:

$$q_s = 441\left[1 + \cos\frac{\pi}{12}(t-12)\right] W/m^2$$

$$q_r = 115 \ W/m^2$$

$$T_{ao} = 20 + 5\cos\frac{\pi}{12}(t-14) \ °C$$

$$T_{ai} = 20\ °C.$$

For α_i is taken 8 W/m² . °C, for α_o 20 W/m² . °C. The absorption factor of the surface area was assumed to be 0.8.

On the results, stated in the table, some remarks can be made.

The maximum value of the temperature of the surface of the constructions 1, 2 and 3 diverges little: 55 to 56°C. This is because in all these constructions, with total thermal resistances between 0.84 and 0.94 W/m² . °C, by far the greater part (90 to 95%) of the solar heat absorbed by these roof surfaces is transferred to the outdoor environment. If the heat gain through these constructions would be nil, the maximum roof temperature would be approximately 58°C.

97

RESULTS FOR SOME ROOF CONSTRUCTIONS

No.	α_0 W/m².°C	r	d cm	λ W/m.°C	a m²/h×10³	α_i W/m².°C	Construction	Surface temperatures (°C) and heat fluxes (W/m²) (t in hours)
1	20	0.1	0.3 7.0 10.0	0.35 0.08 1.7	1.25 0.35 3.0	8	roofing felt wood wool slabs concrete	$T_{wo} = 33.5 + 22.6 \cos \omega(t-12.7)$ $q_{wo} = 12.6 + 48.3 \cos \omega(t-9.6)$ $T_{wi} = 21.6 + 1.1 \cos \omega(t-19.7)$ $q_{wi} = 12.6 + 8.6 \cos \omega(t-19.7)$
2	20	0.1	0.3 10.0	0.35 0.12	1.25 0.34	8	roofing felt reinforced wood wool slabs	$T_{wo} = 33.4 + 21.7 \cos \omega(t-12.8)$ $q_{wo} = 13.9 + 73.9 \cos \omega(t-9.7)$ $T_{wi} = 21.7 + 1.9 \cos \omega(t-18.2)$ $q_{wi} = 13.9 + 15.0 \cos \omega(t-18.2)$
3	20	0.1	0.3 4.0 10.0	0.35 0.05 1.7	1.25 0.23 3.0	8	roofing felt insulation concrete	$T_{wo} = 33.4 + 22.8 \cos \omega(t-12.6)$ $q_{wo} = 13.5 + 38.5 \cos \omega(t-10.0)$ $T_{wi} = 21.7 + 1.3 \cos \omega(t-18.6)$ $q_{wi} = 13.5 + 10.4 \cos \omega(t-18.6)$
4	12	0.1	0.3 4.0 10.0	0.35 0.05 1.7	1.25 0.23 3.0	8	roofing felt insulation concrete	$T_{wo} = 41.7 + 33.7 \cos \omega(t-12.6)$ $q_{wo} = 21.8 + 56.9 \cos \omega(t-10.0)$ $T_{wi} = 22.7 + 1.9 \cos \omega(t-18.4)$ $q_{wi} = 21.8 + 15.4 \cos \omega(t-18.4)$
5	12	0.4	2.0 0.3 4.0 10.0	0.35 0.35 0.05 1.7	1.0 1.25 0.23 3.0	8	gravel roofing-felt insulation concrete	$T_{wi} = 21.4 + 1.2 \cos \omega(t-19.4)$ $q_{wi} = 11.0 + 9.9 \cos \omega(t-19.4)$

Here the value of α_o and the absorption factor of the surface $(1 - r = 0.9)$ are of great influence.

If α_o would be 12 W/m^2 . °C (calm weather), as assumed for construction 4, the maximum value of the outside surface temperature T_{wo} becomes 75°C. A rise in the outside air temperature T_{ao} by 5°C (maximum value T_{ao} = 30°C) results in a rise in the outside surface temperature of approximately $4\frac{1}{2}$°C. So outside surface temperatures of about 80°C, seem to be possible. A rather uncertain factor lies in the value for α_o. This applies both to the average value of α_o for the entire roof surface and to the local differences.

The outside roof surface temperature can be reduced by decreasing the absorption factor of the surface and/or by increasing the heat accumulation in the upper layer. Both effects can be obtained by applying a few cm's thick layer of white gravel as in case of construction 5.

The maximum value of T_{wo} occurs at t = 12.6 to 12.8 hours solar time for constructions 1, 2, 3, 4 and 5, for construction 6 half an hour later.

The heat flow at the inside of the constructions, determining the heat gain, varies in magnitude and phase.

Assuming $r = 0.1$ and $\alpha_o = 20$ W/m^2 . °C the maximum value ranges between 21 W/m^2 in case of construction 1 ($\Sigma\rho d$ = 278 kg/m^2) to 29 W/m^2 in case of construction 2 ($\Sigma\rho d$ = 153 kg/m^2). Of interest are the increase to 37 W/m^2 when α_o = 12 W/m^2 . °C and the considerable decrease when applying a layer of gravel (thickness 2 cm, absorption factor $1 - r = 0.6$).

The time delay in the constructions varies between 5.5 hours in case of the relative light weighed construction 2, and 7.0 hours in case of construction 1. This time delay is, of course, of importance for the momentary heat gain during the operation period of the air conditioning installation.

For practical use the table on page 98 has to be extended, for which purpose the described computer program can be used. We refer also to [13] and [14].

5 — FINAL REMARKS

The matrix method might also be applied for the calculation of the heat storage effects of the inside layers of building constructions, with the possibility to find more accurate values for the well known heat storage factors.

The method can be extended to cylindrical and spherical layers. In this connection, for comparison the work of Kusuda [15] is interesting, who generalised the method of Mitalas and Arseneault [16] for this purpose.

Furthermore we refer to a publication of Muncey and Spencer [17], who applied the matrix method for combinations of walls.

ACKNOWLEDGMENT

The programming was carried out by Mr. G. Brouwer and Mr. F. ten Brug, both of the Institute of Applied Physics TNO-TH, Delft. The authors thank the commissioning Stichting Bouwresearch, The Netherlands, who have consented to publication

SYMBOLS

T_{ao} outdoor air temperature (°C);
T'_{ao} equivalent outdoor temperature (°C);
T_{ai} indoor temperature (°C);

T_{wo}	outside surface temperature (°C);
T_{wi}	inside surface temperature (°C);
q_s	solar radiation intensity (W/m^2);
q_r	net heat radiation flux density (W/m^2);
q_{ao}, q_{wo}	heat flux density outside surface (W/m^2);
q_{ai}, q_{wi}	heat flux density inside surface (W/m^2);
λ	thermal conductivity (W/m . °C);
ρ	density (kg/m^3);
c	specific heat (kJ/kg . °C);
d	thickness (m);
α_o	outside surface film coefficient of convection heat transfer (W/m^2 . °C);
α_i	inside surface film coefficient of heat transfer (W/m^2 . °C);
r	reflectivity for solar radiation;
x	place co-ordinate (m);
t	time (h);
ω	angular frequency (rad/h).

REFERENCES

[1] E.F.M. van der Held, *Physica*, **2** (1935), p. 943.
[2] C.O. Mackay and L.T. Wright. *Trans. ASHVE*, **52** (1946), p. 283.
[3] G. Nehring, *Ges.-Ing.* **83** (1962), p. 253.
[4] R.W. Muncey. *Austr. J. of Appl. Sci.*, **14** (1963), p. 123.
[5] A.W. Pratt. *Int. J. Heat and Mass Transfer*, **8** (1965), p. 861.
[6] H.S. Carslaw and J.C. Jaeger. Conduction Heat in Solids (Oxford 1959), p. 109.
[7] A.H. van Gorcum. *Appl. Sci. Res.*, *A2* (1951), p. 272.
[8] V. Vodicka. *Appl. Sci. Res.*, *A5* (1955), p. 108.
[9] R.W. Muncey, *Appl. Sci. Res.*, *A5* (1955), p. 461.
[10] C.J. Shirtliffe and D.G. Stephenson, Techn. Paper no. 114. Div. Building Res. NRC Canada, Ottawa (1961).
[11] J. Bosma. Report S 505, Lab. for Refrigeration TH, Delft (1967).
[12] G. Brouwer, F. ten Brug and P. Euser, Report 332, Inst. Appl. Physics TNO-TH, Delft (1969).
[13] W. Raiss and J. Masuch. *Ges.-Ing.*, **90** (1969), H3, p. 67.
[14] J. Masuch. *Ges.-Ing.*, **90** (1969) H7, p. 213.
[15] T. Kusuda. *Trans. ASHRAE*, **75** (1969), Part 1, No. 2108, p. 246.
[16] G.P. Mitalas and J.G. Arseneault, DBR Computer Program no. 23, Div. Building Res. NRC, Canada (1967).
[17] R.W. Muncey and J.W. Spencer. *Build. Sci.*, **3** (1969), p. 227.

REMARQUE

J. LEBRUN (Belgique) — A propos de la représentation des échanges convectifs dans une chambre fermée, la température d'air *au centre* n'est que trop peu significative pour en déduire les flux incidents aux différentes parois. Ceci est surtout vrai en ce qui concerne le *plafond* où la température « au large » peut être beaucoup plus élevée que celle à mi-hauteur (il peut même en résulter une inversion des échanges). Sans doute est-il donc nécessaire de rechercher une représentation plus fine et plus complète; c'est dans cette direction que s'orientent nos recherches de simulation au CEDRIC.

DÉTECTION DE DENSITÉ SUPERFICIELLE DE FLUX THERMIQUE PAR SONDES À FAIBLE INERTIE ET HAUTE RÉSOLUTION SPATIALE*

J. LEBRUN et J. HANNAY

Institut de Thermodynamique et CEDRIC
Université de Liège (Belgique)

Detection of the superficial heat flux density via probes of low inertia and high space resolution

SUMMARY: *The principle of measuring flux density consists of interposing, in the interior or at the frontier of a conductive field, a known thermal resistance whose terminals can detect a certain temperature drop.*

Various practical possibilities can be envisaged to this effect: the AA. briefly sum-up the advantages and drawbacks and place particular emphasis on the use of fluxmeters of near-pin-point precision, these meters being wall-mounted. This is a particularly convenient resolve for making rapid in situ measurements.

Conductive disturbances caused through the fluxmeter being wall-mounted have been analogically evaluated.

PRINCIPE DE DÉTECTION

De nombreuses recherches expérimentales, notamment dans le domaine de la climatisation, requièrent la connaissance de densités de flux à la surface de certaines parois (supposées étanches et opaques), qui sont soumises aux échanges convectifs et radiatifs.

— *La densité de flux radiatif* est calculable, moyennant certaines hypothèses (notamment la loi de Lambert) et la connaissance des émissivités et températures superficielles relatives à la paroi et à son rayonnement; il est aussi possible de mesurer cette densité au moyen de deux radiomètres : l'un est braqué sur la paroi, l'autre sur l'environnement.

— *La densité de flux convectif* n'est que rarement calculable; sa mesure est liée à celle du gradient de température du fluide contre la paroi et n'est donc précise que si la couche limite thermique est assez épaisse relativement aux dimensions de la sonde et à l'erreur de localisation.

— Enfin, *la densité de flux conductif* juste derrière la surface est égale à la *densité globale*, soit la somme algébrique des densités radiatives et convectives. Il suffit donc de mesurer le gradient de température à la frontière du milieu solide pour en déduire la densité globale du flux incident.

Trois possibilités pratiques se présentent :

1. mesurer le gradient de température dans la paroi elle-même, *sans altération* (solution idéale);
2. *substituer* une couche fluxmétrique à une couche constitutive de la paroi;
3. *adjoindre* une couche fluxmétrique à la paroi.

(*) Cette communication est extraite d'une étude de synthèse en préparation sous la direction du Professeur G. BURNAY.

Ces trois procédés sont utilisables avec différentes résolutions spatiales; le premier est applicable en régime établi et si la conductivité de la paroi est connue : il suffit alors tout simplement de différencier les températures des deux faces de la paroi (ou d'une couche de celle-ci). Plus fréquemment, on utilise les deux autres procédés, qui peuvent conduire à munir la paroi, soit d'une *peau*, soit de *pastilles* fluxmétriques *encastrées* ou simplement *collées* à la surface de celle-ci (fig. 1).

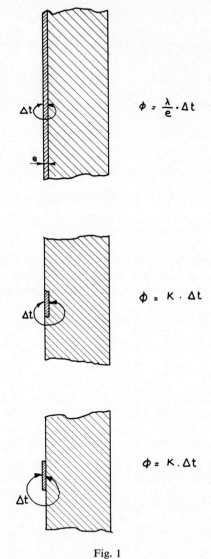

Fig. 1

PEAU FLUXMÉTRIQUE

Cette peau est, par exemple, constituée d'une feuille de frigolite, cousue de thermo-couples cuivre-constantan, qui sont montés en série de façon à mesurer la différence de température entre les deux faces.

Cette solution est notamment utilisée pour la détection de flux résiduels à travers les parois d'une chambre calorimétrique. Il n'y a pratiquement pas de limite à la sensibilité; celle-ci est proportionnelle à la résistance thermique de la peau et au nombre de thermocouples. Chaque élément est étalonné par la méthode de Poensgen, utilisée couramment pour les mesures de conductivité.

Avec ce type de fluxmètre, on peut dresser les bilans thermiques relatifs à une installation complète, avec une précision de l'ordre de 2%.

Fluxmètre encastré dans la paroi

On utilise cette disposition surtout pour la détection *locale* des densités de flux à travers une paroi, éventuellement soumise à un champ convectif très intense. La perméance thermique de cette paroi n'en est pratiquement pas modifiée, pas plus que ses propriétés superficielles.

L'inconvénient est, sans doute, la nécessité d'étalonner les éléments de paroi avec leurs fluxmètres; en effet, la conductivité de la pastille n'est généralement pas la même que celle de la paroi et il en résulte une certaine distorsion des lignes de flux à l'endroit de la mesure.

La nécessité d'introduire le fluxmètre dans une alvéole de la paroi et d'étalonner celle-ci limite l'application de ce procédé aux montages de laboratoire; nous l'avons notamment utilisé pour la constitution de façades légères, dans le cadre de recherches expérimentales sur la climatisation. Le fluxmètre en question est réalisé par Hatfield; il s'agit d'une pastille de Tellurium fritté avec armatures en cuivre : le diamètre et l'épaisseur font respectivement 12 et 1 mm, la résistance thermique $1,2 \times 10^{-3}$ °C.h. m^2/kcal et la sensibilité 0,5 µV/kcal.h.m^2 (ordre de grandeur pour une pastille noyée dans du plexiglas).

L'étalonnage se fait par la méthode de Poensgen comme pour les « peaux fluxmétriques ».

Avec des fluxmètres disposés judicieusement et une densité moyenne d'au moins 100 kcal/h.m^2, il est possible d'atteindre une précision de 4% sur l'ensemble de la façade.

Fluxmètre collé à la paroi

C'est le procédé le plus commode pour de nombreuses applications pratiques. Encore faut-il minimiser la triple perturbation qui en résulte : la présence du fluxmètre modifie les propriétés superficielles de la paroi et donc aussi le flux convectif et radiatif qui en dépendent; le champ conductif est également perturbé.

Il est possible de donner à la surface externe du fluxmètre une émissivité proche de celle de la paroi (tout au moins dans le domaine des infrarouges) et de minimiser ainsi la perturbation radiative; en ce qui concerne le champ convectif, il faut que l'épaisseur du fluxmètre soit la plus petite possible vis-à-vis de celle de la couche limite, ou, à défaut de respecter cette condition, que le fluxmètre soit bien profilé, de façon à minimiser la perturbation des lignes de courant.

Pour cette disposition, nous utilisons un fluxmètre monté sur circuit imprimé cuivre-téflon-cuivre (fig. 2) avec connexions en constantan (fig. 3). Le fluxmètre complet est présenté à la figure 4 : il comprend 110 jonctions, soit 55 couples cuivre-constantan; sa résistance électrique est de 6 Ω. La résistance thermique est très proche de celle de Hatfield, soit $1,5 \times 10^{-3}$ °C.h.m^2/kcal, mais la sensibilité est 6 fois plus grande, soit 3 µV/kcal/h.m^2 [1].

[1] Cette réalisation est due à R. Dupont, Maître-préparateur au service du Professeur G. Burnay.

En raison de la faible résistance électrique du fluxmètre, il n'y a aucune spécification en ce qui concerne l'appareil de mesure; on peut même associer plusieurs fluxmètres en série, de façon à mesurer directement une densité moyenne de flux et ceci avec une sensibilité proportionnelle au nombre des fluxmètres.

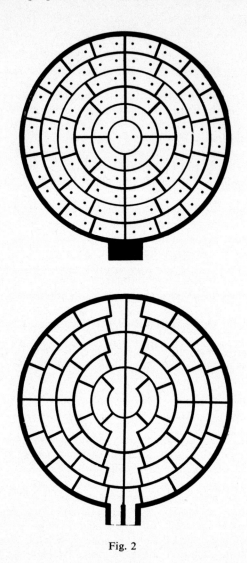

Fig. 2

Si le champ conductif dans la paroi n'est pas sensiblement modifié par la présence du fluxmètre, seule la capacité thermique (C) de celui-ci intervient dans sa constante temps (RC); on a ici

$$C \cong 1 \text{ kcal/m}^2.°C$$

et

$$R = \frac{1}{\alpha + \rho} \text{ en h. m}^2. °C/\text{kcal}$$

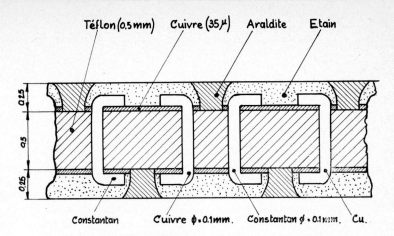

Téflon (0.5mm)   Cuivre (35μ)   Araldite   Etain

0.25   0.5   0.25

Constantan   Cuivre φ=0.1mm.   Constantan φ=0.1mm.   Cu.

Fig. 3

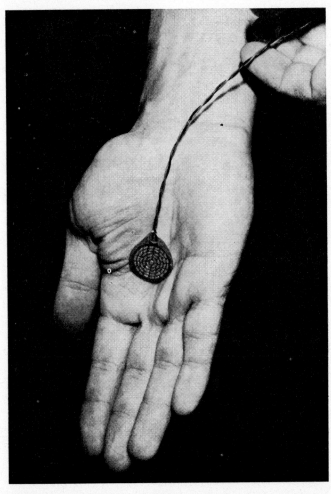

Fig. 4

(la résistance thermique du fluxmètre est négligeable).
Avec

α   coefficient d'échange convectif;
ρ   coefficient d'échange radiatif.

il en résulte que :

$$RC \cong \frac{1}{\alpha+\rho} \text{ en heures.}$$

Ainsi, pour $\alpha+\rho = 10$ kcal/h.m^2.°C, on obtient :

$$RC = 6 \text{ minutes.}$$

On connaît encore mal le domaine d'utilisation de ces fluxmètres dans le champ *convectif*; l'erreur éventuelle diminue en tout cas avec l'intensité de ce champ. Si le champ est très intense (vitesses ou écarts de température élevés), il est prudent d'effectuer un étalonnage dans des conditions comparables quant à la vitesse de l'air au large, l'orientation et l'échauffement de la plaque sur laquelle est collé le fluxmètre (méthode de Poensgen à l'air libre).

Au moins peut-on s'assurer que la perturbation *conductive* (schématisée fig. 5) est vraiment négligeable : ce contrôle est possible par voie analogique.

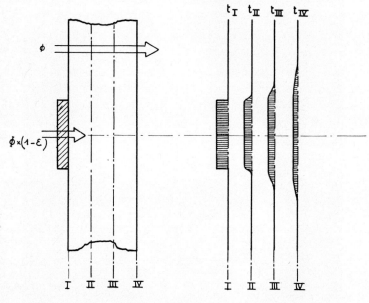

Fig. 5

APPROCHE ANALOGIQUE ET DISCUSSION

Le problème est tridimensionnel, mais la symétrie de révolution autorise une représentation bidimensionnelle : la paroi est découpée en tranches, qui, elles-mêmes, sont découpées en couronnes concentriques; à chacune de ces couronnes correspond un nœud du réseau bidimensionnel.

L'étude a été faite pour une paroi en plexiglas de 6 mm d'épaisseur : le fluxmètre a été représenté par une seule résistance (on y supposait donc l'écoulement unidimensionnel). La paroi a été décomposée en 7 tranches de 7 anneaux, soit 49 éléments représentés par les 49 nœuds d'un réseau de résistances électriques. Les coefficients d'échanges convectifs et radiatifs étaient représentés par d'autres résistances appliquées aux frontières du réseau.

Cette méthode n'est d'une précision acceptable que si la résistance des fluxmètres est relativement importante; aussi ne dispose-t-on de résultats que pour des fluxmètres

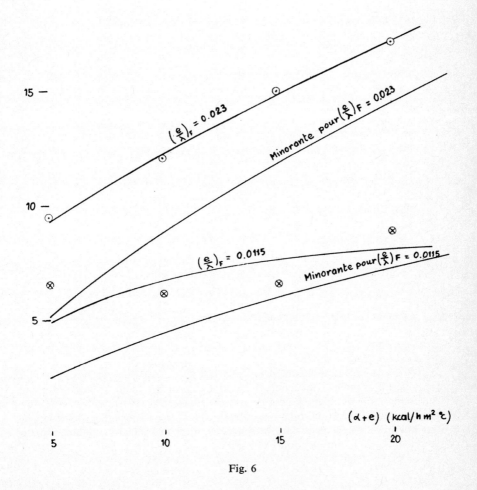

Fig. 6

beaucoup plus résistants que le modèle réel. Ces résultats sont présentés à la figure 6; les notations y sont les suivantes :

ϕ densité de flux traversant la paroi en l'absence de fluxmètre;

ϕ_m densité de flux mesurée par le fluxmètre;

$\dfrac{\phi - \phi_m}{\phi}$ erreur de mesure (en %);

e épaisseur (en m);

λ conductivité (en kcal/h.m.°C);

F   désigne le fluxmètre et P la paroi.

On a également présenté à la figure 6 les courbes *minorantes* de l'erreur, qui sont calculées en considérant le problème comme unidimensionnel; on trouve alors facilement que :

$$\frac{\phi - \phi_m}{\phi} = \frac{(e/\lambda)\,\mathrm{F}}{\dfrac{2}{\alpha + \rho} + \left(\dfrac{e}{\lambda}\right)p}$$

Le décalage relativement important entre courbes expérimentales et minorantes résulte de la déformation des lignes de flux dans la paroi à l'endroit du fluxmètre.

Dans le cas d'une paroi très conductrice, il est évident que l'expression de l'erreur se réduit à :

$$\frac{\phi - \phi_m}{\phi} = \left(\frac{e}{\lambda}\right)_{\mathrm{F}} \times (\alpha + \rho)$$

(Cette expression est aussi une majorante de l'erreur pour une paroi quelconque).

Il ressort de ces résultats que, comme les majorantes et minorantes, les erreurs réelles sont pratiquement proportionnelles à la résistance du fluxmètre : dans le cas courant où

$$\alpha + \rho \cong 10$$

on trouve :

$$\frac{\phi - \phi_m}{\phi} \cong 5 \times \left(\frac{e}{\lambda}\right)_{\mathrm{F}}$$

Pour le fluxmètre dont il est question ici, on trouve donc :

$$5 \times 1,5 \times 10^{-3} = 0,75\,\%$$

et dans le cas de la plaque infiniment conductrice :

$$10 \times 1,5 \times 10^{-3} = 1,5\,\%$$

ce qui reste très acceptable.

On peut donc utiliser le fluxmètre sur une paroi métallique; en pratique, il semble même préférable, lorsque la conductivité de la paroi est faible, d'interposer une rondelle métallique de façon à uniformiser la température superficielle sous le fluxmètre. Sans cette précaution, on a constaté des écarts très importants, probablement dus au fait que l'écoulement dans le fluxmètre n'est pas du tout unidimensionnel comme nous l'avions supposé pour cette étude.

CONVECTION MIXTE LE LONG D'UNE SURFACE VERTICALE. RÉSULTATS EXPÉRIMENTAUX*

P. NUSGENS

Institut de Thermodynamique et CEDRIC
Université de Liège (Belgique)

Mixed air convection along a vertical surface. Experimental results

SUMMARY: *The study in question was carried out in an air environment on an experimental fluxmetric device which was subjected in wind-tunnel to various air velocity and temperature conditions.*

The equipment consisted mainly of an electrically heated plate fitted with a guard ring; a fluxmeter intercepts the residual flux at the back of the plate. The domain investigated covered Reynold's numbers of 0 to 4×10^5 and Grashof's numbers of 1.5×10^6 to 4.3×10^7. In the "open spaces", a turbulence level of about 1% was detected.

The domain (Re, Gr.) referred to was systematically investigated for incidence angles of 0.45° and 90°. Some complementary measurements were made in order to provide better localisation of the critical incidence. The tests were carried out on two types of surface finishes, one smooth and the other rough (a roughened surface of pyramidal form).

INTRODUCTION

Les échanges convectifs au niveau du sol entre l'air et une paroi verticale sont influencés par deux sortes de paramètres, d'une part les paramètres liés à la paroi, à savoir la température de sa surface t_p et son état de surface (forme, rugosité) et d'autre part les paramètres qui caractérisent l'écoulement de l'air, sa température au large t_f, le vecteur vitesse moyenne $\vec{V}$, la turbulence $\tilde{v}$ le long de la paroi. On peut considérer que pour une paroi verticale de bâtiment, le coefficient d'échange convectif au temps τ_i est fonction de ces divers paramètres et écrire

$$\alpha_{\tau i} = f(\vec{V}_{\tau i}, \tilde{v}_{\tau i}, \text{rug}, t_p - t_f)$$

le coefficient de convection moyen $\bar{\alpha}$ durant une période τ valant alors:

$$\alpha_{(\tau m, \; \tau m + \tau)} = \frac{1}{\tau} \int_{\tau m}^{\tau m + \tau} \alpha \tau_i \, \mathrm{d}\tau_i$$

Les grandeurs $\vec{V}\tau_i$ et $\tilde{v}\tau_i$ sont caractérisées par un spectre de variations aléatoires.

Dans le but d'évaluer les relations qui unissent les $\alpha\tau_i$ aux différents paramètres, un dispositf expérimental fluxmétrique a été soumis en soufflerie à diverses conditions de vitesse d'écoulement et d'écarts de température. On a envisagé, en ce qui concerne l'écoulement d'air, des vitesses allant jusque 15 m/s avec un niveau de turbulence de l'ordre du pourcent, des incidences de 0,45 et 90°, et ce pour une paroi lisse et une paroi rugueuse, le domaine de l'écart de température $t_p - t_f$ allant de 2 à 15 deg C.

DESCRIPTION DE L'APPAREILLAGE

Le dispositif expérimental consiste en une plaque active recevant au dos un flux de chaleur produit par un élément chauffant électrique. En régime établi, et pour autant

---

(*) Cette communication est extraite d'une étude de synthèse en préparation sous la direction du Professeur G. BURNAY.

que l'on puisse négliger le flux de fuite périphérique ϕ_f, la puissance électrique se partage en deux, un flux ϕ_1 cédé par la paroi active et un flux de fuite au dos ϕ_2 que l'on contrôle au moyen d'un fluxmètre (fig. 1). Dès lors, on a :

$$\alpha = (\phi - \phi_2)\,[S.(t_p - t_f)]^{-1}$$

et la connaissance de ϕ, ϕ_2, t_p et t_f permet de déduire α.

En fait, le flux ϕ_1 comporte en plus du terme convectif un terme radiatif qui peut être estimé, comme nous le verrons plus loin.

Les dimensions du modèle ont été choisies égales à 30×30 cm afin d'éviter les zones où la veine fluide est perturbée par les bords du tunnel.

Au centre d'une plaque d'aluminium poli de 30 cm de côté et de 2 mm d'épaisseur, un carré de 17×17 cm constituant la paroi active, a été découpé et ensuite resolidarisé par un joint isolant au reste de la plaque, qui forme ainsi l'anneau de garde (fig. 1 et 2). Chacune de ces parties, paroi active et anneau de garde est indépendamment chauffée par une résistance électrique.

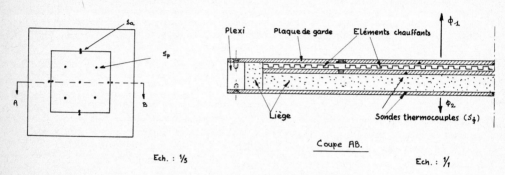

Fig. 1 et 2 — Dispositif expérimental.

En ce qui concerne les éléments chauffants, on s'est efforcé, tout en maintenant une bonne isolation électrique, d'obtenir une faible inertie thermique et un chauffage suffisamment uniforme et d'éviter l'introduction de lame d'air entre les divers éléments du modèle. La solution retenue est de loger les fils de résistance dans les sillons creusés dans une plaque d'amiante et de les y maintenir en refermant le sillon avec un enduit polymérisant. L'enroulement se fait en passant au dos de l'amiante par des trous forés à travers la plaque, un sillon étant réalisé au dos pour le passage transversal (fig. 3 et 4). On obtient ainsi deux éléments en forme de serpentin, un pour la paroi active, l'autre pour l'anneau de garde (fig. 4). Le fil utilisé étant du fil d'acier inoxydable (tophet 30) d'un diamètre de 0,254 mm et de résistivité :

$$\rho = 22{,}144\ \Omega/\text{m}$$

Fig. 3. et 4 — Éléments chauffants.

la résistance centrale vaut 16 Ω et la résistance périphérique 25 Ω. Ces éléments chauffants sont alimentés par une double alimentation continue stabilisée qui fournit une intensité maximale de 2 A sous une tension pouvant varier continûment de 0,5 à 30 V. En ajustant la puissance fournie à la résistance périphérique, on obtient l'annulation du flux de fuite ϕ_f, annulation contrôlée par quatre paires de thermocouples (Cu-Ct-)(Sa) noyés dans l'aluminium et mesurant l'écart de température entre les deux plaques.

La température superficielle t_p de la paroi active est mesurée par cinq thermocouples (S_p) à nouveau noyés dans des sillons creusés au dos de la plaque active et disposés sur les diagonales du carré afin de contrôler l'uniformité de la température (fig. 1).

Le fluxmètre destiné à mesurer le flux au dos est essentiellement constitué d'une plaque de liège entre deux plaques d'aluminium, celle se trouvant près de la paroi active ayant été réalisée également au moyen de deux plaques concentriques reliées par un isolant afin de ne mesurer que le flux ϕ_2 dû à l'élément chauffant central. L'épaisseur du liège a été choisie égale à 10 mm afin d'obtenir dans la lame isolante un gradient de température suffisamment élevé pour atteindre une bonne précision dans la mesure du flux. Quatre thermocouples ($S\phi$ fig. 2) sont logés dans chaque feuille d'aluminium et sont reliés en série de manière à additionner les quatre différences de température existant entre deux sondes opposées.

Les différents éléments constituant le modèle sont empilés et pressés les uns sur les autres par vissage des plaques externes dans un cadre de plexiglas.

LE TUNNEL

Le modèle ainsi réalisé a été placé dans le tunnel aérodynamique de l'Institut de Mécanique et de Thermodynamique de l'Université de Liège. Ce tunnel est alimenté par une soufflerie du type de Prandtl à retour, à aspiration et à fonctionnement continu. L'écoulement est régularisé en amont du tronçon de mesure par un filtre cellulaire formé de mailles de 5 cm de côté et de 10 cm de profondeur. Le tronçon de mesure de 1,44 m de diamètre aval et de 1,2 m de diamètre amont, possède une profondeur de 1,4 m. Le ventilateur est entraîné par un moteur à courant continu alimenté par un groupe Léonard, la vitesse de la veine se réglant avec grande précision en agissant sur les excitations de la génératrice. On a constaté que dans la section droite, la vitesse est constante à moins de 1,5 % près dans un cercle de 1,2 m de diamètre ; elle subit le long de l'axe une diminution de 1 % entre l'entrée et la sortie du tronçon de mesure et l'intensité de la turbulence, suivant l'axe, mesurée avec un anémomètre à fil chaud est de l'ordre du pourcent.

Le modèle expérimental est placé au centre de la veine fluide sur un pivot qui permet de faire varier l'angle d'incidence sur la plaque.

LES TECHNIQUES DE MESURE

Toutes les mesures de température sont faites par l'intermédiaire de thermocouples cuivre-constantan donnant une tension de 40 µV par °C; ces signaux électriques variant entre 0 et 3 mV sont mesurés au moyen d'un appareil potentiométrique permettant des lectures à 5 µV près.

La température de l'air au large de la paroi est donnée également par un thermocouple protégé par un écran d'aluminium poli. La puissance électrique fournie à l'élément central est déterminée par la connaissance de la tension et du courant d'alimentation qui sont mesurés avec le même instrument que les températures, moyennant l'emploi d'un diviseur de tension et d'un shunt.

La vitesse du vent est contrôlée par un anémomètre à fil chaud lorsque la vitesse est inférieure à 3 m/s et par un tube de Pitot relié à un micromanomètre à eau distillée pour les vitesses supérieures.

Une fois que le modèle est orienté suivant l'incidence voulue au centre de la veine fluide, on injecte une certaine puissance à la résistance centrale et on ajuste progressivement la puissance dissipée par la résistance périphérique de manière à obtenir l'égalité de température de la paroi active et de l'anneau de garde.

Quand on a obtenu un régime établi, c'est-à-dire un flux de fuite ϕ_f nul et la stabilité dans le temps de toutes les températures, on répète pendant une demi-heure les diverses mesures.

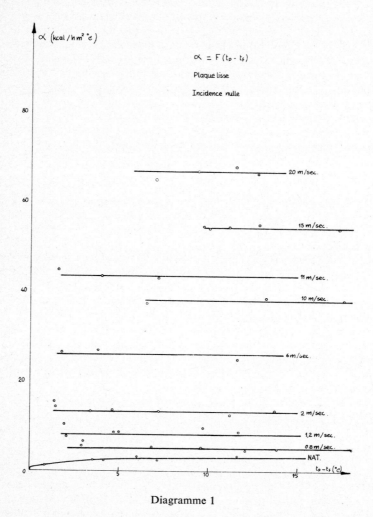

Diagramme 1

En ce qui concerne l'échange radiatif entre la plaque et l'environnement, on a mesuré les températures de surface des parois en regard de la plaque, à savoir des écrans de polystyrène placés pour la circonstance, et constaté qu'elles étaient très proches de celle de l'air (écart de 0,1 à 0,2 deg C). L'émissivité de l'aluminium utilisé a été estimée à $\varepsilon = 0,04$ par comparaison avec les résultats obtenus avec le modèle expérimental recouvert d'une laque noire d'émissivité égale à 0,95.

L'échange radiatif a été calculé pour les différents écarts de température et soustrait de l'échange total dont il ne représente d'ailleurs qu'une faible fraction.

Résultats des essais

Les résultats obtenus lors des essais réalisés ont été regroupés sur des diagrammes illustrant les conclusions que nous pensons pouvoir dégager.

Les diagrammes 1 et 2 expriment le coefficient d'échange convectif d'une plaque lisse en fonction de l'écart de température pour différentes vitesses de vent, dont l'incidence est soit nulle (diagramme 1), soit normale (diagramme 2).

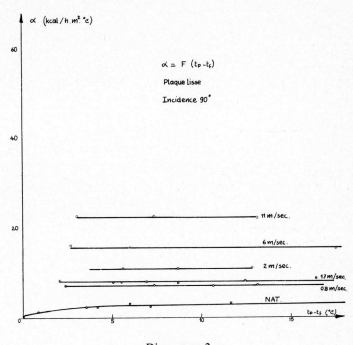

Diagramme 2

En ce qui concerne les essais sous incidence nulle, le domaine des vitesses a été étendu jusqu'à 20 m/s dans le cadre d'une étude sur la climatisation des wagons des Chemins de Fer Roumains. On constate d'une part que la loi obtenue en convection naturelle est proportionnelle à $(\Delta t)^{0,25}$ et d'autre part que pour des vitesses supérieures à 0,8 m/s l'échange convectif ne dépend plus de l'écart de température.

L'influence de l'incidence est montrée par le diagramme 3, déduit des deux précédents, où α est exprimé en fonction de la vitesse. L'échange convectif pour un écoulement parallèle à la paroi est supérieur à celui produit avec un écoulement perpendiculaire de même vitesse. A 11 m/s par exemple, le rapport est égal à 2.

La première série d'essais réalisée systématiquement avec des incidences de 0,45 et 90° a donné des résultats virtuellement semblables pour 45 et 90°. Il semblerait que pour une incidence supérieure à 5°, le coefficient d'échange prendrait déjà la même valeur que pour une incidence de 45 ou de 90° mais ce point demande à être détaillé par une nouvelle série d'essais.

Les essais sur une paroi rugueuse sont faits avec le même modèle expérimental, la paroi active et l'anneau de garde étant remplacés par un dispositif identique réalisé,

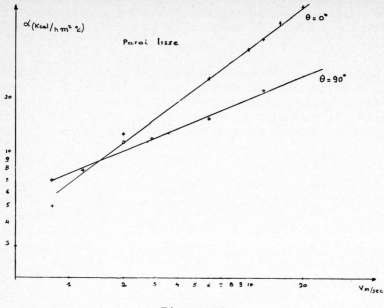

Diagramme 3

cette fois, dans une plaque d'aluminium dont la surface se présente comme une succession continue de pyramides de 1 mm de hauteur et de 1 mm de base. Les résultats sont repris sur le diagramme 4. On constate que la présence d'aspérités a pour effet d'augmenter le coefficient d'échange; cette augmentation, peu sensible pour l'écoulement normal, est très nette pour un écoulement parallèle.

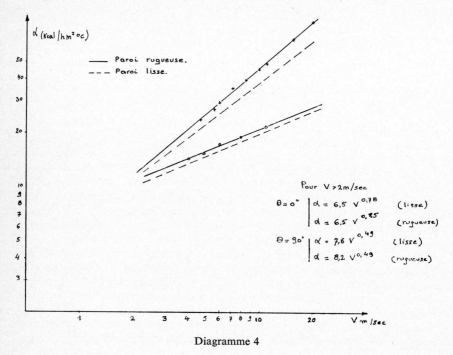

Diagramme 4

114

DISCUSSION

S. KLARSFELD (France) — La régulation de l'anneau de garde par les quatre paires de thermocouples (fig. 1) différentiels se fait-elle pour l'ensemble de l'anneau (intégrant l'effet des fuites sur tout le long du pourtour de la plaque centrale) ou indépendamment pour chacun des quatres côtés de la plaque active ?

P. NUSGENS — Les deux procédés ont été utilisés, sans différence notoire.

H. MARCQ (Belgique) — En fait les mesures faites par M. Nusgens donnent le coefficient *moyen* de convection sur une longueur de 17 cm dans le sens du flux d'air, longueur commençant à 6.5 cm (30-17)/2 du bord d'attaque de la plaque entière (anneau de garde + plaque active). Dans ces conditions, l'épaisseur de la couche limite est encore très faible et le coefficient moyen de convection fort élevé. Pour l'application à la convection forcée sur les parois extérieures des bâtiments, il faudrait étudier l'influence de la longueur de la plaque active et celle de la position de cette plaque par rapport au bord d'attaque pour des longueurs et distances beaucoup plus grandes que dans les essais décrits.

P. NUSGENS — Cette longueur doit effectivement être prise en considération et cette question est à l'étude; il n'est malheureusement pas certain qu'une extrapolation rigoureuse soit toujours possible en ce qui concerne l'estimation des coefficients d'échange relatifs aux façades de bâtiments. Des mesures sur des modèles à échelle différente sont encore nécessaires pour confirmer la validité de cette extrapolation.

UN THERMOANÉMOMÈTRE SIMPLE ET À TRÈS FAIBLE EFFET DIRECTIONNEL*

J. LEBRUN

Institut de Thermodynamique et CEDRIC
Université de Liège (Belgique)

A simple thermo-anemometer of very low directional effect

SUMMARY: *The use of a thermal detector is particularly interesting when measuring low velocities, but it is necessary to minimise the influence of free convection and pure conduction on measurements.*

The detector in question is a thermistor fitted into a Wheatstone bridge; the alternate injection of two electric currents enables the measurement of either the fluid temperature or velocity at any given point. The near-spherical form of the probe virtually eliminates all directional selectivity.

The apparatus is compact and cheap to produce, thus enabling an increase in the number of measurement points in an installation.

DÉTECTION DES FAIBLES VITESSES D'AIR

À l'occasion de recherches ou de contrôles, notamment dans le domaine de la climatisation, on est souvent amené à devoir mesurer des vitesses de l'ordre de 10 cm/s et même moins; ces mesures ne sont accessibles qu'avec un anémomètre thermique et encore ne sont-elles pas sans difficulté, comme on le verra ci-dessous.

Le *principe* de la détection thermique des vitesses d'air est très simple.

On fournit à la sonde une puissance électrique P et on mesure son échauffement Δt :

$$\Delta t = t_{ep} - t_{e0} = \frac{P}{\alpha + \rho}$$

$$t_{e0} = \frac{\alpha t + \rho t_w}{\alpha + \rho}$$

avec

α coefficient d'échange convectif;
ρ coefficient d'échange radiatif;
     (linéarisation de la loi de Stéphan);
t_{e0} température d'équilibre du détecteur à puissance nulle;
t_{ep} température d'équilibre à la puissance P;
t température de l'air;
t_w température moyenne de rayonnement.

Autant que possible, on s'arrange pour que

$$\alpha \gg \rho$$

et comme, dans bien des cas, t diffère peu de t_w, il reste :

$$t_{e0} \cong t.$$

(*) Cette communication est extraite d'une étude de synthèse en préparation sous la direction du Professeur G. BURNAY.

Si donc on connaît t et P, la mesure de t_{ep} suffit à déterminer la vitesse V; en effet

$$\alpha = f(V).$$

En réalité, α n'est univoquement lié à la vitesse que si les forces d'inertie l'emportent nettement sur les forces gravifiques, c'est-à-dire si l'on est bien au-delà de la zone de convection mixte.

Pour une sphère, Yuge [1] trouve les corrélations suivantes :

$$\mathrm{Nu} = 2 + 0,392 \times (\mathrm{Gr})^{0,25} \text{ (convection naturelle)}$$
pour $1 \leqslant \mathrm{Gr} \leqslant 10^5$
et $\mathrm{Nu} = 2 + 0,493 \times (\mathrm{Re})^{0,5}$ (convection forcée)
pour $10 \leqslant \mathrm{Re} \leqslant 1,8 \times 10^3$

avec

Nu nombre de Nusselt;
Gr nombre de Grashof;
Re nombre de Reynolds.

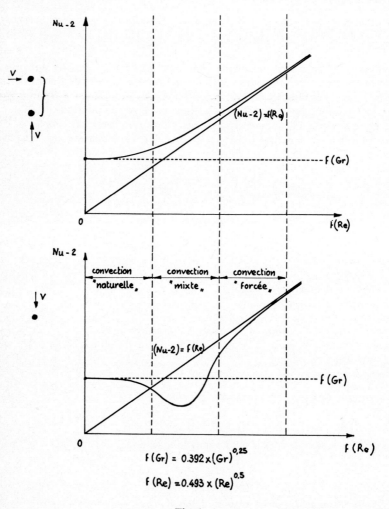

$$f(Gr) = 0{,}392 \times (Gr)^{0,25}$$

$$f(Re) = 0{,}493 \times (Re)^{0,5}$$

Fig. 1.

En convection mixte, l'expression du Nu se complique, car elle dépend notamment de l'*orientation* du vecteur vitesse, comme le montre la figure 1.

L'influence du Gr est donc à minimiser; le critère de Klyachko [2] s'écrit :

$$\frac{Gr}{Re^2} \ll 1$$

ce qui donne ici

$$0{,}033 \times \frac{D\,\Delta t}{V^2} \ll 1 \qquad\qquad (I)$$

(dans l'air à 20 °C) avec D : diamètre de la sphère.

D'autre part, les équations de la convection comportent un terme constant 2, qui n'est autre que le Nu de conduction d'une sphère en milieu infini; si l'on désire avoir le maximum de sensibilité, il faut encore que :

$$2 \ll 0{,}493 \times (Re)^{0{,}5}$$

ce qui peut s'écrire :

$$2{,}47 \times 10^{-4} \ll VD \qquad\qquad (II)$$

(dans l'air à 20 °C).

Enfin, considérant l'influence de la température dans le domaine de 0 à 50 °C, nous pouvons écrire :

$$\frac{1}{\lambda} \times \frac{d\lambda}{d\,t} \cong 0{,}3\,\%/°C$$

et

$$\frac{1}{\nu} \times \frac{d\nu}{d\,t} \cong 0{,}7\,\%/°C$$

λ et ν étant respectivement la conductivité thermique et la viscosité cinématique de l'air.

On en déduit directement :

$$\frac{1}{\alpha}.\frac{d\alpha}{dt} \cong 0{,}3\,\%/°C \text{ pour } VD \ll 2{,}47 \times 10^{-4}$$

et

$$\frac{1}{\alpha}.\frac{d\alpha}{dt} \cong 0 \qquad \text{pour } 2{,}47 \times 10^{-4} \ll VD.$$

On est donc ramené à la condition (II), si l'on désire minimiser les variations de α avec la température.

Les conditions (I) et (II) sont d'autant moins compatibles que la vitesse est faible et Δt doit être maintenu à une valeur suffisamment élevée vis-à-vis de la précision de l'appareil de mesure et surtout vis-à-vis des variations aléatoires de la température du fluide dans le domaine spatio-temporel où s'effectue la mesure de vitesse :

Pour une sonde *différentielle*, la distance séparant les deux détecteurs (« chaud » et « froid ») est une source d'imprécision lorsque le gradient de température du fluide est élevé.

Pour une sonde *simple*, c'est le temps séparant les deux mesures (à « froid » et à « chaud ») qui affecte la précision lorsque la température du fluide varie rapidement.

De tout ceci il résulte que, quelles que soient les caractéristiques choisies pour la sonde, elles ne sont pas immuables, mais uniquement adaptées à un certain domaine d'application.

UTILISATION D'UN DÉTECTEUR À THERMISTANCE N.T.C.

La caractéristique de la thermistance est du type :

$$R = A \cdot e^{B/T} \tag{1}$$

(Exemple : voir fig. 2).

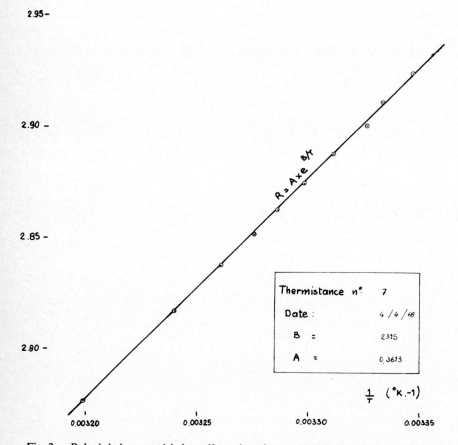

Fig. 2 — Relevé de la caractéristique d'une thermistance N.T.C. (entre 20 et 40 °C).

Cette thermistance est insérée dans un pont de Wheatstone dont les trois autres résistances R_2, R_3, R_4 sont fixes; le pont est alimenté à tension constante U (fig. 3).

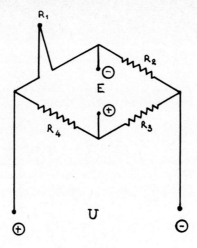

Fig. 3.

Pour la mesure de température (« à froid »), on choisit une tension $U = U_1$ suffisamment faible pour que l'échauffement de la sonde soit inférieur à la précision exigée (par exemple $0,05\,°C$); on peut écrire :

$$\frac{E}{U} = \frac{R_1}{R_1 + R_2} - \frac{R_4}{R_3 + R_4}. \tag{2}$$

On pose :

$$\varepsilon = \frac{E}{U}, \quad \eta_0 = \frac{R_4}{R_3}, \quad \eta = \frac{R_1}{R_2}$$

et on introduit T_0, température d'équilibre du pont.
On obtient ainsi :

$$\varepsilon = \frac{\eta}{1+\eta} - \frac{\eta_0}{1+\eta_0}$$

et

$$\eta = \eta_0\, e^{B\left(\frac{1}{T} - \frac{1}{T_0}\right)}$$

ou, en explicitant par rapport à T :

$$\boxed{\frac{1}{T} = \frac{1}{B}\ln\gamma + \frac{1}{T_0}} \tag{3}$$

121

avec

$$
\begin{cases}
\gamma = \dfrac{1}{\eta_0} \times \dfrac{\varepsilon+\delta_0}{1-(\varepsilon+\delta_0)} \\[4mm]
\delta_0 = \dfrac{\eta_0}{1+\eta_0}
\end{cases}
$$

Par conséquent, ayant choisi η_0 (caractéristique du pont), on peut étalonner la thermistance *avec son pont* et effectuer une régression linéaire sur les variables $1/T$ et $\ln\gamma$; on en tire des tables pour l'utilisation pratique.

Plus simplement encore, on peut, connaissant T_0 par l'étalonnage, chercher parmi les tables, correspondant à cette température et à différentes valeurs de B, celle qui restitue le mieux les autres points de l'étalonnage de la sonde.

Pour la mesure de la vitesse (« à chaud »), on passe à une tension $U = U_2$, telle que l'on obtient un certain échauffement de la sonde; cet échauffement est mesuré comme suit :

$$
\left.
\begin{array}{l}
t_1 = f(E_1/U_1) \\[2mm]
t_2 = f(E_2/U_2)
\end{array}
\right\} \quad \longrightarrow \quad \text{(utilisation des tables)}
$$

$$\Delta t = t_2 - t_1$$

P est facilement calculable :

$$
P = U^2 \times \frac{R_1}{(R_1+R_2)^2} = \left(\frac{U}{R_2}\right)^2 \times \frac{\eta}{(1+\eta)^2}
$$

avec

$$
\eta = \frac{\varepsilon+\delta_0}{1-(\varepsilon+\delta_0)}
$$

et

$$
\delta = \frac{\eta_0}{1+\eta_0} \quad \longrightarrow \quad \text{(caractéristique du pont)}.
$$

On a donc, à une constance près,

$$
P = \frac{\eta}{(1+\eta)^2} = F(\varepsilon)
$$

soit en définitive :

$$
\frac{F(\varepsilon)}{\Delta t} = f(V).
$$

On peut effectuer une régression parabolique sur les résultats de l'étalonnage: d'après la loi de Yuge (cfr. *supra*) on devrait avoir :

$$
f(V) = A + B\sqrt{V}.
$$

En réalité, cette loi ne semble pas assez précise, notamment dans le domaine des très basses vitesses; nous avons donc provisoirement adopté l'équation :

$$V = \alpha_0 + a_1 \cdot \left(\frac{F(\varepsilon)}{\Delta t}\right) + a_2 \cdot \left(\frac{F(\varepsilon)}{\Delta t}\right)^2 + a_3 \left(\frac{F(\varepsilon)}{\Delta t}\right)^3$$

a_0, a_1, a_2, a_3, étant les coefficients de régression obtenus à partir de l'étalonnage.

Pour la pratique, on utilise encore des tables; elles sont cette fois à double entrée : V y est donnée en cm/s pour les couples de valeur $\Delta t = t_2 - t_1$ et $\varepsilon_2 = E_2/U_2$.

UN THERMOANÉMOMÈTRE RÉALISÉ AU CEDRIC[1]

La thermistance que nous avons choisie se présente sous la forme d'une gouttelette; son diamètre moyen est de 0,7 mm.

ETALONNAGE AU TUNNEL (hautes vitesses)

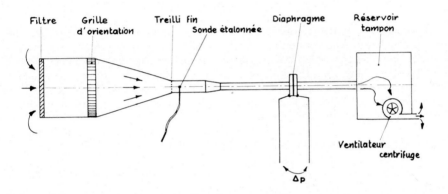

ETALONNAGE SUR CHARIOT (basses vitesses)

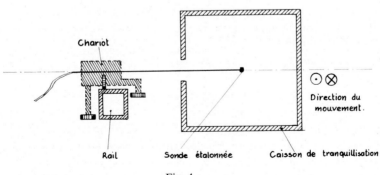

Fig. 4.

[1] Le développement et l'étalonnage de cet appareil ont été menés avec la collaboration de C. ESKENAZI, Ingénieur au CEDRIC.

On a choisi $\Delta t \cong 25\,°C$ à vitesse nulle; il en résulte, en vertu de la condition (I) une certaine influence de la convection naturelle pour les vitesses de l'ordre de quelques cm/s. Mais la condition (II) indique aussi que, dans le même domaine, la conduction est relativement importante, de sorte qu'il n'est pas souhaitable de chercher à encore réduire le diamètre.

La forme quasi sphérique du détecteur supprime pratiquement toute sélectivité directionnelle, si ce n'est évidemment la présence de la tige supportant la sonde.

La constante de temps, qui est fonction décroissante de la vitesse, varie aussi d'une sonde à l'autre; elle est de l'ordre de quelques secondes.

La thermistance présente $1000\,\Omega$ aux environs de $25\,°C$; dans le domaine des températures courantes, les caractéristiques choisies pour le pont sont les suivantes :

$$R_2 = R_3 = 1000\,\Omega$$

$$R_4 = 500\,\Omega\,(\eta_0 = 0,5).$$

Les tensions U_1 et U_2, pour le domaine de 0 à 2 m/s sont :

$$U_1 = 200\,mV \quad (\theta \leqslant 0,05\,°C$$

$$U_2 = 5\,V \qquad (\theta \leqslant 25\,°C).$$

Pour une température comprise, par exemple entre 0 et $50\,°C$, la thermistance peut se situer approximativement entre 1500 et $500\,\Omega$ et la résistance interne du pont va donc de 860 à $600\,\Omega$. L'emploi d'un appareil de mesure à haute impédance d'entrée ($\geqslant 100\,k\Omega$) est donc indispensable; par contre, l'alimentation à tension constante et la longueur des fils de liaison posent peu de problèmes (le pont de Wheatstone, qui ne comporte que des éléments fixes, ne doit d'ailleurs pas nécessairement être ramené près des appareils de mesure).

La simplicité de cet appareil et son faible encombrement permettent de multiplier à peu de frais les points de mesure dans une installation.

Pour l'étalonnage, on dispose d'une double installation (fig. 4) :

— Un tunnel, avec mesure de débit par diaphragme, utilisable au-dessus de 50 cm/s.

— Pour les vitesses allant de 1 à 50 cm/s, un chariot roulant sur un rail rectiligne, parallèlement à un caisson de tranquillisation; ce dernier est rainuré longitudinalement, de façon que la sonde puisse se mouvoir à l'intérieur, en air parfaitement calme.

RÉFÉRENCES

[1] Experiments on heat transfer from spheres including combined natural and forced convection. ASME/C (Aug. 1960), p. 214.
[2] Heat transfer between a gas and a spherical surface with the combined action of free and forced convection. ASME/C (Nov. 1963), p. 355.

DISCUSSION

J. UYTTENBROECK (Belgique) — M. NUSGENS a trouvé que le coefficient α dépend de l'écart de température pour les faibles vitesses. Or dans sa communication M. LEBRUN part de la formule

$$\alpha = f(V).$$

Est-ce que cela signifie que l'anémomètre n'est pas utilisable pour les très faibles vitesses ?

J. LEBRUN — En fait, on cherche précisément à minimiser l'action de la convection naturelle en ce qui concerne l'anémomètre; la formule (1) ((p. 119) montre que l'on doit minimiser le rapport $D\Delta t/V_2$.

Donc, sous une certaine vitesse, et un certain échauffement, l'effet de la convection naturelle s'atténue avec la réduction du diamètre.

Dans notre cas, la sonde a un diamètre inférieur à 1 mm; pour $\Delta t = 25$ deg, l'interaction devient sensible en dessous de 4 cm/sec. Pour la plaque, la longueur caractéristique est environ 200 fois plus grande et le domaine d'interaction se déplace donc certainement vers des vitesses plus élevées, mais la géométrie et les ordres de de grandeur du Re et du Gr sont très différents pour pouvoir appliquer le même critère d'interaction.

W.H. EMERSON (U.K.) — Mr. LEBRUN has drawn attention to the difficulties that attend the measurement of low velocities in fluids, especially where there are significant velocity gradients. There is, however, another circumstance in which the use of hot wire probes, either of the conventional sort or of the kind he has described, becomes problematical because the properties of the fluid are insufficiently well known. An example is a mixture of air and water vapor of which the local composition may not be known and which may even contain water droplets.

A probe to deal with this situation is now being developed at the University of Aberdeen and at the National Engineering Laboratory in the United Kingdom.

It consists of two parallel wires placed across the direction of gas flow and with the second wire lying in the wake of the first. The first wire is essentially a resistance heater and the second a resistance thermometer. A brief, octangular pulse of current is supplied to the first wire and after a short interval the temperature rise in its wake is detected by the second wire, which is incorporated in a bridge circuit. The out-of-balance of the bridge is fed to a double-beam oscilloscope so that the time interval between the pulse and the second wire's response may be recorded. Note that only two measurements are made; the distance between the wires and the time interval. They are all that is needed to compute the velocity, and the physical properties of the fluid are irrelevant. Furthermore, because the time of heating is very brief, the rational convection effects are negligible even at extremely low velocities. The probe is, of course, highly directional.

J. LEBRUN — Ce système est très séduisant, et nous nous sommes aussi intéressés à un appareil de ce genre. Evidement, la nécessité *d'orienter*, la sonde dans l'écoulement la destine à un domaine d'application assez différent de celui de la sonde à thermistance. (omnidirectionnelle).

VISUALISATION D'UN CHAMP CONVECTIF EN CHAMBRE CONFINÉE*

J. HANNAY et J. LEBRUN

Institut de Thermodynamique et CEDRIC
Université de Liège (Belgique)

Visualizing a convective field in a confined space

SUMMARY: *In air conditioning problems, one is confronted with the necessity to make certain estimates of the convective field in a confined enclosure containing one or several heat sources.*

An analytical study of this field being even more difficult, it is necessary to revert to direct experiments in order to obtain, at the least, a qualitative description.

During research into the efficiency of air conditioning equipment, CEDRIC developed certain visualization procedures using first a scaled-down model and later a full-scale test room.

The AA. give some results derived with a test room fitted with a blower working under winter and summer conditions and using two different air jet orientations.

NÉCESSITÉ DE LA VISUALISATION

A propos de l'étude des mouvements d'air en chambre fermée, nous distinguerons trois zones d'observation différant suivant l'échelle et suivant les forces auxquelles l'écoulement y est soumis :

— *Les « jets »* ont une étendue généralement restreinte; l'écoulement y est dominé par les *forces d'inertie*. L'évolution de ces jets, notamment dans leur épanouissement et leur orientation, est aussi affectée par les forces de gravitation (poussée d'Archimède).

— *Les « frontières »* (« couches limites » ou « interfaces ») sont des zones très étroites, les seules où règnent de forts gradients de vitesse et où, de ce fait, les *forces de viscosité* jouent un rôle important. On y trouve parfois aussi de forts gradients de température et donc une influence des forces de gravitation.

— *L'écoulement « au large »*, que les frontières séparent des parois et des jets, occupe la plus grande partie de la chambre; son allure résulte d'une combinaison des forces d'inertie, de gravitation et de pression, dont il est difficile de discerner à priori les importances relatives.

Pas plus que les procédés de calcul, les moyens expérimentaux mis en œuvre ne sont les mêmes pour l'étude de ces zones. Dans les problèmes de climatisation, *l'écoulement au large* est à la fois essentiel du point de vue de l'efficacité et très difficile à approcher théoriquement, ceci en raison de son caractère tridimensionnel, de la multiplicité des forces à considérer et de la complexité des conditions aux limites. D'ailleurs, l'interprétation des résultats expérimentaux, même avec de très nombreux points de mesure de température et vitesses d'air, n'est généralement possible que moyennant une vision, au moins qualitative, de *l'allure du champ convectif*.

La méthode de visualisation, dont il est question ici, peut être adaptée aussi bien à l'étude des jets et des frontières qu'à celle de l'écoulement au large, mais sans doute est-ce pour celui-ci qu'elle est le plus efficace.

(*) Cette communication est extraite d'une étude de synthèse en préparation sous la direction du Professeur G. BURNAY.

Pour connaître le champ vectoriel des vitesses dans un fluide, on peut y introduire des particules, dont on observe les trajectoires; celles-ci sont, moyennant certaines exigences concernant leur finesse et leur poids, identiques en tous points aux vecteurs vitesses, dont on peut ainsi déceler *l'orientation*.

La grandeur de ces vitesses est mesurable, soit en chronométrant le temps de passage d'une particule entre deux repères, soit plus précisément par photographie : sur la pellicule apparaissent des *traits*, dont la longueur est proportionnelle à la vitesse et au temps de pose.

On choisit certains plans de projection, dont la combinaison restitue au mieux l'écoulement tridimensionnel; celui-ci possède d'ailleurs souvent un plan de symétrie, intéressant en raison du fait que l'écoulement y est strictement bidimensionnel. Une rampe d'éclairage à faisceau parallèle permet la sélection du plan de projection, avec une certaine épaisseur de tolérance.

La méthode a tout d'abord été utilisée sur un *modèle réduit* (échelle 0,3), à parois transparentes, de la première chambre d'essais pour radiateurs du CEDRIC. Dans ce modèle, on injecte des paillettes d'aluminium, que l'on éclaire au moyen d'une source à vapeur de mercure; on utilise, pour la photographie, un stroboscope, dont le signal est codé de façon à indiquer l'orientation des vitesses et à éliminer toute confusion, qui serait due au scintillement des paillettes tournant sur elles-mêmes dans le faisceau lumineux.

Ce modèle présente de très nombreuses possibilités d'application; on y a notamment étudié l'influence de la répartition des températures de parois sur le champ convectif.

Pour les essais *en vraie grandeur*, on a aussi réalisé une chambre spéciale, qui est décrite ci-dessous.

Visualisation dans une chambre d'essais

Une nouvelle enceinte légère et transparente a été édifiée à l'intérieur de la chambre d'essais à parois calorimétriques du CEDRIC; cette installation bénéficie d'une infrastructure très complète pour l'essai d'un appareil de climatisation : contrôles et mesures de puissance, de densités superficielles de flux, de températures, de vitesses d'air, etc....[1]

L'enceinte intérieure est parallélépipédique; la base fait 3,80 m sur 4,00 m et la hauteur 2,70 m.

Trois faces sont transparentes : le plafond, au-dessus duquel se trouvent les rampes d'éclairage, et deux parois verticales adjacentes à travers lesquelles se font les observations. Les trois autres faces sont noires, ainsi que tout appareil se trouvant dans la chambre, de façon à ce que seules les particules soient brillantes, sur un fond parfaitement absorbant.

Les rampes sont constituées par juxtaposition de lampes à incandescence, avec déflecteurs paraboliques; pour augmenter le rendement lumineux et réduire l'échauffement du sol, on recouvre celui-ci de miroirs, qui réfléchissent le faisceau lumineux vers le plafond, dans le plan d'observation.

Les particules sont des *flocons de métaldéhyde*, obtenus par sublimation à l'intérieur de la chambre : le métaldéhyde pulvérisé est déposé sur une plaquette, qui est introduite à un endroit tel que la perturbation du champ soit minime (par exemple

[1] L'ensemble a été réalisé et mis en œuvre avec la collaboration de M. Gengler, Maître de Laboratoire au service du Professeur G. Burnay.

dans l'enveloppe de l'appareil de conditionnement); cette plaquette est chauffée électriquement de façon à sublimer le métaldéhyde, qui se cristallise ensuite dans l'air et donne lieu à la formation de flocons blancs emportés par le courant. La vitesse de chute des flocons est inférieure à 1 cm/s, de sorte que cette vitesse est encore mesurable.

Pour faciliter la localisation des prises de vue sur le plan de projection et pour en connaître l'échelle, on a disposé dans la chambre un repérage cartésien gradué de 50 en 50 cm.

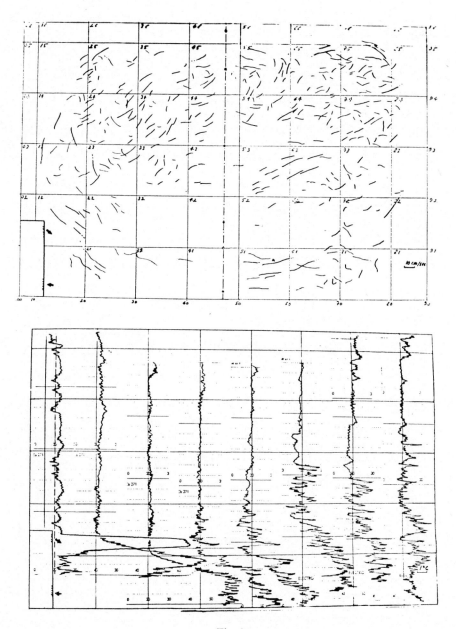

Fig. 1

Le dépouillement peut se faire par projection des négatifs sur un plan, par exemple à l'échelle 1/10, où l'on dessine les trajectoires.

Parallèlement à la visualisation du champ des *vitesses*, on enregistre continûment les températures suivant certaines directions, par exemple des verticales, au moyen de sondes à fil de platine, qui sont montées sur un dispositif de translation automatique. Ces enregistrements complètent fort utilement la visualisation, comme nous le verrons par les exemples qui suivent.

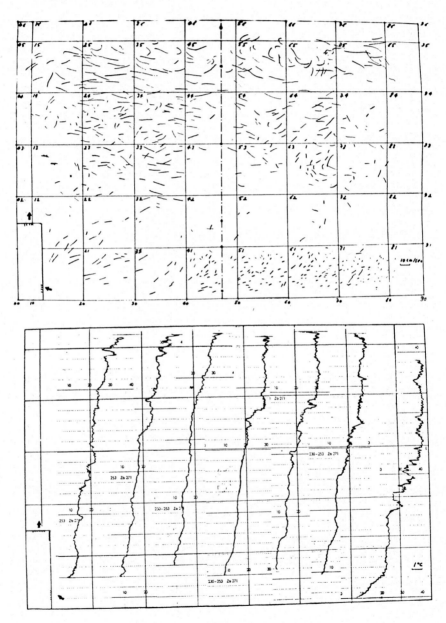

Fig. 2

EXAMPLES

On trouvera ici (fig. 1 à 4) des résultats partiels (dans un seul plan) qui ont été obtenus avec un ventilo-convecteur fonctionnant à puissance réduite, en régime « été » et « hiver » et avec deux orientations du jet de sortie, l'aspiration se faisant toujours de la même façon, à la partie inférieure de l'appareil.

Dans ce cas-ci, le plan médian de la chambre, perpendiculaire à la paroi contre laquelle se trouve l'appareil, est un *plan de symétrie*; le jet est d'ailleurs parallèle à ce plan, qui a, pour cette raison, été retenu pour la visualisation.

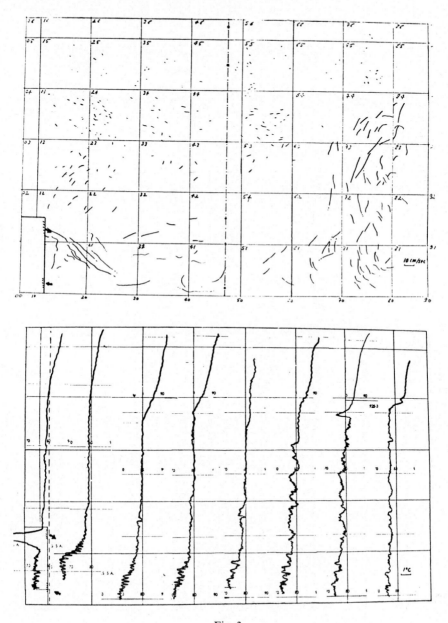

Fig. 3

131

Les enregistrements de température ont été réalisés dans le même plan, par *balayages verticaux* à la vitesse de 2 cm/s; la relative lenteur de ces balayages a permis d'obtenir une indication sur les *oscillations temporelles* de température autour de leur répartition moyenne dans l'espace.

Les caractéristiques des essais sont les suivantes :

— *Régime « hiver »* (fig. 1 et 2) :

Puissance : Q = + 1750 kcal/h

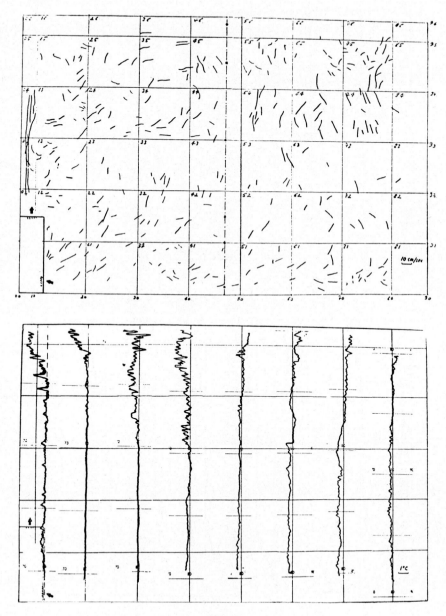

Fig. 4

Ecart de température eau-air : $\Delta t = + 25 \deg C$

— *Régime « été »* (fig. 3 et 4) :

$Q \quad = - 550 \, kcal/h$

$\Delta t \quad = - 8 \deg C$

(Le débit de l'air est sensiblement le même pour les quatre essais, soit environ $500 \, m^3/h$)

Les photos ne montrent pratiquement que l'écoulement au large; pour la compréhension, nous y avons ajouté quatre croquis explicatifs où figurent les *orientations* d'écoulement.

La comparaison entre champs de températures et champs de vitesses fait apparaître, d'une façon générale, que partout où les vitesses sont importantes, les oscillations temporelles de température le sont aussi, relativement à leurs variations spatiales. Dans les zones calmes (fig. 2 près du sol et fig. 3 près du plafond), on trouve de très forts gradients de température, avec de faibles oscillations temporelles.

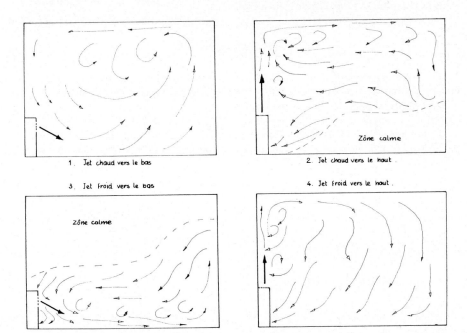

1. Jet chaud vers le bas

2. Jet chaud vers le haut

3. Jet froid vers le bas

4. Jet froid vers le haut

Zône calme

Zône calme

Ainsi, les oscillations de température permettent de suivre le *jet* et même les *courants de retour* qu'il occasionne « au large ».

Les résultats nous paraissent surtout spectaculaires en ce qui concerne l'influence de la température et donc des forces de gravitation, sur l'épanouissement du jet et l'allure des courants de retour :

— Lorsqu'on l'oriente *vers le bas*, le jet *chaud* (fig. 1) gonfle, comme on le voit clairement sur les enregistrements de température, et s'élève devant la paroi opposée, provoquant un courant de retour au ras du plafond et un mouvement tourbillonnaire dans toute la chambre.

Par contre, le jet froid (fig. 3) « colle » davantage au sol et ne remonte que jusqu'à une certaine hauteur ([1]) le long de la paroi opposée et retombe ensuite ([2]),

([1]) Transformation de l'énergie cinétique du jet en énergie potentielle de gravitation.
([2]) Transformation inverse de (1).

provoquant un courant de retour dans la partie basse de la chambre (la retombée est très clairement visible à droite de la figure 3 supérieure).

— Lorsqu'on l'oriente *vers le haut*, le jet *chaud* (fig. 2) « colle » au plafond et redescend le long de la paroi opposée jusqu'à un certain niveau ([1]), puis remonte ([2]) et occasionne un courant de retour confiné dans la partie haute de la chambre.

Par contre, le jet *froid* (fig. 4) diffuse rapidement et retombe en pluie dans toute la chambre.

Il apparaît une certaine analogie respectivement entre les figures 1 et 4 et entre les figures 2 et 3 ; le champ convectif de la figure 2 diffère d'ailleurs peu de celui qui serait induit par un corps de chauffe statique.

Sans entrer dans plus de détails, nous pouvons au moins conclure de ces observations que, toutes autres choses égales, le choix de l'orientation du jet a une influence considérable sur l'allure de l'écoulement au large et sur la champ de température qui lui est associé ; ce choix peut donc être déterminant, en ce qui concerne le *climat* réalisé dans la chambre et l'*efficacité* de l'installation...

DISCUSSION

H. MARCQ (Belgique) — Dans les essais correspondant aux fig. 1 à 4, quelles étaient les parois froides en régime d'hiver et les parois chaudes en régime d'été ? Y avait-il des sources de chaleur intérieures ?

J. HANNAY — Dans aucun des essais il n'y avait de source intérieure, les 6 parois étaient simultanément chauffées ou refroidies suivant le régime (été ou hiver).

([1]) Transformation de l'énergie cinétique du jet en énergie potentielle de gravitation.
([2]) Transformation inverse de ([1]).

MESURE D'UNE TEMPÉRATURE MOYENNE DE PAROI À L'AIDE D'UN RADIOMÈTRE *

J. LEBRUN et J. HANNAY

Institut de Thermodynamique et CEDRIC
Université de Liège (Belgique)

Measuring the average temperature of a wall with the aid of a radiometer

SUMMARY: *The CEDRIC laboratory has developed a thermoelectric radiometer of simple construction for recording average radiation temperatures.*
This receiver may be installed at the far end of a deflector, thus protecting it from air movement and providing better screening of the source examined.
Translating the reading of the radiometer into "true temperature" is only possible through averaging out a certain number of corrections: one must, eventually, take into consideration:
— the influence of the non-aimed *zone via reflection on the deflector;*
— the influence of the "twilight" *effect;*
— the reflection *of surrounding sources;*
— the absorption-emission *of the interposed medium.*

DESCRIPTION DE L'APPAREIL

Le radiomètre, dont il est question ici, peut être utilisé dans un domaine spectral très étendu et notamment dans l'infrarouge lointain, pour autant que l'on n'exige pas une très grande résolution spatiale.

Le récepteur est du type différentiel; il comporte des plages distinctes selon leurs émissivités ou selon leurs expositions et reliées par des couples thermoélectriques; ceux-ci sont disposés en série de façon à constituer une thermopile. Plusieurs dispositions sont possibles, dont celles esquissées à la figure 1 : les deux premières sont du type « à émissivités différentes » et la troisième « à expositions différentes » (le récepteur est alors un simple fluxmètre). Le modèle circulaire à plages concentriques peut être fortement miniaturisé et présenter une faible inertie thermique. Nous l'avons expérimenté sur circuit imprimé cuivre-téflon (épaisseurs 0,035 et 0,025 mm) avec connexions radiales en constantan de 0,08 mm; le circuit en cuivre est présenté à la figure 2, son diamètre hors-tout est ramené à 20 mm [1].

Un tel récepteur est, de préférence, logé dans un boitier de protection et, éventuellement, muni d'un déflecteur dont nous examinerons plus loin l'influence.

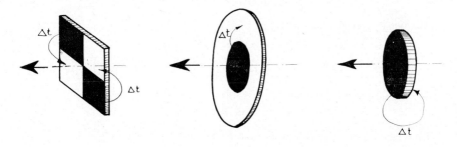

Fig. 1

(*) Cette communication est extraite d'une étude de synthèse en préparation sous la direction du Professeur G. BURNAY.
(1) Cette réalisation est due à R. DUPONT, Maître préparateur au service du Professeur G. BURNAY.

Considérons deux plages réceptrices 1 et 2, dont les émissivités sont ε_1 et ε_2 : soit t_0 la température moyenne hémisphérique de rayonnement et t la température de l'air, relativement à cette face réceptrice, φ_{C1}, φ_{C2}, φ_{R1} et φ_{R2} sont les flux convectifs et radiatifs incidents aux plages 1 et 2.

Fig. 2

Pour le dos du récepteur, on définit une température équivalente t_D (air-paroi) et les échanges globaux φ_{D1}, φ_{D2}.

Enfin, on doit aussi tenir compte du flux conductif φ_{12} entre les plages 1 et 2.

La figure 3 schématise le système considéré : α_1, α_2 sont des coefficients d'échanges convectifs, K_{12} une perméance conductive, K_{D1}, K_{D2} des coefficients de transmission dans l'épaisseur de la plaquette et au-delà.

$$\text{On a}\ (\text{pour } i = 1 \text{ et } 2) \left\{ \begin{array}{l} \varphi_{Ri} = \varepsilon_i\,\sigma\quad (T_0^4 - T_i^4)\ (\sigma = \text{constante de Stéphan}). \\[2mm] \varphi_{Ci} = \alpha_i\quad (t_i - t) \\[2mm] \varphi_{Di} = K_{Di}\quad (t_i - t_D) \end{array} \right.$$

$$\text{et}\qquad \varphi_{12} = K_{12}\quad (t_1 - t_2).$$

Les bilans en 1 et 2 donnent :

$$\varepsilon_1\,\sigma(T_0^4 - T_1^4) = \alpha_1(t_1 - t) + K_{D1}(t_1 - t_D) + K_{12}(t_1 - t_2) \qquad (1)$$

$$\varepsilon_2\,\sigma(T_0^4 - T_2^4) = \alpha_2(t_2 - t) + K_{D2}(t_2 - t_D) + K_{12}(t_1 - t_2). \qquad (2)$$

La conception du récepteur est telle que :

$$\alpha_1 \cong \alpha_2 \text{ et } K_{D1} \cong K_{D2}.$$

En logeant le récepteur dans un boîtier, on peut même obtenir :

$$t_2 \cong t \cong t_D$$

(la plage 2 étant ici celle de faible émissivité).

[2] Ce principe ainsi qu'un exemple de réalisation ont déjà été présentés par V. KORS-GAARD dans l'ASHRE Journal, Section Heating Piping and Air Conditioning (June, July, 1948).

Et comme, $t_1 - t_2 \ll t_0 - t_1$ on obtient en soustrayant (2) de (1) :

$$\boxed{T_0^4 \cong T_m^4 + K(t_1 - t_2)}$$ (3)

avec

$$T_m^4 = \frac{T_1^4 + T_2^4}{2} \text{ et } K = \frac{\alpha_1 + K_{D1} + K_{12}}{\sigma(\varepsilon_1 - \varepsilon_2)}.$$

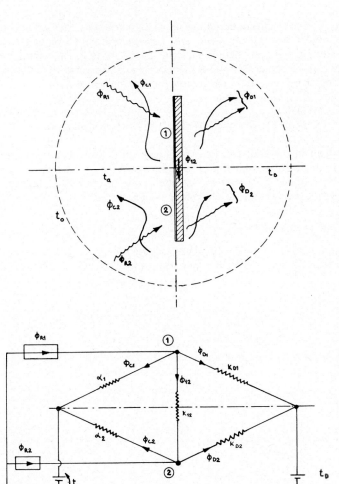

Fig. 3

K_{12} est une constante; K_{D1} aussi, pour autant qu'il y ait un boîtier de protection au dos; enfin, le déflecteur préserve la constance de α_1. On peut donc considérer K comme une constante. (3) se linéarise facilement :

$$\boxed{t_0 = t_m + K'(t_1 - t_2)}$$ (4)

avec

$$K' = \cfrac{K}{4\left(\cfrac{T_0 + T_m}{2}\right)^3}$$

et si la caractéristique thermoélectrique du récepteur est linéaire :

$$t_0 = t_m + K'' E.$$

E est la force électromotrice fournie par la thermopile; K'' est la constante d'étalonnage, obtenue devant un *corps noir*. La température moyenne hémisphérique de rayonnement t_0 est, par définition, celle du corps noir équivalant à l'hémisphère « vu » par le récepteur; il reste à préciser la relation entre cette température fictive et la température vraie de la paroi visée par le radiomètre.

EXAMEN DES CAUSES D'ATTÉNUATION DU SIGNAL

Initialement, on avait été tenté de choisir un déflecteur à surface intérieure très réfléchissante de façon à minimiser son effet sur la mesure et à garder le maximum de sensibilité; comme on va le voir ci-dessous, cette solution est à rejeter dans tous les cas où le radiomètre est assez éloigné de la paroi visée.

Influence de la zone non visée par réflexion sur le déflecteur

Soit, figure 4, une « cible » circulaire 1 de rayon r_1, un récepteur quasi ponctuel 2, un déflecteur conique 3 et l'environnement, qui est représenté par la surface tronconique 4.

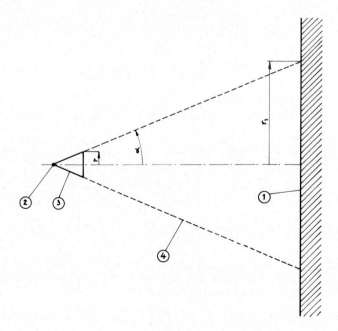

Fig. 4

Supposons provisoirement que toutes les surfaces sont isothermes, que seule la température de la cible diffère de celle des autres et que seule l'émissivité ε du déflecteur peut être inférieure à l'unité.

Une première approximation nous suffit; elle consiste à supposer que le flux réfléchi sur 3 est *diffus* et *uniforme* (comme le flux direct). L'« *atténuation* » A due au déflecteur a été calculée en fonction de l'ouverture γ, de l'émissivité ε et du rapport (r/r_1).

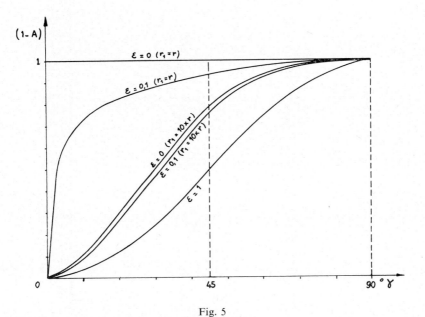

Fig. 5

La figure 5 montre bien que, pour une faible émissivité, les variations d'atténuation avec la distance séparant le radiomètre de la cible (soit avec r/r_1 pour γ fixé) sont inacceptables; par contre, *si le déflecteur est noir* ($\varepsilon = 1$), l'atténuation est sans doute la plus élevée, mais au moins elle ne varie plus avec r/r_1.

À ces considérations s'ajoute que la température de la zone 4 est évidemment plus difficile à estimer que celle du déflecteur.

Influence des dimensions du récepteur

La miniaturisation du récepteur est coûteuse et nuit toujours à sa sensibilité; d'autre part, on doit aussi limiter l'encombrement du déflecteur, d'où l'apparition d'une « *pénombre* » désignée par 4 à la figure 6.

Le facteur de forme F_{24} du récepteur vis-à-vis de cette pénombre vaut :

$$F_{24} = F_{21'} - F_{21}.$$

La figure 7 montre l'importance de F_{24}; autant que possible, on devra donc éviter que la pénombre couvre une zone à température très différente de celle de la cible.

Influence de l'absorption et des réflexions

Considérons encore un environnement isotherme à même température que celle du déflecteur (supposé noir) et que celle du récepteur (infiniment petit); les notations sont celles de la figure 4.

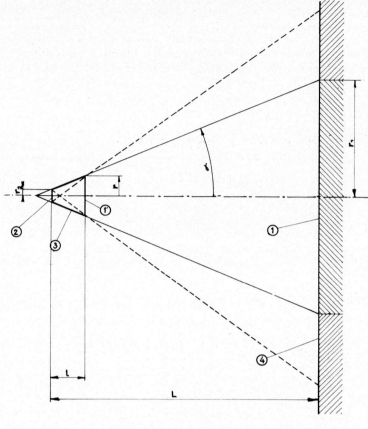

Fig. 6

L'*atténuation* due a l'absorption dans le milieu interposé et aux réflexions sur la cible s'écrit :

$$A = 1 - \varepsilon_1 \, \tau$$

avec

ε_1 émissivité de la cible;
τ facteur de transmission.

Dans l'air et pour les températures courantes, τ est essentiellement lié à la pression partielle de la vapeur d'eau et à la distance parcourue.
En bonne approximation on peut écrire :

$$(1 - \tau) \cong 2{,}3 \times d \times \varphi \times p'$$

Le domaine de validité de cette formule est :

$$0 \leqslant d \leqslant 4$$

$$0 \leqslant \varphi p' \leqslant 0{,}025$$

140

avec

d   distance (en m);
φ   humidité relative (en %);
p' tension de vapeur (en kg/cm²).

Cette formule est linéaire en d; on peut donc y introduire la distance moyenne pour l'ensemble des rayons joignant la cible au récepteur (supposé encore très petit) :

$$d_{moy} = \frac{1}{F_{21}} \times \int F_2 d_1 \times d.$$

$F_2 d_1$ désigne le facteur de forme de 2 vis-à-vis d'une couronne élémentaire de 1; d est la distance correspondante.

On obtient :

$$\boxed{\frac{d_{moy}}{L} = 2 \times \frac{(1 - \cos \gamma)}{\sin^2 \gamma}}$$

L étant la distance normale récepteur-cible (figure 6).

Il en résulte le tableau suivant :

$\gamma(°)$	0	15	30	45	60	75	90
d_{moy}/L	1	1,05	1,08	1,17	1,33	1,60	2

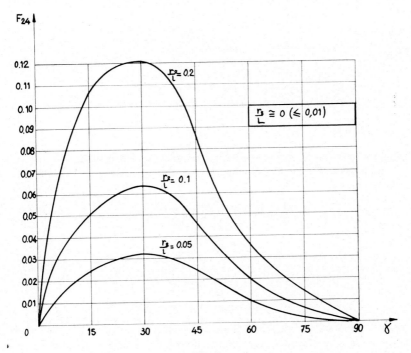

Fig. 7

Ainsi, pour $t = 20\,°C$, $\alpha = 45°$, $L = 1,50$ m, on trouve :

$$\tau = 1 - 0,1\varphi$$

et donc $\tau = 0,9$ pour $\varphi = 1$ (air saturé); d'autre part, on a couramment $\varepsilon_1 \cong 0,9$, de sorte que l'atténuation atteint alors la valeur non négligeable de 20 %.

FORMULES D'APPROXIMATIONS

On peut recourir à trois approximations successives pour définir la température vraie de la cible; nous supposons ici que la linéarisation des échanges radiatifs est acceptable. La pénombre éventuelle est supposée incluse dans la cible :

$$F_{21} + F_{23} = 1 \qquad \text{(notation de la figure 6).}$$

Approximation I

$$\varepsilon_1 = \varepsilon_3 = 1 \qquad \text{(pas de réflexion).}$$
$$\tau = 1 \qquad \text{(pas d'absorption).}$$

L'équation (5) donne avec $t_m = t_2$:

$$\boxed{t_0 = t_2 + K'' E} \qquad (5)$$

d'autre part,

$$t_0 = F_{21} t_1 + F_{23} t_3$$

d'où l'on tire :

$$\boxed{(t_1)_I = t_2 + \frac{1}{F_{21}}\left[(t_2 - t_3) \times F_{23} + K'' E \right]}\,.$$

Approximation II

$\varepsilon_1 \leqslant 1$ c'est-à-dire qu'on tient compte des réflexions sur la cible.

Le radiomètre perçoit une combinaison de la température de la cible avec celle de l'environnement :

$$(t_1)_I = (t_1)_{II} \times \varepsilon_1 + t_5 \times (1 - \varepsilon_1)$$

d'où

$$\boxed{(t_1)_{II} = (t_1)_I + \frac{1 - \varepsilon_1}{\varepsilon_1} \times \left[(t_1)_I - t_5 \right]}\,.$$

t_5 est la température moyenne de rayonnement *vue par la cible*; elle est directement mesurable au moyen d'un second radiomètre sans déflecteur, adossé à la paroi visée (ce radiomètre doit être évidemment assez petit vis-à-vis de la cible).

Approximation III

$\tau \leqslant 1$ c'est-à-dire qu'on tient compte de l'absorption.

On trouve directement :

$$(t_1)_{\text{II}} = (t_1)_{\text{II}} \times \tau + t_6 \times (1 - \tau)$$

d'où

$$\boxed{(t_1)_{\text{III}} = (t_1)_{\text{II}} + \frac{1-\tau}{\tau}\left[(t_1)_{\text{II}} - t_6\right]} \ .$$

t_6 désigne la température moyenne de l'air; $(t_1)_{\text{III}}$ est la *température vraie de la cible*.

MESURES DE TEMPÉRATURES D'AIR EN RÉGIME VARIABLE A PROXIMITÉ DE PAROIS RAYONNANTES*

J. HANNAY

Institut de Thermodynamique et CEDRIC
Université de Liège (Belgique)

Unsteady state air temperature measurements in the proximity of radiating walls

SUMMARY: *The paper presents the results on the development of a resistance thermometric probe having a sensing element of fine platinum wire.*

It is shown how the high convective dissipation properties of the very fine wires enables one to obtain a probe which is little influenced by heat radiation and which has a reduced response time.

This type of probe may be used particularly for the detailed examination of the temperature field inside an air conditioned space.

OBJECTIFS ET DOMAINES D'APPLICATION

La mesure de température d'air en régime variable, dans les jets d'air de conditionnement et à proximité de parois rayonnantes ne peut en général être réalisée au moyen des sondes de température utilisées couramment en laboratoire : en effet, les sondes classiques possèdent souvent une inertie thermique trop grande et une sensibilité au rayonnement qui limitent leur champ d'application si l'on veut obtenir une précision. acceptable.

Unissant le haut pouvoir de dissipation convectif et la faible capacité thermique des fils fins aux caractéristiques bien connues des thermomètres à résistance, il est possible d'obtenir un thermomètre à fil de platine répondant parfaitement aux exigences du problème posé.

Dans le cadre des objectifs poursuivis par la Commission VI, les domaines préférentiels pour l'utilisation de ces sondes portées par un dispositif de translation adéquat sont :

— le relevé de la distribution temporelle et spatiale des températures à proximité des bouches de pulsion d'air conditionné ;

— le relevé du gradient de température dans la couche limite des parois «chaudes» ou « froides » et des corps de chauffe ;

— le relevé rapide suivant des directions préférentielles du profil de température dans tout l'espace d'un local conditionné.

CONDUCTION ET CONVECTION DES FILS FINS

Pour le calcul des échanges de chaleur entre un fil chaud et le fluide environnant, on peut appliquer le principe de superposition; en effet, dans le problème qui nous occupe, les échauffements et les vitesses sont assez faibles.

Le nombre de Nusselt, caractéristique des échanges globaux, s'écrit alors :

$$Nu = Nucd + Nucn + Nucv$$

Relation dans laquelle Nucd, Nucn, Nucv sont respectivement les nombres de Nusselt relatifs aux échanges par conduction, convection naturelle et convection forcée.

(*) Cette communication est extraite d'une étude de synthèse en préparation sous la direction du Professeur G. BURNAY.

Notre but étant de calculer l'échauffement maximum du fil, nous nous placerons dans les conditions les plus défavorables en négligeant l'échange par convection forcée:

$$\text{Nucv} = 0$$

ce qui revient à supposer une vitesse au large égale à zéro.

Les dimensions du fil (surtout son diamétre) étant très petites vis-à-vis de celles de l'environnement, nous considérons le fil comme plongé dans un milieu illimité. Dans ces conditions le nombre de Nusselt relatif à la conduction pure pour un fil de diamètre d et de longueur l s'écrit :

$$\text{Nucd} = \frac{2}{\ln \dfrac{4l}{d}}$$

ce qui revient à admettre que la couche d'air intéressée par les échanges conductifs s'étend pratiquement à une distance du fil égale à sa longueur.

Pour les échanges en convection naturelle, l'analyse dimensionnelle nous montre que le nombre de Nusselt doit se mettre sous la forme :

$$\text{Nucn} = A . \text{Pr}^t . \left(\frac{g\,d^3}{v_2}\right)^q \left(\beta\theta\right)^r$$

Travaillant uniquement dans l'air, nous pouvons négliger la variation du nombre de Prandtl (Pr). De plus, calculant les exposants q et r, différents expérimentateurs ont montré qu'ils prenaient des valeurs sensiblement égales. Il apparaît donc le groupement $\dfrac{g d^3 \beta\theta}{v^2}$ égal au nombre de Grashof (Gr).

En accord avec les résultats de J. Gosse [1], nous écrirons :

$$\text{Nucn} = 0{,}57\,\text{Gr}^{0,125}$$

ÉQUILIBRE THERMIQUE DU FIL

1. *Cas du fil idéal* (sans support).

Soit un fil de platine de diamètre d et de longueur $2l$, parcouru par un courant de mesure d'intensité I, tendu dans l'air à température uniforme t.

Admettons que la température t_f du fil est également uniforme, c'est-à-dire que les gradients de température radiaux et longitudinaux sont nuls au sein du fil, l'équilibre thermique du fil s'écrit :

$$R_0(1+\beta t_f)I^2 = 2\,\alpha\pi\,\mathrm{d}l(t_f - t) \tag{1}$$

La sensibilité de la sonde impose

$$I = \frac{KE}{R_0 \beta} \tag{2}$$

Equations dans lesquelles

R_0 est la résistance électrique du fil à $0\,°C$;

β le coefficient de température du platine*;

(*) Nous avons admis pour le platine une résistivité électrique variant linéairement avec la température.

146

α le coefficient d'échange entre le fil et l'air;

K   une constante qui dépend de la méthode de mesure utilisée;

E   la sensibilité du thermomètre.

Substituant à la longueur du fil une variable auxiliaire sans dimension X caractéristique de son élancement

$$X = \frac{4l}{d} \qquad (3)$$

et considérant uniquement les échanges conductifs

$$\alpha = \frac{2\lambda}{d \ln X} \qquad (4)$$

les équations (1) et (2) deviennent :

$$X^2 = \frac{(KE)^2}{tp-t} A \ln X \qquad (1')$$

$$I = KE \, \beta \, \frac{d}{X} \qquad (2')$$

A et B étant fonction des caractéristiques physiques de l'air et du platine à la température considérée.

Lors de la réalisation de sonde on s'impose généralement :

— le produit KE lié à la sensibilité et à la méthode de mesure;

— l'échauffement maximum $(t_f - t)$;

— le diamètre du fil, dont dépendent la constante de temps et la sensibilité au rayonnement.

Les équations (1') (2') et (3) permettent alors de calculer la longueur du fil et l'intensité de courant.

Le tableau I donne les résultats de calcul pour quelques sondes d'usage courant au laboratoire du CEDRIC avec leurs constantes de temps calculées (RC).

2. Cas de la sonde réelle

En ce qui concerne les sondes thermométriques réalisées au CEDRIC, le fil de platine est soudé par étincelage à l'extrémité de deux tiges d'acier de 0,5 mm de diamètre et de 20 mm de long, elles-mêmes fixées à un support, d'où partent les connections vers le dispositif de mesure.

Pour minimiser l'échauffement en présence de parois rayonnantes, les émissivités des tiges et du support sont rendues aussi faibles que possible.

Pour caractériser la *sensibilité au rayonnement* de la sonde, nous allons d'abord estimer quels sont les échauffements des différentes parties de la sonde lorsque celle-ci est placée devant une paroi rayonnante, cela sans tenir compte des effets de conduction. Soit (fig. 1),

t la température de l'air;

t_p la température de la paroi;

t_1 la température du fil;

t_2 la température des tiges;

t_3 la température du support.

Soit encore

$$\theta_p = t_p - t \qquad \theta_1 = t_1 - t$$
$$\theta_2 = t_2 - t \qquad \theta_3 = t_3 - t$$

Tableau I

Symboles	Valeurs numériques						Unités
10^6 KE			100				V °C^{-1}
θ		0,05			0.10		°C
X		4820			3330		—
10^6d	3	10	30	3	10	30	m
$10^3\, 2l$	7,2	24,1	72,0	5,0	16,6	50,0	m
10^3 I	0,251	0,835	2,51	0,363	1,21	3,63	Amp.
Nucd		0,236			0.246		—
Nucn	0,035	0,054	0.082	0,038	0,059	0,089	—
Nu	0,271	0,290	0,318	0,284	0,305	0,335	—
10^3 RC	0,24	2,5	21	0,24	2,5	21	s.

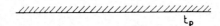

t_p

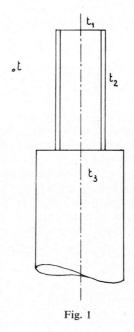

Fig. 1

Nous admettons :

— que la paroi peut être considérée comme un plan noir de dimension infinie;
— que les facteurs géométriques F de chaque partie de la sonde vers la paroi sont tous égaux à 0,5;
— que les facteurs géométriques des parties de la sonde les unes vis-à-vis des autres sont nuls;
— que les émissivités ε_i du fil, des tiges et du support sont toutes égales à 0,08.

Dans ces conditions, l'équilibre entre les échanges radiatifs et convectifs d'une quelconque des parties de la sonde s'écrit :

$$\sigma\,\varepsilon_i(T_p^4 - T_i^4) + \sigma\,\varepsilon i(1-F)\,(T_i^4 - T^4) = \alpha i(t_i - t) \qquad (5)$$

avec la constante de Stephan-Boltzman

$$\sigma = 4,88 \times 10^{-8} \text{ kcal/h. m}^2.\,{}^{\circ}K^4$$

Les écarts θ_i étant relativement faibles, on peut admettre pour le coefficient d'échange par rayonnement une valeur unique

$$\rho = 4\sigma\varepsilon\,F\left(\frac{T_p + T}{2}\right)^3 \qquad (6)$$

L'équation (5) devient

$$\frac{\theta_i}{\theta_p} = \frac{\rho}{\alpha_i} \qquad (5')$$

Tout calcul effectué, on trouve

$$\frac{\theta_3}{\theta_p} \cong 0,06 \qquad\qquad \frac{\theta_2}{\theta_\rho} \cong 0,006$$

d en m	3×10^{-6}	10×10^{-6}	30×10^{-6}
$\dfrac{\theta_1}{\theta_\rho}$	0,00011	0,00035	0,00094

Ces quelques chiffres obtenus rapidement permettent d'apprécier la quasi insensibilité au rayonnement des fils fins. De plus ils mettent en évidence que les parties constituant la sonde peuvent se trouver à des températures différentes. Pour évaluer complètement l'effet du rayonnement sur la sonde, nous devrons donc tenir compte d'un écoulement de chaleur par conduction au sein des tiges et du fil.

A cet effet, considérons un fil de diamètre d et de longueur $2l$ (fig. 2) chauffé par effet Joule et par rayonnement, dont les extrémités sont fixées à des supports massifs dans lesquels l'effet Joule est négligeable mais, qui sont sensibles au rayonnement.

Soit :

— θ_i' l'échauffement du fil par rapport à l'air;
— θ_j' l'échauffement des supports;
— Q la quantité de chaleur fournie à 1 mètre de fil par rayonnement

et par effet Joule

— s la section du fil;

— p son périmètre;

— λ sa conductivité thermique.

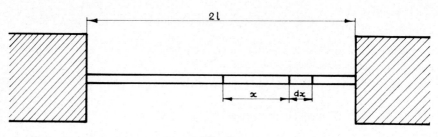

Fig. 2

L'équilibre thermique d'un élément dx du fil s'écrit :

$$s\lambda \frac{d^2\,\theta_i'}{d\,x^2} - p d\,\theta_i' = -Q \tag{7}$$

$$\text{ou} \qquad \frac{d^2\,\theta_i}{d\,x^2} - \frac{p\lambda}{s\lambda}\,\theta_i' = -\frac{Q}{s\lambda} \tag{7'}$$

Cette équation différentielle admet la solution :

$$\theta_i' = A\,\text{ch}\,\sqrt{\frac{p\alpha}{s\lambda}}\,X + \frac{Q}{p\alpha} \tag{8}$$

A, étant une constante d'intégration définie par la condition limite :

$$[\theta_i'(x)]_{x=l} = \theta_j'$$

La température moyenne du fil vaut

$$\overline{\theta_i'} = \frac{1}{l}\int_0^l \theta_i'\,dx$$

Ces formules sont applicables aussi bien aux tiges qu'au fil de la sonde réelle. Pour les extrémités des tiges, les calculs donnent :

$$\theta_2' = 0.023\,\theta_\rho$$

Pour les fils dont les longueurs, les diamètres et les intensités des courants sont données au tableau I, on peut résumer les résultats du calcul des températures moyennes comme suit :

Échauffement par effet Joule :

$\overline{\theta} \cong 0{,}04°C$ pour les fils dont X vaut 4820.

$\overline{\theta} \cong 0.08°C$ pour les fils dont X vaut 3330.

Échauffement supplémentaire par rayonnement direct et par conduction à partir des supports dans le cas où $\theta_p = 10\,^\circ\mathrm{C}$.

$\overline{\Delta\theta} \cong 0{,}010\,^\circ\mathrm{C}$ pour les fils dont le diamètre égale 3 microns.

$\overline{\Delta\theta} \cong 0{,}015\,^\circ\mathrm{C}$ pour les fils dont le diamètre égale 10 microns.

$\overline{\Delta\theta} \cong 0{,}020\,^\circ\mathrm{C}$ pour les fils dont le diamètre égale 30 microns.

CONCLUSIONS

Par la présente communication, nous espérons avoir attiré l'attention des personnes intéressées à l'étude et aux essais des appareils de conditionnement et de chauffage sur les possibilités des sondes thermométriques à fil de platine.

Compte tenu de l'espace qui nous était dévolu, il ne nous a pas été possible de présenter un exemple d'application.

Nous renvoyons les personnes intéressées à la communication « Visualisation d'un champ convectif en chambre confinée » (p. 127) dans laquelle elles trouveront des exemples d'utilisation des ces sondes, intégrés à une méthode générale d'étude des appareils de conditionnement.

RÉFÉRENCE

[1] J. GOSSE. Étude de la convection par les fils aux faibles nombres de Reynolds. Publication scientifique et technique du Ministère de l'Air, (France), n° 322.

PERIODIC HEAT FLOW THROUGH LIGHTWEIGHT WALLS: INFLUENCE OF THE HEAT CAPACITY OF SOLID BODIES UPON ROOM TEMPERATURE

G. GUGLIELMINI and U. MAGRINI,

Istituto di Fisica Tecnica, Università di Genova (Italy)

La propagation des ondes de température dans une ambiance limitée par des parois légères : influence de la capacité calorifique des solides sur la température du local.

RÉSUMÉ : *Une évaluation correcte de la capacité calorifique de l'ambiance, qui participe effective-ment au phénomène thermique oscillatoire, a une grande importance dans la solution des problè-mes qui concernent la transmission de chaleur, en régime périodique stabilisé, à travers les parois extérieures de bâtiments, en particulier dans les cas des constructions légères. En effet l'amplitude et le déphasage de l'onde de température, qui, en ce cas de transmission thermique, s'établit dans l'ambiance, dépendent largement de la capacité calorifique des corps qui y partici-pent.*

La contribution apportée au phénomène oscillatoire par les corps contenus dans l'ambiance dépend notamment du rapport entre leur volume et leur surface exposée à l'air de l'ambiance, et aussi de leurs caractéristiques physiques (conductivité thermique, chaleur et poids spécifiques). Le rapport qui établit, dans certains cas, l'influence prédominante des corps contenus à l'intérieur de l'ambiance en ce qui concerne les oscillations de température, peut varier dans de larges limites.

Dans cette étude, on a examiné les cas les plus intéressants pour la technique, compris entre les limites de variation précitées.

On présente enfin des diagrammes et des relations qui simplifient l'interprétation du phéno-mène et son évaluation pour les applications pratiques.

1. INTRODUCTION

It is well known that under conditions of periodic heat flow, fluid and solid bodies contained inside a room have an influence on the propagation and distribution of thermal waves through the walls and inside the bodies in the room [1].

This is because the heat capacity of these masses causes a time-lag and an amplitude decrease in the room temperature oscillation in comparison to the outdoor air tem-perature oscillation which is taken as the reference.

The heat wave transmitted through the peripheral walls influences all the interior ambient components (air, structures, and contents) which consequently undergo temperature oscillations.

A scheme representing the system can be obtained by considering the heat capacity of each interior element distributed uniformly throughout a room and the temperature of such ambient bodies oscillating with the same period, amplitude and phase. Under these assumptions the heat capacity involved in the phenomenon may be readily evaluated provided that the nature and mass of the room contents are known. This simplified method can be considered sufficiently accurate if we suppose that periodic heat flow is transmitted only to the indoor air by either free convection or a suitably closed circuit ventilation that ensure uniform heat distribution in the room and a single indoor temperature.

This simple procedure however is inadequate inasmuch as the thermal wave transmitted through the walls reaches not only the ambient air but also the solid bodies. It is evident that these heat capacities cannot be considered as uniformly distributed throughout the whole room volume.

The contribution from the room's solid bodies to periodic heat flow depends on the thermal wave penetration depth into the body interiors and also on the extension of the solids boundary surface. The penetration depth of a temperature wave into the interior of a solid depends on the physical properties of the body (e.g. thermal conductivity λ and diffusivity a) and its geometric shape.

Under such conditions the actual heat capacity M involved in the phenomenon cannot be evaluated by the simplified method outlined above. Such heat capacity differs from the so called total heat capacity M_T which takes into account the heat capacity of all the room components (air and solid bodies).

An analytical "exact" solution of the problem was already presented in a preceding paper [3]. The method of calculus used may appear a little cumbersome. A quicker but approximate method involving a "contribution coefficient" of the total heat capacity M_T was developed by D. Faggiani in ref. [4].

In the present communication, the contribution of the room's solid bodies to the periodic heat flow is first evaluated analytically via the above-cited method [3] in terms of the solids' geometrical and physical characteristics.

Next, a straightforward method is presented which provides the amplitude ratio and the time lag of the indoor air temperature wave (in reference to the outdoor temperature oscillation) taking into account the contribution from the room's solid bodies. Such a method involves an "equivalence factor" ε which, multiplied by the total heat capacity M_T, enables the evaluation of a quantity $M_e = \varepsilon M_T$; we call this the "equivalent heat capacity".

This heat capacity M_e, considered as uniformly distributed, enables one to obtain values of amplitude ratio and time lag equal to the true ones. Several values of ε were calculated and are given here for various practical cases.

Once M_e is known then we can easily evaluate the amplitude ratio ρ that the outdoor air temperature wave will undergo in penetrating into the room. This is because the definition of M_e enables the application of simplified methods based on the assumption of uniformly distributed heat capacities among which we mention the analytical methods [1, 3, 5, 6], and those dealing with "electrical analogy" [7, 8, 9, 10].

2. THE CONTRIBUTION OF SOLIDS TO PERIODIC HEAT FLOW IN A ROOM AND ITS DEPENDENCE ON THE PHYSICAL AND GEOMETRIC PROPERTIES OF BODIES.

This refers to the contribution of solid bodies to the phenomenon of periodic heat flow in a room.

It can be seen that the behaviour of a solid mass of total heat capacity M_T under conditions of periodic heat flow will differ on account of the limiting surface exposed to the ambient air provided other variables are kept constant.

As an example let us take a spherical mass. It appears that temperature oscillations will occur only on the surface layer of the body to a depth depending on the heat conductivity λ and diffusivity a of the material; the contribution of the body interior to this phenomenon is negligible.

Let us consider some other bodies with the same mass as the above sphere but differently shaped e.g. prismatic or even finned. Owing to the larger extension of their outer surface, the contribution of such bodies to periodic heat flow will be much stronger than in the case of the spherical mass. To go to an extreme we could say that almost the whole mass of a finned body can be involved in the oscillatory phenomenon provided the fin thickness is sufficiently small.

One can thus conclude that the contribution of a body of given physical properties (λ and a) and total heat capacity M_T will depend upon its geometric characteristics, mainly on its boundary area.

154

We can see, however, that the contribution of flat slabs of equal boundary surface to the oscillatory phenomenon will also depend on the thickness of the plates. To illustrate this we have represented, in figures 1a-1b and 1c, the envelopes of the temperature damped oscillations that may occur under different conditions inside a flat slab.

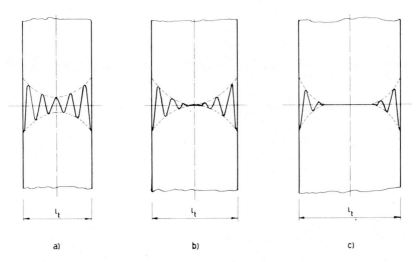

Fig. 1 — Influence of thickness upon the envelopes of the temperature oscillation curves inside a flat slab.

In case a), the temperature waves entering through each side of the slab may influence each other in such a way that even adiabatic conditions can be reached in the mid-plane. For a constant exposed surface S_t, the total heat capacity M_T of the slab increases with increasing thickness l_t and so does the contribution of the plate to the oscillatory phenomenon. For a certain value of l_t (higher than in case a), the temperature oscillating envelope inside the slab takes the form shown in figure 1b.

For slab thickness values much larger than in case b), envelopes like those represented in figure 1c can be found. Under these conditions there is no mutual influence between the temperature waves entering the slab from opposite outer surfaces. An increased total heat capacity M_T, due to the larger value of l_t, is not followed by a larger contribution of the flat body to periodic heat flow; on the contrary, the contribution in case c) is practically equal to that occurring in case b).

Hence it seems interesting to examine what influence the solid bodies thickness l_t has on the amplitude of the temperature fluctuations. To do this we will consider a room containing one partition (i.e. a flat solid body) of thickness l_t. Suppose the characteristics of the room components are as follow:

— Lightweight peripheral wall $l = 0.10$ [m]; $\lambda = 0.03$ [kcal/m.h.°C]; $a = 40 \cdot 10^{-4}$ [m²/h].
— Partition (floor made up of concrete): $\lambda_t = 1.2$ [kcal/m.h.°C];
— $a_t = 30 \cdot 10^{-4}$[m²/h]. The room air heat capacity, taken as uniformly distributed throughout the whole room volume, was given a value of 1 [kcal/m².°C]; this refers to the unit area of the surface S_p of the peripheral wall through which heat flow is transmitted. Let the boundary layer coefficients of heat transfer of the outside and

155

inside surfaces of the peripheral wall be, respectively:

$$\alpha_1 = 18 \left[\frac{\text{kcal}}{\text{m}^2 . \text{h.°C}} \right], \alpha_2 = 8 \left[\frac{\text{kcal}}{\text{m}^2 . \text{h.°C}} \right].$$

The boundary layer coefficient of the partition surface is equal to

$$\alpha_t = 8 \left[\frac{\text{kcal}}{\text{m}^2 . \text{h.°C}} \right].$$

The calculations were carried out according to the method already presented in ref. [3].

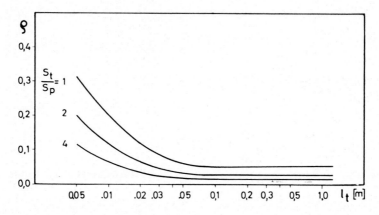

Fig. 2 — Amplitude ratio ρ versus l_t.
  — peripheral wall: $\lambda = 0.03$ (kcal/m. h.°C); $a = 40 \cdot 10^{-4}$ (m²/h)
                  $l = 0.10$ m
  — partition (concrete): $\lambda_t = 1.20$ (kcal/m. h. °C); $a_t = 30 \cdot 10^{-4}$ m²/h
  — room air: $M_a = 1$ (kcal/m². °C)

The results obtained for the above considered systems are represented in figure 2. The amplitude ratio ρ of the room air temperature oscillation was evaluated as a function of the partition thickness l_t, for values of the ratio S_t/S_p, ranging from 1 up to 4.

It may be noted that, for a given ratio S_t/S_p, ρ diminishes with an increasing solid body thickness l_t up to a value called "critical", beyond which ρ becomes independent from l_t. It was noted that this critical value occurred in the situation shown in figure 1b. Figure 2 shows also that, for a fixed l_t, the amplitude ratio ρ decreases with an increasing ratio S_t/S_p.

The critical thickness was evaluated for a range of building materials commonly used in practice.

Table A gives the values for l_t, required to make the temperature oscillation amplitude inside the body equal to 0.10 times the corresponding amplitude of the surrounding air temperature wave. It is necessary to introduce this reference value ($l_{0.10}$) in view of the asymptotic trend, in the envelope of the thermal wave amplitude versus the depth of penetration l_t. Table A also shows both the oscillation amplitude θ_{so} of the solid body surface temperature and the depth $l_{0.33}$ at which the temperature oscillation amplitude inside the body is reduced to 0.33 times the oscillation amplitude of the ambient air temperature.

3. EVALUATION OF THE EQUIVALENCE FACTOR ε

Referring to the use of the equivalence factor $\varepsilon = M_e/M_T$ it was noted that, for the correct evaluation of amplitude and phase of the ambient air temperature wave, it is necessary to define two different equivalence factors: ε_ρ and ε_γ: the former for the amplitude, the latter for the phase.

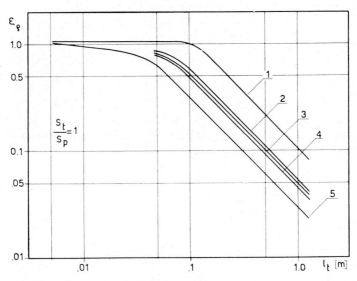

Fig. 3 — Amplitude equivalence factor ε_ρ versus l_t.
— Peripheral wall: $\lambda = 0.03$ (kcal/m. h. °C); $a = 40\cdot10^{-4}$ (m²/h); $l = 0.10$ m
— Partition: 1. Polystyrene foam; 2. Hollow tiles; 3. Eternit; 4. Concrete; 5. Timber.

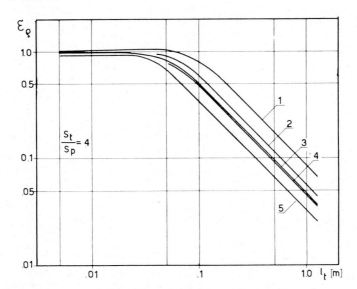

Fig. 4 — Amplitude equivalence factor ε_ρ versus l_t.
— Peripheral wall: $\lambda = 0.03$ (kcal/m. h. °C); $a = 40\cdot10^{-4}$ (m²/h); $l = 0.10$ m
— Partition: 1. Polystyrene; 2. Hollow tiles; 3. Eternit; 4. Concrete; 5. Timber.

This is due to the two degrees of freedom of the equation which defines, analytically, the room ambient air temperature wave.

The evaluation of ε_ρ and ε_γ may be carried out with the same procedure. Due to the main importance of the effects of the thermal wave amplitude reduction, however, the results in this report concern only the first of these factors (ε_ρ).

The results plotted in figures 3 and 4 refer to two systems with a lightweight peripheral wall having five different partitions inside the room.

Let the characteristics of the room components be as follows:

— Peripheral wall:

$$l = 0.10\,[\text{m}]; \quad \lambda = 0.03\left[\frac{\text{kcal}}{\text{m.h.}^\circ\text{C}}\right]; \quad a = 40 \cdot 10^{-4}\left[\frac{\text{m}^2}{\text{h}}\right].$$

The partitions, whose related properties are reported in Table A, are: polystyrene, hollow tiles, Eternit, concrete and timber. The different curves plotted in figures 3 and 4 refer to ratios $S_t/S_p = 1$ and 4, respectively.

Table A

Material	λ	Specific heat	Specific weight	$l_{0.33}$	$l_{0.10}$	θ_{s0}/θ_{e0}
	$\dfrac{\text{kcal}}{\text{m. h. }^\circ\text{C}}$	$\dfrac{\text{kcal}}{\text{kg. }^\circ\text{C}}$	$\dfrac{\text{kg}}{\text{m}^3}$	m	m	
Sandstone masonry	1.50	0.20	1800	0.047	0.262	0.434
Limestone masonry	1.20	0.20	2000	0.045	0.228	0.450
Asbestos - cement	0.80	0.18	1800	0.064	0.229	0.533
Common bricks	0.70	0.20	1800	0.058	0.205	0.537
Plaster	0.65	0.20	1900	0.055	0.193	0.534
Glass	0.60	0.20	2500	0.041	0.157	0.513
Asphalt	0.52	0.22	2200	0.043	0.152	0.537
Gypsum	0.40	0.20	1200	0.078	0.214	0.663
Hollow tiles	0.35	0.22	1000	0.080	0.217	0.689
Asbestos-cement "Eternit"	0.32	0.20	1900	0.051	0.148	0.633
Cellular concrete	0.25	0.22	600	0.102	0.246	0.777
Timber	0.13	0.60	500	0.047	0.116	0.761
Vermiculite	0.12	0.20	350	0.110	0.248	0.878
Cork board	0.085	0.40	500	0.052	0.120	0.832
Glass wool	0.06	0.16	100	0.180	0.386	0.957
Eraclit	0.06	0.40	430	0.049	0.112	0.866
Wood fiber board	0.06	0.60	300	0.048	0.108	0.863
Faesite	0.05	0.45	340	0.049	0.109	0.884
Expanded cork	0.04	0.49	150	0.066	0.143	0.925
Polystyrene foam	0.035	0.31	50	0.140	0.297	0.967

It should be observed that the plotted ε_ρ values were calculated in neglecting the contribution from the ambient air heat capacity which is always uniformly distributed throughout the room. To obtain the equivalent heat capacity of a room and its contents (air and solid bodies) one need only add the ambient air heat capacity M_a to the evaluated M_e value.

4. Remarks

It should be noted that the equivalence factor ε_p may sometimes work out slightly higher than unity. This is because boundary conditions and the most significant variables concerning the contribution from interior solid bodies are vastly different to the situations compared in this paper (i.e. the actual case and that approximated via the assumption of uniform heat capacity distribution).

Comparing the data plotted in figures 3 and 4 shows that the ratio S_t/S_p has a noticeable influence on the variations of ε_p as a function of l_t for polystyrene or timber partitions; there is no appreciable influence in other cases.

The curves evaluated are collated in figure 5. Note that all the representative curves of the various systems considered here (with heavy or lightweight peripheral walls and different partitions) are rather near each other, except those pertaining to systems with timber or polystyrene partitions.

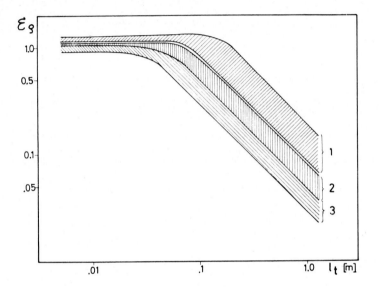

Fig. 5 — Collation of data obtained;
ε_p as a function of l_t for different partitions. Region 1: polystryrene; region 2: hollow tiles, Eternit, concrete; region 3: timber.

The curve $\varepsilon_p \rightarrow l_t$ plotted in figure 5 can be simply represented, with satisfactory accuracy, by the equation:

$$\varepsilon_p\, l_t = C.$$

The values of C given in Table B were evaluated for some materials commonly used in practice and for the above mentioned peripheral walls.

Practical application of the method

Let us examine a system with a peripheral wall having the following properties:

$$l = 0.10\,[\text{m}]\,; \quad \lambda = 0.03\left[\frac{\text{kcal}}{\text{m.h.}^\circ\text{C}}\right]\,; \quad a = 40 \cdot 10^{-4}\left[\frac{\text{m}^2}{\text{h}}\right].$$

159

Exterior wall	Material	Partition		S_t/S_p	$C = \varepsilon_\rho \, l_t$ (*)	for l_t
		λ_t	a_t			
		$\dfrac{kcal}{m.\,h.\,°C}$	$\dfrac{m^2}{h} \cdot 10^4$	m^2/m^2	m	m
	Concrete	1.2	30	1	0.07	0.07
	"	1.2	30	4	0.06	0.07
	Hollow tiles	0.35	16	1	0.075	0.07
	" "	0.35	16	4	0.065	0.07
$\lambda = 0.7$ (kcal/m. h. °C)	Asbestos cement	0.32	16	1	0.07	0.07
$a = 20 \cdot 10^{-4}$ (m²/h)	" "	0.32	16	4	0.06	0.07
$l = 0.10$ m	Timber	0.13	4.3	1	0.05	0.04
	"	0.13	4.3	4	0.04	0.04
	Polystyrene	0.035	22.5	1	0.20	0.15
	"	0.035	22.5	4	0.15	0.15
	Concrete	1.2	30	1	0.051	0.07
	"	1.2	30	4	0.046	0.07
	Hollow tiles	0.35	16	1	0.058	0.07
	" "	0.35	16	4	0.055	0.07
$\lambda = 0.03$ (kcal/m. h. °C)	Asbestos cement	0.32	12	1	0.048	0.07
	" "	0.32	12	4	0.047	0.07
$a = 40 \cdot 10^{-4}$ (m²/h)	Timber	0.13	4.3	1	0.030	0.04
	"	0.13	4.3	4	0.032	0.04
$l = 0.10$ m	Polystyrene	0.035	22.5	1	0.105	0.12
	"	0.035	22.5	4	0.096	0.12

(*) For values of l_t higher than those listed in the last column.

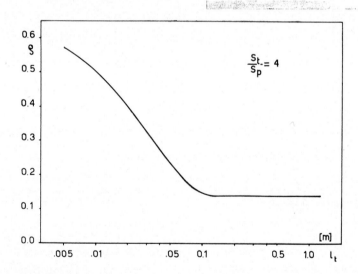

Fig. 6 — Amplitude ratio ρ versus l_t.
— Peripheral wall:
 $\lambda = 0.03$ (kcal/m. h. °C); $a = 40 \cdot 10^{-4}$ (m²/h); $l = 0.10$ m
— Partition (polystyrene):
 $\lambda_t = 0.035$ (kcal/m. h. °C); $a_t = 22.5 \cdot 10^{-4}$ (m²/h)
— Room air:
 $M = 1$ (kcal/m². °C)
— $S_t/S_p = 4$

Let the ambient air heat capacity be 1 [kcal/m² °C], referred to S_p, and suppose that it may be considered as uniformly distributed throughout the whole room volume. Let the properties of the room solid body (polystyrene partition) be as follows:

$$l_t = 0.10\,[\text{m}]\;;\quad \lambda_t = 0.035\left[\frac{\text{kcal}}{\text{m.h.°C}}\right]\;;\quad a_t = 22.5 \cdot 10^{-4}\left[\frac{\text{m}^2}{\text{h}}\right].$$

Assume the ratio $S_t/S_p = 4$ and the partition total heat capacity

$$M_T = c'\,l_t\,\frac{S_t}{S_p} = 15.55 \cdot 0.1 \cdot 4 = 6.22\left[\frac{\text{kcal}}{\text{m}^2.°\text{C}}\right].$$

Suppose that the curve of the amplitude ratio ρ versus M had been plotted for the peripheral wall under the assumption of uniformly distributed heat capacity. The equivalence factor will be:

$$\varepsilon_\rho = \frac{M_e - M_a}{M_T} \cong \frac{0.096}{l_t} = \frac{0.096}{0.1} = 0.96 \quad \text{(see Table B)}.$$

One therefore obtains:

$$M_e = \varepsilon_\rho M_T + M_a = 0.96 \cdot 6.22 + 1 = 6.97\left[\frac{\text{kcal}}{\text{m}^2.°\text{C}}\right].$$

Inserting the curve ρ versus M with $M_e = 6.97$ we get an amplitude ratio of $\rho = 0.14$ which agrees with the value we should obtain by evaluating ρ as a function of l_t (fig. 6).

SYMBOLS

a	thermal diffusivity of the peripheral wall;
a_t	thermal diffusivity of the partition;
l	thickness of the peripheral wall;
l_t	thickness of the partition;
$l_{0.10}$	depth at which the temperature oscillation amplitude inside the body is reduced to 0.10 times the oscillation amplitude of the ambient air temperature;
$l_{0.33}$	depth at which the temperature oscillation amplitude inside the body is reduced to 0.33 times the oscillation amplitude of the ambient air temperature;
M	heat capacity of unit surface area of the peripheral wall;
M_a	heat capacity of the room ambient air;
M_e	equivalent heat capacity;
M_T	total heat capacity;
S_p	surface area of the peripheral wall;
S_t	surface area of the partition;
α_1	boundary surface coefficient of heat transfer of the peripheral wall outside surface;
α_2	boundary surface coefficient of heat transfer of the peripheral wall inside surface;
α_t	boundary surface coefficient of heat transfer of the partition;

γ	time lag between the room air temperature oscillation and the outdoor air thermal wave;
$\varepsilon = M_e/M_T$	equivalence factor;
ε_ρ	amplitude equivalence factor;
ε_γ	phase equivalence factor;
θ_e	outdoor air temperature;
θ_{e0}	amplitude of the outdoor air temperature oscillation;
θ_i	room air temperature;
θ_{i0}	amplitude of the room air temperature oscillation;
θ_{s0}	amplitude of the partition surface temperature oscillation;
λ	thermal conductivity of the peripheral wall;
λ_t	thermal conductivity of the partition;
$\rho = (\theta_{i0}/\theta_{e0})$	amplitude ratio between the room air temperature oscillation and the outdoor air thermal wave.

REFERENCES

[1] U. MAGRINI, « Influenza della capacità termica dell'ambiente sul regime periodico stabilizzato di temperatura ». *Ricerche di Termotecnica*, n° 11, (1962).

[2] D. FAGGIANI and U. MAGRINI, « Isolamento termico ed onde di temperatura nelle pareti sottili ». *Il Giornale del Genio Civile*, n° 7-8, (Jul.-Aug. 1966).

[3] U. MAGRINI, « Aspetto tecnico della trasmissione di un'onda termica attraverso una parete ». *La Termotecnica*, n° 5, (May 1965).

[4] D. FAGGIANI, « Le onde di temperatura nei moderni edifizii in relazione al benessere ». *Atti del IX Congresso Naz. ANDIL Rimini*, (June 1969).

[5] D. FAGGIANI, « Criteri di caratterizzazione delle pareti nei riguardi dell'isolamento di onde di temperatura ». *La Termotecnica*, n° 10, (Oct. 1967).

[6] D. FAGGIANI, U. MAGRINI and G. REALE, « Note sulla propagazione di onde di temperatura negli edifici, *Quaderni di Fisica Applicata — Istituto di Fisica Tecnica dell'Università di Genova*, n° 4, Publ. Scient. d'Ing.—Genova, (1966).

[7] L. MATTAROLO, « Analogia elettrica e metodi dei quadripoli in problemi di trasmissione termica », *La Termotecnica*, n° 8, (Aug. 1962).

[8] M. SOVRANO and G. ZORZINI, « Il coefficiente di trasmissione « K » ed il tempo di risposta « τ » nello studio del comportamento termico delle pareti ». *Ricerche di Termotecnica*, n° 16, (1966).

[9] I. BARDUCCI, « Attenuazione termica attraverso pareti multistrati in regime periodico -Misura delle reti elettriche equivalenti ». *Quaderni di Fisica Tecnica dell'Ist. di Fis. Tecn. della Univ. di Roma Sett. 1964.*

[10] A. SACCHI, « Sul calcolo dell'attenuazione termica di pareti in regime periodico ». *La Termotecnica*, n° 2, (Feb. 1966).

SYSTEM FOR THE CALCULATION
OF TRANSIENT THERMAL RESPONSES OF ROOMS

W. GROSSMANN, P. KÖNIG and K. H. SCHEUNEMANN

Institut für Luft- und Kältetechnik, Dresden (East Germany)

Système de calcul de la réponse thermique des locaux

Résumé : *L'Institut für Luft- und Kältetechnik de Dresde, a mis au point des méthodes de calcul de la réponse thermique des locaux, qui utilisent à la fois des calculateurs analogiques et digitaux. Les méthodes s'appuient sur la solution de l'équation différentielle de Fournier. En introduisant des incréments différentiels finis pour les coordonnées topologiques, on réduit l'équation aux dérivées partielles à une équation différentielle ordinaire du temps. Au moyen de la transformation de Laplace, celle-ci est présentée comme fonction de transfert.*

Dans le calcul analogique, cette dernière fonction est simulée directement par les composants du circuit. Le calcul digital est conduit au moyen de la réponse complexe de fréquence.

Les calculateurs permettent de déterminer la charge thermique des locaux en conditions optimales et la température de l'air des locaux ventilés mécaniquement ou naturellement. Des considérations pratiques, dans chaque cas, permettent de réduire dépense et précision.

Modern building designs and the use of new building materials have given the transient thermal response of buildings the status of an important technical characteristic; hence, the interest of air-conditioning engineers, architects and sanitation engineers in this parameter. The transient thermal response of a building defines the reaction of a structure to such influences of climate as temperature, radiation and wind inside and outside of the building.

In the following pages the calculation of the thermal behaviour of cooled and air-conditioned buildings by computing the thermal load and the determination of the temperature in ventilated rooms is described. The problem consists of a simple solution for general use of the differential equations describing the transient thermal transfer through building walls.

Papers dealing with this subject have been published already in U.S.A. during the forties (see Mackey, Wright, Stewart *et al.*). The increasing use of electronic data processing equipment for scientific investigations and practical project planning also provides a greater number of solution methods of the present problem. A noteworthy circumstance in this connection seems to be that calculating methods for such a specific purpose as the calculation of thermal load or room temperature with digital or analogue computers are familiar, as are the methods of approximation for "hand calculation", but which are each based on a different algorithm for solution. This makes it difficult for the project engineer to understand and handle the various methods for the suggested purposes.

This paper gives a calculating system of the transient thermal response of rooms where every partial solution of the thermal load and temperature calculation is based on a uniform statement for use with digital or analogue computers, or suitable for "hand calculation".

This method is the result of the work carried out in recent years in the "Institut für Luft- und Kältetechnik", Dresden. It is intended as a fundamental for planning ventilating equipment with the aid of computers in the air-conditioning and ventilation branch of industry of the GDR. The problem centers on the determination of the heat transfer through heat storing components which is not constant but periodic and develops through the influence of various transient conditions. The solution method of the system has been worked out by Grossmann [1, 3, 8, 9] and is based on the definition of system time behaviour with the aid of translation functions familiar from control engineering.

Every component of the building is divided in homogeneous layers, i.e., layers consisting of the same material. Every layer is resolved into a certain number of elements for which the heat balance is determined. The premise, however, is that heat storage is concentrated in the centre of the element in a plate of infinite thinness and that heat transfer is unidimensional only.

The heat balance equations given in figure 1, after translation into the Laplace range, can be written as a matrix (with $T_i = R_i C_i$)

$$\begin{pmatrix} \vartheta_{a_I} \\ \\ q_{a_I} \end{pmatrix} = \begin{pmatrix} A_{I11} - R_I A_{I12} \\ \\ -\dfrac{1}{R_I} A_{I21} \quad A_{I22} \end{pmatrix} \begin{pmatrix} \vartheta_{e_I} \\ \\ q_{e_I} \end{pmatrix} \qquad \text{with}$$

$$A_{I11} = A_{I22} = \tfrac{1}{2} T_I S + 1$$
$$A_{I12} = \tfrac{1}{4} T_I S + 1 \quad (1)$$
$$A_{I21} = T_I S$$

For arranging the n elements in series to a layer the law of matrix multiplication applies, thus:

$$\begin{pmatrix} \vartheta_{W_I} \\ \\ q_{W_I} \end{pmatrix} = a_I^n \begin{pmatrix} \vartheta_{W_A} \\ \\ q_{W_A} \end{pmatrix}, \qquad a^n = a = \begin{pmatrix} A_{11} \quad -R_0 A_{12} \\ \\ -\dfrac{1}{R_0} A_{21} \quad A_{22} \end{pmatrix} \qquad (2)$$

$$R_0 = R_1 \cdot n; \quad C_0 = C_1 \cdot n \qquad T_0 = T_1 \cdot n^2$$

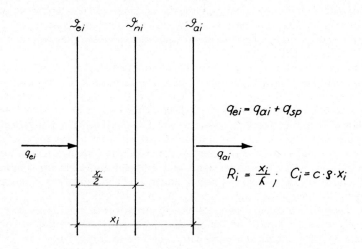

$$q_{ei} = q_{ai} + q_{sp}$$

$$R_i = \frac{x_i}{k}; \quad C_i = c \cdot \mathcal{g} \cdot x_i$$

$$\tilde{\vartheta}_{ai} = \frac{R_i C_i}{2} \cdot \frac{d\tilde{\vartheta}_{ei}}{dt} + \tilde{\vartheta}_{ei} - R_i \frac{R_i C_i}{4} \cdot \frac{dq_{ei}}{dt} + q_{ei}$$

$$q_{ai} = -\frac{1}{R_i} R_i C_i \frac{d\tilde{\vartheta}_{ei}}{dt} + \frac{R_i C_i}{2} \frac{dq_{ei}}{dt} + q_{ei}$$

Fig. 1

164

where

A_{11} to A_{22} represent functions of the form of differential equations of the order of n.

With multi-layer walls an analogous calculation is given with the transfer matrix (m = number of layers) to

$$a = a_1 \cdot a_2 \cdot \ldots a_m.$$ (3)

On this transfer matrix the definition of the thermal behaviour of a room is constructed with respective consideration of the prevailing boundary conditions.

These boundary conditions are defined as follows:

heat flow on the wall surfaces

$q_{WA} = q_{KA} + q_{SA}$ indexes: W total

                                             K convection

                                           A outside

$q_{WJ} = q_{KJ} - q_{SJ}$ J inside

                                           S radiation

The convection component q_K includes the heat transfer between air and wall as well as the radiation exchange due to the different surface temperature between wall and surroundings calculated to a fictitious heat transfer value. The radiation component q_S only contains the radiation of the sun and the diffuse radiation of the sky. Solar radiation and temperature can then be combined to express solar air temperature for the outer walls and for radiated inner walls in the familiar manner by:

$$\vartheta_S = \vartheta + \frac{a \cdot q_S}{\alpha}.$$ (4)

Heat transfer of the windows is considered stationary and expressed in the conventional way.

The energy release of interior heat sources, such as machines, lighting and people, is differentiated to components of convection and radiation.

For heat stored in the air of a room the following statement applies:

$$q_L = Cp_L \cdot \rho_L \cdot V_R \cdot \frac{d\vartheta_L}{dt}.$$ (5)

From the mentioned equations and the matrix of transient heat transfer the heat transfer q_{KJ} of a wall is expressed by the relation:

$$q_{K_I}(S) = W_1 \frac{1}{r_D} \vartheta_{S_A}(S) - W_2 \frac{1}{r_D} \vartheta_I(S) + W_3 \frac{r_z}{r_D} q_{S_I}(S).$$ (6)

Then the transfer functions W_1 and W_3 are given by:

$$W_1 = \frac{1}{\frac{r_A}{r_D} A_{11} + \frac{r_o}{r_D} A_{12} + \frac{r_I r_A}{r_D r_o} A_{21} + \frac{r_I}{r_D} A_{22}}$$

165

$$W_2 = \frac{\dfrac{r_A}{r_o} A_{21} + A_{22}}{\dfrac{r_A}{r_D} A_{11} + \dfrac{r_o}{r_D} A_{12} + \dfrac{r_I r_A}{r_D r_o} A_{21} + \dfrac{r_I}{r_D} A_{22}}$$

$$W_3 = \frac{\dfrac{r_A}{r_z} A_{11} + \dfrac{r_o}{r_z} A_{12}}{\dfrac{r_A}{r_D} A_{11} + \dfrac{r_o}{r_D} A_{12} + \dfrac{r_I r_A}{r_D r_o} A_{21} + \dfrac{r_I}{r_D} A_{22}} \ .$$

The already mentioned transfer matrix components are expressed with A_{11} and A_{22}, resistance values are thus:

$$r_z = r_A + r_{o'}. \tag{8}$$

For calculating the temperature ϑ_I of a room with differing walls and the temperature ϑ_{Re} of the air entering the room, Grossmann [2] derived the following relation:

$$\vartheta_I = \frac{\sum\limits_z (K_1 W_1 \vartheta_{S_A}(S)) z}{K_4 W_4 + \sum\limits_z (K_2 W_2) z} + \frac{\sum\limits_z (K_3 W_3 q_{S_t}) z}{K_4 W_4 + \sum\limits_z (K_2 W_2) z} +$$

$$+ \frac{K_4}{K_4 W_4 + \sum\limits_z (K_2 W_2)_z} \vartheta_{Re} \frac{1}{K_4 W_4 + \sum\limits_z (K_2 W_2)_z} Q_1. \tag{9}$$

Because of the coupling of Z walls, the transfer functions are also coupled. Every wall input has its own transfer function:

$$W_4 = 1 + \frac{M_R}{m_L} S$$

$$K_1 = K_L \cdot \frac{F}{r_D} \qquad K_2 = \frac{r_z \cdot F}{r_D} \tag{10}$$

$$K_4 = m_L \cdot C \rho_L .$$

Prior to the computation of the transfer functions, the number of incremental steps (number of elements) for certain layers of the wall have to be determined. In order to keep an acceptable relation between input tolerance and calculation errors Grossmann [1] has shown that transfer equations of the 2nd to the 5th order are necessary for the input of the thermal ventilation engineering calculations, i.e., solar air temperature, outer air and room temperature, and for the transfer properties of conventional walls and roofs. To obtain a better accuracy the polynomial coefficients of the mentioned order can be determined by reduction of a statement of a higher order [1].

In figure 2 an example of a calculation diagram is given for the cooling load of a wall from the ILK-programme "ANAREMA" (see [1, 8]), which can be derived directly from the transfer equation.

166

For the digital calculation according to the ILK-programme "ELDIREMA"the transfer functions W(S) are translated into frequency responses $F(j\omega)$ by replacing the Laplace operator $s = x + jy$ with the Fourier operator $S = j\omega$ (see [3]). This frequency response defines the relation between initial value and input of a system in built-up state on sinusoidal fluctuation of the input.

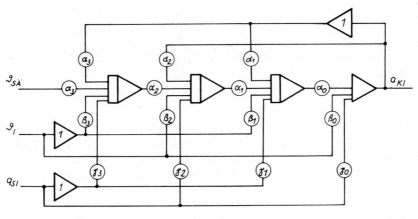

Fig. 2

Frequency response is a complex function and in this case takes the form:

$$F(j\omega) = \frac{R_z(j\omega) + j\,J_z(j\omega)}{R_N(j\omega) + j\,J_N(j\omega)}. \tag{11}$$

Amplitude frequency response and phase frequency response can be separated by resorting to Euler's formulas, thus:

$$H(j\omega) = \frac{\sqrt{R_z^2(j\omega) + J_z^2(j\omega)}}{\sqrt{R_N^2(j\omega) + J_N^2(j\omega)}} = H(j\omega)\,e^{\varphi(j\omega)}$$

$$\tag{12}$$

$$\varphi(j\omega) = \arctan\frac{J_z(j\omega)}{R_z(j\omega)} - \arctan\frac{J_n(j\omega)}{R_n(j\omega)}.$$

Amplitude attenuation and phase displacement of a sinusoidal input when passing the system is here expressed with H and φ, respectively.

In practice every input has consequently to be developed into a Fourier series by analysis of the harmonics, with the basic frequency, for instance, representing the cycle of one day. With the computer the various harmonics are separately calculated with respect to amplitude attenuation and phase displacement and combined to represent the total initial value.

To reduce computation effort and to provide a possibility for "hand calculation" of cooling load and room temperature, König [4] used the simplified statement of Mackey and Wright [5]. This operates with only one average value, that is independent of frequency, for each of the periodic inputs of the various harmonics separated by analysis as to amplitude attenuation and phase displacement. It is relatively simple to determine these data for the respective walls, because in many instances the amplitude

167

attenuation and phase displacement of the basic harmonic can be utilized. For "hand calculation" these can also be listed in tables of the most frequently used wall structures. Then, the calculation of the heat transfer q_{KJ} of one wall is made with the simple equation:

$$q_{K_I}(t) = \frac{1}{r_D}\,\vartheta_{S_{Am}} + \frac{r_z}{r_{K_I}r_D}\,\vartheta_{S_{Im}} - \frac{1}{r_{K_I}}\,\vartheta_{Im} +$$

$$+ \frac{1}{r_D}\,f_1\{\vartheta_{S_A}(t+\varphi_1/\omega) - \vartheta_{S_{Am}}\} + \frac{r_z}{r_{K_I}r_D}\,f_3\{\vartheta_{S_I}(t+\varphi_3/\omega) - \vartheta_{S_{Im}}\} -$$

$$- \frac{1}{r_{K_I}}\{\vartheta_I(t) - \vartheta_{Im}\}. \qquad (13)$$

The stationary component is here given by the first line, while the second represents the component that changes with time and is written in the form of a phase displaced and attenuated frequency, i.e., phase displacement φ and amplitude attenuation f, respectively.

Equations for calculating room temperature can also be translated in a similar manner [4].

The introduction of this approximation theorem results in a marked reduction of calculating effort. Nehring [6] has pointed out the error that can occur in this mode. For this reason it is feasible to employ this approximation only where the advantage of simplified calculation is not offset by the inherent inaccuracies. Such cases are, for example, the calculation of the energy consumption with the aid of typical daily cycles [7] and the "hand calculation" of the relatively simple problem of cooling load, and primarily of the room temperature in ventilated rooms.

The described system for the calculation of the thermal response of a building gives the following advantages:

— By modifying the uniform method, heat load and temperature of a room can be calculated with a digital or an analogue computer.

— With simplification and approximation statements based on a uniform method, the calculation effort can be adapted to the desired problem solution. This is, for example, the frequent computation of daily cycles with a computer for determination of energy consumption or even the "hand calculation" of the simpler cases of determination of room temperature of ventilated rooms.

— The presentation of relations with transfer functions permits the clear analysis of the various influences of input data and the consideration of their optional periodic time functions. There is no restriction, for instance, to a constant interior room temperature.

— The possibility of the "hand calculation" of temperatures in ventilated rooms provides a means for the planner of ventilating equipment to decide between ventilating equipment or air-conditioning systems for a specific case.

LIST OF SYMBOLS

a	absorption factor
A	matrix link matrix
c	specific heat

C	storage capacity
f	amplitude attenuation
F	area
H	amplitude frequency response
K	transfer factor
m	number of wall layers
m_L	flow
M_R	volume of air in a room
n	number of elements
q	heat flow
q_K	heat flow from convection
q_L	heat stored in the air of a room
q_S	heat flow from radiation
q_{sp}	stored amount of heat
q_w	heat flow on wall surface
R	resistance
r	resistance per unit of area
S	Laplace operator
t	time
T	time constant
V_R	volume of a room
W	transfer function
x	length
α	heat transfer factor
ϑ	temperature
ϑ_L	air temperature
ϑ_{Re}	room inlet temperature
ϑ_W	surface temperature of a wall
ϑ_s	solar air temperature
λ	heat transfer coefficient
ρ	density
ω	rotational frequency (radian frequency)
P_n	phase displacement
φ	phase frequency response

Indexes

a	output/outlet
A	outer side
D	passage
e	entry/input/inlet
i	running numerator
I	inner side
k	convection
L	air
m	average value/mean value
o	wall
z	running numerator

REFERENCES

[1] W. GROSSMANN, Calculation of transient thermal transfer of walls. *Vortrag auf der Fach-tagung Luft- und Kältetechnik*, Dresden (April 1968).
[2] W. GROSSMANN, Calculation of room temperature with transfer functions (unpublished).

[3] W. Grossmann, Calculation of heat and cooling loads with consideration of the heat storage in space enclosing components with digital computers — Transfer function programme — Mathematical model and rough programme process (unpublished).

[4] P. König, Simplified calculation of cooling load and room temperature with transfer functions (unpublished).

[5] C.O. Mackey and L.T. Wright, Periodic heat flow — homogeneous walls and roofs. *A.S.H.V.E. Transactions*, **50** (1944), S. 293.

[6] G. Nehring, Heat flow through outer walls and roofs into air-conditioned rooms due to periodic daily cycles of specific meteorological elements. *Gesundheitsingenieur*, München, **83** (1962), 7, S. 185/89; 8, S. 230/42; 9, S. 253/69.

[7] P. König, Methods of calculating energy consumption of ventilating and air-conditioning plants. *Luft- und Kältetechnik*, Berlin, **5** (1969), S. 59/66.

[8] W. Grossmann, Calculation of heat and cooling loads with consideration of the heat storage in space enclosing components with analogue computers (unpublished).

[9] W. Grossmann, Calculation of the cooling loads of a hotel with consideration of the heat storage of space enclosing components. *Luft- und Kältetechnik*, Berlin, **3** (1967), 6, S. 263/67 and 4 (1968), 1, S. 8/12.

REMARQUE

F. M. CAMIA (France) — Il serait excellent, croyons-nous, que soit faite une synthèse des travaux effectués relativement aux échanges thermiques des locaux et de l'ambiance, en se référant à diverses sources. Pour notre part nous connaissons la SFT et le COMPLES.

L'utilisation de la fonction de transfert pour les échanges calorimétriques est faite à notre connaissance par le Prof. Petit à l'INSA de Lyon, par le Dr Zielen Kiewicz à Varsovie, et par le Dr Tappe à Berlin Est. Enfin, la fréquence de la variation thermique imposée à un mur doit entrer au nombre des paramètres. Une variation périodique de longue période est très pénétrante.

ON COMPUTING THE TEMPERATURE OF A ROOM
UNDER SOLAR RADIATION

J. CHYSKY

Polytechnical School, Prague (Czechoslovakia)

Calcul de la température d'un local soumis au rayonnement solaire

RÉSUMÉ : *Lorsqu'on conçoit de nouveaux bâtiments, il faut souvent décider si une installation de conditionnement d'air est nécessaire pour l'été. Dans la littérature, ce problème est souvent étudié en fonction des variations prévues de la température du local pendant toute une journée. En agissant ainsi, les auteurs comptent sur des variations cycliques régulières des charges thermiques, ce qui ne se vérifie pas en pratique, même approximativement.*

Ce rapport tente de définir les conditions réelles en déterminant la température moyenne d'un local en fonction de la charge thermique moyenne quotidienne et en substituant à la charge thermique réelle soit une charge harmonique, soit une charge constante pendant la période d'insolation qui entraîne une certaine variation de température par rapport à la température moyenne. Des conditions thermiques ainsi calculées, on devrait alors tirer des données qui permettraient de savoir si une installation de conditionnement d'air est nécessaire ou non.

Comme exemple de ce calcul, on cite le nouveau bâtiment de l'Université Technique de Prague, dont 50 % de la surface est vitrée. Les résultats des calculs sont comparés à des mesures faites par la suite.

Ces mesures ont aussi montré l'influence de la hauteur du bâtiment sur la répartition des températures, l'importance de la ventilation nocturne des locaux, de même que l'influence de persiennes extérieures et intérieures sur la variation de température dans les locaux ne disposant pas de conditionnement d'air.

Recently, repeated discussions have been held concerning the necessity of air conditioning in civil buildings under Central European climatic conditions (comfort purposes) and the adequacy of air cooling, shading from solar radiation being carried out by current means or by special glass.

Under Central European conditions the thermal load on the windows, at maximum air temperatures, may be higher than in more southerly areas due to the mutual position of the sun and the wall. Figure 1 gives the values for solar radiation through a window facing south and south-east at 30 and 50°N latitude in the month of July. With the window facing south in our geographic latitudes the maximum irradiation in July is 3.5 times and, with the window facing south-east, 1.4 times larger.

Apart from the maximum solar radiation, the overall climatic conditions are also affected by the frequency of summer days with sunshine. Their number varies considerably under climatic conditions in Czechoslovakia. However, even in an average year their number is such that it must be taken into account in solving the thermal conditions (e.g., in 1968 the number of hours of sunshine amounted to 225 in June, 241 in July and 180 hours in August). As the passage of solar radiation through windows is instantaneous and, as in summer one can only exceptionnally compensate this load by ventilation, it is clear that buildings must be designed, even as far as the construction is concerned, with a view to the summer thermal load.

The thermal load through the irradiated windows and the thermal properties of the external and internal walls have a predominant effect upon the temperature conditions in the rooms. The temperature conditions in buildings, which are defined by these conditions, are the main reason for using comfort air-conditioning equipment. One must also consider other reasons, e.g., the impossibility of opening windows because of the high level of dust in large cities, the noise level and the intense air flow due to aerodynamic conditions and to the temperature differences.

In order to establish an objective scale as regards the necessity for air conditioning in buildings, some authors have proposed certain criteria (mentioned in the references).

After testing their applicability, extensive measurements of summer temperature conditions in the buildings of the Mechanical Engineering Faculty in Prague were made. In the following the most important results will be given and an analysis of these conditions will be made. If not mentioned otherwise, the measurements were carried out with the windows closed without any protection against radiation. Al-

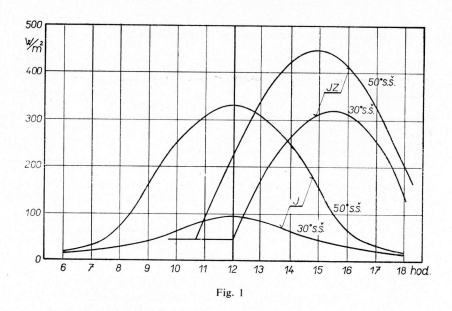

Fig. 1

Fig. 2

172

though the conditions were observed in all the rooms. e.g., drafting rooms, auditoria, only the values observed in the offices of the individual departments will be mentioned.

The building investigated was of concrete, of rectangular cross-section with nine storeys, facing south-east and north-west (fig. 2). The windows were double-glazed in metal frames, occupying about one half of the facade. The internal walls were of hollow bricks and the thickness of the walls, inclusive of plaster, was 12 cm.

CONCLUSIONS DRAWN FROM THE OBSERVATION RESULTS

The average internal temperatures in the offices were higher, in all cases, than the average outside temperatures; e.g., in a room on the 9th floor the average temperature, without internal heat sources, was 29.4 °C over a period of 24 hrs when the outside temperature was 22.8 °C. The average solar temperature, defined as $t_s = t_e + \varepsilon I / \alpha_e$, nearly coincided with the average temperature in the room. Absolute agreement would be random, because under the same conditions the temperatures in the lower floors were lower. Deviations were mainly due to varying infiltration.

The delay in the temperature variation in the rooms with respect to the variation of the external thermal load was comparatively small, mostly less than 2 hrs. The temperature variation of the wall surfaces could only be properly measured on the internal surfaces of the outside wall and it did not exceed 1.5–2 deg C. The increase in the temperature of the walls had an adverse effect on the resultant temperature; this was mostly higher than the air temperature by 1 deg C. Long-term temperature variations (cooling, warming) became apparent in the average temperature values after a delay of two days approximately.

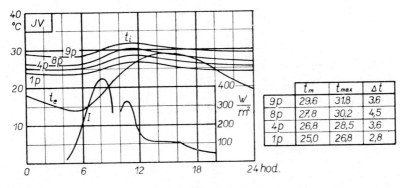

	t_m	t_{max}	Δt
9p	29,6	31,8	3,6
8p	27,8	30,2	4,5
4p	26,8	28,5	3,6
1p	25,0	26,8	2,8

Fig. 3

The effect of the height of the building on the temperatures in the individual floors has not, to date, been treated sufficiently in literature. The results of our observations are given in figure 3. The differences observed between the individual floors are comparatively large. If one disregards the ninth floor which is anomalous, and which has a flat roof, the differences in the maximum temperatures between the 8th floor and ground level amounted to 3.4 deg C in the rooms facing south-east and the differences in average temperatures amounted to 2.8 deg C. On an average, this means an increment of a maximum temperature of 0.5 deg C per floor. These conditions are mainly due to the cooling of the rooms on the lower floors by the night infiltration of cool air. The corridors between the individual floors were not separated by doors.

Effect of orientation. In the rooms facing south-east, the mean temperature was somewhat higher than in the rooms facing north-west. The difference (on identical floors) was, on an average, 1 deg C. The difference in solar temperatures for the two directions mentioned was slightly higher: 2 deg. With the rooms facing north-west, it was a convenient factor that maximum temperature was reached after working hours, at 6.30 p.m. On the other hand the temperature in the rooms on the 9th floor, facing south-east, exceeded 30 °C between 7 a.m. and 2 p.m., i.e., during nearly all the working hours.

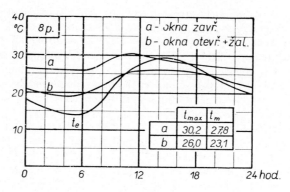

Fig. 4

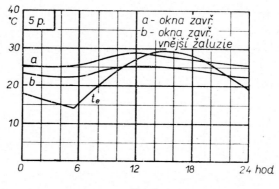

Fig. 5

The infiltration into the 9th floor was investigated informatively. Under calm air conditions it had an updraught direction (average internal temperatures were higher than external). The observed values were in a wide range covering multiples of 0.5—2 of the air-change volume in the room.

The possibilities of decreasing the inside temperatures in the rooms were also investigated. Figure 4 shows the temperature variations in two neighbouring offices facing south-east. In room *b* the window was left permanently open and the outside blinds were pulled down during the day. The temperatures in this room were lower by 4.7 deg C on the average and the maximum diurnal temperature was 4.2 deg lower than in the neighbouring room *a* with its windows closed. The effect of the outside blinds can be seen in figure 5. The difference between the maximum temperatures of otherwise identical rooms amounted to 3.5 deg; the average temperatures to 3 deg. The internal blinds had no noticeable effect on the air temperature but they acted favourably by shading the working space and decreasing the surface temperature of irradiated areas in the room.

ANALYSIS OF THE THERMAL LOAD AND CONDITIONS IN A TEST ROOM (ROOM 815)

In order to test the possibilities of computing the temperature in the rooms under different conditions, observed values of the intensity of solar radiation, the appropriate temperatures and air changes were used for a computing procedure relevant to the test room. The average amount of heat entering the room amounted to 433 W and the amount leaving the room amounted to 343 W. The total thermal balance on the day of the investigation was positive: $+90$ W. This would correspond to a temperature increase of 0.5 deg in all the walls of the room. However, this value is uncertain due to some of the chosen quantities. Especially, the value for dirty panes is problematic. For determining the average diurnal temperature in the room, t_{im}, the following relation was derived:

$$t_{im} - t_{em} = \frac{\dfrac{k_w}{\omega}\, \Delta t_{sm} + \sum q_{Rm}}{\dfrac{k_w}{\omega} + k_0 + \dfrac{M_L c_L}{S_0}},$$

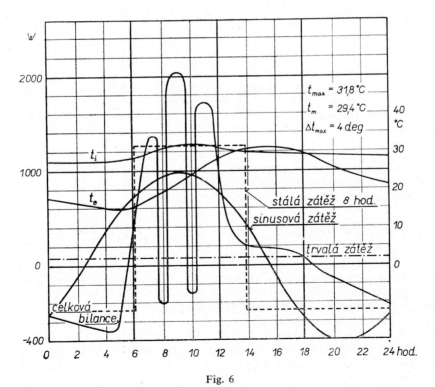

Fig. 6

where:

k_w and k_o are the heat transfer coefficients for the walls and the windows respectively;

$\Delta t_{sm} = t_{sm} - t_{em}$ is the increment of the average solar temperature with respect to the air temperature;

q_{Rm} is the mean specific thermal load due to solar radiation through the windows;

ω is the relative measure of glazing of the external walls;

S_o is the window area, and

$M_L c_L$ is the water equivalent of the air entering the room.

After substituting numerical values, the average internal temperature in the room came to 29.5 °C, which is nearly identical with the observed value.

The variation of the internal temperatures was computed in two ways:

a) under the assumption of a harmonic variation of the thermal load. The real variation of the load was transformed to a harmonic variation according to a sinusoid under the following conditions: the axis of the sinusoid is identical with the value of the average load, the area limited by the real load was substituted by the same area under the sinusoid. From this condition follows the amplitude of the variation load, 890 W (fig. 6). According to Šklověr's simplified relation,

$$\theta = \frac{\sum Q}{\sum B_i S_i},$$

the temperature variation comes to ± 2.5 deg C. B is the coefficient of thermal absorptivity of the walls of the room. The surface S was considered to be the area of all the walls inside the room without subtracting the areas taken up by furniture; on the other hand, the fitting of room furniture was not taken into account. The difference between the maximum and minimum temperatures was computed to be 5 deg, the observed value being 4 deg.

b) The second method was carried out by means of an idealization of the variable thermal load by a constant load of 1 130 W, lasting for 8 hrs, covering the period of direct irradiation of the window. This corresponds to a difference between the maximum and minimum temperatures (with a view to the same source as in the first case) of 3.5 deg. In this case the result was 0.5 deg lower than the observed value.

Both the relations used are considerably complicated as regards computation and practically very difficult to apply. A simpler approach seems to be the simplified criterion relationship. From the dimensional analysis

$$t_{imax} - t_{imin} = \frac{\sum I}{M_L c_L} f\left(\frac{a\tau_s}{\delta^2}, \frac{\alpha\delta}{\lambda}\right) = \frac{\sum I}{M_L c_L} A \, \mathrm{Fo}^m \, \mathrm{Bi}^n,$$

where:

I is the thermal load during the period of solar irradiation after subtracting the losses due to ventilation;

$M_L c_L$ is the thermal capacity of the brickwork per room;

δ is the thickness of the walls per room;

τ_s is the period of solar irradiation.

The criterion relationship would have to be determined from a series of observations. In the case investigated, the product $\mathrm{Fo}^m \, \mathrm{Bi}^n$ does not differ too much from 1.

CONCLUSION

It is important to compute the temperature conditions in non-air-conditioned rooms in order to be able to decide whether it is necessary to use air-conditioning equipment. The theoretical relationships, obtained by idealizing the thermal load and the thermal properties of buildings, are too elaborate for computation with a view to their comparatively low accuracy, because the initial theoretical assumptions are usually not even approximately satisfied. The author is of the opinion that satisfactory results may be achieved with the help of an average thermal balance and of the said criterion relationship. The effect of the Biot criterion would be smaller than that of Fourier's. Since the values for Fo need not differ substantially if current building material is used, the quite simple relation for characterizing temperature variation would be applicable in some cases,

$$t_{i\,max} - t_{i\,min} = C \cdot \frac{\sum \cdot}{M_L c_L},$$

from which it follows that the increase in the internal temperature is indirectly proportional to the thermal capacity of the mass per single room and directly proportional to the thermal load. A similar assumption was considered in the case of the Carrier computing data.

The results were also compared with the criteria for air conditioning according to Barcs, Duta and Windisch. In considering both references, it is necessary to use air conditioning. Neither of the references, however, respect the average internal temperature nor is the substitution of the simple maximum values of the loads correct. The author considers it more suitable to judge the necessity of air conditioning with regard to expected internal temperatures and to their variation.

REFERENCES

[1] ŠHKLOVĚR, Teploperedacha pri periodicheskikh teplovӯkh vozděistviyakh (Moscow 1961), p. 160.
[2] V. BARCS, The influence of an up-to-date architectural arrangement of dwellings and public buildings on energy demand for heating and possible cooling, *Paper presented at the Symposium on Problems of Optimum Economics.* (Prague 1966).
[3] G. DUTA and K. WINDISCH, Zur Ermittlung von Kriterien für die Notwendigkeit der Klimatisierung, *Luft u. Kältetechnik* (1967), pp. 176-179.

ELECTRICAL ANALOG MODEL
OF A SYSTEM WITH TRANSIENT RADIATION HEAT TRANSFER

F. DE PONTE and G. ZORZINI

*Istituto Fisica Tecnica e Laboratorio per la Tecnica
del Freddo dell'Università di Padova (Italy)*

Modèle analogique électrique pour l'étude d'un système avec transmission de chaleur par rayonnement, en régime variable

RÉSUMÉ : *On décrit l'emploi d'un modèle analogique réalisé par un réseau passif et des éléments linéaires et non linéaires d'un calculateur analogique.*

Le circuit permet l'étude du comportement en régime variable de deux parois d'épaisseur s entre lesquelles il y a seulement échange de chaleur par rayonnement.

Pour commenter l'emploi du modèle, on a mesuré les temps de réponse de quelques types de parois, dans des conditions déterminées.

1. INTRODUCTION

This work has been chosen because of some problems arising out of the interpretation of experimental data regarding tests on multi-layer cavity walls.

In an air cavity, thermal exchange takes place by conduction, convection and radiation. It is well known that by considering only the last way of exchange, thermal resistance occurs which is a function of the temperatures of the cavity surfaces. As a consequence, tests under non-steady state conditions aimed at measuring the response time* yield different results according to the test procedures themselves.

2. FORMULATION OF THE PROBLEM

Consider the diagram in figure 1 where two indefinite walls A and B are defined by the parallel planes S_1, S_2, S_3, and S_4 with abs. temperatures T_1, T_2, T_3, and T_4 respectively.

Let surfaces S_2 and S_3 be "grey" and a_1 and a_2 the respective absorptivities.

Suppose that thermal exchange occurs by mutual radiation through the empty space s_1, by conduction through walls A and B and by convection and radiation between face 4 and the fluid at temperature T_5 in contact with it. The study of the system's non-steady state behaviour is under consideration.

Non-steady state heat transfer problems with non-linear boundary conditions have already been approached by means of analogical methods.

By way of example, references [2] and [3] in the bibliography relate to the use of non-linear resistors or diode function generators. Both notes assume constant the temperature of the medium radiating towards the extreme surfaces of the system examined, or towards which such surfaces radiate at varying temperature. In the above-quoted non-linear elements, the current is only a function of the potential difference of the terminals and does not depend on the single values of such potentials.

This study, instead, is related to an electric circuit that still obeys laws similar to those for mutual thermal radiation and enables the realization of boundary conditions free from the limitations stressed in this paragraph.

(*) The definition of response time is given further on.

3. DESCRIPTION OF THE ANALOG MODEL

In the model, two conventional resistance and capacity networks correspond to walls A and B, that are structurally marked with only two T-shaped cells.

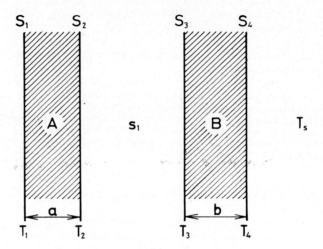

Fig. 1

The voltages corresponding to the abs. temperatures T are marked with a V, while:

$$i_r = K(V_2^4 - V_3^4) \, [\text{ampère}] \qquad (1)$$

indicates the current corresponding to the specific heat flow exchanged by the two surfaces S_2 and S_3:

$$q_r = \frac{T_2^4 - T_3^4}{\dfrac{1}{\sigma_2} + \dfrac{1}{\sigma_3} - \dfrac{1}{\sigma_0}} \left[\frac{\text{kcal}}{\text{h. m}^2} \right] \qquad (2)$$

where

$$\sigma_2 = a_2 \sigma_0 \, ; \quad \sigma_3 = a_3 \sigma_0 \, ; \quad \sigma_0 = 4.96 \cdot 10^{-8} \, \frac{\text{kcal}}{\text{h. m}^2 \, {}^\circ \text{K}^4} .$$

If $a_2 = a_3 = a$, we have:

$$q_r = \sigma_0 \frac{a}{2-a} (T_2^4 - T_3^4). \qquad (3)$$

Let us now consider the block diagram in figure 2.

Block s_1 stands for the electric circuits capable of drawing a current $i_r = K(V_2^4 - V_3^4)$ from node 2 and feeding the same into node 3. Since i_r is a function of V_2 and V_3, these signals must enter into block s_1. The relative inputs 7 and 8, however, must have practically infinite impedance so that the transfer of current i_r from the output of A to the input of s_1 and from the s_1 output to the B input may occur with negligible error.

Block s_1's internal structure is schematically shown in figure 3. The individual blocks A_1, A_2, M_1, M_2, M_3, K_1, K_{-1}, η_1, η_2 are obtained by means of an analogue computer and they represent respectively:

a) A_1 and A_2, two dc amplifiers with gain $A = 2$ and an input impedance of about 1,000 MΩ;

b) M_1, M_2, M_3, three non-linear elements with two inputs (A_i, B_i) and an output (C_i), which perform respectively operations $A_1^2 + B_1^2/100$, $A_2^2 - B_2^2/100$ and $A_3 B_3/100$;

c) K_1, K_{-1}, two linear elements gaining respectively K_1 and -1;

d) η_1, η_2, two voltage-current converters in which the output current $i = (-v/R\eta)$ corresponds to the input voltage v.

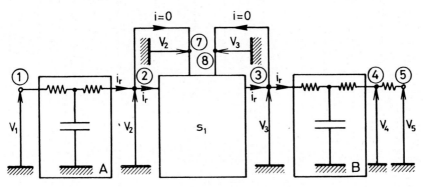

Fig. 2

4. CIRCUIT CHECK AND EXPERIMENTAL RESULTS

Let us consider the physical system in figure 1 and suppose 0.1 m be the thicknesses a and b of the walls A and B. The conductivity λ of the constitutive material be equal to 1 kcal/h. m. °C and $c = 0.2$ kcal/kg. °C and $\rho = 1800$ kg/m³ be its specific heat capacity and its density respectively. Let us lastly put $a_2 = a_3 = 0.8$ and $\alpha = 10$ kcal/m². h. °C as the combined convection and radiation coefficient between the surface S_4 and the air at T_5 temperature.

Let us call $R_{12} = R_{34}$, R_r and R_t the thermal resistances corresponding to one m² of surface in relation to the conduction into the homogeneous layers A and B, to the radiation into the empty space s_1 and to radiation and convection between the surface S_4 and the room at temperature $T_5 = 300$°K. The same symbols will be used later on to indicate the electric resistances corresponding to the analogue model.

For $\tau < 0$ suppose that the system in figure 1 be in a steady state condition with $T_1 = T_2 = T_3 = T_4 = T_5 = 300$°K.

The initial R_r resistance can be analytically estimated and it is equal to:

$$[R_r] \text{ initial} = 0.28 \text{ Fou. m}^2$$

Suppose that at $\tau = 0$, T_1 changes from 300°K to 600°K, while T_5 keeps its initial value (300°K). As the new steady state condition is reached, temperatures T_1, T_2, T_3, T_4, T_5 will assume the following values analytically inferred:

$$T_1' = 600°K ; \quad T_2' = 517.5°K ; \quad T_3' = 465.1°K ;$$

$$T_4' = 382.5°K ; \quad T_5' = 300°K$$

181

while the resistance to radiation takes the final value:

$$[R_r]_{final} = 0.0636 \text{ Fou. m}^2$$

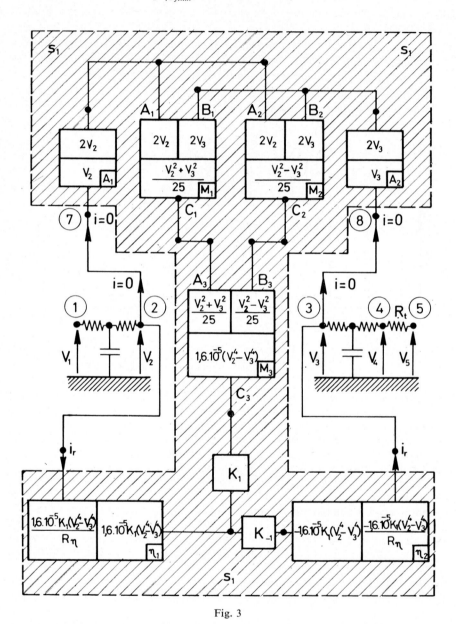

Fig. 3

After having built the electric model with a ratio R_e/R_t between an electric resistance R_e and the corresponding thermal resistance R_t equal to 5 MΩ/Fou. m^2, with a single cell capacity equal to 0.34 μF, and after having chosen a ratio V/T equal to 1/15 volt/°K, the following experimental values were recorded in relation

182

to the voltages at the points corresponding to the surfaces S_2, S_3, S_4:

$$V_2' = 34.50 \text{ volt}; \quad V_3' = 31.00 \text{ volt}; \quad V_4' = 25.50 \text{ volt}.$$

It follows that:

T_2' (experim.) $= 517.5\,°K$
T_3' (experim.) $= 465.0\,°K$
T_4' (experim.) $= 382.5\,°K$.

It has been further experimentally found:

$$[R_r]_{final}^{experim.} = 0.0634 \text{ Fou. m}^2.$$

These values are in perfect agreement with the computed values; it is to be noted that they are supplied by a computer in its best working condition.

It is obvious that similar tests have been also performed in correspondance with values for a temperature drop T_1 to T_5 and resistances $R_{1,2} = R_{3,4}$ that differ from those considered above. All tests draw the conclusion that the circuit works, for all practical purposes, in a satisfactory manner.

Models like the one in figure 2 with electrical resistances $R_{1,2} = R_{3,4}$ equal to 100, 250, 500, 1000 KΩ were then taken into consideration. This means that 0.02, 0.05, 0.1, 0.2 Fou. m^2 values were taken as the ratio between the $a = b$ thicknesses of the walls A and B and their conductivity λ.

Starting from steady state conditions, V_1 was varied, as a step function, from the initial values to the final ones recorded in table 1, where the respective temperature T_1 variations are reported.

Table 1

T_1		V_1		T_1		V_1	
From °K	*to* °K	*From* V	*to* V	*From* °K	*to* °K	*From* V	*to* V
300	375	20	25	450	300	30	20
375	300	25	20	300	600	20	40
300	450	20	30	600	300	40	20

For the sake of brevity we shall, from now on, deal with "direct" tests when voltage V_1 goes from $V_{10} = 20$ volt to a higher value and "inverse" tests when from the steady state condition with $V_1 > 20$ volt we have V_1 going back to 20 volt.

We shall call "response time" of the model at point 4 (fig. 2) the following expression:

$$\tau_0^* = \int_0^\infty \left[\frac{V_1(\tau) - V_{1i}}{V_{1f} - V_{1i}} - \frac{V_4(\tau) - V_{4i}}{V_{4f} - V_{4i}} \right] d\tau \qquad (4)$$

in which i and f indicate, in the same order, the initial and final values.

If V_1 shows a step variation, the value of τ_0^* is:

$$\tau_0^* = \int_0^\infty \frac{V_{4f} - V_4(\tau)}{V_{4f} - V_{4i}} d\tau. \qquad (5)$$

183

For a linear system, formula (4) is a constant. It does not depend therefore on the values of the input variable as this can easily be brought about by the transfer function of the system assigned in Laplace variable domain.

If the system is not a linear one, expression (4) becomes a function of the input variable also, so that it would not be altogether correct to call it a response time. At any rate it enables comparison of the transient behaviour of one system in nonsteady state condition with another system affected by identical inputs.

The results of experiments are recorded in table 2. As a function of the step of V_1 and of the electrical resistance R_e per cell (in $K\Omega$) of the quadripoles corresponding to the homogeneous layers A and B, the following quantities may be observed in it:

1) The electrical resistances R_{rin} and R_{rf} (in $K\Omega$) equivalent to thermal resistances to radiation in steady state condition when the system is in the initial and final conditions of the direct tests respectively (columns 1 and 2). (In inverse tests, final resistances turn into initial ones and vice-versa);

2) The percentage shifting of R_{rf} from R_{rin} (column 3);

3) The average value of R_{rf} and R_{rin} (column 4);

4) The equivalent electrical resistance $R_{r\,eq}$, defined as the constant resistance that being substituted in the analogue model to the non-linear calculus elements that simulate radiation, i.e., placed between points 2 and 3 in figure 2, allows in point 4 the same response time related to a direct test (column 5). (It is worth noting that R_{eq} cannot be correlated with R_{rin} and R_{rf} in a simple form);

5) Values τ_{0d}, τ_{0i}, τ_{0in}, τ_{0f} and τ_{0m} expressed by:

$$\tau_0 = K \int_0^\infty [V_{4f} - V_4(\tau)] \, d\tau \quad (K, \text{ const.}) \qquad (6)$$

in the case of a direct test, inverse test and direct test when between the terminals 2 and 3, in figure 2, R_{rin}, R_{rf} and R_{rm} are substituted respectively to the computation elements which simulate radiation (columns 6, 7, 9, 10 and 12);

6) The percentage shiftings of τ_{0i} from τ_{0d}, τ_{0f} from τ_{0in}, τ_{0m} from τ_{0d} and from τ_{0i} respectively (columns 8, 11, 13 and 14).

Expression (6) differs from (5) only for the multiplicative constant K $(V_{4f} - V_{4i})$. It is obvious therefore that a direct comparison among the values of τ_{0d}, τ_{0i}, τ_{0in}, τ_{0f} and τ_{0m} has a meaning only if the difference $V_{4f} - V_{4i}$ is equal.

If the electric model is in steady state condition we have:

$$V_4 - V_5 = \frac{R_t}{R_{1,2} + R_r + R_{3,4} + R_t} (V_1 - V_5) \qquad (7)$$

and by introducing into the formula the resistance values corresponding to the initial and final conditions, the difference $V_{4f} - V_{4i}$ is then determined (it should be remembered that in direct tests $V_{4i} = V_5$ and in inverse tests $V_{4f} = V_5$). Such operations have been done in every line in table 2 for τ_0, so that their comparison with τ_{0d} of the same line and with each τ_0 belonging to the same line have been made possible.

Incidentally, it should be kept in mind that in a linear model the response time does not depend upon the input. It follows that τ_{0in}, τ_{0f} and τ_{0m} assume equal values for direct and inverse tests respectively starting from identical initial and final conditions.

Finally, if the real response time is of any interest, the connection between the times of the thermal phenomena and those of the electric ones must be determined.

We shall now analyze the response time of the model in figure 2. Since this question is of a relative nature it is sufficient to know a number proportional to it, i.e. τ_0, which, however, has the advantage of not being affected by implicit errors in determining the scale factors.

We will begin by noting that from the starting instant down to the final steady state condition of a direct test, the resistance R_r, which simulates heat transfer by radiation, undergoes very high drops, going from 80 to 30% of the initial values. (See columns 1, 2, and 3 in table 2.)

The response time was then checked by substituting a constant resistance of equal value to R_{rf} and R_{in} respectively to the analogue circuit for radiation simulation.

In the columns 9, 10, and 11 data, one may observe that the response time reduction oscillates from 40 to 5%. In other words, a percentage variation given to R_r understandably leads to a percentage variation inferior to τ_0, being R_r only partly the circuit's total resistance.

It is further understandable that the response time is an increasing function of the circuit's total resistance. If, therefore, for the variable resistance R_r we substitute respectively the min. value R_{rf} and the max. value R_{rin} that R_r can take up in the unsteady state condition, one can get, respectively, the min. and max. response time obtainable with any other input defined by the same range of the voltage variation for V_1.

An example is given by the response time of the direct and inverse tests where, against the output quantity, the input value undergoes equal and opposite variations. τ_{0i} can assume greater or lesser values than τ_{0d}. The first assumption materialises with small values of R_e and the second with high values of R_e.

It is, however, always true that $\tau_{0f} < \tau_{0i} < \tau_{0in}$ and that $\tau_{0f} < \tau_{0d} < \tau_{0in}$ with some shifting of τ_{0i} in relation to τ_{0d} which seldom exceed 6% and never 10%.

As a last statistical remark one may say that τ_{0m} takes up greater or lesser values than τ_{0d} or τ_{0i}, prevailing the former ones in inverse tests while the latter ones do so in direct tests. However, they never differ more than 5% from τ_{0d} and from τ_{0i}.

SOME CONCLUSIONS

In concluding this study it is worthwhile noting that the analogue circuit in figure 3 enables one to determine the corresponding physical system's non-steady state behaviour, no matter what the input is, if it can be reproduced by electrical means.

Besides radiation, there is no difficulty in also simulating convection and conduction in cavity s_1, provided that the above-mentioned phenomena can be taken as linear. By the same token, it is easy to simulate some internal or superficial generation phenomena by feeding in electrical currents, however varying, into some points of the circuit. This can be obtained by function (voltage) generators followed by voltage-current converters.

It should be noted that such problems do not have an easy analytical solution [3].

Finally it is to be noted that the manifold use of passive networks and active elements of an analogue computer for the simulation of physical processes gives much greater flexibility than the exclusive use of the former ones. It gives, furthermore, considerable simplicity and economy in respect to the simulation on an exclusively analogue computer. This procedure has already been used on other occasions [4] and has yielded some interesting results.

Table 2

V_1	$R_{e/cell}$	1 R_{ri}	2 R_{rf}	3 $\dfrac{R_{rf}-R_{ri}}{R_{ri}}100$	4 R_{rm}	5 R_{req}	6 τ_{0d}	7 τ_{0i}	8 $\dfrac{\tau_{0i}-\tau_{0d}}{\tau_{0d}}100$	9 τ_{0in}	10 τ_{0f}	11 $\dfrac{\tau_{0f}-\tau_{0in}}{\tau_{0m}}100$	12 τ_{0m}	13 $\dfrac{\tau_{0m}-\tau_{0d}}{\tau_{0d}}100$	14 $\dfrac{\tau_{0m}-\tau_{0i}}{\tau_{0i}}100$
20–40 and 40–20	20	1400	255	−81	823	550	61.8	67.4	+9.1	78.8	47.3	−40	70.0	+13.3	+3.9
	50	1400	290	−79	845	750	65.9	64.9	−1.5	74.8	54.8	−26	67.5	+ 2.4	+4.0
	100	1400	317	−77	859	910	67.0	62.9	−6.1	71.6	58.6	−18	66.0	− 1.5	+4.9
	200	1400	350	−75	875	1050	65.5	61.2	−6.6	67.9	59.8	−12	63.5	− 3.1	+3.8
20–30 and 30–20	20	1400	558	−60	979	1050	27.2	27.7	+1.8	29.8	23.1	−22	26.5	− 2.6	−4.3
	50	1400	587	−58	994	1200	29.0	28.0	−3.5	30.2	25.4	−16	27.8	− 4.1	−0.7
	100	1400	630	−55	1020	1250	30.9	28.3	−5.7	30.2	27.0	−11	28.6	− 4.7	+1.1
	200	1400	673	−52	1040	1310	29.9	28.5	−4.7	30.1	27.9	− 7	29.0	− 3.0	+1.8
20–25 and 25–20	20	1400	895	−36	1150	1200	11.4	11.4	+0.2	12.0	10.6	−12	11.3	0.7	−0.9
	50	1400	920	−34	1160	1250	12.3	12.1	−2.1	12.8	11.7	− 9	12.1	− 1.6	+0.5
	100	1400	935	−33	1170	1300	13.2	12.8	−2.7	13.6	12.7	− 7	13.1	− 0.6	+2.2
	200	1400	950	−32	1180	1350	14.0	13.5	−3.2	14.3	13.6	− 5	13.8	− 1.2	+2.1

QUADRIPOLES B AND C ARE DIVIDED IN 5 CELLS; CAPACITY PER CELL: 0.34 μF; $\dfrac{V}{T}=\dfrac{1}{15}\ \dfrac{\text{Volt}}{^\circ\text{K}}$. $R_t=500$ KΩ; $R\eta=2,000$ KΩ (all these data are approximated at the 3rd figure).

186

REFERENCES

[1] P. Di Filippo and G. Zorzini, Misura del coefficiente di trasmissione H e del tempo di risposta τ_0 delle pareti, Lavoro presentato al XXIII Congresso Nazionale A.T.I., Bologna (Sept. 1968).

[2] D.I. Lawson and J.H. McGuire, The solution of transient heat - flow problems by analogous electrical networks. Department of scientific and industrial research and Fire Offices' Committee, Joint Fire Research Organisation (Oct. 1952).

[3] R.D. Zerkle and J. Edward-Sunderland, The transient temperature distribution in a slab subject to thermal radiation, *Journal of Heat Transfer* (Feb. 1965).

[4] G. Zorzini and A. Cavallini, Simulatore di processo per un impianto di evaporazione a multiplo effetto, *La Termotecnica*, n° 1 (Jan. 1969).

PROBLEMS OF ROOM HEAT-UP IN SUMMER

K. GERTIS

Institut für Technische Physik
Aussenstelle Holzkirchen (W. Germany)

Problèmes de l'échauffement des locaux en été

RÉSUMÉ : *L'échauffement des locaux en été est dû à l'influence combinée de l'énergie solaire introduite dans le local et de l'absorption de chaleur par les éléments de la construction entourant le local, ainsi que de la chaleur éliminée par les appareils de conditionnement d'air. Les entrées de chaleur solaire ne sont pas constantes dans le temps mais suivent plutôt (suivant l'heure, la saison et la largeur des fenêtres) une certaine fonction instable de temps donnée par la nature. Il faut donc, lorsqu'on calcule l'échauffement des locaux, utiliser les équations — relativement compliquées — de l'écoulement de chaleur instantané, dans lesquelles, outre d'autres facteurs thermiques, on introduit toutes les capacités spécifiques des éléments de construction entourant le local.*

Jusqu'ici on n'a pas suffisamment prêté attention à l'influence de la capacité thermique des éléments de construction sur l'échauffement des locaux en été. Comme l'a prouvé l'examen de divers cas, la capacité thermique des matériaux dont sont composés les éléments de construction entourant le local a une grande importance qui, pour son influence sur l'échauffement des locaux, peut être considérée comme égale aux dimensions des fenêtres et à la sorte de vitres choisie. Les éléments de construction intérieure lourds à grand volant thermique ont une influence favorable. Il faut, en revanche, considérer comme risquées les constructions légères et les grandes surfaces vitrées. D'une façon générale, il est cependant possible de créer, sans dépenses techniques excessives, des conditions climatiques saines, même dans les murs-rideau modernes. Pour cela, il faut équilibrer de façon appropriée la maçonnerie et les surfaces vitrées.

1. INTRODUCTION

The heating-up of rooms in summer is due to the joint influence of the solar energy supplied to a room, the heat absorbed by the room envelope, and the heat taken away by means of conditioning devices, respectively. The solar heat gains depend, on the whole, on the radiation behavior of the window construction. These gains are not constant in relation to time, but rather follow a certain function of time given by nature (according to hour and season and to the directions of the windows).

Therefore, in computing the heat-up of rooms, the—relatively complicated—instantaneous heat flow equations must be used into which are fed, together with other thermostatic factors, above all the specific capacities of the building elements enveloping the room.

The following is a brief account on the heat transfer through glazing materials into a room, and of the thermal effect due to energy transmission upon the heat-up of rooms.

2. HEAT TRANSMISSION THROUGH FENESTRATION AREAS

Under the aspect of radiation, glasses differ from other building materials by their property of not only absorbing and reflecting, but also of transmitting at a considerable rate, the radiation that falls upon them. This is due to their transparency, and the ratio of the transmitted, absorbed, and reflected radiation depends on the sort of glass used and the angle of incidence of solar radiation. On the whole, only the transmitted and the absorbed radiation contribute to the heat-up of a room, the latter because it leads to a heating of the glass and thus brings about an additional (*secondary*) heat supply by the pane into the room by convection and long wave radiation.

189

Extensive investigations into radiation processes and room heat-up resulting from them were undertaken at the Holzkirchen Research Station. To mark the supply of energy by glasses, the so-called *identification value of glass* (G) was introduced which could be successfully used for the calculation of room heat-up*. The ID-value of glass G represents the percentage of solar energy which falls into a room through a certain sort of glazing, and during one day of radiation. (The continuous change of the angle of incidence of the solar radiation process is taken into consideration here.) According to this definition, the G-values thus vary between G 0 for a—fictious—window that won't let through a bit of energy, and G 100 for an ideal energy transmitting pane. Ordinary clear glass double glazing has an ID-value of about G 60. Shading glasses rank in value between G 25 and G 50. Glazing with effective shading devices can give ID-values below G 25, while clear glass single glazing is characterised by ID-values above G 60.

3. ROOM HEAT BALANCE

The right insight into the conditions of room heat-up in summer is granted by consideration of the room heat balance. The room heat balance results from the joint influence of the supply of solar energy into a room (marked by the G-values), the heat absorption by the building envelope, and the heat taken away by means of conditioning devices, respectively. The theoretical and practical evaluation of such a balance consideration meets with particular difficulties in the present case, because the solar radiation is not constant in relation to time, but rather (according to hour and season and to the direction of the window) follows a certain unsteady function of time given by nature. Therefore, in computing the heat-up of room envelopes, the relatively complicated instantaneous heat flow equations must be used, into which are fed—aside from the thermal conductivities of the building materials— especially their specific heat capacities.

For the mathematical registration of the phenomena of instantaneous heat conduction in building envelopes there exist three possibilities; building envelopes are treated mathematically as:

a) storage cores, i.e. as materials without any heat conduction resistance, however with specific heat capacity. In this case, thermal conductivity is assumed to be infinitely high, whereby anytime during the instantaneous process the same temperature prevails at any point of the enveloping element;

b) semi-infinite solids, very thick walls for instance, of finite thermal conductivity and capacity;

c) flat slabs of finite thickness that are supplied with heat from both sides. This approach can be regarded as the norm for building elements secluding rooms from each other inside building complexes since rooms of one building are usually built in the same way and all rooms limited by the same façade are heated according to the same function of time.

If the increase of indoor temperature caused by solar radiation through a window due south is figured out for a certain room according to the three methods listed, the curves shown at the foot of figure 1 are arrived at. We can see that for the storage

(*) H. KÜNZEL and C. SNATZKE: Neue Untersuchungen zur Beurteilung der Wirkung von Sonnenschutzgläsern auf die sommerlichen Temperaturverhältnisse in Räumen. Glastechnische Berichte 41 (1968), H. 6, pp. 315-325.

core approach, where the heat conduction resistance is neglected, temperature rises are found which are considerably smaller than for the mathematical treatment of the enveloping elements as flat slabs because practically no heat is held back in the zones near the surface and heat can flow off freely into the interior of the building elements.

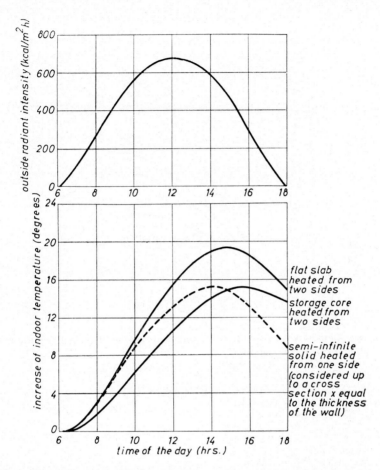

Fig. 1 — Increase of indoor temperature computed according to 3 different methods (room-enclosing elements considered as storage cores, as semi-infinite solids, and as flat slabs). Above the temporal course of outside radiant intensity is indicated.
It was gone out from the following basis:
Dimensions of the room: 4 m × 4 m × 2,5 m
Room-enclosing elements: 20 cm cellular concrete
Window-area : 4 m²
Glazing: G57; window due south
(No heat exchange by ventilation)

This is caused by the immense thermal conductivity of the flat slab ($\lambda = \infty$). The semi-infinite solid (dotted line) shows at first (during the morning hours) the same results as the flat slab, then, however, the heat flow into the interior of the solid becomes conspicuous so that the temperature rise, for this approach, is seen to be considerably lagging in comparison to the increase of temperature computed according to the slab method as the day is advancing. The temperature peak which is of crucial consequence for the heat-up of a room can thus be computed only after the flat slab approach where the instantaneous processes of heat flow are taken care of. This method, however, is somewhat more extensive. In the following the results of some practical examples are discussed in which this slab method was used as the basis of computation.

4. SOME PRACTICAL APPLICATIONS

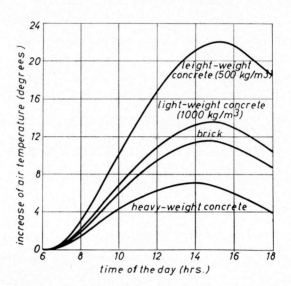

Fig. 2 — Temporal curve of the air temperature of a room inside a building complex whose room-enclosing elements are 15 cm thick and are made up of different building materials. Radiation of the sun at equinox.

Dimensions of the room: 4 m × 4 m × 2,5 m (standard room)
Glazing: Window area: 4 m² (due south)
           Type of glass: clear glass; double glazing G 57

Weights of the building materials:
light-weight concrete: 500 kg/m³
                         1000 kg/m³
brick: 1500 kg/m³
heavy-weight concrete: 2500 kg/m³
(No heat exchange by ventilation)

The heat capacity of the building materials the room envelopes are made from has an essential significance which, in its effect on the heat-up of rooms, must be considered equal to the importance of the size of the windows and the sort of glass chosen. This is made obvious in figure 2. Here, the courses of the indoor temperature rises during a summer day with rich radiation are compared to each other; while the sizes of the rooms and the constructions of the windows are the same, the room envelopes are made of different building materials. Rooms of light-weight construction (i.e. made up of elements out of light-weight concrete) undergo relatively high temperature rises due to the small heat capacity of light materials, while rooms out of heavy building elements of a higher ability to store heat remain much cooler. To give an example, the maximum temperature rise in rooms out of heavy-weight concrete will reach only about one-third of the maximum temperature to be expected in a room out of light-weight concrete elements weighing 500 kg/m³.

In spite of this aspect of heavy indoor constuctions which is unquestionable under the thermostatic aspect, light type elements are preferred —especially in offices and school buildings— for reasons of fitting the parts together and possibly displacing them at some time. In this case it must be thought about means to diminish the heating-up of rooms in summer by reducing the solar energy supplied to a room. Therefore, either the fenestration area can be reduced or a sort of glass and glazing, respectively, can be chosen which transmits less radiation, i.e. which of the two possesses a smaller ID-value G. It is shown in figure 3 where the maximal increase of indoor temperature

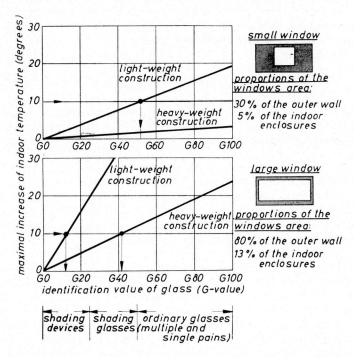

Fig. 3 — Maximal increase of indoor temperatures for a room of light-weight construction and one of heavy-weight construction. Each room has a window due south, different from the other by fenestration area and glazing.

Dimensions of the room: 4 m × 4 m × 2,5 m

Light-weight construction: 20 cm light-weight concrete (500 kg/m³)

Heavy-weight construction: 20 cm heavy-weight concrete (2500 kg/m³)
(Radiation of the sun at equinox; no heat exchange by ventilation)

for a small and a large window are depicted depending on the G-value, that for light-weight and heavy-weight constructions the heating up of rooms increases in a linear way after an increase of the G-value. If, for instance, in a room having a large window due south whose area amounts to 80% of the outer wall (see lower diagram), an increase in temperature of 10 deg C is considered permissible, it is possible to keep within that limit, for heavy-weight construction, by means of a shading glass glazing with an ID-value of G 42; for a room of the light-weight type, a glazing with shading device (ID-value G 12) should be selected. In the case of a small window with an area of 30% of the outer wall (see upper diagram) it has to be dealt with different conditions. Here the permissible limit is never exceeded in a room of the heavy-weight type regardless of what kind of glazing is used. In a room out of light-weight elements it is still possible to choose a G 52 glass which is a glazing that protects from the sun only a little better than an ordinary clear glass double glazing.

5. Conclusions

It can be seen from the foregoing statements that one has at hand a number of purely architectural means, e.g. the choice of interior building materials, use of special glass and shading elements, etc., to solve satisfactorily the problem of the heat-up of rooms in summer. The installation of air conditioning devices should be reasonably accorded with these architectural means. A cooling load calculation for which the listed architectural and mechanical feasibilities must be taken into account and which has to be carried out while a building is only at the planning stage, will help to avoid uncomfortably high indoor temperatures in summer.

DISCUSSION

J. ROTHSCHILD (Israel) — What size of a window is permissible in order to have the same cooling capacity required for a room, taking into account the time of highest heat transfer from the wall and the change in outside temperature?

K. GERTIS — That depends on the material of the room envelope. In practice, when you have heavy indoor elements, the permissible size of a window is about 10% of the area of elements enclosing the room (included ceiling and soil). When you choose a size of window, which surpasses this limit, the indoor thermal conditions become critical. Lightweight indoor constructions must be considered generally as critical, even though you may have a small window area.

G. DE LEPELEIRE (Belgique) — M. GERTIS a calculé les températures de l'air telles qu'elles résultent de différentes constructions des parois et fenêtres. Sans doute, les conclusions pertinentes de l'auteur seraient encore plus convaincantes, si l'on visait non pas la température sèche, mais la température résultante. Celle-ci en effet est déterminée e.a. par le champ de rayonnement des parois qui sera bien plus important (et gênant) avec les constructions légères par exemple, ou les grandes surfaces vitrées.

K. GERTIS — Le calcul de la température d'air des chambres avec des fenêtres assez grandes représente un premier pas. Quant à la température résultante, nous avons quelques hésitations, parce qu'on a dans des chambres avec des grandes baies vitrées, des différences très grandes de température entre les surfaces vitrées et les surfaces des parois ou du sol ou du plafond. En été les baies sont assez chaudes, les parois assez fraîches. En hiver cette situation est renversée. Mais dans les deux cas la température résultante de tout l'environnement est environ la même malgré ces asymétries. Quoi qu'il en soit, M. de LEPELEIRE, je vous consens, que la considération de la température résultante est importante, mais je crois, jusqu'aujourd'hui que ces critères fixes, qui permettraient d'estimer exactement les conditions asymétriques, manquent.

Voir aussi — *See also*: DISCUSSION pp. 215-216.

LA CAPACITÉ THERMIQUE DES BÂTIMENTS

J. KOZIERSKI

Institut de la Physique Technique et de l'Equipement Technique des Bâtiments,
Varsovie (Pologne)

Heat capacity of buildings

SUMMARY: *In designing building air-conditioning—especially for summer operation—it is usually accepted that excess heat in the rooms is absorbed by the conditioned air. Generally it is not taken into account that the walls surrounding the conditioned room absorb heat gains that vary during the day; via this way, the walls decrease the air-conditioning cooling loads.*
These calculations are explained and, according to the A's observations in a building in Warsaw, large savings were realised with the air-conditioning.
The report also describes a method for checking the calculations with a computer.

Quand on projette la climatisation d'un bâtiment, on admet que *le débit maximal de l'installation du refroidissement* doit être *égal à la somme des apports de chaleur maximaux.*

Ces apports maximaux de chaleur sont calculés en fonction des paramètres maximaux de l'air dans les locaux (température de l'air t_i max $+26\,°C$, et humidité de l'air :

$$HR\ i\ max \leqslant 55\%).$$

Les apports de chaleur proviennent des sources suivantes :

— des occupants;
— de l'éclairage;
— des installations mécaniques;
— de la conduction de chaleur par les parois;
— de l'insolation par les parois opaques et les parois transparentes.

De ces sources de chaleur, les trois premières ne dépendent pas du climat extérieur.
Ces trois sources peuvent donc être définies comme *sources à caractère constant.*
Les sources restantes dépendent du climat extérieur et des changements de ce climat. On peut définir ces sources comme *sources à caractère périodiquement variable.*
La chaleur émise par convection constitue la charge thermique de l'air de climatisation immédiate et effective.
Cette chaleur constitue une réelle demande de refroidissement de l'air dans le local examiné.
Par contre la chaleur émise ou transmise par rayonnement réchauffe les objets solides (murs, planchers, plafonds, meubles, équipements, matériaux, etc.).
Ensuite cette chaleur est transmise à l'air, si la température des objets est plus élevée que celle de l'air.
En conséquence, la chaleur transmise par rayonnement ne constitue pas une immédiate charge d'air climatisé et cette chaleur parvient à cet air indirectement, quelque temps après et souvent en quantité réduite.
Sur la base de ces constatations un projet de climatisation fût exécuté en 1964 dans un grand magasin à Varsovie.
Le principe de l'accumulation de chaleur fût contrôlé pendant l'été 1965.
La figure 1 présente les variations, par 24 heures, des températures de l'air extérieur, des températures de l'air intérieur et des températures des parois dans une pièce climatisée, ainsi que les changements de quantité de chaleur accumulée par ces parois.

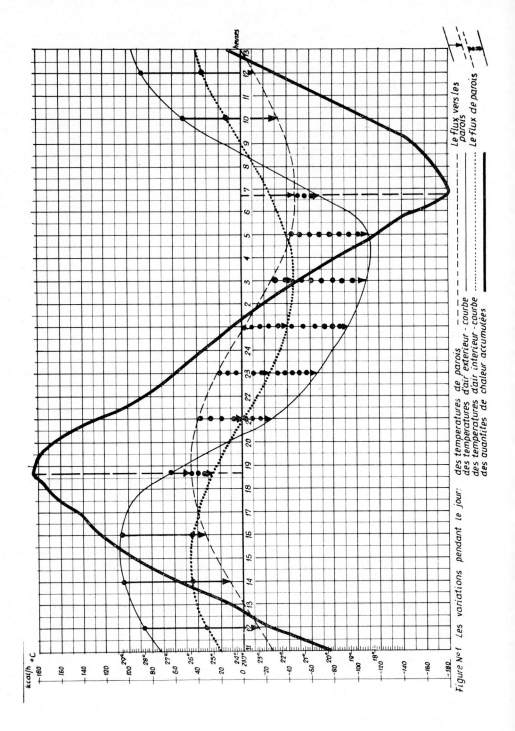

kcal/h °C

Figure N°1 Les variations pendant le jour: des temperatures de parois
des temperatures d'air exterieur - courbe
des temperatures d'air interieur - courbe
des quantites de chaleur accumulées

Le flux vers les parois
Le flux de parois

De ces considérations il résulte, pour les projets de climatisation, que le principe, consacré par l'usage, suivant lequel tous les gains de chaleur sont immédiatement transmis à l'air, et ensuite évacués par l'air de climatisation, est en désaccord avec la réalité.

Selon les données citées par O. O. Mackey et N. R. Gay [3], la chaleur transmise se présente comme suit :

— pour les occupants : 50 % par convection et 50 % par rayonnement;
— pour l'éclairage : 32,9 % par convection et 67,1 % par rayonnement (20 % absorbés par le plafond et 47,1 % par le plancher);
— pour les équipements mécaniques, on évalue la part transmise par *convection à* 90 %, la part transmise par *rayonnement à* 10 %;
— la chaleur transmise par conduction, à travers les parois externes, et par insolation à travers les parois opaques, fait intervenir le coefficient de transmission $a_t = 7$ kcal/m². h. °C (normes polonaises). De cette valeur, le coefficient de convection se définit :

$$a_c = 1,57 \sqrt[4]{\Delta t} = \text{ca. } 2,5$$

le coefficient de rayonnement se définit : $a_r = $ca. 4,5. En conséquence la part de chaleur transmise par *convection* est de 36 %, par *rayonnement de* 64 %;
— enfin la chaleur transmise par l'insolation à travers les parois transparentes, est d'abord absorbée par les objets solides presque exclusivement par rayonnement. En somme, pour tous les gains de chaleur, on peut établir, en se basant sur quelques exemples, que *la part* de ces *gains par convection*, est d'environ 25 %, et la part *par rayonnement* d'environ 75 %.

Si on admet, de façon erronée, que le débit de refroidissement de ces installations doit être égal aux gains momentanés maximaux de chaleur, on est conduit à réaliser des installations frigorifiques de trop grandes dimensions et trop coûteuses.

Évidemment tous les gains de chaleur d'un local, admis en 24 heures, doivent être absorbés par la climatisation.

Cependant, comme la plus grande partie de ces gains est accumulée par les parois et les objets qui se trouvent dans le local, l'installation frigorifique peut être dimensionnée non sur la valeur maximale, mais seulement sur une valeur moyenne du jour.

En mettant à profit le phénomène d'accumulation de chaleur par les parties solides

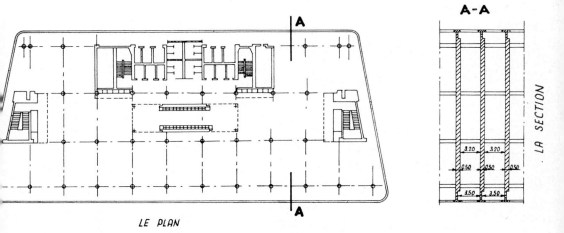

Fig. 2 — Le plan et la section verticale d'un étage des grands magasins à Varsovie — climatisé selon la méthode proposée.

d'un local, on peut faire des économies, en organisant le refroidissement d'un bâtiment pendant la nuit (entre 23 et 6 heures) quand la température de l'air extérieur est plus basse que la température moyenne du jour. On réalise le refroidissement de ces parties solides en augmentant durant cette période la ventilation mécanique du bâtiment, alimentée par l'air froid extérieur. L'installation est beaucoup moins coûteuse d'investissement et d'exploitation.

En s'appuyant sur ces observations, on a élaboré le projet de climatisation d'un étage dans un grand magasin à Varsovie.

La figure 2 représente le plan et la section verticale de cet étage.

Dans ce magasin, édifié il y a une quinzaine d'années, on a réalisé une ventilation mécanique sans refroidissement d'air. Comme le bâtiment est amplement vitré la température intérieure de l'air pendant l'été dépassait souvent la zone de confort.

La réalisation d'une climatisation centralisée dans un bâtiment existant serait très difficile et très coûteuse. Aussi, nous avons projeté le refroidissement de l'air dans la salle examinée, à l'aide de 7 climatiseurs individuels.

Outre cela on a projeté une adaptation de la ventilation existante, avec un taux de renouvellement d'air de 2 volumes par heure (V = 1 600 m^2 × 3,2 m × 2 = 10 240 m^3. h), ce qui donne pour 240 personnes dans la salle, un volume d'air de 10 240 : 240 = 40 m^3 par heure et par personne.

Le tableau 1 contient les calculs de base du bilan thermique de la climatisation discutée.

La méthode de calcul pour ce projet se présentait comme suit :

1.   On a admis les températures maximales de l'air extérieur en été à Varsovie pour une période de 24 heures (t_{et}, rubrique 2 du tableau 1).

La figure 3 présente les diagrammes de variation de la température maximale de l'air extérieur pendant le jour selon deux alternatives : 1^0 selon les données météorologiques (première courbe) et 2^0 en partant du principe que ces changements sont exactement harmoniques, selon la relation : $t_{et} = t_{em}$ · sin A (deuxième courbe);

$$t_{em} = +23,7°C$$

A : angle fonction de l'heure du jour (= 0 à 9 h du matin).

Comme on le voit sur cette figure, les deux courbes ne diffèrent que d'une façon insignifiante; on peut donc admettre les variations de la température de l'air extérieur comme harmoniques.

2.   Ensuite nous calculons les variations de la température de l'air maximales (t_{it}) dans la salle, en fonction des éléments suivants :

a)   on admet que la température moyenne de l'air intérieur est la même que la température maximale moyenne extérieure :

$$t_{itm} = t_{etm} = +23,7°C$$

b)   on admet comme température maximale de l'air intérieur :
$t_{itmx} = +26°C$, température assurant un confort satisfaisant dans un magasin ;

c)   on admet que les variations de la température de l'air intérieur sont harmoniques, conformément à la relation

$$t_{it} = t_{itm} + (t_{itmax} - t_{itm}) · sin B$$

Selon nos observations, la température t_{it} est maximale à 15 heures comme la température t_{et} de l'air extérieur.

De ces observations, il résulte que l'angle B = 0 à 9 heures, et = 90° à 15 heures.

198

Tableau N° 1
les calculs des dispositifs frigorifiques pour la climatisation d'une salle dans les magazins à Varsovie

N°	temp. d'air extérieur t_{et} °C	temp. d'air intérieur t_{it} °C	la différence $\Delta t = t_{et}-t_{it}$ °C	les gains de chaleur — par transmission Q_t kcal/h	par ventilation Q_v kcal/h	par insolation Q_i kcal/h	du peuplement Q_p kcal/h	des dispositifs mechaniques Q_d kcal/h	éclairage électrique Q_e kcal/h	la somme des gains Q_h kcal/h	le rendement des dispositifs frigorifique kcal/h	les variations du flux de chaleur ΔQ kcal/h	la chaleur accumulée kcal	temperat. des parois interieur t_{ip}°C	la différence $\Delta t' = t_{it}-t_{ip}$ °C
1	2	3	4	5	6	7	8	9	10	11	12	13	14	15	16
1	19.2	21.7	-2.5	-4950	-7950	-	-		+3600	-9300	-18570	-27870	+10525	+23.7	-4.5
2	18.6	21.5	-2.9	-5750	-9200	-	-		+3600	-11350	-18570	-29920	-19395	+23.1	-4.5
3	18.3	21.4	-3.1	-6150	-9850	-	-		+3600	-12400	-18570	-30970	-50365	+22.5	-4.2
4	18.2	21.5	-3.3	-6550	-10450	-	-		+3600	-13400	-18570	-31970	-82335	+22.1	-3.9
5	18.2	21.7	-3.5	-6950	-11050	-	-		+3600	-14400	-18570	-32970	-115305	+21.7	-3.5
6	19.1	22.1	-3.0	-5950	-9500	-	-		+3600	-11850	-18570	-30420	-145725	+21.5	-2.4
7	21.0	22.5	-1.5	-2950	-4750	+17700	-		+3600	+13600	-37140	-23540	-169265	+21.4	-0.4
8	22.6	23.1	-0.5	-1000	-1600	+21000	-		+3600	+22000	-37140	-15140	-184405	+21.5	+1.1
9	24.5	23.7	+0.8	+1600	+2550	+35500	+13700	+6900	+3600	+63850	-37140	+26710	-157695	+21.7	+2.0
10	26.2	24.3	+1.9	+3750	+6000	+41900	+13700	+6900	+3600	+75850	-37140	+38710	-118985	+22.1	+4.1
11	27.3	24.9	+2.4	+4750	+7600	+43700	+13700	+6900	+3600	+80250	-37140	+43110	-75875	+22.5	+4.8
12	28.0	25.3	+2.7	+5350	+8550	+40200	+13700	+6900	+3600	+78300	-37140	+41160	-34715	+23.1	+4.9
13	28.5	25.7	+2.8	+5550	+8900	+43700	+13700	+6900	+3600	+82350	-37140	+45210	+10495	+23.7	+4.8
14	28.9	25.9	+3.0	+5950	+9500	+42200	+13700	+6900	+3600	+81850	-37140	+44710	+55205	+24.3	+4.6
15	29.1	26.0	+3.1	+6150	+9850	+35500	+13700	+6900	+3600	+75700	-37140	+38560	+93765	+24.9	+4.2
16	29.1	25.9	+3.2	+6350	+10150	+24600	+13700	+6900	+3600	+65300	-37140	+28160	+121925	+25.3	+3.8
17	28.8	25.7	+3.1	+6150	+9850	+17600	+13700	+6900	+3600	+57800	-37140	+20660	+142585	+25.7	+3.1
18	28.0	25.3	+2.7	+5550	+8550	+26500	+13700	+6900	+3600	+64600	-37140	+27460	+170045	+25.9	+2.1
19	26.5	24.9	+1.6	+3150	+5050	+19100	+13700	+6900	+3600	+51500	-37140	+14360	+184405	+26.0	+0.5
20	24.6	24.5	+0.3	+600	+950	-	+13700	+6900	+3600	+25750	-37140	-11390	+173015	+25.9	-1.3
21	22.5	23.7	-1.2	-2400	-3800	-			+3600	-2600	-37140	-39740	+133275	+25.7	-3.2
22	21.5	23.1	-1.8	-3550	-5700	-			+3600	-5650	-37140	-42790	+90485	+25.3	-4.0
23	20.5	22.5	-2.0	-3950	-6350	-			+3600	-6700	-18570	-25270	+65215	+24.9	-4.4
24	19.8	22.1	-2.3	-4550	-7300				+3600	-8250	-18570	-26820	+38395	+24.3	-4.5
	568.8 :24= =23.7°C	568.8 :24= =23.7°C	-27.6 +27.6 $\Sigma\Delta t$	-54700 +54700 =0	-87500 +87500 =0	+409200	+164400	+82800	+86400	742800	742800	+368810 -368810 =0		568.8 :24= =23.7°C t_{ipm}	-40.8 +40.8 $\Sigma\Delta t'=0$

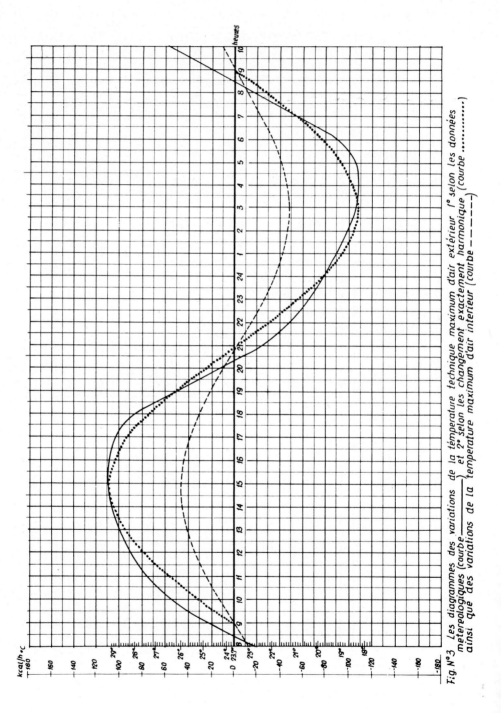

Fig. N°3 Les diagrammes des variations de la température technique maximum d'air extérieur 1° selon les données
météorologiques (courbe ⎯⎯⎯) et 2° selon les changement exactement harmonique (courbe ....)
ainsi que des variations de la température maximum d'air intérieur (courbe ⎯ ⎯ ⎯ ⎯)

d) sur la figure 3, la troisième courbe représente les variations de température de l'air intérieur (t_{it}) pendant 24 heures; dans le tableau 1 les valeurs de t_{it} sont données dans la rubrique 3.

3. Après l'établissement des variations de températures de l'air extérieur et intérieur, on calcule les gains de chaleur de diverses sources, à savoir :

a) les gains de chaleur de conduction par les parois extérieures; dans notre salle toutes ces parois sont vitrées doublement (coefficient de transmission K = 3, selon les normes polonaises); comme la surface des parois extérieures est $F_e = 200 \times 3,3 = 660 \text{ m}^2$, les gains de chaleur de cette source se chiffrent à :

$$Q_t = F_e \cdot K(t_{et} - t_{it}) = F_e \cdot K \cdot \Delta t = 660 \cdot 3 \cdot \Delta t = 1980 \cdot \Delta \ ,$$

les valeurs de Q_t pendant 24 heures sont données dans la rubrique 5 du tableau 1;

b) on a admis pour la salle une ventilation d'extraction mécanique et l'arrivée d'air frais directement de l'extérieur, par les ouvertures des parois externes; le taux de renouvellement d'air admis est de deux volumes par heure; comme le cube de la salle est :

$$V_i = 1600 \text{ m}^2 \times 3,2 \text{ m} = 5120 \text{ m}^3,$$

les gains de chaleur de cette source se chiffrent à :

$$Q_v = 2V_i \cdot c(t_{et} - t_{it}) = 2 \cdot 5120 \cdot 0,31 \cdot \Delta t = 3170 \cdot \Delta t;$$

où $c = 0,31$ — chaleur spécifique de l'air comptée par m^3 d'air. Les valeurs de Q_v pendant 24 heures sont données dans la rubrique 6 du tableau 1;

c) dans la rubrique 7 de ce tableau, sont donnés les gains de chaleur par insolation des *parois externes transparentes*; la salle ne possède pas de parois externes opaques; ces gains ont leur maximum à *13 heures*;

d) les rubriques 8, 9 et 10 du tableau 1 contiennent les gains de chaleur provenant des sources suivantes :

rubrique 8 — occupants de la salle (chaleur sensible), entre 8 et 20 heures, avec une valeur moyenne maximale de 240 personnes [7] :

$$Q_p = 240 \times 57 \text{ kcal/h} = 13\,700 \text{ kcal/h};$$

rubrique 9 — équipement mécanique, entre 8 et 20 heures; 14,3 kW [7] :

$$Q_d = 14,3 \times 860 \cdot \cos F \cdot 0,7$$

$$= 14,3 \times 860 \cdot 0,8 \cdot 0,7 = 6\,900 \text{ kcal/h};$$

rubrique 10 — éclairage électrique; cet éclairage fonctionne tout le jour sur une surface de 300 m^2 avec une puissance moyenne maximale 14 W/m^2 de plancher :

$$Q_e = 300 \text{ m}^2 \times 14 \times 0,860 = 3\,600 \text{ kcal/h};$$

e) la rubrique 11 du tableau 1 contient la somme des gains de chaleur pour chaque heure du jour; la somme maximale est :

$$Q_{h\ max} = 82\,350 \text{ kcal/h à 13 heures.}$$

La somme des gains de chaleur pour tout le jour est :

$$Q_j = 742\,800 \text{ kcal.}$$

4. Après avoir calculé la somme des gains de chaleur pour tout le jour Q_j, nous pouvons déterminer la puissance frigorifique des équipements pour le refroidissement de l'air.

Cette puissance est minimale si l'on admet que ces équipements vont travailler pendant tout le jour :

$$Q_{r\,min} = Q_j : 24 \text{ heures} = 742\,800 : 24 = 30\,950 \text{ kcal/h}.$$

Si nous admettons un temps de travail des équipements plus court, la puissance de ces équipements devra être plus grande, par exemple :

a) pour un temps de travail de 18 heures

$$Q_r = Q_j : 18 = 742\,800 : 18 = 41\,300 \text{ kcal/h};$$

b) pour un temps de travail de 12 heures

$$Q_r = Q_j : 12 = 74\,800 : 12 = 61\,900 \text{ kcal/h}.$$

5. Pour le temps du travail des équipements pour le refroidissement d'air nous admettons une méthode spéciale.

Dans la salle examinée, nous avons admis que les variations de la température de l'air pendant le jour étaient périodiques et harmoniques, depuis $+21,4\,°C$ jusqu'à $+26\,°C$; la température moyenne de cet air se définit à $t_{itm} = \pm 23,7\,°C$. En conséquence, l'amplitude des variations de la température maximale de l'air intérieur est :

$$A_t = 26 - 23,7 = 23,7 - 21,4 = 2,3\,°C \text{ par jour}.$$

Selon la théorie du flux discontinu de chaleur, élaborée par A. M. Schklover [4 et 5], la relation entre les variations du flux de chaleur A_q et les variations de la température de l'air intérieur A_t dépend de trois facteurs :

a) avant tout des qualités thermiques des parois, et notamment des coefficients d'accumulation de chaleur par la surface de ces parois (B kcal/m². h. °C), des coefficients d'absorption de chaleur par la surface des couches extérieures de ces parois (Y kcal/m². h. °C) et des coefficients d'absorption de chaleur par les matériaux qui forment ces couches particulières :

$$S = \sqrt{2\pi K_m c_m g_m : Z} \text{ kcal/m². h. °C}$$

où K_m définit la conductibilité du matériel en kcal/m. h. °C; c_m chaleur spécifique, en kcal/kg. °C; g_m son poids spécifique en kg/m³ et Z, la période de variation en heures (Z = 24 heures);

b) ensuite de la capacité thermique de l'air intérieur ($V_a \cdot C_a$) où V_a est le cube d'air dans la salle en m³ et C_a la chaleur spécifique de l'air (en kcal/m³), pour une unité de période de variation harmonique (2 $V_a C_a$: Z);

c) enfin, si le local est ventilé, notre relation ($A_q : A_t$) dépend de la capacité thermique du volume d'air renouvelé pendant une heure (produit L_a m³/h $\cdot C_a$ kcal/m³. °C où L_a est le volume d'air renouvelé pendant une heure et C_a la chaleur spécifique de cet air).

Selon Schklover, la relation entre les variations du flux de chaleur (A_q) échangé entre l'air du local et les parois, et les variations de la température de cet air se présente comme suit :

$$A_q : A_t = P = \Sigma BF + \frac{2\pi}{Z} \cdot V_a \cdot C_a \cdot i + L_a \cdot C_a \qquad (1)$$

où $i = \sqrt{-1}$; F, la surface de chaque paroi; Σ BF, la somme des produits des coefficients B et des surfaces F, pour toutes les parois. Le coefficient B, coefficient d'accumulation thermique de la surface de la paroi, se définit comme suit d'après la théorie élaborée par Schklover :

$$B = \frac{Y\sqrt{i}}{a_p + Y\sqrt{i}}; \qquad \text{où} \qquad \sqrt{i} = \sqrt{\sqrt{-1}}; \qquad (2)$$

a_p : coefficient de transmission de chaleur entre l'air intérieur et les parois.
Aux relations 1 et 2, s'applique la solution de l'équation différentielle de Fourier pour le flux de chaleur discontinu et unidimensionnel, à savoir :

$$a \cdot \frac{\partial^2 t}{\partial x^2} = \frac{\partial t}{\partial z} \qquad \text{où} \qquad a = k/g \cdot c. \qquad (3)$$

La solution de cette équation fut donnée :

a) en France par L. Nisolle en 1931;

b) en Russie par A. M. Schklover en 1952 [4];

c) aux États-Unis par C. O. Mackey et L. T. Wright en 1954.

On a adopté la méthode de *Schklover simplifiée*, où les relations principales se présentent comme suit :

$$A_q : A_t = P = BF + \frac{2\pi}{Z} V_a C_a + L_a C_a \qquad (4)$$

$$B = m \cdot a_p \cdot Y : (a_p + Y) \qquad \text{où} \qquad m = 1,02 \text{ jusqu'à } 1,08. \qquad (5)$$

Pour les parois intérieures d'une structure lourde, ce qui est le cas dans les salles des grands magasins (plafonds massifs par suite de grandes charges, murs intérieurs massifs vu la sécurité contre l'incendie), le coefficient Y (absorption de chaleur par la surface extérieure des parois) est égal au coefficient $S = \sqrt{2\pi c\,\overline{kg} : z}$. (6)

6. En définitive, on obtient une installation avec les caractéristiques suivantes :
— variations de la température de l'air intérieur :

$mu - A_t = +2,3\,°\text{C}$ (depuis $t_{imim}^{\daleth} = 21,4\,°\text{C}$ par

$t_{im} = +23,7\,°\text{C}$ jusqu'à

$t_{imx} = +26\,°\text{C}$ a 15 heures, en même temps que

$t_{emx} = +29,1\,°\text{C}$)
(rubrique 2 du tableau 1)
— variations du flux de chaleur maximale :

$A_{q\,max} = 45\,210$ kcal/h à 13 heures

(rubrique 13 du tableau 1);

— absorption de chaleur par les parois et par l'air de la salle de 8 à 19 heures :
$+184\,405$ kcal, perte de chaleur, de 19 à 8 heures : $-184\,505$ kcal
(rubrique 14 du tableau 1);

— variations de la température des parois intérieures t_{ip} avec une valeur maximale de cette température $t_{ip\,max}$ à 19 heures; comme la chaleur dans la salle est surtout absorbée par rayonnement, la température des parois intérieures est au moins égale à la température de l'air intérieur; d'où :

$$t_{ip\,max} \geqslant t_{ia\,max} = +26\,°C$$

$$t_{ip\,min} \geqslant t_{ia\,min} = +21,4\,°C$$

(rubrique 15 du tableau 1 et figure 1);

— outre les équipements frigorifiques (7 climatiseurs individuels) on a projeté pour le local une ventilation mécanique d'extraction, avec un taux de renouvellement d'air de 2 volumes par heure.

* * *

Pendant l'été 1965, on a suivi le fonctionnement de la salle et on a constaté ce qui suit :

— la température maximale de l'air intérieur est atteinte au même moment que la température maximale de l'air extérieur et pendant l'insolation, le plus souvent à 15 heures;

— la température maximale des parois intérieures a lieu à la fin de l'absorption de chaleur par ces parois; elle est au moins égale à la température maximale de l'air intérieur;

— les jours où les gains de chaleur dans la salle examinée sont plus grands, la température moyenne de l'air intérieur et des parois intérieures augmente; dans les cas inverses, ces températures moyennes diminuent.

En outre, on a calculé, à l'aide d'une calculatrice électronique, les variations des températures extérieures d'insolation, des températures intérieures de l'air et des parois intérieures. Comme résultat de ce calcul, on a trouvé que la température maximale des parois intérieures a lieu à 19 heures, ce qui est conforme aux résultats de nos évaluations renfermées dans le tableau 1.

Cette constatation a démontré que la méthode admise pour le projet de climatisation est bien fondée.

RÉFÉRENCES

[1] F.C. HOUGHTEN, J.L. BLACKSHAW, E.M. PUGH, and P. Mc DRMOTT, Heat transmission as influenced by heat capacity and solar radiation. *Transactions of A.S.H.V.E.* (1932) vol. 38, p. 231.
[2] C.E.A. WINSLOW, L.P. HERRINGTON, and R.J. LORENZI, "The influence of heat capacity of walls in interior thermal conditions and heat economy". *Transactions of A.S.H.V.E.* (1945), vol. 51, p. 197.
[3] C.O. MACKEY and N.R. GAY, "Heat gains are not cooling Loads". *Transactions of A.S.H.V.E.*, (1949), vol. 55, p. 415.
[4] A.M. SCHKLOVER, « Tieploustrojcziwost zdanij », « La statique thermique des bâtiments ». *Académie d'Architecture de l'URSS*, Moscou 1952.
[5] A.M. SCHKLOVER, B.F. WASILIEFF et F.B. USCHKOFF, « Osnowy stroitielnoi tieplotiechniki », « Les principes de la technique thermique des bâtiments. *Académie d'Architecture de l'URSS*, Moscou 1956.
[6] Institut des Installations aux Bâtiments de l'École Polytechnique de Varsovie — Le projet des dispositifs frigorifiques et de ventilation dans les grands magasins centraux à Varsovie.
[7] A. DESPLANCHES, *Guide chauffage, ventilation, conditionnement d'air*, Paris 1959.

DISCUSSION pp. 215-216.

ACTUAL COOLING LOADS FROM SOLAR RADIATION THROUGH WINDOWS OF AIR-CONDITIONED ROOMS

B. TODOROVIĆ

The University, Belgrade (Yugoslavia)

Besoins de froid réels, dus au rayonnement solaire à travers les fenêtres des locaux climatisés

RÉSUMÉ : *Pour déterminer avec précision les besoins de froid réels, il faut tenir compte de l'accumulation de chaleur dans la construction.*
L'expression mathématique de ce problème s'appuie sur le processus physique de transfert de chaleur selon lequel les rayons solaires transmis à travers les fenêtres frappent la surface des parois intérieures, le sol, le plafond, etc., où ils sont en partie absorbés et en partie réfléchis. La quantité de chaleur absorbée provoque une élévation de la température superficielle de la paroi par rapport à la température interne de la paroi et à l'air ambiant. Les différences de température provoquent un écoulement de chaleur par conduction dans les parois et par convection dans l'air ambiant.

La représentation mathématique de ce phénomène s'appuie sur les points suivants :
1. *La paroi observée sépare deux locaux climatisés qui sont à la même température constante t_i.*
2. *On a pris le cas où les deux surfaces de la paroi observée absorbent des quantités égales de chaleur, dues à l'insolation.*
3. *Les coefficients de transfert de chaleur par convection et par conduction α et λ sont constants.*
4. *La paroi observée est constituée par un matériau homogène.*
5. *La conduction de chaleur dans la paroi suit une direction perpendiculaire à la surface de la paroi.*
6. *L'épaisseur de la paroi observée est 2δ.*
7. *Le rayonnement solaire frappant les surfaces de la paroi (fig. 2) est représenté par les relations suivantes :*
Pour une surface de paroi $x = 0$

$$F_{(\tau)} = f \cdot I_d \cdot \sin \omega_d (\tau - \tau_d) + f \cdot I_D \cdot \sin \omega_D (\tau - \tau_D)$$

Pour une surface de paroi de $x = \delta$

$$F_{(\tau)} = 0$$

(I_d — rayonnement solaire diffus et I_D — rayonnement solaire direct à travers les vitres; f — partie du rayonnement direct et diffus absorbée par la surface de la paroi.)

INTRODUCTION

In air-conditioning design, it is a special problem to calculate actual cooling loads from solar radiation through windows, taking into account the effect of heat storage in the building structure. There is no existing method which would be acceptable for general application in air-conditioning design.

In this direction, some well known firms have done a lot of work but their practice is based on some suppositions which are unsatisfactory and represent only a rough simplification, so that they cannot be accepted without reserve.

The above mentioned remarks primarily refer to the following:
1. The coefficients of heat transfer by conduction are taken as infinite.
2. It has been assumed that all parts of internal walls, floors and ceilings absorb equal quantities of solar radiation energy.

The first of these suppositions neglects the temperature gradient in buildings materials. The second is not based on fact since the distribution of solar energy passing through windows (even if the windows are fitted with screens) depends primarily upon reciprocal geometrical forms of the room and the window. Besides, the absorbed part of the energy depends upon the physical properties of the materials of walls, floors and ceilings. (The author is currently measuring this distribution in geometrically different rooms; some results are available.)

The aim of this work is to provide factual coefficients of thermal conductivity and to give calculation methods. Via experience through measuring the distribution of solar energy in rooms, these data would theoretically and practically represent a satisfactory way of calculating cooling loads.

PHYSICAL EXPRESSION OF THE PROBLEM

The physical process consists of heat transfer in which the solar rays transmitted through windows strike surface of internal walls, floors, and ceilings, where they are partly absorbed and partly reflected. The absorbed quantity of heat gain increases the temperature of the walls (floors, ceilings) surface above the temperature inside the wall and the surrounding air. The established differences in temperature cause heat flow by conduction into the wall material and by convection into the surrounding air.

The mathematical representation of this phenomenon for internal walls is based upon the following:

1. The observed internal wall separates two air-conditioned rooms which have equal and constant temperatures (fig. 1).

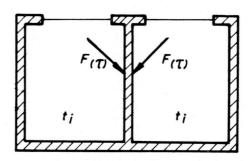

Fig. 1 — Position of observed wall.

2. In the treated case both surfaces of the observed wall absorb equal quantities of heat from solar radiation $F_{(\tau)}$.

3. The coefficients of heat transfer by conduction and convection (λ and α) are constant.

4. The observed wall is made of a homogeneous material.

5. The solar radiation which is absorbed by the surfaces of the wall is, in general, represented by the relationship:

$$F_{(\tau)} = f \cdot I_d \cdot \sin \omega_d(\tau - \tau_d) + f \cdot I_D \cdot \sin \omega_D(\tau - \tau_D)$$

206

with time limits :

$$\tau_d \leqslant \tau \leqslant \tau_d + \frac{T_d}{2}$$

$$\tau_D \leqslant \tau \leqslant \tau_D + \frac{T_D}{2}$$

6. The heat conduction into the wall is in one direction and normal to the wall surface.
7. The thickness of the observed wall is 2 δ.

MATHEMATICAL EXPRESSION OF THE PROBLEM

The basic equation is Fourier's differential equation (2) for heat conduction under unsteady state :

$$\frac{\partial t}{\partial x} = a^2 \frac{\partial^2 t}{\partial x^2} \tag{2}$$

The boundary condition for surface $x = 0$ is (fig. 2):

$$-\lambda \frac{\partial t}{\partial x} + \alpha(t - t_i) \bigg|_{x=0} = F_{(\tau)} \tag{3}$$

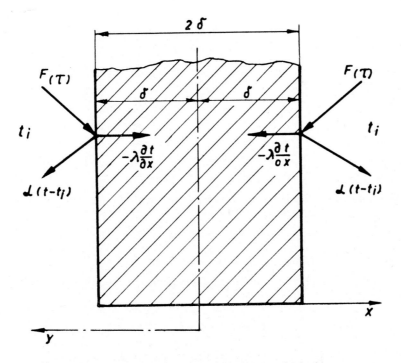

Fig. 2 — Observed wall with heat balance on both sides.

The second boundary condition is taken for $x = \delta$, based on the symmetry of the problem :

$$-\lambda \frac{\partial t}{\partial x}\bigg|_{x=\delta} = 0 \tag{4}$$

The initial condition is based on the assumption that the temperatures in the rooms on both sides of the observed wall are constant during the whole 24 hours. Thus, it can be assumed that, at the commencement of solar radiation ($\tau = \tau_0 = \tau_d$), the temperatures inside the observed wall are equal to the temperature of the rooms and are independent of the coordinate x:

$$t_{(x, \tau_0)} = t_i = \text{const.} \tag{5}$$

The solution*

In order to obtain a simple proper function, instead of the variable x, y was introduced, to give the dependency:

$$y = \delta - x \tag{5'}$$

The transformed equation (2) is now:

$$\frac{\partial t}{\partial \tau} = a^2 \frac{\partial^2 t}{\partial y^2} \tag{6}$$

and boundary conditions (3) and (4) after transformation and arrangement become:

$$\frac{\partial t}{\partial y} + b \cdot t \bigg|_{y=\delta} = \psi_{(\tau)} + k \tag{7}$$

$$\frac{\partial t}{\partial y}\bigg|_{y=0} = 0 \tag{8}$$

where:

$$\psi_{(\tau)} = \frac{F_{(\tau)}}{\lambda} \tag{9}$$

$$k = \frac{\alpha}{\lambda} t_i = b \cdot t_i \tag{10}$$

The solution for equation (6) including conditions (7), (8), and (9) are given in the form:

$$t_{(y,\tau)} = U_{(y,\tau)} + V_{(y,\tau)} \tag{11}$$

The determination of function $U(y, \tau)$

The function $U(y, \tau)$ has to accomplish boundary conditions (7) and (8), and should be in the following form:

$$U_{(y, \tau)} = C \cdot y^2 [\psi_{(\tau)} + k] \tag{12}$$

(*) The author is grateful to Mr. Djordje Djukić for assisting him in mathematical treatment of the problem.

208

By substituting (12) in (7) the constant C becomes:

$$C = \frac{1}{b\delta^2 + 2\delta} \tag{13}$$

For further calculation the following form of $U(y, \tau)$ was introduced:

$$U_{(y, \tau)} = C \cdot \phi_{(\tau)} \cdot y^2 \tag{14}$$

where:

$$\phi_{(\tau)} = \psi_{(\tau)} + k \tag{15}$$

The determination of function $V(y, \tau)$

The function $V(y, \tau)$ has to accomplish both boundary and initial conditions. Substituting (14) into (11), the following is obtained:

$$t_{(y, \tau)} = C \cdot \phi_{(\tau)} \cdot y^2 + V_{(y, \tau)} \tag{16}$$

Differentiating (16) and substituting in Fourier's equation (6) we get:

$$\frac{\partial V}{\partial \tau} = a^2 \frac{\partial^2 V}{\partial y_2} + 2a^2 \cdot C \cdot \phi_{(\tau)} - C \cdot \phi'_{(\tau)} \cdot y^2 \tag{17}$$

In order to determine function $V(y, \tau)$, the supposed form of solution (11) has to be substituted into boundary conditions (7) and (8), in which case new boundary conditions are obtained and which have to be satisfied by function V:

$$\left. \frac{\partial V}{\partial y} + b \cdot V \right|_{y=\delta} = 0 \tag{18}$$

$$\left. \frac{\partial V}{\partial y} \right|_{y=0} = 0 \tag{19}$$

The initial condition, which completes (18) and (19), is obtained from the initial condition (5) and the supposed form of solution (11):

$$t_{(y, \tau_0)} = U_{(y, \tau_0)} + V_{(y, \tau_0)} = t_i$$

$$V_{(y, \tau_0)} = t_{(y, \tau_0)} - U_{(y, \tau_0)} = t_i - U_{(y, \tau_0)} = t_i - \frac{k}{b\delta^2 + 2\delta} y^2 \tag{20}$$

To determine $V(y, \tau)$ one may choose from equations (17), (18), (19), and (20). The first to be treated is the homogeneous equation:

$$\frac{\partial V}{\partial \tau} = a^2 \frac{\partial^2 V}{\partial y^2} \tag{21}$$

and a solution is sought with the help of Fourier's method of separating the variables, starting from the form:

$$V_{(y, \tau)} = \sum_{n=1}^{\infty} T_{1(\tau)} \cdot Y_{(y)} \tag{22}$$

209

where the function $V(y, \tau)$ has to satisfy homogeneous boundary conditions (18) and (19).

After differentiating expression (22) and its substitution in (21), one obtains:

$$\frac{T'_{1(\tau)}}{a^2 \cdot T_{1(\tau)}} = \frac{Y''_{(y)}}{Y_{(y)}} = -\mu \tag{23}$$

Equating $-\mu$ shows that the expression on the left side of equation (23) is a function depending only upon τ and the expression on the right, only upon y. This is possible only if both expressions are equal to the constant μ, where $\mu > 0$.

From (23) the characteristic equation is obtained:

$$Y''_{(y)} + \mu Y_{(y)} = 0 \tag{24}$$

and using boundary conditions (18) and (19), bearing in mind that $T_1 \not\equiv 0$, it follows that:

$$Y'_{(\delta)} + b \cdot Y_{(\delta)} = 0 \tag{25}$$

$$Y'_{(0)} = 0 \tag{26}$$

The solution of equation (24), i.e., proper functions from function $V(y, \tau)$ have the form:

$$Y_{n(y)} = A \cdot \cos \sqrt{\mu}\, y + B \cdot \sin \sqrt{\mu}\, y \tag{27}$$

Changing the expression for Y_n (27) into (25), $B = 0$ so that the proper functions have the final form:

$$Y_{n(y)} = A \cdot \cos \sqrt{\mu}\, y \tag{28}$$

Substituting (28) into (25), and considering that $A \neq 0$, the following equation can be used to determine characteristic values for μ:

$$\operatorname{tg} p = \frac{q}{p} \tag{29}$$

where roots of equations are $p_n = \sqrt{\mu_n}\,\delta$; $q = b \cdot \delta$ — and characteristic values:

$$\mu_n = \frac{p_n^2}{\delta^2} \qquad (n = 1, 2, 3, \dots) \tag{30}$$

As the proper functions (28) form an orthogonal set, using the conditions of orthogonality, the constant A (now written as A_n) can be determined:

$$A_n = \frac{1}{\sqrt{\dfrac{\delta}{2} + \dfrac{\sin 2\sqrt{\mu_n}\,\delta}{4\sqrt{\mu_n}}}} \tag{31}$$

The solution of the non-homogeneous equation (17) can be represented in the form of an infinitive set:

$$V_{(y, \tau)} = \sum_{n=1}^{\infty} T_{n(\tau)} \cdot Y_{n(y)} \tag{32}$$

After differentiating (32) and substituting it in (17), the following is obtained:

$$T'_{n(\tau)} \cdot Y_{n(y)} - a^2 \sum_{n=1}^{\infty} T_{n(\tau)} \cdot Y''_{n(y)} = 2a^2 \cdot C \cdot \phi_{(\tau)} - C \cdot \phi'_{(\tau)} \cdot y^2 \qquad (33)$$

As the proper functions Y_n satisfy equation (24), equation (24) is used to obtain the relationship:

$$Y''_{n(y)} = -\mu_n Y_{n(y)} \qquad (34)$$

With this relation (34) expression (33) takes on a new form:

$$\sum_{n=1}^{\infty} Y_{n(y)} [T'_{n(\tau)} - a^2 \cdot T_{n(\tau)} \cdot \mu_n] = 2a^2 \cdot C \cdot \phi_{(\tau)} - C \cdot \phi'_{(\tau)} \cdot y^2 \qquad (35)$$

If equation (35) were multiplied with $Y_{m(y)}$ and integrated within limits of $[0, \delta]$, using the following condition of orthogonality:

$$\int_0^\delta Y_{n(y)} \cdot Y_{m(y)} \, dy = 1 \qquad \text{for } n = m$$

it will be:

$$T'_{n(\tau)} - a^2 \cdot T_{n(\tau)} \cdot \mu_n = \int_0^\delta [2a_{\dot{4}}^2 \cdot C \cdot \phi_{(\tau)} - C \cdot \phi'_{(\tau)} \cdot y^2] Y_{n(y)} \cdot dy \qquad (36)$$

The solutions of the integrals in (36) are:

$$\int_0^\delta 2a^2 \cdot C \cdot \phi_{(\tau)} \cdot Y_{n(y)} \cdot dy = A_n \cdot R_n \cdot \phi_{(\tau)} \qquad (37)$$

$$\int_0^\delta C \cdot \phi'_{(\tau)} \cdot y^2 \cdot Y_{n(y)} \cdot dy = A_n \cdot S_n \cdot \phi'_{(\tau)} \qquad (38)$$

Equation (35) takes the form of a linear equation:

$$T'_{n(\tau)} - a^2 \cdot \mu_n \cdot T_{n(\tau)} = A_n \cdot R_n \cdot \phi_{(\tau)} + A_n \cdot S_n \cdot \phi'_{(\tau)} \qquad (39)$$

and has the following solution:

$$T_{n(\tau)} = e^{-a^2\mu_n\tau} [C_n + A_n \cdot R_n \int_{\tau_0}^\tau \phi_{(\tau)} e^{a^2\mu_n\theta} \, d\theta + A_n \cdot S_n \int_{\tau_0}^\tau \phi'_{(\tau)} e^{a^2\mu_n\theta} \, d\theta] \qquad (40)$$

Determination of the constant C_n

Constant C_n in (40) is determined from initial condition, changing $\tau = \tau_0$:

$$T_{n(\tau_0)} = C_n \cdot e^{-a^2\mu_n\tau_0}$$

$$C_n = T_{n(\tau_0)} \cdot e^{a^2\mu_n\tau_0} \qquad (41)$$

From (32)

$$V_{(y,\,\tau_0)} = \sum_{n=1}^{\infty} T_{n(\tau_0)} \cdot Y_{n(y)}$$

$$T_{n(\tau_0)} = \int_0^\delta V_{(y,\,\tau_0)} \cdot Y_{n(y)} \cdot dy \qquad (42)$$

211

Changing $V(y, \tau_0)$ from (20) into an expression for T_n (42):

$$T_{n(\tau_0)} = \int_0^{\delta} \left[t_i - \frac{k}{b\delta^2 + 2\delta} \cdot y^2 \right] Y_{n(y)} \cdot dy \tag{43}$$

The solution for $T_{n(\tau_0)}$, using constants R_n and S_n (37) and (38) will be:

$$T_{n(\tau_0)} = A_n \left[t_i \frac{\sin \sqrt{\mu_n} \delta}{\sqrt{\mu_n}} + k \cdot S_n \right] = A_n \cdot M_n \tag{44}$$

Substituting (44) into (41):

$$C_n = A_n \cdot M_n \cdot e^{a^2 \mu_n \tau_0} \tag{45}$$

The general solution

The general solution for $t(y, \tau)$, based on form (11), now involves (14), (15), (28), (32), and (40):

$$t_{(y, \tau)} = \frac{\psi_{(\tau)} + k}{b\delta^2 + 2\delta} \cdot y^2 + \sum_{n=1}^{\infty} T_{n(\tau)} \cdot A_n \cdot \cos \sqrt{\mu_n} y \tag{46}$$

Going back to variable x, with the help of dependency $y = \delta - x$, the solution is:

$$t_{(x, \tau)} = \frac{\psi_{(\tau)} + k}{b\delta^2 + 2\delta} (\delta - x)^2 + \sum_{n=1}^{\infty} e^{-a^2 \mu_n \tau} \left[M_n \cdot e^{a^2 \mu_n \tau_0} + \right.$$

$$\left. + R_n \int_{\tau_0}^{\tau} \phi_{(\tau)} \cdot e^{a^2 \mu_n \theta} \, d\theta + S_n \int_{\tau_0}^{\tau} \phi'_{(\tau)} \cdot e^{a^2 \mu_n \theta} \, d\theta \right] A_n^2 \cdot \cos \sqrt{\mu}(\delta - x) \tag{47}$$

THE GENERAL SOLUTION FOR WINDOWS FACING SOUTH

Since windows facing south in the first and third periods of solar radiation (fig. 3) only possess diffuse radiation and in the second period both diffuse and direct radiation, the solution for (47) has to cover each period of solar radiation and also the night period. Due to this, equation (47) is integrated between limits which depend upon the period for which the solution is to be determined:

Solution for 1st period ($\tau_0 \leqslant \tau \leqslant \tau_D$)

$$t_{(x, \tau)} = \frac{\psi_{(\tau)} + k}{b\delta^2 + 2\delta} (\delta - x)^2 + \sum_{n=1}^{\infty} \left[\left(N_n - \frac{bt_i}{a^2 \mu_n} P_n - T_{2n} \right) e^{-a^2 \mu_n (\tau - \tau_0)} + \frac{bt_i}{a^2 \mu_n} P_n + \right.$$

$$\left. + T_{1n} \sin \omega_d(\tau - \tau_0) + T_{2n} \cos \omega_d(\tau - \tau_0) \right] \cos \sqrt{\mu_n}(\delta - x) \tag{48}$$

where

$$\psi_{(\tau)} = \frac{f \cdot I_d}{\lambda} \sin \omega_d(\tau - \tau_d)$$

$$\tau_0 = \tau_d$$

212

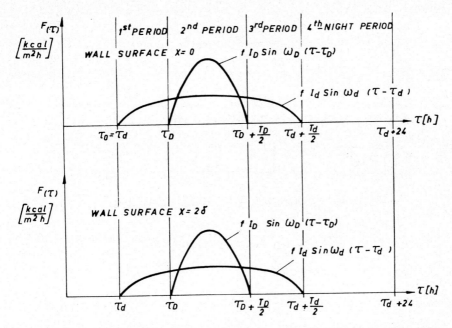

Fig. 3 — The variation in solar radiation absorbed by wall surfaces in rooms with windows facing south.

Solution for 2nd period $[\tau_D \leqslant \tau \leqslant \tau_D + (\tau_D/2)]$

$$t_{(x,\tau)} = \frac{\psi_{(\tau)} + k}{b\delta^2 + 2\delta}(\delta - x)^2 + \sum_{n=1}^{\infty}\left[\left(N_n - \frac{bt_i}{a^2\mu_n}P_n - T_2\right)e^{-a^2\mu_n(\tau - \tau_0)} - \right.$$

$$- T_{4n}\,e^{-a^2\mu_n(\tau - \tau_D)} + \frac{bt_i}{a^2\mu_n}P_n + T_{1n}\sin\,\omega_d(\tau - \tau_0) + T_{2n}\cos\,\omega_d(\tau - \tau_0) +$$

$$\left. + T_{3n}\sin\,\omega_D(\tau - \tau_D) + T_{4n}\cos\,\omega_D(\tau - \tau_D)\right]\cos\sqrt{\mu_n}(\delta - x)\quad(49)$$

where

$$\psi_{(\tau)} = \frac{f \cdot I_d}{\lambda}\sin\,\omega_d(\tau - \tau_d) + \frac{f \cdot I_D}{\lambda}\sin\,\omega_D(\tau - \tau_D)$$

$$\tau_0 = \tau_d$$

Solution for 3rd period $[\tau_D + (\tau_D/2) \leqslant \tau \leqslant \tau_d + (\tau_d/2)]$

$$t_{(x,\tau)} = \frac{\psi_{(\tau)} + k}{b\delta^2 + 2\delta}(\delta - x)^2 + \sum_{n=1}^{\infty}\left[\left(N_n - \frac{bt_i}{a^2\mu_n}P_n - T_{2n}\right)e^{-a^2\mu_n(\tau - \tau_0)} - \right.$$

$$- T_{4n}\,e^{-a^2\mu_n(\tau - \tau_D)} - T_{4n}\,e^{-a^2\mu_n[\tau - (\tau_D + T_D/2)]} + \frac{bt_i}{a^2\mu_n}P_n +$$

$$\left. + T_{1n}\sin\,\omega_d(\tau - \tau_0) + T_{2n}\cos\,\omega_d(\tau - \tau_0)\right]\cos\sqrt{\mu_n}(\delta - x)\quad(50)$$

213

where

$$\psi_{(\tau)} = \frac{f \cdot I_d}{\lambda} \sin \omega_d(\tau - \tau_d)$$

$$\tau_0 = \tau_d$$

Solution for night period $[\tau_d + (\tau_d/2)) \leqslant \tau \leqslant \tau_d + 24]$

$$t_{(x, \tau)} = \frac{k}{b\delta^2 + 2\delta}(\delta - x)^2 + \sum_{n=1}^{\infty}\left[\left(N_n - \frac{bt_i}{a^2\mu_n}P_n - T_{2n}\right)e^{-a^2\mu_n(\tau - \tau_0)} - \right.$$

$$- T_{4n}e^{-a^2\mu_n(\tau - \tau_D)} - T_{4n}e^{-a^2\mu_n[\tau - (\tau_D + (T_D/2))]} -$$

$$\left. - T_{2n}e^{-a^2\mu_n[\tau - (\tau_d + (T_d/2))]} + \frac{bt_i}{a^2\mu_n}P_n\right]\cos\sqrt{\mu_n}(\delta - x) \quad (51)$$

where

$$\psi_{(\tau)} = 0$$

CONCLUSIONS

Equation (47) represents the general solution for the temperature field of an internal wall which absorbs, on both sides, equal quantities of solar radiation. For the coordinate $x = 0$, this solution gives temperature variations on the surface of the wall. Using these values, it is possible to calculate the actual cooling loads with the help of the following formula:

$$q = \alpha\left[t_{x=0} - t_i\right] \qquad \left[\frac{\text{kcal}}{\text{m}^2 \cdot \text{h}}\right] \qquad (52)$$

The solutions which are given in this work refer to homogeneous materials and can be used for most internal walls. For floors and ceilings which consist of more varied materials, a similar mathematical treatment would reveal the solution but the treatment would be longer and more complicated due to more boundary conditions which depend upon the number of the different material layers.

APPENDIX

τ_0	initial time
τ_d	initial time of diffuse radiation
τ_D	initial time of direct radiation
T_d	the period of diffuse radiation
T_D	the period of direct radiation
2δ	thickness of observed wall (m)

214

$$a^2 = \frac{\lambda}{c \cdot \gamma}$$

$$R_n = 2a^2 \frac{\sin \sqrt{\mu_n}\delta}{\sqrt{\mu_n}} \cdot \frac{1}{b\delta^2 + 2\delta}$$

$$S_n = -\left[\frac{\delta^2 \sin \sqrt{\mu_n}\delta}{\sqrt{\mu_n}} + \frac{2\delta \cos \sqrt{\mu_n}\delta}{\mu_n} - \frac{2 \sin \sqrt{\mu_n}\delta}{\mu_n \sqrt{\mu_n}}\right] \frac{1}{b\delta^2 + 2\delta}$$

$$N_n = t_i \left[\frac{\sin \sqrt{\mu_n}\delta}{\sqrt{\mu_n}} + \frac{b \cdot S_n}{b\delta^2 + 2\delta}\right] \frac{1}{\dfrac{\delta}{2} + \dfrac{\sin 2\sqrt{\mu_n}\delta}{4\sqrt{\mu_n}}}$$

$$P_n = \frac{R_n}{\dfrac{\delta}{2} + \dfrac{\sin 2\sqrt{\mu_n}\delta}{4\sqrt{\mu_n}}}$$

$$Q_n = \frac{S_n}{\dfrac{\delta}{2} + \dfrac{\sin 2\sqrt{\mu_n}\delta}{4\sqrt{\mu_n}}}$$

$$T_{1n} = \frac{f \cdot I_d/\lambda}{\omega_d^2 + a^4 \mu_n^2} (a^2 \mu_n \cdot P_n + \omega_d^2 \cdot Q_n)$$

$$T_{2n} = \frac{f \cdot I_d/\lambda}{\omega_d^2 + a^4 \mu_n^2} (a^2 \mu_n \cdot Q_n \omega_d - \omega_d \cdot P_n)$$

$$T_{3n} = \frac{f \cdot I_D/\lambda}{\omega_D^2 + a^4 \mu_n^2} (a^2 \mu_n \cdot P_n + \omega_D^2 \cdot Q_n)$$

$$T_{4n} = \frac{f \cdot I_D/\lambda}{\omega_D^2 + a^4 \mu_n^2} (a^2 \mu_n \cdot Q_n \omega_D - \omega_D \cdot P_n)$$

DISCUSSION

H. MICHEL (France) — 1. Quelle est a variation de température acceptée dans la pièce? L'hypothèse de la température constante n'est-elle pas théorique?

2. La surface insolée est variable au cours de la journée. En a-t-il été tenu compte dans le calcul?

B. TODOROVIĆ — 1. The temperature was taken as a constant with time (in the room), and the temperature of the wall varied because of solar radiation which was absorbed in the observed wall which creates the various cooling loads.

2. The mathematical treatment in my work was calculated on 1 m^2 of the surface, so it is equal if we have a window with or without screens. But, in practice, the air conditioned rooms always have creens. Of course, whether with screens or without, solar radiation passes through the window and this is a variable with time. In both cases solar radiation can be represented with sinus functions.

R. LANDSBERG (Israel) — The three papers, each of them most valuable, fail to take into account work done earlier in similar directions: research done at Newcastle University. Directions given by Carrier's comprehensive guide papers published by Abramovitz and myself at the Madrid conference and in Kältetechnik, 1967, etc. The I.I.F. might well try to coordinate the various approaches to the problems.

K. GERTIS (W. Germany) — I think also that it is necessary to coordinate the different works on cooling load calculations. Every work contains certain approximations and assumptions. Especially it is necessary to explain these relatively complicated heat transfer processes in a manner, which can be understood by the architects because the architects often design facades out of glass without any regard on the thermal consequences happening later in the rooms behind such facades.

216

EXPERIMENTS ON STATIONARY AND OSCILLATING HEAT TRANSFER IN LARGE WALLS

P. BONDI, C. CODEGONE, V. FERRO and A. SACCHI

Istituto di Fisica Tecnica del Politecnico, Turin (Italy)

Expériences sur la transmission de chaleur au travers de grands murs en régimes permanent et cyclique

RÉSUMÉ : *On décrit un nouvel appareil de mesure de la transmission de chaleur à travers des échantillons de murs.*

La mesure comprend l'effet de résistances thermiques superficielles, car les échantillons sont chauffés et refroidis par de l'air maintenu dans des chambres fermées et recirculé à l'aide de ventilateurs. Les valeurs des températures, fixes ou cycliques, sont déterminées par des systèmes de commande adéquats.

L'appareil peut être utilisé pour des parois multiples dont les surfaces externes ne sont n planes ni parallèles, c'est-à-dire pour tout modèle de mur préfabriqué.

On donne les résultats d'expériences sur des échantillons de murs contenant des parois vitrées.

En régime stable, on étudie l'effet de différentes vitesses d'air (c'est-à-dire les résistances thermiques superficielles) et en régime cyclique, on examine une gamme complète de périodes (3, 6, 12 et 24 heures).

On a mis au point une méthode de calcul théorique des paramètres caractéristiques, que l'on a comparée aux mesures expérimentales.

INTRODUCTION

After the completion of an apparatus for measurements of thermal conductance on large wall specimens [1] and systematic tests on many structural elements with the said apparatus [2], a new apparatus for measuring the thermal transmittance of the same large walls was built.

The term "thermal conductance" means the heat transfer coefficient without boundary thermal resistances, i.e., heat flow per unit area and a temperature difference between opposite surfaces of the homogeneous wall; the term "thermal transmittance" means the heat transfer coefficient including film thermal resistances, i.e., heat flow per unit area and an ambient temperature difference between the rooms separated by the homogeneous wall.

If the wall is composite, these coefficients must be intended as equivalent [3] or apparent [4].

The new apparatus is different from those already built [5], in that it does not repeat unidirectional heat flow from the hot chamber, protected by a peripherical guard, towards a single cold chamber but actuates bi-directional, symmetrical distribution. The thermal flow is transmitted from a central hot room through two identical wall specimens to two external rooms symmetrical to the first.

DESCRIPTION OF THE APPARATUS

The apparatus can be used to measure the thermal transmittance of walls from 3×1 m up to 3×3 m.

It consists of a steel frame completely covered by an insulating material (fig. 1). Two rails hold the two wall specimens in such a way as to provide a hot room in the central zone of the apparatus; there are two cold rooms on each side of the hot room.

The large dimensions of the frame enable the testing of specimens of prefabricated walls or entire modules of walls, glass surfaces, ordinary doors or similar building elements. Variations in the thickness of walls or other elements do not hinder the execution of tests; the tests can simulate conditions as would be found in a real

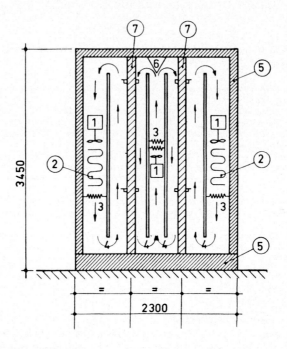

Fig. 1 — Scheme of the apparatus:
1. Fans; 2. Refrigerating coils; 3. Electrical resistances; 4. Light-reflecting screens; 5. Heat insulation; 6. Air partition; 7. Wall specimens.

building. The insulating material used was expanded polystyrene for the external surfaces of the refrigerating chambers and the floor. For the outer surfaces of the hot chamber, polyurethane foam was expanded directly onto the peripheric surface. Apart from giving good thermal insulation, good impermeability to hot air dispersions was also achieved because polyurethane foam adheres to any surface and fills cavities.

HOT ROOMS

The hot room is the most important part of the apparatus since heat flow through the specimens is measured in this enclosure.

Heat transfer through the lateral walls of this room can produce errors in the measurements and in order to limit this, a thin light-reflecting aluminum sheet was applied to the insulation on both the interior and exterior walls. In order to evaluate thermal dispersion due to lateral walls, thermocouples were placed in suitable positions since the thermal conductivity of the insulating material is well known. The error in the final values for heat flow through the tested walls was small. The heat from the circulating fans was taken into account in the sensible heat load. The air speed in the hot room was adjusted in order to get a thermal film coefficient of 7 to 8 $W/m^2.°C$.

Air velocities of 0.28 to 0.30 m/s at an average air temperature of $\sim +20°C$ were chosen (according to Rowley and Algren [6]).

Heat flux was provided by electrical resistances placed in line with the fans and regulated from the exterior; the heat input was measured through a wattmeter.

In the hot chamber between the wall specimens two light-reflecting screens are spaced about 30 cm apart and have the same surface dimensions as the specimens; fans and heaters are placed between the screens. A deflector directs the air flow into two paths down past the specimens. At a distance of 10 cm from each wall specimen, nine thermocouples in rows of three and at different heights measure and control the temperature distribution in the heated zone; these couples are screened. A hot wire anemometer is placed before each specimen in order to measure air velocity and control the value of the thermal film coefficient. The head of the anemometer is placed in a horizontal plane at 1.50 m height and is movable in this plane in order to plot the velocity profiles.

COLD ROOMS

The cold rooms are placed symmetrically to the hot chamber, similar to the flat plate apparatus used in this Institute [1]. Each room is divided by a light-reflecting screen into two intercommunicating compartments. The outer compartment contains the fans, the evaporating units and electrical resistances for the regulation of temperature. The second compartment carries the thermocouples and an anemometer, similar to those in the hot chamber.

Particular care was taken to obtain a good seal between the specimens and the frame and the hot and cold air sections. Apart from the tested walls, all surfaces were covered with reflecting aluminum sheet in order to limit radiation heat transfer between parts of the system; greater repeatability and a better comparison of measurements can be obtained via these precautions.

Air circulation was calculated to give a heat transfer coefficient of 9 to 20 W/m^2.°C with air velocities 1 to 4 m/s and a temperature $\sim +7°C$ (Rowley and Algren [6]).

CONTROL SYSTEM

A system having proportional, integral and derivative action controls, through a magnetic amplifier, the electrical heat input to the hot zone at a preset constant temperature. An analogous system closely regulates the temperature in the cold zone. The refrigerating unit is controlled by an on-off regulator which presets the expansion temperature. The air velocity is manually controlled. Considering the "prototype" nature of the actual apparatus it was decided not to automate the ventilating system. A safety cut-out stops the power input to the resistances and fans should the temperature in the rooms attain a pre-established maximum value.

EXPERIMENTAL MEASUREMENTS

The usual expression for calculating thermal transmittance is:

$$W = H \cdot S \cdot (T_h - T_c) \quad [W] \tag{1}$$

and is valid under stationary conditions for a plane, homogeneous, indefinite slab with isothermal surfaces parallel to the external surfaces and thermal flow lines perpendicular to them.

Definitions

W	Thermal flow rate through one of the specimens (1/2 of the electrical input into the hot chamber including the fans)	[W]
H	Thermal transmittance	[W/m².°C]
S	Frontal surface of specimen	[m²]
T_h	Mean temperature of hot zone	[°C]
T_c	Mean temperature of cold zone	[°C]
T_m	Mean temperature of the wall	$T_m = (T_h + T_c/2)$ [°C]
v	Air velocity in cold chambers	[m/s]

In the case of non-homogeneous wall made from layers of different materials and having thermal bridges, projections, and other irregularities on its external surface, the formula still holds valid (1) but H assumes the meaning of "equivalent thermal transmittance" H_e, i.e., equivalent to that of a flat homogeneous wall which transmits the same heat flux per unit of frontal area for a temperature difference of $T_h - T_c$.

It has already been mentioned that the measurement for heat input is the summation of the electrical input to the resistances and the fans; the measurement was effected by means of a precision wattmeter (class 0.2). The temperatures were measured by a 100 channel data logger.

Tests were made on two identical specimens of a glass wall by varying the temperature of the internal room and air velocity in the external one and steadying the air velocity in the internal room and temperature in the external one. In figure 2 the equivalent thermal transmittance is plotted against mean temperature between the hot and cold rooms; three straight lines correspond to the value for air velocity in the cold

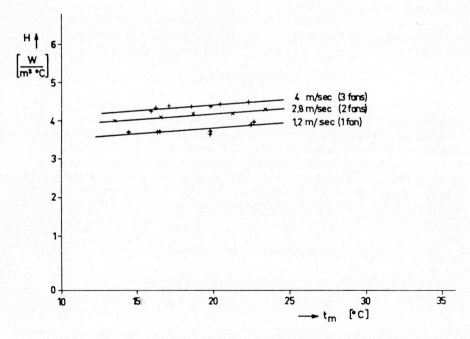

Fig. 2 — Thermal transmittance of tested specimens versus mean temperature.

zones. Transmittance can be made a function of temperature as follows:

$$H_1 = 4.14 + 0.016 \, T_m \, [W/m^2.°C] \qquad \text{for} \qquad v = 4 \quad m/s$$

$$H_2 = 3.88 + 0.016 \, T_m \, [W/m^2.°C] \qquad \text{for} \qquad v = 2.8 \, m/s$$

$$H_3 = 3.50 + 0.016 \, T_m \, [W/m^2.°C] \qquad \text{for} \qquad v = 1.2 \, m/s$$

and in general:

$$H = H_0 + 0.016 \, T_m.$$

The constant H_0 can be written as a velocity function:

$$H_0^{\overline{s}} = 3.30 \cdot v^{0.16},$$

and then the general expression follows:

$$H = 3.30 \, v^{0.16} + 0.016 \, T_m \, [W/m^2.°C],$$

which agrees with experimental values to within $\pm 2\%$.

OSCILLATION EXPERIMENTS

A wall [7, 8] subjected to thermal perturbations of small amplitude so as to hold constant the transfer parameters, behaves like a linear, passive four-terminal network. If the thermal perturbations are periodical for each harmonic component of v frequency, the following relations will be valid:

$$\begin{vmatrix} \theta_{0v} \\ \varphi_{0v} \end{vmatrix} = \begin{vmatrix} A_v & B_v \\ C_v & D_v \end{vmatrix} \cdot \begin{vmatrix} \theta_{nv} \\ \varphi_{nv} \end{vmatrix} \qquad (2)$$

Constants A_v, B_v, C_v, and D_v, as with temperatures and fluxes, should be considered complex numbers.

A similar expression can be derived by considering the transfer phenomenon between external and internal rooms:

$$\begin{vmatrix} \theta_{ev} \\ \varphi_{ev} \end{vmatrix} = \begin{vmatrix} E_v & F_v \\ G_v & H_v \end{vmatrix} \cdot \begin{vmatrix} \theta_{iv} \\ \varphi_{iv} \end{vmatrix}. \qquad (3)$$

Here also, E_v, F_v, H_v, G_v are complex numbers.

Both the matrices of coefficients in (2) and (3) are unimodular. If the wall has a symmetrical disposition of layers we get $A_v = D_v$ and if boundary layer coefficients are also equal the result is: $E_v = H_v$.

PRINCIPLES OF THE METHOD

The following experiments were made:

a) Measurement of E_v

From expression (3) one derives:

$$E_v = \left(\frac{\theta_{ev}}{\theta_{iv}} \right)_{\varphi_i = 0} = \left(\frac{\theta_{ev}}{\theta_{iv}} \right)_{C_i = 0}. \qquad (4)$$

This condition can be realized by making nil or constant, the value for heat flux to the air capacity of the internal chamber. An apparatus can be so arranged that the thermal flux transmitted from the internal chamber is related to the difference of temperature between the upper and lower zone of the channels facing the same wall.

Imposing the constancy of the aforesaid difference of air temperatures by means of a suitable automatic controller for the heaters and coolers we get $\varphi_{iv} = 0$ and $C_i = 0$. (This last condition is not absolutely correct; indeed C_i is equal to both the thermal capacity of air held in the channels and that from the lateral walls of the same channels, a lesser value in relation to the thermal capacity of the test walls.)

The mean temperature of air in the channel will therefore oscillate in relation to the temperature oscillations of the external air and θ_{iv} will represent one of its harmonics. In the present experiment we introduced perturbations of a single harmonic θ_{ev} (of v frequency) and by a constant term constituting the average value of a resulting diagram. The mean values of temperatures and fluxes are bound by the steady state coefficient of the examined wall (under the hypothesis of linearity of the system and thus the validity of d'Alembert's principle).

b) Measurement of F_v

From equation (3) we get:

$$F_v = \left(\frac{\theta_{ev}}{\varphi_{iv}}\right)_{\theta_{iv}=0} = \left(\frac{\theta_{ev}}{\varphi_{iv}}\right)_{C_i=\infty}. \tag{5}$$

Such a condition can be actuated by holding constant the temperature of air in the internal chamber via an automatic controller and measuring the harmonic φ_{iv} of flux given off by the controller.

c) Measurement of H

By inverting the direction of thermal propagation within the specimens, i.e., by exchanging external and internal surfaces, we get:

$$\begin{vmatrix} \theta^*_{ev} \\ \varphi^*_{ev} \end{vmatrix} = \begin{vmatrix} H_v & F_v \\ G_v & E_v \end{vmatrix} \cdot \begin{vmatrix} \theta^*_{iv} \\ \varphi^*_{iv} \end{vmatrix} \tag{6}$$

and:

$$H_v = \left(\frac{\theta^*_{ev}}{\theta^*_{iv}}\right)_{\varphi_{iv}=0} = \left(\frac{\theta^*_{ev}}{\theta^*_{iv}}\right)_{C_i=0} \tag{7}$$

The experimental disposition is identical to that used for measuring E_v. If the wall is symmetrical, as with the wall specimens we examined, it is sufficient to exchange boundary layer resistances between the two of them.

d) Measurement of G_v

From the unimodularity relation:

$$H_v \cdot E_v - G_v \cdot F_v = 1 \tag{8}$$

comes:

$$G_v = \frac{H_v \cdot E_v - 1}{F_v}. \tag{9}$$

222

In the theory of heat transfer under steady or oscillating conditions the temperatures of the internal and external chambers are supposed uniform. In practice, however, there is a difference between the upper and lower zones sufficient to require average values in the calculations.

This is somewhat in contrast to theory but it simulates a condition that is much more approximate to actual application.

Thus the thermal transmittance and characterizing parameters under oscillating conditions are much nearer to reality than with the more rigorous theoretical considerations and the measurements for thermal conductance and oscillating parameters A_v, B_v, C_v, D_v [7]. It is furthermore to be noted that relations (2), (3), and (8) require no hypothesis other than the four-terminal system to be linear and passive, which is valid in our experiments if variations in the parameters θ and φ are small.

Parameters E_v, F_v, G_v and H_v denote a factor similar to thermal transmittance. Thus our method characterizing the parameters has a direct derivation and therefore is not aggravated by further numerical calculations.

REGULATING APPARATUS

The scheme of the automatic regulating apparatus is indicated in figure 3.

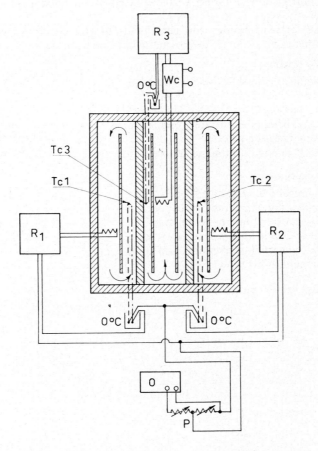

Fig. 3 — Scheme for regulating apparatus realizing $C_i = \infty$.

223

An oscillator produces a sinusoidal output voltage at the required frequency which, having suitably been reduced with a partition resistance, is opposed to the e.m.f. of thermocouples T_{c1} and T_{c2} situated on the surface of the specimens in order to evaluate the mean temperature of the air in the external room. The total signal is sent into regulators R_1 and R_2 which vary the heat input into the external chambers until the difference between the thermocouple and oscillating signal is constant.

The maximum amplitude of the temperature oscillation is set by varying the value of this constant, i.e., the mean value of oscillations, while regulating the output of the oscillator. For the internal zone the mode of regulation depends upon the condition required. For infinite equivalent capacity the layout in figure 3 is used in which regulator R_3 holds constant the temperature measured by couple T_{c3}.

Electrical power necessary to attain this condition will vary with time, following temperature oscillations in the external chambers. (W_c records the sum of electrical input to R_3 and to the fans in the internal chambers.)

The condition of nil equivalent capacity ($C_i = 0$) can be obtained with a re-arranged version of figure 3 (see ref. [12]).

EXPERIMENTAL RESULTS

The experiments were performed on the same glass-containing specimens described previously.

Recordings of temperatures and heat inputs to give conditions of equivalent internal thermal capacity equal to zero and infinity are as below:

1. For infinite capacity:
   Period: nominal 6 hrs effective 5 hrs 22 mn, $v = 51.75~\mu\text{Hz}$;
   mean temperature of external oscillation: $+7.0\,°\text{C}$;
   internal temperature: $+17.9\,°\text{C}$;
   amplitude of external temperature oscillation: $7.3\,°\text{C}$;
   mean internal heating power: 522 W;
   amplitude of internal heating power oscillation: 371 W.

2. For zero capacity:
   Period: nominal 6 hrs effective 5 hrs 34 mn, $v = 49.87~\mu\text{Hz}$;
   mean temperature of external oscillation: $+7.5\,°\text{C}$;
   mean temperature of internal oscillation: $+17.7\,°\text{C}$;
   amplitude of external temperature oscillation: $3.5\,°\text{C}$;
   amplitude of internal temperature oscillation: $2.9\,°\text{C}$;
   mean internal heating power: 441 W

From these values one can calculate:

$$|E_v| = 3.5/2.9 = 1.207$$

$$\underline{|E_v} = 0.550 \text{ rad}$$

$$|F_v| = \frac{7.3 \cdot 11.32}{371} = 0.223 \frac{°\text{C.m}^2}{\text{W}}$$

$$\underline{|F_v} = 0.348 \text{ rad}$$

since the surfaces of the walls exchanging heat are $S = 11.32 \text{ m}^2$.

224

Thermal transmittance as referred to mean values for temperature and heating power is for $C_i = \infty$

$$\mathcal{H} = \frac{522}{(17.9 - 7.0) \cdot 11.32} = 4.23 \, \frac{W}{m^2 \cdot {}^\circ C}$$

for $C_i = 0$

$$\mathcal{H} = \frac{441}{(17.7 - 7.5) \cdot 11.32} = 3.82 \, \frac{W}{m^2 \cdot {}^\circ C}.$$

Results for periods of 3, 6, 12 and 24 hrs were recorded.

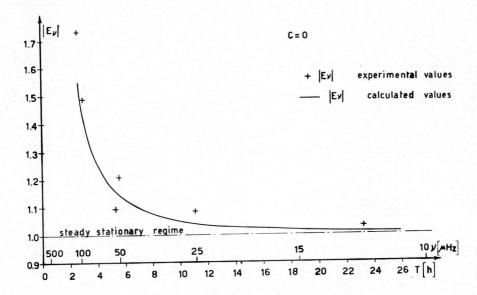

Fig. 4 — Modulus E_y versus frequency.

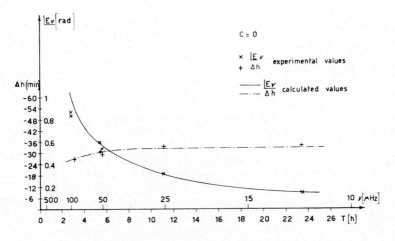

Fig. 5 — Phase E_y versus frequency.

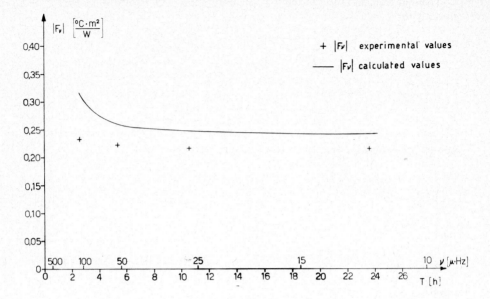

Fig. 6 — Modulus F_v versus frequency.

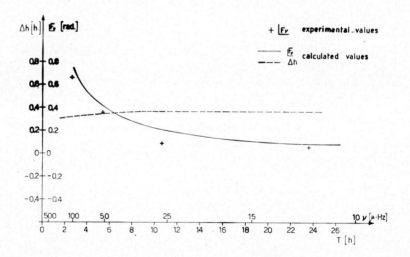

Fig. 7 — Phase F_v versus frequency.

In figures 4, 5, 6 and 7 are plotted parameters E_v and F_v versus frequency; $\mathscr{H}$ values versus frequency were also plotted. Good agreement was found with steady stationary results.

To measure the value of H_v (since the wall was symmetrical) it was sufficient to exchange boundary heat transfer coefficients by lowering the air velocity on the external surface to 0.30 m/s and correspondingly raising the internal one to 4 m/s. Realizing condition $C_i = 0$ the internal boundary coefficient does not influence the

226

phenomenon as the flux transmitted to the internal surface is zero. Hence the value of 0.30 m/s was held constant (i.e., velocity of internal air) to eliminate the necessity of using circulators with higher rating and electrical consumption in the internal chamber since this would have meant a rise in the operating temperature and in the temperature differences.

In figures 8 and 9 parameter H_v is plotted against frequency.

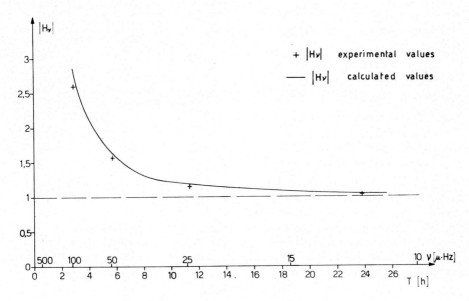

Fig. 8 — Modulus of H_v versus frequency.

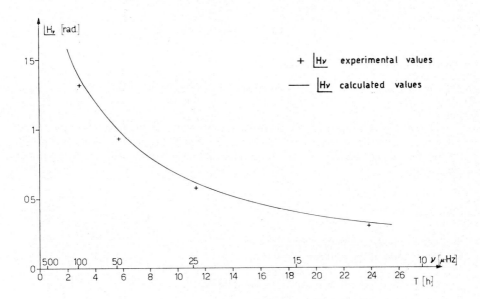

Fig. 9 — Phase of H_v versus frequency.

227

Finally we calculated the values for the same parameters using the theoretical calculations [8, 9, 10], supposing that the wall was built of two glass panes 6 mm thick and with an air gap 13 mm wide.

Thermal bridges have been taken into account by correcting values for thermal conductivity, density and specific heat of the layers in proportion to the ratio of glass surfaces to the frontal surface of thermal bridges.

Values calculated were:

Glass surface 11.30 m^2
Frontal surface of thermal bridges 0.016 m^2.

Equivalent values for equivalent glass layer:

density 2,600 kg/m^3
specific heat 0.84 kJ/kg. °C
thermal conductivity 0.998 W/m. °C.

Equivalent values for equivalent air gap:

density 5.21 kg/m^3
specific heat 0.999 kJ/kg. °C
thermaı conductivity 0.334 W/m^2. °C.

The values for E_v, F_v, G_v, and H_v were calculated on the basis of assumed values and for boundary coefficients of 7 W/m^2. °C for the interior and 20 W/m^2. °C for the exterior, for periods of T = 24, 12, 6, 3 hrs [9].

The agreement between experimental and calculated values is obvious.

The value of equivalent thermal impedance $Z_{i,eq}$ of the central internal channel and channels connected to thermocouples T_{c4} and T_{c5} can be calculated from temperature and heat flux recordings viz:

$$Z_{i,eq} = \frac{(\theta_i)_{c_i = 0}}{(\varphi_i)_{c_i = 0}}$$

since the thermal flux exchanged with the specimens is zero (i.e., the oscillating part).

$$Z_{i,eq} = (0.08 - 0.488\,j)\,\frac{°\text{C}.\text{m}^2}{\text{W}}$$

and therefore:

$$Y_{i,eq} = (0.345 + 2.11\,j)\,\frac{\text{W}}{\text{m}^2.°\text{C}}.$$

This means that apparent thermal dispersion from the chamber at a considered frequency (T = 6 hrs nominal) is 0.345 W/m^2. °C and the equivalent thermal capacity of the same equals 6.72 kJ/°C. m^2, both referred to 1 m^2 of the surface of the tested specimen.

SYMBOLS

θ	amplitude of temperature oscillation	[°C]
φ	amplitude of thermal flux oscillation	[W]
A, B, C, D, E, F, G, H	complex coefficients of the matrix expressing the behavior of a linear passive four-terminal network equivalent to a tested wall	
ν	frequency	[μHz]

228

C_i	equivalent thermal capacity of the internal chamber	$[\text{kJ/m}^2.°\text{C}]$
$\theta*$	amplitude of temperature oscillation (inverse heat flow)	$[°\text{C}]$
$\varphi*$	amplitude of thermal flux oscillation (inverse heat flow)	$[\text{W}]$
S	specimen's surface involved in heat transfer	$[\text{m}^2]$
$\mathcal{H}$	thermal transmittance	$[\text{W/m}^2.°\text{C}]$
T	period of oscillation	$[\text{h}]$
Z	thermal impedance	$[°\text{C m}^2/\text{W}]$
Y	thermal admittance	$[\text{W/m}^2.°\text{C}]$
Δh	phase lag in hours	$[\text{hr}]$

BASES

o	related to the external surface of the wall
n	related to the nth surface of the wall (usually the last or internal one)
ν	related to frequency
i	related to internal chamber
e	related to external chamber
eq	equivalent

REFERENCES

[1] C. CODEGONE and V. FERRO, « Un apparecchio per la determinazione della conduttanza termica di pareti prefabbricate». *La Termotecnica* (1965), vol. 19, p. 704.

[2] P. BONDI, V. FERRO, C. LOMBARDI and A. SACCHI, « Prove sistematiche di conduttanza termica eseguite su pareti prefabbricate per l'edilizia», *Atti e Rassegna Tecnica Soc. Ing. e Arch. in Torino* (1967), vol. 21, p. 119.

[3] C. CODEGONE, « Rassegna di metodi di misura della conduttività termica dei materiali da costruzione», *Atti e Rassegna Tecnica Soc. Ing. e Arch. in Torino* (1955), vol. 9, p. 297.

[4] G. B. WILKES, "Heat Insulation", J. Wiley, New York (1950), p. 37.

[5] Thermal Conductance and Transmittance of Built-up Sections by Means of the Guarded Hot Box-A.S.T.M., designation C 236-66 adopted 1960, revised 1966.

[6] F. B. ROWLEY and A. B. ALGREN, "Thermal Conductivity of Building Materials", *Engineering Experimental Station*, Bull. no. 12, University of Minesota (1937), p. 21.

[7] V. FERRO and A. SACCHI, « Oscillazioni in pareti composte», *Ricerche di Termotecnica* n° 16 (1966), pp. 22-27.

[8] C. CODEGONE, « Corso di Fisica Tecnica », vol. II, Termocinetica, Parte 2, p. 535 (Appendice n° 7 a cura dell'Ing. A. Sacchi), Giorgio, Torino 1967.

[9] A. SACCHI, « Il calcolo numerico dei parametri caratteristici in regime termico periodico di pareti composte» — Pubblicazioni dell'Istituto di Fisica Tecnica del Politecnico di Torino (Dec. 1967), memoria n° 333.

[10] C. BOFFA, V. FERRO and A. SACCHI, «Tabelle numeriche per il calcolo rapido dei parametri caratteristici di pareti composte», *Ricerche di Termotecnica* n° 16 (1966), pp. 52-62.

[11] C. CODEGONE and V. FERRO, « Una nuova apparecchiatura per prove di trasmittanza termica di pareti di grandi dimensioni». *La Termotecnica* (1967), vol. XXI, pp. 463-467.

[12] P. BONDI and A. SACCHI, « Oscillazioni termiche di pareti di grandi dimensioni». *Atti e Rassegna Tecnica Soc. Ing. e Arch. in Torino* (1968), vol. XXII, pp. 65-71.

EXAMINATION OF CONVECTIVE HEAT TRANSPORT IN AN AIR-PROOF CASED PLATE OF MINERAL WOOL

I. PALJAK

The National Institute for Materials Testing, Stockholm (Sweden)

Examen du transfert de chaleur par convection dans une plaque de laine minérale placée dans une enveloppe étanche à l'air

RÉSUMÉ : *Ce rapport a pour objectif, d'une part, de contribuer à déterminer la quantité de transfert de chaleur par convection pour une plaque de laine de verre placée dans une enveloppe étanche à l'air et, d'autre part, de prouver qualitativement l'existence de la convection en mesurant la répartition de la température à l'intérieur de la plaque d'essai.*

L'examen doit spécialement indiquer si, dans la mesure courante du coefficient K de la laine minérale, il est nécessaire de redouter des courants de convection dans la plaque d'essai.

INTRODUCTION

The present work, at one hand, aimed to contribute to the elucidating of the size of convection heat transport for an air-proof cased plate from glass wool and on the other hand, qualitatively prove the existence of convection by measuring the temperature distribution inside the test plate.

As a special task, the examination should state if, by routine measuring of the *k*-value of mineral-wool in a hot plate apparatus it is necessary to fear convection currents inside the test plate.

CHOICE OF TESTING MATERIAL AND TESTING CONDITIONS

Former examinations have shown that the convection in porous materials with open cells cased between air-proof layers is increasing with:

a) the increasing air permeability of the material;
b) the increasing thickness of the layer of the material;
c) the reduction of the height of a vertical plate till the height equals the thickness;
d) the increasing temperature gradient through the specimen;
e) the decreasing mean temperature of the specimen.

Since the extent of the examination referred to had to be limited it was decided to choose a combination of a material of common occurrence and testing conditions which, according to the conditions mentioned above were expected to cause a high convection part in the heat transport.

After having made an inventory of the existing mineral-wool qualities in Sweden, glass wool of the type "Gullfiber 3004" with the nominal bulk density of 16 kg/m^3 was chosen. Two testing plates of the material were made with the following bulk density and dimensions.

Plate 1: Bulk density 16.7 kg/m^3 and dimensions:
$600 \times 600 \times 95$ mm.

Plate 2: Bulk density 15.9 kg/m^3 and dimensions :
$600 \times 600 \times 100$ mm.

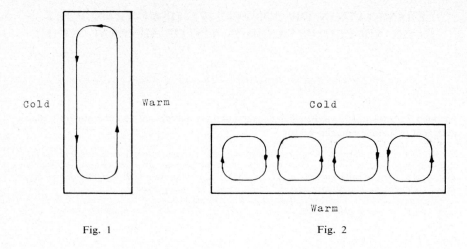

Cold   Warm   Cold

Warm

Fig. 1   Fig. 2

The following testing conditions were chosen: temperature difference over the test plate was 40 deg C at which the surface temperature of the specimen was $+30°C$ and $-10°C$, the mean temperature of the specimen being $+10°C$.

The temperature difference and the surface temperature chosen are extreme values in regard to the usual temperature gradients in the walls of a building and danger of water vapor condensation in the specimen during measurement.

MEASURING PRINCIPLE

The convection part of the heat transport was examined for two orientations of specimen, namely the vertical and the horizontal ones with the heat flow turned upwards. The examination was done by comparing thermal conductivity numbers at the two positions with thermal conductivity number in horizontal position with the heat flow turned downwards. The measurement was done on a specimen by a guarded hot plate of the laboratory which can be rotated in different positions in relation to the ground plane [8] (see fig. 3). The apparatus is used in a one-sided form, at which the heat flow from the hot plate is directed through the test plate by means of a special compensating hot plate (see fig. 4). The edges of the test plate were sealed against airflow by plastic film.

In order to map the air flow pattern inside the material, the temperature distribution of test plate 2 was measured by 25 thermocouples of iron-constantan, wire thickness 0.1 mm. The thermocouples were placed in a section through the centre of the specimen according to figure 5. The thermocouple wires were placed in the test plate in such a way that they couldn't contribute to temperature-equalization in the measuring points in the directions of the convection current or of the heat flow.

If convection exists in a plate with vertical arrangement a temperature distribution in the cut can be expected, in principle, which is evident from figure 6, provided that the plate has got uniform structure as seen from the viewpoint of heat transfer. Unfortunately, this condition has not been realized for glasswool which causes distortion of the isotherms an difficulties to illustrate conformities in the temperature distribution.

If the convection is weak, the difficulties are increasing still more.

In order to illustrate small convection currents in inhomogenous material it is possible to proceed in the following way: the temperature distribution in the section is determined with the plate arranged sometimes horizontally with the heat flow turned

Fig. 3

downwards, sometimes vertically and sometimes [vertically but rotated 180° in relation to its former position. The convection currents at the two vertical arrangements are counter-directed (see fig. 7). The temperature distributions of the section at the three arrangements are introduced as relative values compared to the temperature distributions at a horizontal arrangement of the plate. If the temperature curves of the three arrangements are drawn on the same figure it is, in case of convection, possible to expect a configuration of principally the following shape (see fig. 8). In the figure the temperature curves marked "*a*" correspond to horizontal arrangement, "*b*" to the first vertical arrangement, and "*c*" to the second one. Curves for five rows of measurement points according to figure 5b are given.

233

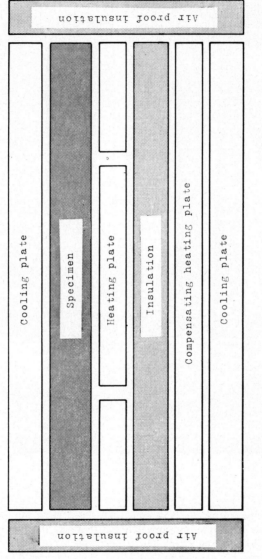

Fig. 4

234

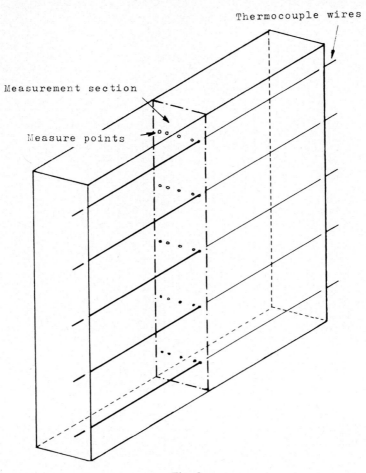

Fig. 5a

TEST RESULTS

Determination of thermal conductivity for the specimen at different positions. Material: glasswool "Gullfiber 3004".

<table>
<tr><td colspan="4">Series No. 1</td></tr>
<tr><td>Bulk density:
Thickness:</td><td>16.7 kg/m³
95 mm</td><td colspan="2">(1.04 lb/cu.ft)
(3.74 in.)</td></tr>
<tr><td>Measurement No.</td><td>1</td><td>2</td><td>3</td></tr>
<tr><td>Direction of heat flow through
  the specimen</td><td>↓</td><td>→</td><td>←</td></tr>
<tr><td>Mean temperature, °C</td><td>+9.2</td><td>+9.4</td><td>+9.7</td></tr>
<tr><td>Temperature difference, °C</td><td>38.3</td><td>38.6</td><td>38.7</td></tr>
<tr><td>Thermal conductivity
  kcal/m hr.°C
  Btu in./sq. ft. hr.°C</td><td>0.0317
0.256</td><td>0.0317
0.256</td><td>0.0316
0.255</td></tr>
</table>

235

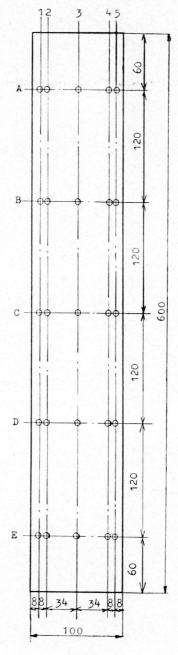

Fig. 5b

236

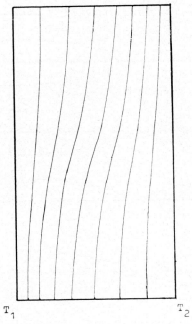

Isothermes in a section of a
homogenous specimen with convection

$$T_1 > T_2$$

Fig. 6

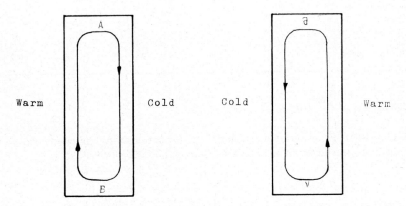

Fig. 7

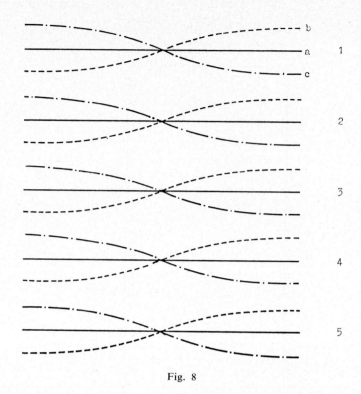

Fig. 8

Series No. 2

Bulk density:		16.0 kg/m³		(1.0 lb/cu.ft)	
Thickness:		100 mm		(3.94 in.)	

Measurement No.	1	2	3	4
Direction of heat flow through the specimen	↓	↑	→	←
Mean temperature, °C	+11.5	+11.6	+11.5	+11.6
Temperature difference, °C	41.0	41.5	41.2	41.6
Thermal conductivity				
kcal/m hr.°C	0.0317	0.0315	0.0317	0.0315
Btu in./ sq.ft.hr.°F	0.256	0.254	0.256	0.254

Determination of air permeability of the material

For a closer definition of the material the air permeability, perpendicular to the fibre direction was determined according to a method described in [9].

TEST RESULT

Pressure difference over the specimen: 0.5 mm of water.
Air permeability: 5.3 m³/m.hr.mm of water.

238

Determination of temperature distribution of the specimen

After the temperature values being treated according to the method stated above, the result of the temperature measurement appears from figure 9.

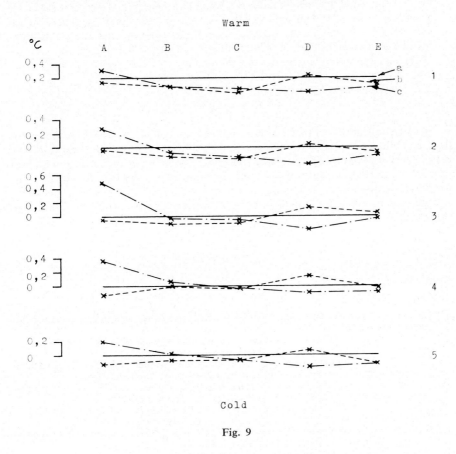

Fig. 9

COMMENTS ON THE MEASUREMENT RESULTS

The size of the convective heat transport

As appears from the test result above, no difference between the measured thermal conductivity values for various directions of heat flow through the test plate was obtained within the reliability range of the method. This means that any measurable convection part cannot be indicated in the heat transport through air-proof sealed test plates of light glass-wool (bulk density 16 kg/m³, 1 lb/cu.ft and air permeability 5.3 m³/m.hr.mm of water), dimensions $600 \times 600 \times 100$ mm at a temperature difference of 40 deg C and a mean temperature of $+10\,°\text{C}$.

Temperature distribution inside the test plate

When studying the relative temperature distribution curves according to figure 9, the following observations can be made:

a) The two curves for vertical arrangement of the test plates (*b* and *c*) are only a little different from the curve for horizontal arrangement (*a*). The difference is approaching the size of the error for temperature measurement, which is estimated to be $\pm 0.1\,°\text{C}$.

b) On the other hand the form of the curves looks like the expected temperature curves according to figure 8. This indicates that convection currents of the type sketched before probably exist in the material. As appears from the quantitative determination (see above) the current is not sufficient to increase the thermal conductivity of the material in any degree worth mentioning.

CONCLUSION

Since a combination of material type and exterior conditions had been chosen for the examination causing the *conditions* of a high convection part in the heat transport the conclusion of the result will be the following. The influence of convection on heat transfer in insulation of mineral wool in buildings can be neglected provided that the insulation is cased without adjoining air space in the wall construction. This conclusion is confirmed by other test results [4, 5, 6, 10] which also have included storage-high specimens of the material.

REFERENCES

[1] E. GRIFFITHS, Heat Insulators, London (1929).
[2] E.A. ALLCUT and F.G. EVANS, Thermal conductivity of insulating materials, *Canad. J. Res.*, A 17 (1939).
[3] G. LORENTZEN and E. BRENDENG, The influence of free convection in insulated vertical walls, *Insulation Rev.*, 4 (1960).
[4] W.F. CAMMERER, Der Konvektions-einfluss auf die Wärmeleitfähigkeit von Wandisolierungen aus Mineralfasern. *Allg. Wärmetechnik*, 11 (1962).
[5] H. ZEHENDNER, Einfluss der freien Konvektion auf die Wärmeleitfähigkeit einer leichten Mineralfasermatte bei tiefen Temperaturen, *Kältetechnik*, Band 16 (1964).
[6] S. WOLF, K.R. SOLVASON, and A.G. WILSON, Convective Air Flow Effects with Mineral Wool Insulation in Wood-Frame Walls, *ASHRAE Transactions* (1966).
[7] I. HÖGLUND and T. HANSSON, Inre konvektion i byggnadskonstruktioner, Meddelande 37 från KTH (1964).
[8] B. GÖSTRING, Plattapparat för bestämning av värmeledningsförmåga hos byggnadsmaterial. Rapport från Byggforskningen, no. 1667.
[9] Metod för bestämning av luftgenomsläpplighetstal för isolermaterial enligt "skyddsringprincipen" SP VVS-2-68, Stockholm (1968).
[10] W.A. LOTZ, Heat and air transfer in cold storage insulation, *ASHRAE transactions*, vol. 70 (1964).

DISCUSSION

D. FOURNIER (France) — L'expérimentation de M. PALJAK est voisine de celle que nous avons effectuée nous-mêmes et que nous présentons dans un autre rapport soumis à cette réunion de Liège. Il nous semble tout à fait normal que quantitativement il n'ait pu mettre en évidence la présence de convection dans la laine de verre soumise à ses conditions d'expérience. Nous ne sommes par contre pas d'accord avec le fait qu'il estime que, malgré tout, des courants de convection « existent probablement » dans le matériau fibreux soumis aux conditions d'expérience particulières choisies par lui. En effet un travail très approfondi, à la fois sur le plan théorique et sur le plan expérimental, effectué récemment en France dans le cadre d'un contrat de recherche « Euratom » relativement à la convection naturelle dans les isolants fibreux en espace confiné (référence [8] par exemple, citée dans notre rapport p. 331) a montré qu'il existe un *seuil* de déclenchement des mouvements convectifs dans un matériau fibreux. Ce seuil ainsi que le degré d'importance de la convection étant liés à deux nombres sans dimensions :

1. Le nombre de Rayleigh de filtration où figurent les paramètres thermiques et ceux caractérisant le milieu fibreux (en particulier la perméabilité).

2. L'allongement de la cellule c'est-à-dire le rapport $\dfrac{hauteur}{épaisseur}$.

Dans les conditions d'expérimentation de M. PALJAK ces résultats montrent que la convection ne peut exister, même à l'état de traces.

I. PALJAK — In answer to Mr. FOURNIER's second statement I would recall that the qualitative determination of the free convection by temperature measurement on a section inside the specimen in fact indicated a configuration of the relative isotherms on the specimen which was principally similar to the *expected* configuration of the relative isotherms. It is agreed that the difference between the reference curve and the relative curves for both the vertical orientations are small and within the measurement error.

W. MOLNAR (U.K.) — I should like to make the following comments on convective heat transfer. My work has been involved with insulation between 90 and 290°K where convection currents can be disastrous and increase the ideal heat inleak by a factor of at least 3. My predecessor at British Oxygen, Mr. L. L. Katan, found that using a lightweight mineral wool 100 mm thick there was no change in effective conductivity as the height of the insulation increased from 300 to over 1,200 mm whereas one would expect a decrease if the "race track" theory shown in figure 1 were true. Because of the random nature of fibres in the insulation the air currents are broken up and result in eddy diffusion within the insulation so that height to thickness ratio is 1/1. This has been found by Prof. Haselden at Leeds University. Thus I am not surprised that he did not obtain his theoretical curves. His conductivities are about 50% higher than that for still air and this could be accounted for by radiation or convection. Convection could account for this increase in heat flow but its total contribution is difficult to assess. The question I would like to ask is whether the warm surface had been put on the top to cut out convection entirely and if so, what was the effect? Also had he tried a more dense grade of wool say 60-70 kg/m³ to see whether he obtained a lower conductivity?

I. PALJAK — As an answer to Mr. MOLNAR, I would say the following.

a) An investigation of Zehendner no. [5] in the reference table, p. 240, shows that measurable free convection for mineral wool (8 kg/m³) starts at about 250°K for the temperature gradient used in my work.

b) The influence of height of the specimen is discussed, among others, by Cammerer [4] and he has found that convection increases with reduction of height of a vertical plate till the height equals the thickness.

c) It is very probable that the air flow in the specimen has the shape of a "race track" because the smallest resistance to air flow is in the direction of fibres in the specimen, e.g. parallel to the long sides.

d) A k-value for a horizontal position with the warm surface upwards was used as a reference for determination of the existence of convection for the other positions of the specimen.

e) Mineral wool with a bulk-density higher than 16 kg/m³ was not examined as the probability for convection in this case is still smaller.

TRANSFERT DE CHALEUR DANS UNE ISOLATION POREUSE AVEC DIFFUSION DE GAZ À TRAVERS LES PAROIS

G. SAINT GIRONS

AFICO, Lausanne (Suisse)

Heat transfer in a porous insulation with gas diffusion through the walls

SUMMARY: *When a room is insulated with cellular materials, a slight excess pressure produces a diffusion of the gas through the walls.*

Assuming some simplifications in the calculation, equations are established showing the heat transfer respectively by conductivity and by gas diffusion.

This kind of "dynamical" insulation results in economic advantages over a conventional "statical" insulation, subject to the condition that the gas is not recycled but dispersed in the surrounding atmosphere after the diffusion. In such a case, which corresponds to any refrigerating system using cryogenic fluids such as liquid nitrogen, that part of the apparent heat remaining available in the gas is directly recovered.

If the cryogenic fluid is also used, besides cooling the room, for other refrigerating purposes such as freezing merchandise within the room, that part of the fluid required for balancing the heat losses through the walls is still further reduced.

Savings in fluid requirements are calculated in two concrete applications using liquid nitrogen.

Theoretically, the "dynamical" insulation is also applicable to an air conditioning system where a part of the air is permanently renewed in open circuit.

Les isolations en matériaux cellulaires sont par nature plus ou moins poreuses et un léger excès de pression suffit pour engendrer un transfert permanent de gaz à travers la paroi. Un tel système est appelé « Isolation dynamique » et peut dans certains cas présenter des avantages techniques et économiques par rapport à l'isolation statique classique [1].

On se propose de calculer le bilan des transferts de chaleur respectivement dûs à la conduction et à la diffusion du gaz dans une isolation dynamique, puis d'en déduire les économies réalisables dans certaines applications.

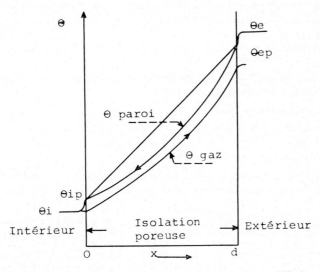

Fig. 1 — Gradient des températures dans une paroi poreuse avec diffusion du gaz vers l'extérieur à travers la paroi.

La figure 1 représente le gradient des températures réelles et la figure 2 correspond à l'hypothèse simplificatrice adoptée pour le calcul, c'est-à-dire en supposant que les températures intérieures θi et extérieures θe relatives à l'ambiance, à la paroi et au gaz peuvent être respectivement confondues, ce qui implique que la vitesse de diffusion du gaz à travers l'isolation poreuse est suffisamment faible pour obtenir en chaque point l'égalisation complète des températures.

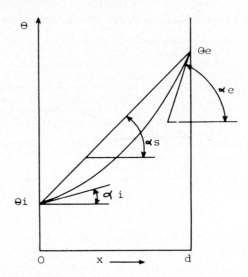

Fig. 2 — Représentation du même gradient de température suivant les hypothèses simplificatrices utilisées dans le calcul.

On admet de plus que la conductibilité λ du matériau d'isolation n'est pas modifiée par le phénomène de diffusion du gaz.

On appelle :
— p le débit du gaz par unité de surface de la paroi et c sa chaleur spécifique ;
— d l'épaisseur de l'isolation en prenant la face intérieure de la paroi pour origine de l'axe des x ;
— α l'angle avec l'axe des x de la tangente à la courbe représentant le gradient de température ;
— w le flux d'énergie transférée sous forme de chaleur soit par conduction soit par suite de la diffusion du gaz.

et on adopte les indices suivants :
— i et e pour les faces respectivement intérieure et extérieure de la paroi ;
— g pour le gaz ;
— s pour l'isolation statique classique.

On considère que la paroi isolée est suffisamment grande pour que la température d'un plan intermédiaire d'abscisse x soit uniforme, une fois le régime stationnaire établi.

Dès lors on peut écrire que la différence entre les quantités de chaleur entrant par conduction à travers les 2 faces du solide limité par le plan x et le plan infiniment voisin, soit :

$$\lambda \frac{d^2 \theta}{dx^2}$$

est égale à la quantité de chaleur captée par le gaz qui traverse l'isolation soit :

$$pc \frac{d\theta}{dx}.$$

D'où l'équation :

$$\lambda \frac{d^2\theta}{dx^2} - pc \frac{d\theta}{dx} = 0$$

soit en intégrant et compte tenu des conditions aux limites :

$$\theta = (\theta e - \theta i) \frac{e^{\frac{pcx}{\lambda}} - 1}{e^{\frac{pcd}{\lambda}} - 1} + \theta i.$$

On remarque que l'exposant de e dans cette équation dépend du rapport r entre :

— la quantité de chaleur captée par le gaz

$$Wg = pc(\theta e - \theta i)$$

— et la quantité de chaleur transmise par conduction dans une isolation statique classique de même épaisseur d et même conductibilité λ :

$$Ws = -\frac{\lambda}{d}(\theta e - \theta i).$$

Posons :

$$R = -r = -\frac{Wg}{Ws} = \frac{pcd}{\lambda}$$

et l'équation devient

$$\theta = (\theta e - \theta i) \frac{e^{R(x/d)} - 1}{e^R - 1} + \theta i.$$

Le flux de chaleur transmis par conduction à travers le plan d'abscisse x est égal à :

$$W = -\lambda \frac{d\theta}{dx}$$

soit

$$W = -\frac{\lambda}{d}(\theta e - \theta i) \frac{Re^{R(x/d)}}{e^R - 1}$$

qu'on peut aussi écrire :

$$W = -pc(\theta e - \theta i) \frac{e^{R(x/d)}}{e^R - 1}.$$

Au niveau des faces de la paroi on aura

$$Wi = -pc(\theta e - \theta i) \frac{1}{e^R - 1}$$

$$We = -pc(\theta e - \theta i) \frac{e^R}{e^R - 1}$$

ou encore :

$$W_i = -W_g \frac{1}{e^R - 1} \qquad (1) \qquad \text{et} \qquad W_e = -W_g \frac{e^R}{e^R - 1} \qquad (2)$$

et finalement :

$$(-W_e) = (-W_i) + W_g$$

comme représenté sur la figure 3 où l'on voit comment la quantité de chaleur entrant par la face extérieure (We) sert en partie à réchauffer le gaz extrait de la paroi par diffusion et comment la chaleur entrant par la face intérieure (Wi) est donc réduite d'autant.

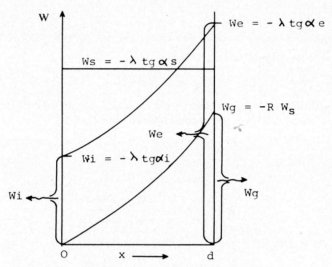

Fig. 3 — Bilan des transferts de chaleur respectivement par conduction et par diffusion du gaz.

1. Comparaison avec l'isolation statique classique

On a vu que le flux de chaleur transmise par conduction dans la même paroi d'épaisseur d utilisée comme isolation statique classique était :

$$W_s = -\frac{1}{R} W_g$$

d'où :

$$W_i = W_s \frac{R}{e^R - 1} \qquad \text{et} \qquad W_e = W_s \frac{Re^R}{e^R - 1} \qquad (3)$$

et l'on a toujours :

$$(-W_i) < (-W_s) < (-W_e) \qquad (4)$$

avec des différences d'autant plus grandes entre les membres de l'inégalité que

$$R = \frac{pcd}{\lambda}$$

246

est plus élevé, c'est-à-dire l'isolation meilleure ou le débit et la chaleur spécifique du gaz eux-mêmes plus élevés.

Sachant que

$$Tg\,\alpha = -\frac{W}{\lambda}$$

on a la même inégalité

$$Tg\,\alpha i < Tg\,\alpha s < Tg\,\alpha e$$

comme on s'en rend compte sur la figure 2.

Il résulte de l'inégalité (4) que l'isolation dynamique ne présentera un avantage sur l'isolation classique que si le gaz diffusant à travers la paroi est évacué à l'extérieur sans recyclage.

Il est clair que l'isolation dynamique permet alors de récupérer une partie de la chaleur sensible du gaz encore disponible entre les températures d'ambiance extérieure et intérieure pour réduire dans une certaine proportion la quantité de chaleur qui pénètre effectivement dans l'enceinte, c'est-à-dire celle qui traverse la paroi intérieure de l'isolation.

2. Enceintes refroidies par évaporation d'un fluide cryogénique

Dans de telles enceintes, le fluide évaporé est évacué à l'extérieur sans recyclage d'où l'intérêt d'une isolation dynamique.

2.1 *Débit de fluide évaporé avec une isolation dynamique d'épaisseur* d *pour maintenir une température* θi.

On appelle q la quantité de chaleur absorbée par l'évaporation de l'unité de poids du fluide et par son réchauffement jusqu'à la température θi et l'on désigne par K le rapport :

$$K = \frac{q + c(\theta e - \theta i)}{q}$$

où $c\,(\theta e - \theta i)$ représente la part de chaleur sensible récupérable, d'où :

$$\theta e - \theta i = \frac{q}{c}(K - 1).$$

On aura

$$(-Wi) = pq$$

et

$$Wg = pc(\theta e - \theta i) = pq(K - 1)$$

donc, en appliquant (1) :

$$e^R = K$$

d'où le débit de fluide d'après la définition de R :

$$p = \frac{\lambda}{dc}\,Lg\,K. \tag{5}$$

2.2. *Economie de fluide par rapport à l'isolation classique*

Avec une isolation classique, le poids de fluide évaporé serait :

$$p_s = -\frac{Ws}{q} = \frac{\lambda}{dq}(\theta e - \theta i) = \frac{\lambda}{dc}(K-1).$$

L'économie de fluide s'exprime par

$$\frac{p_s - p}{p_s} = 1 - \frac{Lg\ K}{K-1} \tag{6}$$

d'autant plus élevé que K, c'est-à-dire la part de chaleur sensible récupérable, est plus élevée.

2.3 *Accroissement de l'économie de fluide nécessaire à la compensation des déperditions dans le cas où le débit total de fluide est augmenté*

On a supposé jusqu'ici que la totalité du fluide évaporé servait uniquement à compenser les déperditions internes $(-Wi)$. Si au contraire une quantité supplémentaire est évaporée, par exemple pour congeler des produits, le débit effectivement consommé pour compenser les déperditions internes de l'isolation dynamique sera encore réduit par rapport à une isolation classique.

Soit le nouveau débit total de fluide évaporé :

$$p' = np$$

dont une partie seulement soit p'_u sert à compenser les déperditions internes. On a encore d'après (1)

$$W'i = -\frac{W'g}{e^{R'}-1}$$

avec

$$W'g = p'q(K-1)$$

et

$$p'_u = -\frac{W'i}{q}$$

d'où

$$p'_u = p'\frac{K-1}{e^{R'}-1}$$

comme

$$e^{R'} = (e^R)^n$$

et

$$e^R = K$$

En définitive :

$$p'_u = np\frac{K-1}{K^n-1}.$$

248

L'économie de fluide due à l'isolation dynamique devient :

$$\frac{p_s - p_u'}{p_s} = 1 - n \frac{pcd}{\lambda} \frac{1}{K^n - 1}$$

ou :

$$\frac{p_s - p_u'}{p_s} = 1 - \frac{\mathrm{Lg}\, K^n}{K^n - 1} \tag{7}$$

puisque

$$\frac{pcd}{\lambda} = R = \mathrm{Lg}\, K$$

Comme le terme $\mathrm{Lg}\, K^n / K^n - 1$ décroît toujours si n augmente, on voit que l'économie réalisée sur la consommation de fluide nécessaire pour compenser les déperditions augmente avec le débit total de fluide évaporé.

Ce fait résulte logiquement de l'augmentation de la quantité de chaleur sensible disponible pour la récupération par diffusion du gaz dans l'isolation.

2.4 *Calcul d'application à des enceintes refroidies par évaporation d'azote liquide*

On considère deux cas :

a) Enceinte à $-196\,°C$ c'est-à-dire à la température d'évaporation de l'azote sous la pression atmosphérique où l'on a sensiblement :

$$K_A = 2$$

b) Enceinte à $-30\,°C$ où l'on a sensiblement :

$$K_B = \frac{10}{9}.$$

En appliquant la formule (7), on obtient le tableau suivant :

	Économies de fluide pour les déperditions	
	Cas A : $\theta i = -196\,°C$	Cas B : $\theta i = -30\,°C$
$n = 1$	31 %	4,6 %
$n = 2$	54 %	10 %
$n = 3$	70,4 %	14,6 %
$n = 4$	81,6 %	19,1 %
$n = 5$	88,9 %	23,5 %
$n = 6$	93,4 %	27,8 %

3. APPLICATION AU CONDITIONNEMENT D'AIR

Dans les installations de climatisation comportant un renouvellement d'air permanent en circuit ouvert et où une légère surpression pourrait être admise, il serait théoriquement possible d'évacuer l'air vicié par diffusion à travers les parois et de bénéficier ainsi des économies décrites.

Si par exemple le taux de renouvellement d'air est tel que la puissance frigorifique nécessaire au refroidissement de l'air frais prélevé à l'extérieur soit égale aux déperditions statiques, on aura :

$$Wg = -Ws$$

et par conséquent $R = 1$

On tirera de (3) l'économie réalisée sur les déperditions, soit :

$$\frac{Ws - Wi}{Ws} = 1 - \frac{R}{e^R - 1} = 41\%.$$

On peut vérifier que de telles conditions sont cohérentes et correspondraient à un taux de renouvellement d'air de l'ordre de 1 fois par heure.

RÉFÉRENCE

[1[N. A. GOLOVKIN and N. N. KOSHKIN, New cold store insulation system. *Annexe 1960-3, Bull I.I.R.*, pp. 443-447. Commission V, Marseille.

DISCUSSION

E. EMBLIK (Suisse) — Pour obtenir le débit d'air à travers l'isolation dynamique, il faut maintenir une pression élevée dans la chambre. Est-que l'énergie nécessaire pour le ventilateur donnant cette surpression est comprise dans les calculs d'économie ?

G. SAINT-GIRONS — Il n'a pas été tenu compte, dans le calcul, de l'énergie nécessaire pour maintenir la surpression qui engendre elle-même la diffusion du gaz à travers l'isolation mais, dans le cas des fluides cryogéniques, cette énergie n'est pas en cause. En ce qui concerne l'application au conditionnement d'air, il existe des cas où une surpression est volontairement maintenue en tout état de cause, pour d'autres raisons. Si au contraire une dépression permanente est nécessaire, par exemple par sécurité, la diffusion pourrait s'effectuer de l'extérieur vers l'intérieur et un calcul analogue du bilan des échanges thermiques pourrait être établi.

C. F. KAYAN (U.S.A.) — A brief question on the order of magnitude of the pressure drop associated with the cross-insulation fluid flow.

(I have been interested in an electrical analogue study of this type of problem, carried out some years ago.)

G. SAINT-GIRONS — N'ayant pas fait d'expériences personnelles je peux seulement répondre au Prof. KAYAN en me référant à la communication présentée par MM. N. A GOLOVKIN et N. N. KOSHKIN à la réunion I.I.F.-Com. V de Marseille en 1960 d'après laquelle :
— la surpression nécessaire pour obtenir un débit d'air de 1,5 à 2,5 kg/m².h par diffusion à travers l'isolation dans une application pratique était de 2 à 6 mm de colonne d'eau.

W. MOLNAR (U.K.) — I agree with the conclusions in Mr. SAINT-GIRONS' paper but the difficulty of obtaining an insulant with sufficient open porosity to allow gas transmission without setting up convection currents may be very great. Such convection currents would negate the advantages achievable with his theory. The soundness of his theory has been proven because I have used the principle

of utilizing the refrigeration in the gas for the storage of liquid helium. At 100 %
efficiency loss rate decreases by a factor of 70 can be obtained and we have obtained a
reduction of a factor of 25.

G. SAINT-GIRONS — D'après les réponses faites aux questions du Prof.
EMBLIK et du Prof. KAYAN, la surpression nécessaire est faible et le matériau d'isolation à considérer ne devrait pas présenter un coefficient de porosité plus élevé que
les isolations cellulaires usuelles.

De plus si l'on évoque les phénomènes séparés de conduction, radiation et convection interne qui prennent place à l'intérieur d'un matériau pour concourir à
former ensemble sa conductibilité globale λ, ils sont tous pris en compte dans le
calcul présenté.

On peut même penser que les courants de convection interne sont diminués par
suite de l'écoulement du gaz et ceci pourrait expliquer la valeur réduite de conductibilité apparente Ω, trouvée expérimentalement par MM. GOLOVKIN et KOSHKIN
d'après leur communication I.I.F.-Marseille déjà citée.

H. JUNGNICKEL (East Germany) — Only a small remark on Mr. SAINT-
GIRONS' paper. In practice, insulations at -200 °C radiation should not be neglected.
Therefore one must install radiation shields which are of metal and make it impossible
to use this diffusion-cooling effect.

But this effect has long since been used under the name of "transpiration
cooling" in rocket technology. You remember that it is useful for protecting the
combustion chambers of rockets and it is done much work in this direction.

G. SAINT-GIRONS — La théorie exposée ne s'applique bien entendu qu'aux
isolations cellulaires poreuses et non aux isolations par le vide ou par multicouches.
Comme souligné, l'isolation dynamique n'a d'intérêt que s'il peut s'établir un courant
de diffusion du gaz en circuit ouvert et les applications pratiques éventuelles ne
sauraient être envisagées que cas par cas.

J.C. MARÉCHAL (France) — J'ajouterai que la terminologie utilisée peut
prêter à confusion. Que l'on cherche à utiliser les gaz froids détendus qui seraient
perdus en les utilisant dans un échangeur qui se trouve être la paroi poreuse, c'est
légitime sur le plan économique. Il ne semble pas correct de considérer ce passage
forcé sous le vocable de «transfert de chaleur avec diffusion»—surtout que dans la
proposition des conditions on veut ignorer les variations du matériau.

ON THE WATER VAPOR TRANSMISSION THROUGH POROUS MATERIALS*

V. FERRO and A. SACCHI

Istituto Fisica Tecnica del Politecnico di Torino (Italy)

Diffusion de la vapeur d'eau à travers des matériaux poreux

RÉSUMÉ : *On expose quelques considérations sur le phénomène de la diffusion de la vapeur d'eau à travers des matériaux poreux en conditions stationnaires et, sur la base de certaines hypothèses simplifiées, on développe le problème en partant de la loi de Fick.*

On fait la description d'un appareil réalisé pour la mesure de la perméabilité à la vapeur d'eau à travers des matériaux poreux, appareil qui comprend deux chambres à température et degré hygrométrique contrôlés, séparées par le panneau du matériel en essai.

L'appareil est pourvu d'un dispositif qui mesure la masse de vapeur d'eau transmise à travers le panneau.

On examine ensuite les résultats de certains essais faits sur des matériaux tels que le liège expansé émulsionné avec bitume, mousse phénolique et polystyrènes expansés.

1. The diffusion of atmospheric water vapour through porous materials used in buildings and especially in cold stores, can be dealt with easily if it is considered as the flow of a single gaseous fluid through a porous material in steady state and isothermal conditions, subjected only to the partial pressure gradient of the vapour.

In this condition, the flow can be considered as resulting by molecular (or Knudsen) and laminar (or Poiseuille) flow.

In reality, however, the phenomenon is much more complicated. Even if the steady state and isothermal conditions are left unchanged, it must more generally be considered as an ordinary molecular and laminar diffusion flow of a binary mixture of two gaseous fluids, air and water vapour, between two rooms with different concentrations, and hence with different partial pressures, separated by a porous wall.

In the case under examination it should also be observed that the surface of most porous building materials has a certain affinity for water molecules. The corresponding molecular attraction forces binding water molecules to this surface, rapidly decrease with increases in the mean free path of the molecules and thus, under isothermal conditions, with decreases in ambient relative humidity ψ.

This means that the thickness of the molecular film of water on the surface in equilibrium with surrounding increases with increases in ψ.

Near the saturation point, small pores are totally filled with water, while larger pores are partially filled.

In function of the values of relative humidity of the two rooms, the migration of water inside a porous material having an affinity for water, will take place through series and parallel capillary flows of the vapour and liquid phases, the influence of the latter increasing with increases in the value of ψ. These two phases cannot be dealt with separately since they are strictly interconnected through the flow paths as a result of alternating evaporation and condensation.

Further complications of the phenomenon are also introduced by the presence of electrical potentials and of salts in the porous material.

All these difficulties mean that a simplified treatment of the problem is desirable. Here it has been assumed that the Fick's law applies to the phenomenon from a macroscopical point of view.

(*) This work was sponsored by C.N.R. under the contract 115/0284/0/4653.

2. The equation used in the transport of water vapour through a slab consisting of homogeneous, isotropic and isothermal material, is based on Fick's law, which in unidimensional form is written:

$$dG = -\mu dS \frac{dp}{dx} d\tau \qquad (1)$$

where:

G	mass of water vapour transported through the material	[kg]
S	frontal surface area perpendicular to flow direction	[m^2]
p	partial pressure of the vapour	[mbar]
x	distance, in the direction of flow	[m]
τ	time	[hr, sec]

The coefficient μ, which relates the vapour flow per unit of surface area, $(dG/d\tau).(1/S)$, to the vapour partial pressure gradient, dp/dx, is called the "permeability to water vapour" and is a characteristic property of the transport of this fluid into building materials.

This parameter will obviously be dependent on the thermodynamic condition of the water-vapour air mixture, on the affinity of water vapour for the porous material, on the size and shape of the pores and on the interconnection of these pores within the material. Its value is usually given in kg.m/hr.m^2.mbar or in kg.m/hr.m^2.mmHg.

If in a given material, whose porosity is assumed to be homogeneously and isotropically distributed, all points having the same partial vapour pressure are assumed to be connected, a series of isobaric surfaces will be formed and, by analogy with thermal conduction, one can define the flow lines and the flow tubes of the vapour in the material.

The analogy with thermoconduction is only formal, since the two phenomena concern the first the energy transport and the latter the mass transport. In the last case a changement of aggregation state may happen.

It may be remarked that, as given by equation (1), the water vapour permeability is valid, under isothermal conditions, for similar thickness, density, porosity and disposal of the specimen as well as for constant relative humidity values.

For the same slab, with given orientation, in steady state and isothermal conditions, permeability is therefore a function of the degree of humidity and hence of the partial vapour pressure.

We can therefore write [1]:

$$\frac{\int_{p_2}^{p_1} \mu dp}{p_1 - p_2} = \bar{\mu}, \qquad (2)$$

where $\bar{\mu}$ is a mean coefficient of permeability. The relationship between μ and p is generally non-linear and may be expressed by an empirical expression of the type:

$$\mu = a + bp + cp^2 + \dots \qquad (3)$$

Equation (2) can be expanded to give:

$$\bar{\mu}(p_1 - p_2) = \int_{p_2}^{p_1} \mu dp = \int_{p_2}^{p_1} (a + bp + cp^2 + \dots) dp$$

$$= a(p_1 - p_2) + \frac{b}{2}(p_1^2 - p_2^2) + \frac{c}{3}(p_1^3 - p_2^3) + \dots \qquad (4)$$

If $\overline{\mu}$ values are determined for a pre-arranged value of p_1 and for three different values of p_2, equation (4) can be used to evaluate the constants a, b and c. Equation (3) is then employed to determine μ in function of p with a sufficient degree of approximation within the experimental field p_1, p_2.

3. The transport of water vapour in terms of equation (1) under isothermal conditions is of somewhat limited application, since isothermal diffusion of atmospheric water vapour through the outer walls of buildings is rarely encountered in practice. Radiation or superficial and internal evaporation of fluid in the walls, or the more common case of a wall separating two rooms at different temperatures, are accompanied by the heat transport, which interferes with the mass transport.

When an initially isothermal, homogeneous and isotropic wall, containing a uniformly distributed mass of water vapour, is subjected to a temperature gradient, the vapour tends to migrate from the higher to the lower temperature zone. If the vapour contained in the wall has initially a certain relative humidity value, it will flow through the pores in the direction mentioned. The value of ψ will gradually increase and there may be condensation in the direction of the cold zone according to the value of the temperature difference through the wall. The diffusion flow may thus go through the wall wholly or partly in the liquid phase; this phenomenon is also dependent on the relative humidity of the colder room, especially if the wall is very porous.

The diffusion of water vapour in a porous slab under the combined effect of heat and mass transfer has been studied by several authors [2, 5]. However no quantitative relationships valid for open systems in general, such as the external walls of buildings, have been determined [6], and therefore the phenomenological coefficients of mutual interference are difficult to evaluate.

A first approximate formulation can be obtained by linearisation of the terms appearing in more complex relationships, such as given in [2, 3, 5]. For the specimen under study, one can write:

$$M = -A\frac{dp}{dx} - B\frac{dT}{dx} \qquad (5)$$

where $M = G/S\tau$ represents the mass transferred per unit of time and of frontal area of the specimen, expressed in $kg/hr.m^2$, and A and B are parameters derived from the linearisation of the problem.

From (5) it follows:

$$\mu^* = -\frac{M}{dp/dx} = A + B\frac{dT/dx}{dp/dx} \qquad (6)$$

where μ^* can be considered as an equivalent permeability, which is a function of the mean temperature and of its gradient. Equation (1) can be derived from (6) for $dT/dx = 0$, and one obtains $A = \mu$. Therefore:

$$\mu^* = \mu + B\frac{dT}{dp} \qquad (7)$$

4. The apparatus, we set up for these measurements, is formed of two cubic isothermal rooms (fig. 1), each with approximately. 5 m sides separated by test slabs of surface area $.50 \times .50$ m and with a thickness varying from 1 to 100 mm.

The slab edges were sealed with resin size. In the first room relative humidity is maintained at 100%; in the latter, values are adjustable to a constant level in the 20%-80% range by means of an electrically-operated lithium chloride hygrometer pre-set to a given humidity value. The instrumental error is $\pm 1\%$ of the full scale and

255

the sensitive element is calibrated by comparison with a dewpoint hygrometer. The reading of hygrometer is the input of a controller which switches on the humidity-absorbing element of silica gel. The relative humidity value is thus maintained at about the pre-set level to the same degree of accuracy as the sensitive element.

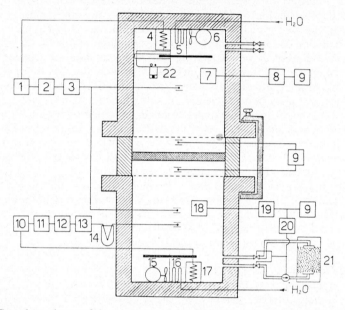

Fig. 1 — Complete scheme of the apparatus. 1-10 Magnetic amplifiers; 2-11 PID temperature controllers; 3-12 Deviation amplifiers; 4-17 Heating elements; 5-16 Refrigerating exchangers; 6-15 Fans; 7-18 Li Cl humidity sensors; 8-19 RH on-off controllers; 9 RH and temperature recorders; 13 Set point unit; 14 Reference thermocouple joint at 0°C; 20 Power relays; 21 Silica gel dryer; 22 Water reservoir.

The water vapour mass transmitted through the test material under steady state conditions is detected by the corresponding increase in mass of the silica gel.

Variations in temperature and humidity in the two rooms during the experiments are determined by means of a multipoint recording potentiometer, which has accuracy $\pm$ .2% and sensitivity .03% of the full scale.

In both rooms a thermal control system consists of a small cooling radiator placed between a fan and an electrical heater, with a chromel-alumel thermocouple connected to a regulation line made by the following components: a set point unit, a deviation amplifier, a P.I.D. controller, a magnetic amplifier power circuit with internal feed back to cut out the effects of variations in the supply voltage; this amplifier is used to feed the electrical heater placed before the radiator.

To avoid radiation effects the specimen under test and thermocouple junctions are shielded with respect to thermal components inside the rooms.

With the described control systems it is possible to maintain the two rooms at a pre-set temperature with an accuracy of $\pm$ .1 °C.

Periodical weighing of the absorbent material on a balance with a 5 kg full scale and a sensitivity of 25×10^{-6} kg is carried out approximately every 12-24 hr (according to the permeability of the material) to determine the point at which the flow of vapour into the test material reached steady state condition.

Steady state times varies from 6 to 15 days in relation to the material permeability.

Figure 2 is an outside view of the apparatus with the measuring and regulating instruments.

Fig. 2 — Photographic view of the apparatus.

5. Four series of tests were run. Two were aimed at determining the permeability of expanded cork with asphaltic binder and expanded polystyrene slabs versus density under isothermal conditions ($+20°$ C), with relative humidity values of 100% and 50% in the upper and lower rooms respectively. The results are given in figure 3 and table I.

The third series of tests was run on the same specimen of expanded phenol resin (50 mm thickness, density 48.5 kg/m^3) at 20° C with a relative humidity range of 20%-

Table I

	Ψ_1	Ψ_2	ρ	$\bar{\mu}$
	%	%	kg/m^3	$\dfrac{10^{-6} \text{ kg.m}}{\text{hr.m}^2\text{.mbar}}$
Expanded cork with asphaltic binder	100	50	98	3.64
	”	”	125	3.15
	”	”	150	2.36
	”	”	171	2.17
	”	”	202	1.54
Expanded polystyrene	100	50	12	2.7
	”	”	16	2.0
	”	”	20.6	1.6
	”	”	26	0.9
	”	”	31.5	0.6

257

Table II

EXPANDED PHENOLIC RESIN $\rho = 48.5$ kg/m^3

Ψ_1	Ψ_2	p_1	p_2	Δp	$\bar{\mu}$
%	%	mbar	mbar	mbar	$\dfrac{10^{-6} \text{ kg.m}}{\text{hr.m}^2.\text{mbar}}$
100	70	23.37	16.36	7.01	7.5
”	60	”	14.01	9.36	6.5
”	50	”	11.68	11.69	5.7
”	40	”	9.34	14.03	5.0
”	30	”	7.01	16.36	4.5
”	20	”	4.69	18.68	4.1

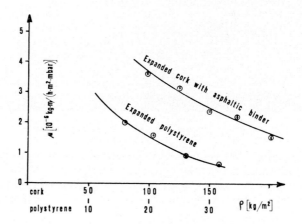

Fig. 3 — Vapor permeability of expanded cork with asphaltic binder and of expanded polystyrene versus density at $\Psi_1 = 100\%$, $\Psi_2 = 50\%$ and $t_1 = t_2 = 20\,°C$.

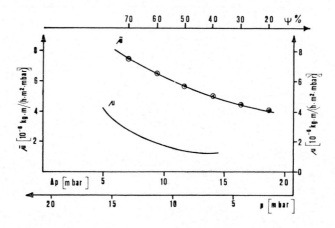

Fig. 4 — Vapor permeability μ and mean vapor permeability $\bar{\mu}$ of expanded phenolic resin (density = 48.5 kg/m^3) versus vapor partial pressure at $t_1 = t_2 = 20\,°C$.

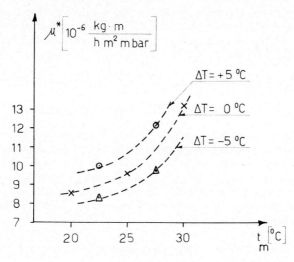

Fig. 5 — Vapor permeability μ^* of expanded phenolic resin (density 56.3 kg/m^3) versus mean temperature $t_m = (t_1 + t_2)/2$ at different values of temperature gradient dT/dx but at constant vapor pressure difference $p_1 - p_2 = 12.9$ mbar.

Table III

EXPANDED PHENOLIC RESIN, $\rho = 56.3$ kg/m^3

t_1	Ψ_1	t_2	Ψ_2	p_1	p_2	Δp	G/τ	μ^*	$t_m = \dfrac{t_1 + t_2}{2}$
°C	%	°C	%	mbar	mbar	mbar	gr/hr	$\dfrac{10^{-6}\ \text{kg.m}}{\text{hr.m}^2.\text{mbar}}$	°C
20	100	20	45	23.37	10.52	12.85	0.46	8.59	20
20	100	25	33	23.37	10.45	12.92	0.45	8.36	22.5
25	100	20	80	31.66	18.70	12.96	0.54	10.00	22.5
25	100	25	59	31.66	18.68	12.98	0.52	9.61	25
25	100	30	44	31.66	18.66	13.00	0.53	9.78	27.5
30	100	25	93.3	42.42	29.54	12.88	0.65	12.11	27.5
30	100	30	70	42.42	29.69	12.73	0.70	13.20	30

Table IV

t_m	A	B
°C	$\dfrac{10^{-6}\ \text{kg.m}}{\text{hr.m}^2.\text{mbar}}$	$\dfrac{10^{-6}\ \text{kg.m}}{\text{hr.m}^2.°\text{C}}$
22.5	9.0	2.2
25	9.6	2.45
27.5	10.9	2.7

70% in the lower room. The mean permeability values ($\bar{\mu}$) are given in figure 4 and in table II; figure 4 also shows the behaviour of μ versus p as determined by means of equation (4). In the field of experimental measurement, equations $\bar{\mu}\,(\Delta p)$ and $\mu\,(p)$ can be expressed analytically as:

$$\bar{\mu} = 11.49 - 6.91 \cdot 10^{-1}\,\Delta p + 1.83 \cdot 10^{-2} \cdot \Delta p^2 - 1.32 \cdot 10^{-4}\,\Delta p^3 \qquad (8)$$

and

$$\mu = 2.438 - 3.19 \cdot 10^{-1} \cdot p + 1.79 \cdot 10^{-2} \cdot p^2 + 5.28 \cdot 10^{-4}\,p^3 \qquad (9)$$

the permeabilities being expressed in 10^{-6} kg.m/hr.m^2.mbar and Δp and p in mbar.

The fourth series of tests investigate the influence of the temperature gradient with constant vapour pressure difference between the two rooms. The specimen is expanded phenolic resin 56.3 kg/m^3 density and 60 mm thickness.

The results are reported in table III and plotted in figure 5.

Table IV gives the values of $\mu = A$ and B deduced by figure 5 and equation (7) in the temperature range $22.5 \div 27.5\,°C$.

REFERENCES

[1] C. CODEGONE. « Sur la détermination des coefficients de température de la conductivité thermique des matériaux isolants », *Annexe 1960-2 Bull. I.I.F.* pp. 103-105, Commission 2, Belgrade.
[2] J. R. PHILIP and D. A. DE VRIES. « Moisture Movement in Porous Materials under Temperature Gradients », *Trans. Am. Geophys. Union*, **38** (1957), pp. 222-232.
[3] D. A. DE VRIES, « Simultaneous Transfer of Heat and Moisture in Porous Media », *J. Geophys. Res.*, **67** (1958), pp. 909-916.
[4] W. WOODSIDE and J. B. CLIFFE : "Heat and Moisture Transfer in Closed Systems of Two Granular Materials", *Soil Sci.*, **87** (1959), pp. 75-82.
[5] S. A. TAYLOR and J. W. CARY : "Linear Equations for the Simultaneous Flow of Matter and Energy in a Continuous Soil System", *Soil Sci. Soc. Am. Proc.*, **28** (1964), pp. 167-172.
[6] T. AMBRUS and J. KOMOLY : "Moisture Movement under the Combined Effects of Temperature and Partial Vapor Pressure Gradient", *J. Refrign*, **11** (1968), pp. 182-183, 197-202, 236-238, 253-255.

DISCUSSION

H. MYNCKE (Belgique) — 1. Personnellement je regrette que dans la présente publication le symbole μ soit utilisé pour la perméabilité à la vapeur d'eau. En général ce symbole indique le facteur de résistance à la diffusion à la vapeur d'eau, tandis que la perméabilité est indiquée par δ.

2. De quelle façon la différence de la pression de vapeur d'eau fut-elle mesurée ? A-t-on tenu compte des coefficients de transition à la surface de l'échantillon ?

V. FERRO — 1. For vapor permeability we used the notation introduced by ASHRAE, but we had no difficulties in adapting different symbols.

2. We did not introduce the film resistance between ambient air and specimen surfaces for the following reasons:

a) the air velocity near the surfaces is very high;
b) the flow resistance of the specimen is much higher than that of the boundary layer;
c) the authors who studied this phenomenon did not introduce a flow resistance similar to the film thermal resistance;
d) in our measurements, with the sensibility of our apparatus and with the specimen used, we did not find functional dependence between thickness and flow resistance. This fact shows the small influence of the resistance of boundary layer or its own absence.

URETHANE RIGID FOAMS:
A REVIEW OF THE FACTORS
AFFECTING THE ATTAINMENT AND RETENTION OF
OPTIMUM THERMAL CONDUCTIVITY AND
THEIR SIGNIFICANCE IN SERVICE APPLICATIONS

G.W. BALL, R. HURD and M.G. WALKER

Imperial Chemical Industries Ltd, Dyestuffs Division
Manchester (United Kingdom)

Mousses rigides de polyuréthane : Facteurs qui influent sur l'obtention et le maintien de la conductivité thermique optimale et leur signification pratique

RÉSUMÉ : *On passe en revue les données de laboratoire, les données pratiques relatives à la conductivité thermique des mousses rigides de polyuréthane; ces résultats ont été obtenus par les auteurs et de nombreux autres chercheurs. On étudie la conductivité thermique de ces mousses, sur la base d'une relation maintenant bien établie, qui peut être utilisée pour calculer la conductivité de l'isolant, d'après des valeurs supplémentaires imputables au transfert de chaleur par conduction, radiation et convection pour les phases gazeuses et solides de l'isolant poreux. On attache une attention particulière à l'aspect le plus important, à savoir le vieillissement. On compare les résultats des essais de laboratoire et des essais pratiques aux relations théoriques de l'évolution du coefficient λ en fonction du temps, qui ont été décrites dans un récent travail de Norton. En pratique, il est confirmé que, au contact de l'air, les échantillons accusent une certaine perte de gaz intracellulaire, le trichloromonofluorométhane. Cependant ces données de laboratoire ne contredisent pas les précisions théoriques de Norton et confirment que les mousses rigides de polyuréthane gardent toujours, à température ambiante, leurs avantages, en ce qui concerne la conductivité thermique, par rapport aux isolants remplis d'air, pour des temps supérieurs aux durées techniques intéressantes.*

1 — INTRODUCTION

Easy processing characteristics combined with unique physical properties —some still not yet fully explored— continue to create novel applications for rigid urethane foams [1, 2]. Whilst their low initial thermal conductivity, low water vapour transmission rate and high strength-to-weight ratio have rarely been disputed, the permanence properties —particularly their ability to retain their low values of thermal conductivity— have been occasionally criticized, and this has prompted a number of authors to examine this question both practically and theoretically [3-17]. This present paper notes the factors determining initial conductivity and compares existing with new thermal conductivity ageing data obtained using the Guarded Hot Plate (G.H.P.), Heat Flow Meter (H.F.M.) and Double Line Probe methods [18, 19, 20].

The thermal conductivity of cellular materials can be sub-divided into four components associated with the three mechanisms of heat tranfer for the solid and gaseous phases (*Equation 1*) [9, 21, 22, 23].

$$\lambda = \lambda_G + \lambda_S + \lambda_R + \lambda_C \tag{1}$$

Where

λ the conductivity of the cellular material;
λ_G the conductivity component associated with conduction through the gaseous phase;
λ_S the conductivity component associated with conduction through the solid phase;

261

λ_R the conductivity component associated with radiative transfer across the cells; and

λ_C the conductivity component associated with convective transfer through the gaseous phase.

λ_G is by far the most significant term in equation 1. The gaseous phase occupies approximately 97 % of the volume of foams in the 28-48 kg/m^3 density range and these materials depend for their technical superiority over air-filled insulants upon their ability to retain low-conductivity gases in their closed cells. The values of λ_S and λ_R, unlike λ_G, are characteristics of the polymer network and do not appear to change during the ageing process. These factors have been fully discussed elsewhere [3, 5, 10, 11, 12, 14, 23, 24, 25, 26, 31]. In practical terms only the gases listed in Table 1 are of commercial significance, and it can be seen that the use of trichloromonofluoromethane and analogous compounds confers a substantial λ_G advantage. λ_C is zero [3, 10, 24, 25].

Table 1

Thermal conductivity data for gases used to expand commercially available plastic insulating foams

Insulant	Cell gas	λ value at 0 °C (kcal/m.h.°C) 10^2
Urethane	Trichloromonofluoromethane	0.695
Urethane	Carbon dioxide	1.265
Urethane	Dichlorodifluoromethane	0.83 (at 24 °C)
Others	— Hydrogen sulphide	1.08
	— Oxygen } Air	
	— Nitrogen }	2.08

2 — AGEING

The dominance of λ_G in equation (1) naturally leads to concern about its permanence, particularly at the low value associated with the use of trichloromonofluoromethane in the cells. Changes in λ_G occur when foams are exposed to air, air diffusing into the cells and carbon dioxide (if present) rapidly diffusing out.

In previous papers some factors affecting the rates of gas permeation were discussed [3, 5]. It was shown that in practice carbon dioxide diffuses out very rapidly. Air diffuses into the foam more slowly and the loss of trichloromonofluoromethane is so slow that it is not technically significant. If, however, all the diffusion processes were to go to completion, for a closed-cell rigid foam, its conductivity would change with time in a similar way to that depicted in figure 1. This curve can be divided into five regions, each bounded by characteristic times T_A, T_B, T_C, T_D obtained by the extrapolation of the linear regions to their points of intersection. The region up to T_A is an induction period of short duration which can have a negative slope, if carbon dioxide is present in the cells, due to its rapid loss.

T_A to T_B is the next stage when sufficient air is entering the foam to increase significantly the λ value owing to dilution of the trichloromonofluoromethane. The characteristic time T_B is the time at which the partial pressure of air in the cells of the foam reached atmospheric pressure.

The region from T_B to T_C is present only when the temperature of measurement of the thermal conductivity of the foam is below 24 °C. In this region the thermal conduc-

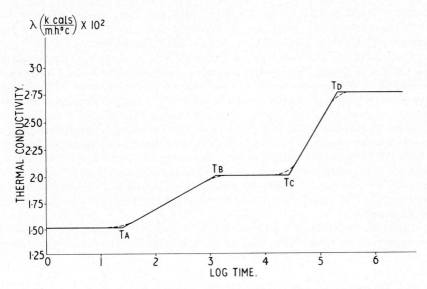

Fig. 1 — Thermal conductivity changes under conditions which allow rapid diffusion.

tivity remains constant since the partial pressure of trichloromonofluoromethane remains constant at the appropriate saturated vapour pressure, any excess being present as condensed liquid (fig. 2). An increase in thermal conductivity cannot take place until the reserve of condensed liquid has been exhausted by evaporating into the gaseous phase to replace the trichloromonofluoromethane being lost by diffusion. The characteristics time T_C is, therefore, defined as the estimated time at which all the condensed trichloromonofluoromethane in the cells has vaporised.

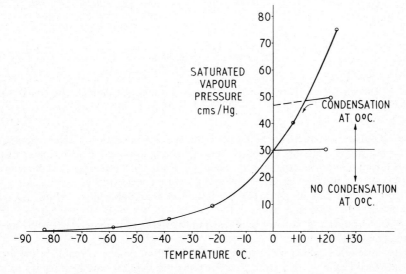

Fig. 2 — Saturated vapour pressure of $CFCl_3$ as a function of temperature.

The region T_C to T_D is one of rising conductivity due to the gradual loss of gaseous trichloromonofluoromethane from the foam. At times greater than T_D the foam contains only air in the cells and hence has a constant conductivity.

In practice, with cut-foam samples 50 mm thick, aged at ambient temperature in the authors' laboratory, the ageing process has never proceeded to the T_D end-point. One set of samples is of particular interest: these were prepared some seven years ago to evaluate the effect of cell dimensions on conductivity and are identical in all other. respects. An updated set of results, which were first given in the paper by Buist *et al.* in 1965, is displayed in figure 3 [5]. The λ values were determined at a mean temperature of 0 °C. Line D is the curve predicted using Norton's theory of conductivity ageing [12, 27, 28, 29]. The same data are displayed in figure 4 using a logarithmic time scale. The closeness of the agreement between the predicted curve and that given by the fine-celled sample (C) is remarkable. All four lines give very similar values for T_B and have similar slopes for the T_A—T_B region. It should be noted that Norton's theory takes no account of cell diameter.

From the similarity of T_B values it can be surmised that the ageing rate, with respect to air entering the foam, does not increase as the number of barriers is reduced

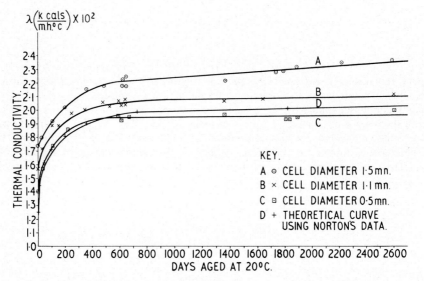

Fig. 3 — Thermal conductivity changes with time at room temperature: effect of cell size. (Corrigendum: A, B, C, read: cell diameter 1.5 mm; 1.1 mm; 0.5 mm.)

because the effect is balanced by the increased thickness of the barriers. Further experimental proof of this is required. Moreover, it is interesting to note that with respect to loss of trichloromonofluoromethane, a T_C value is indicated in figure 4 for the coarse material (A) but not the finer materials (B and C). From this evidence a value for T_C in excess of 7 years can be predicted for foams of reasonably fine texture. By comparison with the predicted behaviour it appears that a final conductivity value of 2.75×10^{-2} kcal/m.h.°C will not be reached until after approximately 200 years' ageing for the fine-celled material.

Very little other evidence of ageing for this length of time is available in the literature. Benford has, however, given data for polyester-based foams aged for up to nine years at 24 °C and Schmidt shows data up to 5 years [16, 17]. No thickness is quoted

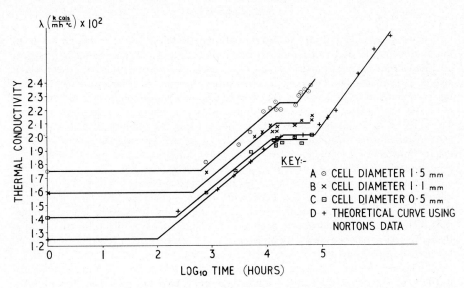

Fig. 4 — Thermal conductivity changes with time for samples aged at 20 °C: effect of cell size

in Benford's data but it seems reasonable to assume from the article that this is 2 in. (50.8 mm) or less. His data are reproduced in figure 5. For a conductivity-measuring temperature of 24 °C, no equilibrium region T_B—T_C is expected. In this case, however, T_C is depicted at a λ value of approximately 2×10^{-2} kcal/m.h.°C when a distinct increase in the slope of the λ/log time curve should occur. This change in slope is not seen in figure 5 and therefore is must be concluded that the appropriate T_C values exceed nine years.

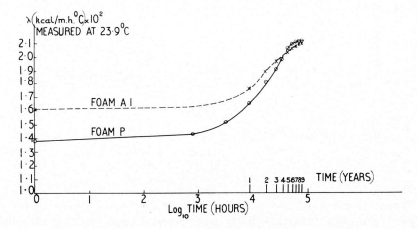

Fig. 5 — Ageing characteristics of polyester-based foam samples of unspecified thickness between 3/8″ and 2″ aged at 23.9 °C. (Plotted from figure 1, reference 16).

3 — Discussion

That it is possible to design *in situ* insulation on the basis of the maintenance of the initial λ value has been substantiated many times, provided always that the foam is anclosed within barriers impermeable to air. Levy has confirmed this for water vapour also [23]. It follows that λ can only be reduced by reducing the contribution of the opertieve components in equation (1). The choice of trichloromonofluoromethane is determined by many aspects of the technology as well as its λ value, and a change is unlikely. The value of λ_S could clearly be reduced by reducing the foam density, but this can lead to dimensional instability [4]. The value of λ_R can be minimized by the use of efficient nucleating mixing heads and surfactant additives so as to create a foam containing small cells [5]. The value of λ_C is zero [3, 10, 24, 25].

Various factors provide the means for minimizing the rate of ageing. Temperature and sample thickness effects have been dealt with elsewhere [30, 31]. The influence of the polyol on the diffusion coefficients, and the impedence offered by surfacing materials such as paper should not be overlooked. Figure 6 clearly illustrates that a polythene-treated paper can delay the ageing of 1 in. (25.4 mm) thick paper-faced foam laminate considerably. Polyol A offers better ageing characteristics than Polyol B.

It has been shown elsewhere that expansion of the foam by trichloromonofluoromethane only produces higher T_C values than foam expanded by mixed blowing techniques [31]. It follows, therefore, that the amount of carbon dioxide used in a mixed blown system should be kept to a minimum, consistent with other technological requirements.

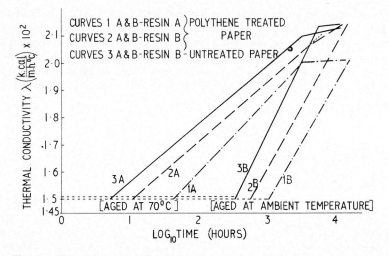

Fig. 6 — Thermal conductivity ageing characteristics of paper-faced laminates.

4 — Practical implications

There are still several factors (for example, effect of cell size) that prevent very accurate theoretical predictions being made. Practical ageing data are still needed. T_C values for 20 °C ageing for freely expanded trichloromonofluoromethane-containing samples 2 in. (50.8 mm) thick can be confidently expected, from the authors' laboratory data, to be in excess of 7 years. T_D values, that is, the time when the values rises to that of air-filled insulants, will be at least ten times longer and certainly too long to be of technological and commercial significance. Norton's theory has been used to calculate a theoretical value of T_C which applies to the author's data [31]. This value was found to be nine years. Norton, in his paper, estimated a T_C value of twenty years for 2 in. (50.8 mm) thick samples aged at ambient temperature [12]. This apparent discrepancy is fully discussed elsewhere and arises solely from the different measuring and ageing temperatures and trichloromonofluoromethane cell gas pressures used [31]. Schmidt's comparable estimate for T_C is a hundred years [9]. Benford's experimental data are based on polyester foam but the data do indicate T_C values in excess of nine years.

Low-temperature applications will give very much larger values of T_C owing to the strong temperature-dependence of the gas diffusion coefficients, not to mention the additional impedence to gaseous diffusion offered by the unavoidable use of vapour barriers in such applications. There are available, however, some special foam types that may be used at temperatures up to 150°C for applications such as pipe insulation [30]. In this case, a mean temperature of 70°C is a reasonable assumption with the pipe insulation being subjected to a temperature gradient. The exact T_C values appropriate to pipe insulation machined from large blocks of foam remain to be established experimentally. Moreover, it is also common practice to prepare these materials in the moulded form to create surface skins, which were shown by Schmidt to delay diffusion considerably [9].

Therefore, it is clear that urethane foams can continue to provide, in service, the thermal conductivity advantages over air-filled insulants for which they are renowned, for times exceeding those of technological interest.

REFERENCES

[1] G. W. BALL, A. M. WOOLER, and C. F. KENYON. *J. Cell. Plast.*, **5**,2 (1969), 85.
[2] R. HURD., *Annex Bull. I.I.R.* 1957-1, pp. 195-214.
[3] D. J. DOHERTY, R. HURD, and G. R. LESTER. *Chemistry and Industry*, **30** (1962), 1340.
[4] D. J. DOHERTY, and G. W. BALL. *Trans. Plast. Inst.*, Lond., **35** (1967), 345.
[5] J. M. BUIST, D. J. DOHERTY, and R. HURD. *Progress in Refrigeration Science and Technology* (1965), 271.
[6] R. H. HARDING and B. F. JAMES. *Modern Plastics*, **39**,7 (1962), 133.
[7] R. E. KNOW. Paper presented to the Plastics Foam Insulation Conference of S.P.I. (April 1962).
[8] Current literature for "Hartmoltopren" (Bayer), Figures 10.5 to 10.9 (1969).
[9] W. SCHMIDT. *App. Phys.*, **114** (1968), 19.
[10] R. H. HARDING. "Progress Design and Development", *Ind. Eng. Chem.*, **3**,2 (1964), 117.
[11] R. H. HARDING. *J. Cell. Plast.*, **1**,3 (1965), 385.
[12] F. J. NORTON. *J. Cell. Plast.*, **3**,1 (1967) 23.
[13] R. H. VARLAND. *S.P.E. Journal*, **22**,11 (1966), 34.
[14] K. C. FRISCH. *J. Cell. Plast.*, **1**,2 (1965), 321.
[15] J. M. BUIST, R. HURD, and R. L. STAFFORD. Advances in Polyurethane Technology, Chapter 6; published by Maclaren & Sons (1968).
[16] A. E. BENFORD. Paper presented at the 2nd International Cellular Plastics Conference, New York (Nov. 1968).
[17] W. SCHMIDT. *Kältetechnik*, **2** (1968), 387.
[18] British Standard 874.
[19] G. W. BALL. Paper presented to the N.P.L. Int. Conference on Thermal Conductivity U.K. (1964).
[20] VOS. Inst. of Refrig. Meeting, Cambridge (Sept. 1961).
[21] O. R., McINTIRE and R. N. KENNEDY. *Chem. Eng. Prog.* **44**,9 (1948), 727.
[22] H. G. NADEAU, P. H. WASZECIAK, and A. A. R. SAYIGH. Paper presented at the 22nd Technical Conference of S.P.E. (1966).
[23] M. M. LEVY. *J. Cell. Plast.*, **2**,1 (1966), 37.
[24] M. E. STEVENSON, Jr. and M. MARK. *ASHRAE Journal*, **3**,2 (Feb. 1961).
[25] R. E. SKOTCHDOPOLE. *Chem. Eng. Progress*, **57**,10 (1961), 55.
[26] G. W. BALL. (Paper to be published.)
[27] A. B. NEWMAN. *Trans. Amer. Inst. Chem. Engrs.*, **27** (1931), 203.

[28] A.B. Newman. *Ibid*, (1931), 310.
[29] A.B. Newman. *Chem. and Met. Eng.*, **38** (1931), 710-713.
[30] G.W. Ball, G.A. Haggis, R. Hurd, and J.F. Wood. *J. Cell. Plast.*, **4**,7 (1968), 248.
[31] G.W. Ball, R. Hurd, and M.G. Walker. (Paper to be published.)

REMARK

I. PALJAK (Sweden) — As a complement to Mr. Hurd's description of ageing of the urethane foams, I would like to mention our experience at the National Institute of Materials Testing in Stockholm as follows: we have found, that when ageing a urethane quality (bulk density 1.9 lb/ft^3), the k-value rises considerably faster when conditioning the specimen at a high relative humidity (90%). See the curve p. 269. Two specimens, the k-values of which were the same at the beginning were conditioned in different ways (refer to the figure). After 250 days the k-values for the both specimens were 0.0226 and 0.0214 kcal/m.h.°C respectively. The investigations are going on, especially in order to explain the phenomenon physically.

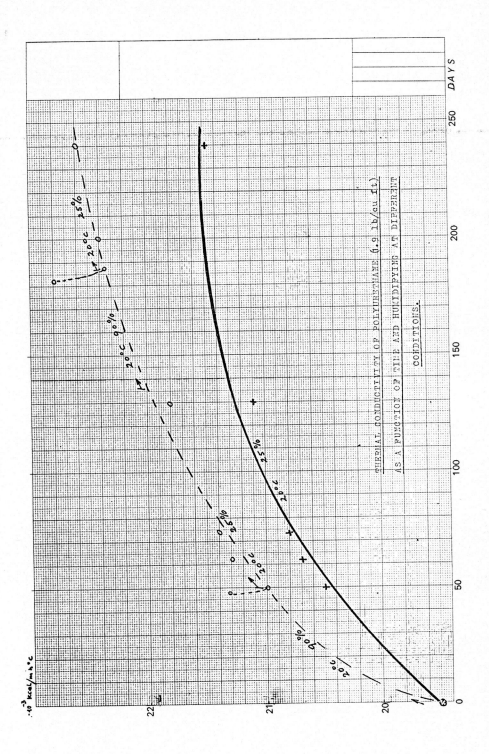

THERMAL CONDUCTIVITY OF POLYURETHANE (1.9 lb/cu.ft)
AS A FUNCTION OF TIME AND HUMIDIFYING AT DIFFERENT
CONDITIONS.

DAYS

269

A SIMPLE TRANSIENT METHOD FOR THE RAPID COMPARISON OF THE THERMAL DIFFUSIVITIES OF POOR CONDUCTORS

E. G. LOMAX and D. R. SCOTT

National College of Refrigeration,
Air Conditioning and Fan Engineering
London (United Kingdom)

Méthode simple en régime transitoire pour comparer rapidement les diffusivités thermiques des conducteurs faibles

RÉSUMÉ : *Le maintien de conditions climatiques tolérables dans les usines souterraines et les mines de charbon profondes, au cours de la dernière décennie, au Royaume-Uni, a conduit à étudier la diffusivité thermique de roches sèches ou humides. Pour ces dernières, l'expérience doit être courte pour éviter les erreurs dues à l'évaporation et à la migration de l'eau.*

Diverses méthodes en régime transitoire ont été utilisées. Les techniques de Van Rest, Billington, de Chung et Jackson sont longues. Celles de Clarke et Kingston et celle de Carte peuvent donner des difficultés dans la préparation des échantillons. La méthode de la sonde chauffée électriquement, de Hooper et Lepper et de Blackwell est suffisamment rapide, mais la résistance de contact à l'interface rend les résultats difficiles à interpréter. Dans une méthode rapide due à Kitto, une température déterminée est appliquée soudainement à l'extrémité d'une carotte de roche en équilibre thermique avec l'ambiance. On compare, pour différentes roches, le temps au bout duquel l'autre extrémité atteint la même température. Les valeurs semblent ainsi obtenues avec une déviation de 7% environ.

La diffusivité thermique peut aussi être déduite indirectement des mesures de conductivité si on connaît la chaleur massique et la masse volumique. La conductivité est généralement mesurée en régime stationnaire et, avec l'appareillage ordinaire, l'équilibre thermique n'est atteint qu'après plus d'une heure. Le rapport décrit une méthode rapide en régime transitoire utilisant un appareil simple.

The engineering problems associated with the maintenance of tolerable climatic conditions in underground factories and in deep coal mines have, over the past decade, directed interest in this country to the determination of the thermal diffusivities of rocks in both the dry and moist condition. When testing a wet porous material such as a natural rock, it is desirable that the experiment should be of short duration both to save time and to prevent errors due to the evaporation and migration of moisture. Such considerations apply equally to thermal insulants.

Various transient methods have been used to measure the diffusivities of bad conductors. Of these, techniques adopted by Van Rest [1], Billington [2] and Chung and Jackson [3] are lengthy, whilst the methods of Clarke and Kingston [4] and Carte [5] can give trouble in the preparation of specimens.

The electrically heated probe as developed by Hooper and Lepper [6] and Blackwell [7] is sufficiently rapid, but the contact resistance at the interface makes the results difficult to interpret.

In a rapid method due to Kitto [8], a known fixed temperature is suddenly applied to one end of a core of rock initially in thermal equilibrium with its surroundings. The method is a comparative one, the times for the other end of different specimens to reach the same temperature being compared. No theoretical treatment is given, and the method is subject to a "zero" error. The standard deviation of the values of thermal diffusivity obtained seems to be about 7%.

The thermal diffusivity may also be deduced, indirectly, from measurements of conductivity if values of specific heat and density are available. Thermal conductivity is usually measured in the steady state and, with the conventional divided bar apparatus, the time to reach thermal equilibrium is in excess of one hour.

271

The present paper describes a rapid transient method using simple apparatus for the determination of diffusivity.

When calibrated against standard specimens, an accuracy of $\pm 2\frac{1}{2}\%$ is attained, which is sufficient for many practical purposes.

The method adopted is similar to that used by Kitto, in that a sudden temperature rise takes place at one end of a core of rock in thermal equilibrium with its surroundings.

In the present case, however, the rate of temperature rise at the further end of the core is determined, and the result is interpreted by theoretical analysis.

Experimental technique

The method is a comparative one, the apparatus being calibrated by means of standard cores of dry rock of the same size and under identical conditions, the thermal diffusivity of these cores being accurately known. Advantages of the method are that the cores of rock are easily and quickly prepared, and each determination takes only a few minutes to perform, which allows the testing of moist rocks. In addition, the many heterogeneous rocks found in the coal measures can be studied as the samples are large enough to be representative.

Methyl alcohol vapour was used to apply the fixed temperature rise. The liquid has a boiling point of 65 °C, and for this temperature the vapour pressure gradient of the water in a moist specimen is insufficient to appreciably affect the heat transfer.

A core length of about $1\frac{1}{2}''$ was found to be the most suitable for this temperature, as the temperature rise was then sufficiently fast to avoid errors due to the migration of moisture, and yet not so rapid that it prevented accurate measurements. The core

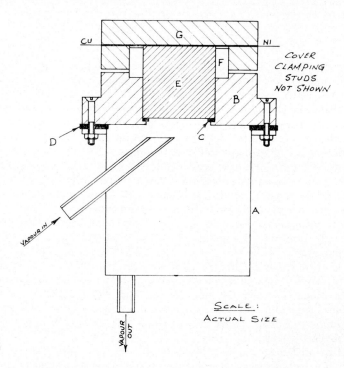

Fig. 1 — A. Copper Container — B. Tufnol Support — C. Rubber Ring — D. Cardboard Gasket — E. Rock Specimen — F. Perspex Ring — G.Tufnel Cover.
Sectional drawing of thermal diffusivity. Apparatus testing assembly.

272

samples were cut into cylinders by means of an impregnated diamond slitting wheel on a special assembly [13]. There was no necessity to reface the samples after slitting, as the variation in length was never greater than ± 0.005 in 1.445 inches.

One end of the core sample to be tested was always treated with Bostik dissolved in toluene to render it impervious to the methyl vapour. The paste was made as thick as was conducive to satisfactory application, for if too much toluene was used, the paste became permeable, probably due to the miscibility of toluene and methyl alcohol.

The core was insulated at the further end and along its length to minimise heat losses.

During experiment, the rock core was held in the testing assembly shown in figure 1. This consisted of a copper can 3 inches in diameter and 3 inches high, fitted with $\frac{1}{4}''$ bore inlet and outlet tubes, the inlet tube being set at an angle as shown. To a lip on the top edge of the can a machined Tufnol support was screwed, the joint being sealed by a cardboard gasket. Into this support the rock specimen fitted tightly, with its impervious face above the end of the inlet tube. A rubber ring was placed between the rock face and a step in the Tufnol support to prevent access of vapour up the sides of the specimen. The vapour was thus supplied directly to practically the whole lower face of the specimen core.

A perspex ring $\frac{5}{8}''$ long and having an external diameter of $2''$ was housed in a groove cut in the upper side of the support. It was a tight fit over the specimen core, and provided thermal insulation for the slides.

The temperature-measuring element was a calibrated Cu-Ni thermocouple carried in a groove cut diagonally across an end cover made from Tufnol, the groove being slightly less in depth than the diameter of the wire, so that when the end cover was screwed down tightly on the core, the thermocouple was in good contact with the top end face. The thermocouple was made by silver-soldering the nickel and copper wires end to end to avoid overlapping, so that only the temperature at the centre of the core end face was measured. It was decided to locate the thermocouple at the centre of the end face instead of a point along the circumference to the core because, in the latter case, it was difficult to ensure good thermal contact with the core and also the insulation surrounding it was disturbed. The Tufnol end cover was made a good fit over the perspex ring, thus ensuring that the core was well insulated, and that the thermocouple was always located at the centre of the face. The cover was secured by means of four studs from the Tufnol support, these passing through holes in the cover and fitted with wing nuts.

The nickel wire from the thermocouple was passed to a vacuum flask containing oil where a soldered junction was made with another copper wire. This constituted a junction of constant temperature over the period of the test. The copper wire from the thermocouple and the copper wire from the junction were then passed to a sensitive Cambridge spot galvanometer. This was such that, on setting the galvanometer at zero before an experiment, a full-scale deflection of 100 divisions corresponded to a temperature rise at the end of the rock core of about 11 °C. When using methyl alcohol, it was sufficient to take readings up to 50 divisions, i.e., a temperature rise of about $5\frac{1}{2}$°C which is suitable for work with moist specimens.

The methyl alcohol was boiled in a copper can. The outlet from the can was connected to a large three-way glass tap, one arm of which was connected to the testing assembly, and the other arm passed to a condensate jar closed by a rubber bung. The outlet pipe from the testing assembly carried a Moh's clip and was also connected to the jar. A third tube connected the condensate jar to a water cooled condenser. Plastic tubing was used for all connections, as methyl alcohol attacks rubber.

Before testing, each rock specimen was put in place in the testing assembly and left for about half an hour with the perspex ring and cover fitted. This served to ensure that the rock core was in equilibrium with its surroundings and was indicated by no movement on the galvanometer. The temperature of the cold junction was taken to be the temperature of the assembly. The liquid was then heated, with the three-way tap open to the condensate jar, but closed to the testing assembly. Similarly, the Moh's clip was kept closed. A sheet of asbestos was located between the heating can and the testing assembly and cold junction. These precautions prevented the equilibrium of the specimen from being disturbed.

When the liquid was boiling, the galvanometer was adjusted to zero at the left hand end of the scale, and the Moh's clip was opened. The three-way tap was then altered to close the connection to the jar and to open the one to the testing assembly, thus putting the vapour in immediate contact with the lower face of the core and, at the same time, the stop-clock was started. Readings of the time were taken at every five divisions on the galvanometer up to the 50 divisions mark.

The apparatus was so designed that the measuring conditions, such as the insulation and the position of the thermocouple, were always the same for all experiments. The only important variable, apart from the type of rock, was the ambient room temperature. The room in which these experiments were carried out was partly below ground, and was deliberately chosen for the very steady temperature conditions within, so that, in practice, the difference between the ambient and cold junction temperatures was negligible.

The changes in barometric pressure were insufficient to appreciably affect the boiling point of the methyl alcohol used.

From a previous calibration, galvanometer readings were directly convertible to Centigrade degrees. A typical set of results are plotted in figure 2. It will be seen that a distinct straight line portion AB exists on either side of which the graph has an opposite curvature.

The thermal diffusity can be obtained from the gradient of the straight line portion according to the following theory.

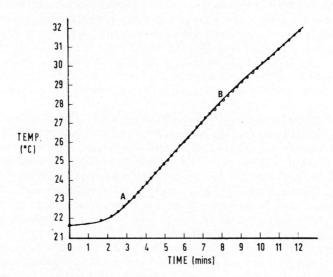

Fig. 2 — A typical curve obtained from the thermal diffusivity apparatus.

Theoretical investigation

The mathematical model considered is a uniform cylinder of radius $r = a$ and length $z = l$ which is initially at zero temperature. A temperature V_0 is suddenly

274

applied at the end $z = 0$. The end $z = l$ and the curved surface $r = a$ are assumed to be perfectly insulated.

It is required to find the relation between temperature and time at $z = l$.

As there is no heat loss from the sides, we require a solution of the reduced equation

$$K \frac{\partial^2 v}{\partial z^2} = \frac{\partial v}{\partial t} \tag{1}$$

subject to the restrictions:

Initial conditions

$$v = 0 \quad \text{when} \quad t = 0 \quad \text{and} \quad 0 < z \leqslant l \tag{2}$$

Boundary conditions

$$v = V_0 \quad \text{when} \quad z = 0 \quad \text{and} \quad t \geqslant 0 \tag{3}$$

$$\frac{\partial v}{\partial z} = 0 \quad \text{at} \quad z = l \tag{4}$$

Solution

The solution is readily found to be

$$\bar{v} = \frac{V_0 \coth q(l-z)}{p \coth ql} \tag{5}$$

where $\bar{v}$ is the Laplace transform of the temperature v and

$$q^2 = \frac{p}{K}$$

The required rate of temperature rise at $z = l$ is consequently given by

$$\bar{G} = \left(\frac{\partial v}{\partial t} \right)_{z=l} = \frac{2 V_0}{e^{ql} + e^{-ql}} \tag{6}$$

For small values of time the solution may be expanded in terms of negative exponentials and we obtain the series:

$$\bar{G} = 2 V_0 (e^{-ql} - e^{-3ql} + e^{-5ql} - \dots) \tag{7}$$

and inverting we have finally

$$G = \frac{V_0 l}{\sqrt{\pi K t^3}} \{ e^{-l^2/4 Kt} - 3e^{-9l^2/4 Kt} + 5e^{-25 l^2/4 Kt} - \dots \} \tag{8}$$

The series is rapidly convergent and for the practical range considered only the first term need be retained.

The point of inflection is then given by $dG/dt = 0$ or

$$\frac{Kt_i}{l^2} = \frac{1}{6} \tag{9}$$

275

and substituting in (8) above we have

$$G_i = K V_0 \frac{12}{l^2} \left(\frac{3}{2\pi}\right)^{\frac{1}{2}} e^{-3/2} \tag{10}$$

from which we conclude that G_i, the gradient at the point of inflection, is directly proportional to the diffusivity K.

In practice, there will be departures from the conditions assumed for the simple mathematical model: thus heat losses will occur from the surface of the specimen and the temperature different V_0 is not instantaneously applied. However, the experimental curve in figure 2 shows the general features of the analysis, and the gradient of the straight line portion is found to be linearly related to the thermal diffusivity as predicted.

Numerical data

Length of core	$l = 3.67$ cm
Radius of core	$a = 1.82$ cm
Minimum value of conductivity	$k_o = 3.39 \times 10^{-3}$ c.g.s.
Minimum value of diffusivity	$K_o = 6.50 \times 10^{-3}$ c.g.s.
Maximum value of conductivity	$k_m = 13.1 \times 10^{-3}$ c.g.s.
Maximum value of diffusivity	$K_m = 21.2 \times 10^{-3}$ c.g.s.
Applied temperature difference	$V_o = 45°C$

The point of inflection on the time-temperature graph occurred in the region of 300 seconds.

Errors of measurement

(a) *Length of core.* The length l in equation (10) occurs squared. The actual variation in length was never greater than ± 0.005 inches in 1.445 inches, so that the expected error in the gradient due to his cause was

$$\frac{\pm 2 \times 0.005 \times 10^{-2}}{1.445} = \pm 0.69\%$$

(b) *Radius of core.* More detailed analysis indicates that any error in the measurement of the radius "a" of the core gives rise to a term of the second order only. Further, any departure of the thermocouple from centre should result in an error of similar magnitude. Precautions were taken in designing the apparatus to ensure accurate centering of the thermocouple so the errors resulting from these causes should be very small.

(c) *Temperature.* From (10), the gradient G_i is directly proportional to V_0, the temperature excess, so that any variation in this quantity must be allowed for by correcting to a standard value. The probable error in measurement of temperature in the author's experiments is $\pm 1\%$.

(d) *Time.* As the method involves the comparison of *rates* of temperature change in the vicinity of a point of inflection, the "zero" error in Kitto's method is avoided where the times to reach a certain temperature are compared. The errors in the experimental measurement of time are considered negligible.

(e) *Overall error.* The overall errors of measurement and dimension are estimated as $\pm 1\frac{1}{2}\%$. These estimates are borne out by the practical results obtained in calibrating the apparatus.

Calibration

The thermal diffusivity apparatus was calibrated by testing a number of dry rocks of known thermal diffusivity under the same conditions as those used for the normal measurements. The thermal diffusivity value of each of the rocks used for the calibration by the divided bar method, its specific heat by the adiabatic calorimeter and its bulk density. Full details are given elsewhere [9].

In the thermal diffusivity apparatus, the applied temperature was 65°C. The thermal conductivity values were determined at a mean temperature of approximately 28°C, and the specific heat over a range of 65°C→26°C.

The samples for the calibrations were selected from homogeneous rock and the cores were cut from the same specimens as those from which the discs for the divided bar apparatus were prepared.

In the accompanying table, the results are given of the known thermal diffusivity values and the measured gradients for each of the rock samples used in the calibration.

TABLE I

Type of rock	Inclination to bed	Thermal diffusivity $K \times 10^3$ (c.g.s.)	Gradient G galvo divs/min.
Fine Sandstone	‖el	16.2	16.1
Coarse Sandstone	⊥′r	16.1	15.8
Cavernous Dolomite	⊥′r	13.8	13.7
Fine Sandstone	⊥′r	13.4	13.1
Compact Limestone	⊥′r	10.2	10
Compact Limestone	⊥′r	9.27	9.54
Dark Grey Shale	⊥′r	8.29	8.62
Red Silty Sandstone	⊥′r	7.52	7.67

The above results may be represented by the formula

$$K = 1 \cdot 00 \times 10^{-3} \times G$$

with a standard deviation of 2.29 % in K.

CONCLUSIONS

The apparatus affords a cheap and easy method for the determination of the thermal diffusivity of rock and other poor conductors. The specimens are quick to prepare and, as the actual determination takes only a few minutes, moist materials may be tested.

The time is lengthened by the half hour required for the apparatus to achieve thermal equilibrium with its surroundings before the actual test but duplication of the simple testing assembly avoids this delay.

When the apparatus is calibrated against eight standard specimens of known diffusivity in the range

$7.52 \times 10^{-3} \rightarrow 16.2 \times 10^{-3}$ c.g.s., the results are shown to have a standard deviation of $\pm 2\frac{1}{2}\%$.

The estimated error due to errors in measurement is $\pm 1\frac{1}{2}\%$ which is in good accord. Measurement errors could be reduced by carrying out the experiment in a constant-temperature room.

In the case of materials which are not homogeneous in composition and those with large grain size, the method will probably give a better result than the divided bar, as the diffusivity is measured over a greater length of core.

REFERENCES

[1] E. D. VAN REST. "A Method for the Measurement of the Thermal Diffusivity of Poor Conductors". *Proc. Phys. Soc.*, **46** (1934), p. 537.
[2] N. S. BILLINGTON. "The Thermal Diffusivity of Some Poor Conductors". *J. Sci. Instrum.*, **26** (1949), p.20.
[3] P. K. CHUNG and M. L. JACKSON. "Thermal Diffusivity of Low Conductivity Materials". *Ind. Eng. Chem.*, **46** (Dec. 1954), p. 2563.
[4] L. CLARKE and R. S. T. KINGSTON. "Equipment for the Simultaneous Determination of Thermal Conductivity and Diffusivity of Insulating Materials by a Variable State Method". *Australian J. Appl. Sci.*, **1** (1950), p. 172
[5] A. E. CARTE. Thermal Conductivity and Mineral Composition of some Transvaal Rocks". *Amer. J. Sci.*, **253** (Aug. 1955), p. 482.
[6] F. C. HOOPER and F. R. LEPPER. "Transient Heat Flow Apparatus for the Determination of Thermal Conductivities". *Heat. Pip. Air. Cond.*, **22** (Aug. 1950), p. 129.
[7] J. H. BLACKWELL. "A Transient Flow Method for Determination of Thermal Constants of Insulating Materials in Bulk. Part 1—Theory". *J. Appl. Phys.*,**25**, 2 (Feb. 1954), p. 137.
[8] P. H. KITTO. "The Measurement of Rock Conductivity and Diffusivity". *Min. Vent. Soc. South Africa Bull.*, **7**, 1 (Jan. 1954).
[9] G. LOMAX. "The Thermal Constants of Coal Measure Rocks". Nottingham University Ph. D. Thesis (1956).

REMARK

R. P. TYE (U.S.A.) — The paper contains a bibliography concerning measurements of thermal conductivity of rocks but I would like to make reference to the extensive work by W. H. Semerton and his various colleagues at the University of California.

SECTION 3

MESURES DE CONDUCTIVITÉ THERMIQUE

MEASUREMENT OF THERMAL CONDUCTIVITY

ON THE DESIGN OF A GUARDED HOT PLATE APPARATUS

E. BRENDENG and P. E. FRIVIK

Norges Tekniske Høgskole, Trondheim (Norway)

Conception d'un appareil à plaque chaude gardée

RÉSUMÉ : *On a construit, à l'Institut du Froid de l'Université Technique de Norvège, un appareil à plaque chaude gardée en vue de la mesure de la conductivité thermique des isolants. Le courant de chaleur est unidirectionnel et la température de la plaque chaude peut, au stade actuel de la mise au point, varier de 20° à −70° C. Des températures de la plaque froide de −180° C sont prévues avec un matériel supplémentaire.*

La section de mesure est de 200 × 200 mm²; la largeur de la plaque de garde est de 100 mm et l'épaisseur maximale de l'échantillon expérimental est de 100 mm. Les éléments chauffants sont en fils chauffants gainés d'acier inoxydable soudés directement dans les tôles de cuivre de l'assemblage de la plaque chaude. On a réduit l'influence des conditions thermiques ambiantes sur la précision à l'aide d'une enveloppe refroidie. Le fonctionnement est automatique, avec utilisation d'amplificateurs d'écarts, de régleurs de température et de stabilisateurs de tension. L'appareil est placé dans un récipient, permettant des essais sous vide ainsi qu'à des pressions atteignant 3 bars.

On a calculé les isothermes et l'on a amélioré quelques points de la conception à l'aide de la méthode de relaxation.

On a vérifié la précision de l'appareil par rapport à l'appareil à plaque chaude du « Norges Byggforskningsinstitutt » de Trondheim et on a trouvé qu'elle était satisfaisante. On a trouvé que les valeurs obtenues dans les essais de cet appareil concordaient étroitement avec les valeurs du NPL et du NBS.

INTRODUCTION

The conventional type of guarded hot plate apparatus seldom permits measurements of thermal conductivity over a wide temperature range to very low temperatures. Further, measurements are usually restricted to atmospheric pressure. Since conduc-

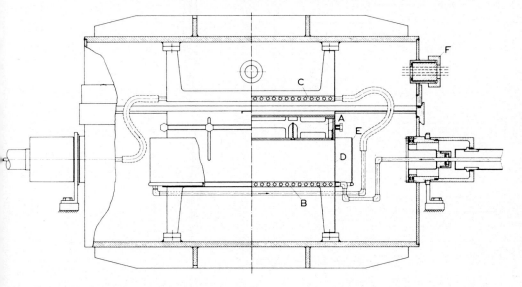

Fig. 1 — General arrangement.
  A. Hot plate assembly; B. Cold plate; C. Cold guard plate; D. Stainless steel gradient guard; E. Second gradient guard; F. Thermocouple bushing.

tivity values for insulating materials at low temperatures are of increasing importance, and valuable knowledge on the nature of heat transfer through fibrous materials can be obtained through measurements at low temperature and low pressure, a guarded hot plate apparatus of suitable type for these purposes was designed at the Institute of Refrigeration, The Technical University of Norway.

DESCRIPTION OF THE APPARATUS

The apparatus is of the Jacobs-type, with uni-directional heat flow, and is placed inside a pressure vessel, permitting tests under vacuum as well as under pressure up to 3 bars with different gases. The vessel is supported on bearings, and measurement may be taken in any position (fig. 1). The metering section is 200×200 mm^2, and the size of the test sample is 400×400 mm^2, i.e. the width of the guard ring is 100 mm. The maximum thickness of the test sample is 100 mm.

The hot plate assembly consists of the main plate, the guard ring, the secondary guard ring and the guard plate. 4 mm copper sheet is used in these parts, and all heating elements are 1 mm stainless steel sheathed heating wire directly soldered into milled slots in the plates (fig. 2).

Two cold plates are used, both made of 12 mm copper with drilled channels for the coolant. The upper plate is used in order to compensate for the heat leakage when working with hot plate temperatures below ambient temperatures. This plate

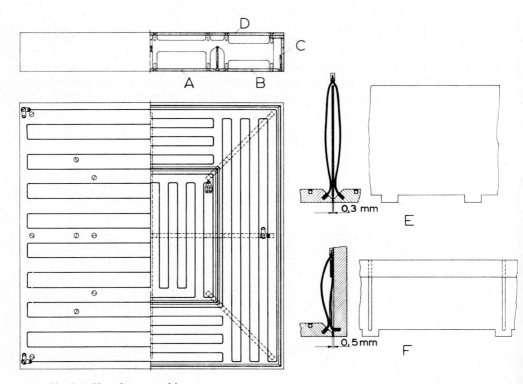

Fig. 2 — Hot plate assembly.
A. Main plate; B Guard ring; C. Secondary guard ring; D. Guard plate; E. Thermopile main plate — guard ring, with support; F. Thermopile guard ring — secondary guard ring, with support.

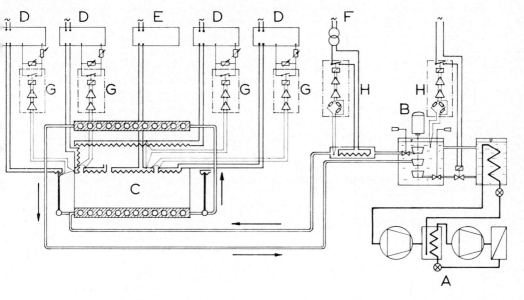

Fig. 3 — Control diagram.
A. Cascade refrigeration plant; B. Coolant pump; C. Hot plate assembly; D. Programmable DC power supplies; E. Stabilized DC power supply; F. Variac; G. Deviation amplifiers; H. Temperature controllers.

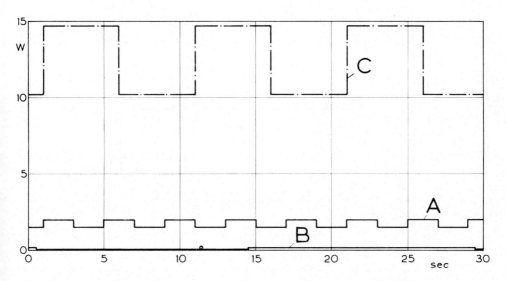

Fig. 4 — Control characteristics.
A. Power input, guard ring; B. Power input, guard plate; C. Power input, secondary guard ring.

is kept in place in the lid of the pressure vessel by means of a stainless steel frame, and connected to the cooling system by means of flexible tubes.

In order to reduce the influence of ambient temperature conditions, a 1 mm stainless steel temperature gradient guard is soldered to the cold plate, and kept in contact with the secondary guard ring with screws. Thus the temperature difference across the gradient guard corresponds to the temperature difference across the test sample, and linear isotherms within the test area should be ensured. The gradient guard also serves as a frame for the insulation under test, and keeps the hot plate assembly in place.

At low temperature levels the temperature gradient in the gradient guard becomes unlinear due to heat leakage from ambient, and a second gradient guard made of 0.5 mm brass and placed 40 mm outside the first gradient guard was found to be necessary .This second gradient guard is soldered to a cooling tube at the lower end.

A heater wire is placed along the upper edge of the second gradient guard, the heating to be used at higher temperature levels. The surfaces of the cold and hot plate facing the test sample are painted dull black, to give a well defined radiant heat transfer.

COOLING SYSTEM

The coolant used at present is alcohol, cooled by a small cascade refrigerating system, with $-70\,°C$ as the lowest obtainable temperature. Later, propane or another suitable coolant will be used, with liquid nitrogen or a Philips cryogenerator equipped

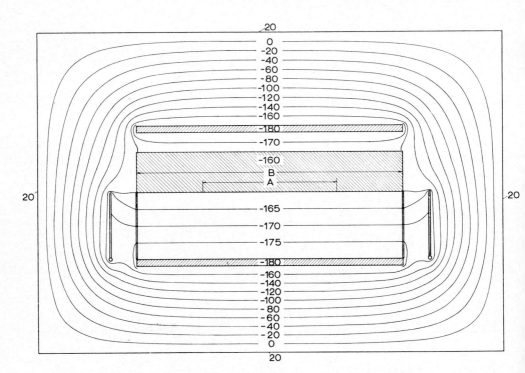

Fig. 5 — Calculated isotherms at low temperature conditions.
A. Main plate; B. Area with linear isotherms.

with a pressure head as the primary refrigeration system. The coolant is circulated by means of an adjustable capacity pump.

The power input to the metering section of the hot plate should be continuous in order to permit the highest accuracy in the measurement of the heat flow through the insulation under test. Thus the voltage is set manually, which also permits the necessary overshoot to reach stable conditions as soon as possible. In order to obtain the desired hot plate temperature some adjustment may be necessary during the stabilization period. DC-current is used, the voltage being stabilized to 0.01 per cent.

Apart from the manually set voltage to the main plate, the control of the apparatus is automatic. A thermopile consisting of 32 copper-constantan thermocouples in series is installed between the main plate and the guard ring. The output from the thermopile is fed to a precision deviation amplifier, with low drift and high sensitivity. The output from the deviation amplifier is again fed to a power amplifier with a relay, this relay controlling a programmable DC-power supply. By means of two low wattage potentiometers the voltage can be set to switch from a higher to a lower value by means of the relay (fig. 3).

The same arrangement is used for the control of the voltage to the secondary guard ring and the guard plate and for the heater along the upper edge of the second gradient guard. However, a less costly type of deviation amplifier is used for these parts of the control system. The thermopile between the guard ring and the secondary guard ring consists of 40 thermocouples, and the thermopile between the hot plate and the guard plate consists of 16 thermocouples.

The time constants in the control system are low and the switching from high to low voltage is quick, with intervals of a few seconds (fig. 4). At a cold plate temperature of $20 - °C$ stable readings are obtained after 24 hours.

The temperature of the coolant is maintained by means of two temperature controllers, one of which controls the temperature in the main bath by switching on and off a solenoid valve feeding cold alcohol from the low temperature alcohol reservoir. The other temperature controller switches on and off the current to a low inertia heater of 1 mm stainless steel sheathed heating wire, placed directly into the flow of alcohol. Thermistors are used as temperature transducers.

The positions of the isotherms in a cross-section of the whole apparatus were calculated by means of the successive over-relaxation method (fig. 5). The influence of different materials and sheet-thicknesses in the two gradient guards was tried out at different temperature levels, and the best combination at the most interesting temperatures could be chosen.

The influence of the ambient temperature conditions on the linearity of the isotherms within the insulation in the guard ring area is shown in figure 6.

Temperatures are measured with 0.15 mm or 0.08 mm copper-constantan thermocouples on the cold and hot plates, on the gradient guards and on the guard plate, as well as in the coolant circuit. The temperature difference across the insulation is measured with a thermopile, consisting of 8 thermocouples in series.

All temperatures are recorded by means of a data logger with resolution 1 μV, with a paper tape punch as output device. The calculations are completed on a computer. As a control temperatures may also be measured with an automatic precision potentiometer. The power to the main plate is measured by means of a multirange Volt/Amperemeter, calibrated to better than 0.2 per cent. Separate conductors to the heater for voltage measurements are used, to avoid voltage drop in the current leads. To obtain better accuracy at low temperature differences a precision potentiometer may be used for voltage and current measurements.

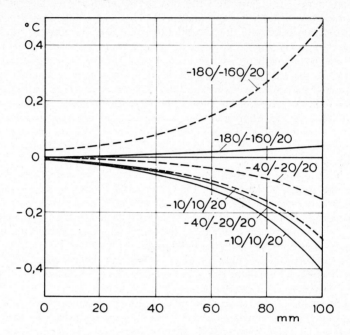

Fig. 6 — Calculated deviation from linearity of mean isotherm in guard ring area.
Fully drawn lines: Second gradient guard 0.5 mm brass.
Dotted lines: Second gradient guard 1 mm stainless steel.
Cold plate/ hot plate/ ambient temperatures varying from − 10/10/20 to − 180/ − 160/20.

The following instruments were used:
DC-power supplies:
    Oltronix, stability ±0.01 per cent, ripple 0.5 mV, programmable.
Deviation amplifiers with specially built power amplifiers:
    Main plate—guard ring: Honeywell deviation amplifier, zero stability 0.5 μV/ 8 hours. Guard ring—secondary guard ring, main plate—guard plate, and gradient guard—second gradient guard: Knick JV, zero stability 5 μV/month.
Data logger:
    Electro Instruments, 400 channels, resolution 1 μV, accuracy ∼ 3 μV.
Volt/amperemeter:
    AEG multimeter class 0.2 and ESI potentiometric Voltmeter class 0.02.

THICKNESS MEASUREMENTS

The thickness of test sample is measured indirectly as the distance from the top of the hot plate assembly to the top of the gradient guard. The thermal contraction

286

of the stainless steel gradient guard is approximately 0.30 mm at $-180\,^{\circ}$C, and since the temperature distribution is known, the contraction can easily be accounted for. When making conductivity measurements of board insulations, the thermal contraction of these materials may exceed the contraction of the stainless steel frame, and special precautions must be taken to ensure firm contact between the insulation boards and the hot and cold plates.

ACCURACY

The performance of the apparatus has been checked against the apparatus at Norges Byggforskningsinstitutt (NBI, Norwegian Building Research Institute) through conductivity measurements on the same test samples at identical temperature conditions. During the international comparative measurements of thermal conductivity in 1967, the test figures from NBI were found to be in close agreement with the values from NPL and NBS.

The NBI-apparatus is designed according to ASTM standard C177-63, with bi-directional heat flow. Two fibre glas boards of 46 mm thickness were used as test samples, and the thermal conductivity were measured by NBI at three mean temperatures.

A difference from the reference value of 1.8 to 2.4 per cent was found when measurements were made with one test sample. With two test samples laid together to form a slab with total thickness 92 mm, the disagreement in the test figures increased to 3 to 4 per cent (fig. 7).

The thermal contact between the secondary guard ring and the stainless steel gradient guard is less efficient than the thermal contact between the cold plate and the gradient guard. At higher temperature levels this leads to a too low temperature in the gradient guard, and the measured conductivity is too high. Through manual

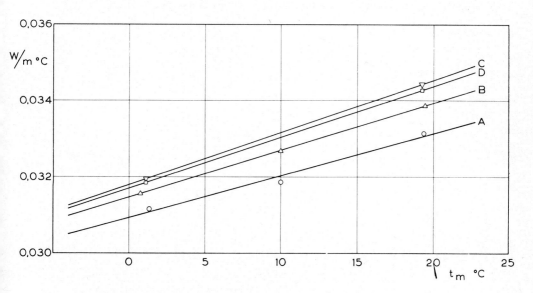

Fig. 7 — Comparison between conductivity measurements.
    A. Test figures from NBI.
    B. Conductivity measured with one sample with thickness 46 mm.
    C. Conductivity measured with two samples together with total thickness 92 mm, with automatic control of second gradient guard heater.
    D. As B, but with manual control of second gradient guard heater.

control of the second gradient guard heater, the temperature distribution in the stainless steel gradient guard can be influenced, and linear isotherms established in the test specimen.

With the temperature at the upper edge of the second gradient guard increased about 20°C over the hot plate temperature, the measured conductivity was reduced with about 0.3 per cent. Thus, the accuracy is not too dependent on the temperature of the second gradient guard, or on ambient conditions.

The apparatus is very sensitive to thermal unbalance between the main plate and the guard ring, due to the rather narrow gap between these parts. With a test sample thickness of 92 mm, the measured conductivity increased with 2 per cent for an unbalance of 0.025°C. Since this feature may lead to inaccurate measurements due to a slightly uneven temperature distribution in the hot plate, an increase in the gap width is considered.

Corresponding measurements with an unbalance of 0.1°C between the main plate and the guard plate increased the measured conductivity with 2 per cent. Thus, a slight unbalance between these two parts do not seriously effect the accuracy.

CONCLUSION

The hot plate apparatus described herein is convenient in operation, since most of the controls are automatic. It is almost independent on ambient temperature conditions, and the accuracy is satisfactory for sample thicknesses up to 50 mm. The sensitivity to thermal unbalance between main plate and guard ring should be reduced, and the accuracy be improved for a sample thickness of 100 mm. Further investigation are necessary to establish the accuracy at low temperatures.

DISCUSSION pp. 317-319.

INFLUENCE DU MODE DE CONSTRUCTION DES PLAQUES CHAUFFANTES À ANNEAU DE GARDE ET DE L'ISOLATION LATÉRALE SUR LA MESURE DE LA CONDUCTIVITÉ THERMIQUE DES MATÉRIAUX ISOLANTS

R. DOUSSAIN*

Laboratoire National d'Essais, Paris (France)

Influence of the mode of construction of guard-ring heating plates on the thermal conductivity measurements of insulating materials

SUMMARY: *Whilst the guard-ring heating plate method is virtually a proven reference method for measuring the thermal conductivity of insulating materials, it is nevertheless necessary to point out the difficulties encountered in applying this technique, particularly when taking measurements below the ambient temperature.*

These difficulties are essentially associated with the control of lateral heat losses. This control is a function of the method of lateral heat insulation and the ambient temperature level, as well as heat regulating conditions which are directly influenced by the mode of construction of the heating plate.

The paper first briefly deals with current tendencies with equipment and in relation to the influence of the thermal resistance, lateral insulation and the ambient temperature. Particularly referred to is the usefulness of the environmental characteristic index recently proposed by W.H. Orr.

The paper then quotes the results of tests carried out on four types of plates having the minimum of constructional differences. These results have been interpreted according to Woodside's analysis which reveals the possibilities of important divergencies.

In concluding, the author proposes that future inter-laboratory preoccupations should not only be overall, but they should include a systematic study of lateral insulation and environmental conditions as well as the mode of construction of heating plates so that one may appreciate the actual working limits of this equipment.

I — INTRODUCTION

La méthode de la plaque chauffante à anneau de garde utilisée en régime permanent constitue, actuellement, une méthode de référence pratiquement incontestée dans le domaine des mesures de conductivité thermique appliquées aux matériaux isolants ou faiblement conducteurs.

On sait qu'il existe déjà quelques normes ou réglementations tendant à fixer les conditions de mesures et il paraît probable qu'on assistera, dans un prochain avenir, au développement sur un plan international de cette tendance à la normalisation.

La première étape dans la recherche d'une harmonisation des conditions de mesure se trouve, évidemment, dans la réalisation d'essais simultanés d'échantillons identiques par les différents laboratoires intéressés et, de ce point de vue, les résultats récemment publiés par l'I.I.F. représentent une contribution importante.

Il est cependant à craindre que de telles confrontations, de par leur caractère global, ne risquent de donner une vue trop optimiste de la méthode et ne mettent pas suffisamment en garde contre les difficultés d'application de la méthode dès que l'on s'écarte sensiblement des conditions de mesure voisines de l'ambiance.

Ce risque était limité pour la comparaison interlaboratoire de l'I.I.F. puisque le domaine de température pour lequel les échantillons de fibre de verre avaient été expérimentés était compris entre 0 et 40 °C.

(*) Laboratoire National d'Essais, 1, rue Gaston Boissier, Paris-XVe, France.

En fait, les besoins croissants en isolation frigorifique font souhaiter que ce type de confrontation puisse être étendu vers les basses températures et éventuellement jusqu'à −100 °C.

Dans cette zone de température, les difficultés d'application deviennent notables puisque le principe même de cette méthode basée sur le contrôle de la déperdition thermique latérale se trouve mis en cause.

On sait que ce contrôle est une fonction complexe des données suivantes :

I — 1. *Conditions d'environnement* qui mettent en jeu le mode d'isolation thermique latérale de l'empilage ainsi que le niveau de température ambiante.

I — 2. *Conditions de réglage de la surchauffe,* ces conditions étant, en fait, gouvernées par le mode de réalisation de la plaque chauffante.

II — EXAMEN DE L'INFLUENCE DES CONDITIONS D'ENVIRONNEMENT

Dans la pratique, on semble actuellement s'orienter vers deux types principaux de dispositifs :

II — 1. Dans une première variante, le procédé utilisé consiste à introduire près de la surface latérale du dispositif et à l'intérieur de l'isolation thermique latérale, un échangeur thermique qui absorbe le flux thermique de réchauffage provenant de l'ambiance [1]. Il est alors possible sous réserve de réglage précis, de maintenir un flux de chaleur positif dans les échantillons et l'on peut utiliser la surchauffe latérale pour redresser les isothermes.

II — 2. Dans une deuxième variante [2], l'empilage classique constitué par les échantillons, la plaque chauffante et les plaques froides isothermes est introduit dans un caisson à parois isolantes. L'air contenu dans le caisson est maintenu à des conditions déterminées de température et, éventuellement, d'humidité. Ce type de dispositif a été décrit dans les publications du N.B.S. [3].

Un appareillage assez comparable a également été réalisé par le Laboratoire National d'Essais [4], pour des échantillons carrés de 50 × 50 cm expérimentés entre 100 et −80 °C. Dans cet appareil, le caisson enveloppant le dispositif est conditionné en température par évaporation d'azote liquide, la température du milieu fluide en contact avec les parois latérales du dispositif d'essai étant constamment maintenue à environ 10 °C au-dessous de la température des plaques froides isothermes. En principe, un dispositif de ce type est automatiquement replacé dans les conditions théoriques de la plaque chauffante à anneau de garde, de telle sorte que la déperdition thermique puisse être contrôlée par le réglage habituel de surchauffe.

Les contrôles effectués au moyen de ce type d'appareillage sur des échantillons connus, ont permis de souligner qu'une mesure correcte de la conductivité thermique dépendait de façon complexe de la résistance thermique latérale appliquée à l'empilage, du niveau de température du milieu ambiant et, vraisemblablement, de la nature des échantillons ainsi que des gradients de température existant dans l'empilage, ainsi que du mode de construction de la plaque chauffante.

II — 3. Un travail récent de W.H. Orr [5] a particulièrement mis en valeur les risques importants d'erreur qui apparaissent dans ce type d'expérimentation. W.H. Orr a utilisé un appareillage NBS comportant un caisson à ambiance interne réglable appliqué à des échantillons d'environ 20 × 20 cm. L'expérimentation a été réalisée sur des échantillons de polyuréthane et de silicone et elle a permis de justifier la possibilité d'utiliser des échantillons d'épaisseur importante, de 75 à 150 mm d'épaisseur, sous réserve de réaliser une combinaison optimale entre le niveau de température ambiante et la résistance thermique de l'isolation latérale.

Cette première conclusion est déjà particulièrement importante puisqu'elle permet d'espérer une extension des conditions d'application de la méthode au-delà des limites fixées par les normes existantes [6]. De plus, l'étude de W.H. Orr a le mérite de proposer un paramètre définissant, tout au moins en première approximation, l'influence des conditions de température du système constitué par l'empilage et le milieu ambiant. Cet indice dénommé par l'auteur « Ambiant Temperature Index » est défini comme suit :

$$\text{ATI} = (\theta_a - \theta_f)/(\theta_c - \theta_f)$$

où θ_a est la température ambiante supposée uniforme; θ_c et θ_f les températures des faces chaude et froide des échantillons.

Nous nous bornerons à reproduire l'un des graphiques (fig. 1) figurant dans l'étude de W.H. Orr.

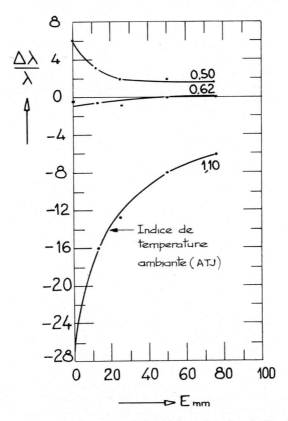

Fig. 1 — D'après W.H. Orr. – Erreur relative $\Delta\lambda/\lambda$ en fonction de l'épaisseur d'isolation latérale et de la température d'environnement (indice ATI) – Échantillon de mousse de polyuréthane de $200 \times 200 \times 85$ mm – E → épaisseur d'isolation latérale de mousse de polyuréthane.

Ce graphique représente l'erreur relative sur la valeur de la conductivité thermique λ en fonction de l'épaisseur d'isolation thermique latérale (de 0 à 100 mm de mousse de polyuréthane), pour trois valeurs caractéristiques de l'indice de température ambiante (0,5, 0,62 et 1,10).

Il est remarquable de constater une erreur maximum de -28% sur la valeur de la conductivité lorsque l'épaisseur d'isolation latérale est négligeable, pour une valeur de l'indice ATI de 1,10.

Par contre, on observe une zone optimale d'isolation latérale caractérisée par des épaisseurs comprises entre 40 et 80 mm, l'indice ATI ayant une valeur d'environ 0,62.

II — 4.   On ne doit pas perdre de vue que cette expérimentation a été conduite dans un domaine limité. En effet, la température moyenne de l'échantillon restait fixée à 24°C, l'écart de température entre faces chaude et froide était également maintenu à environ 24°C, tandis que les températures ambiantes étaient réglées à 3 niveaux distincts (24°C, 38°C et température ambiante du local).

Quoiqu'il en soit, il est certain que cette méthode d'analyse présentera un grand intérêt si l'on veut réaliser des appareillages véritablement adaptés au fonctionnement à basse température et, à plus forte raison, si l'on veut s'orienter vers des confrontations plus rationnelles entre les laboratoires.

III — INFLUENCE DU MODE DE CONSTRUCTION DES PLAQUES CHAUFFANTES

Notre intention est également de souligner les risques d'erreur qui peuvent résulter de différences minimes dans le mode de réalisation des plaques chauffantes respectant les conditions fixées par les normes existantes.

III — 1.   CARACTÉRISTIQUES HABITUELLES DE CONSTRUCTION

La plupart des plaques chauffantes connues comportent les éléments suivants :

III — 1.1.   Un élément chauffant constitué par un matériau électriquement isolant (amiante, mica, ciment réfractaire, alumine, etc...) enrobant un réseau de résistances électriques.

III — 1.2.   Deux plaques métalliques, placées de part et d'autre de l'élément chauffant et destinées à uniformiser le flux thermique délivré par cet élément.
Ces plaques comportent un système de fente créant une discontinuité de conduction thermique entre une zone centrale de mesure et la zone constituant l'anneau de garde.

III — 1.3.   Un système de détection du gradient de température au niveau de l'entrefer ci-dessus, ce système de détection étant généralement constitué par deux feuillets minces supportant des thermopiles formées de thermocouples disposés symétriquement de part et d'autre de l'entrefer.
Ces deux thermopiles sont généralement raccordées en série. Elles peuvent être placées de deux façons distinctes, soit entre l'élément chauffant et la face interne des plaques d'uniformisation, soit entre la face externe de ces plaques et les faces «chaudes» des échantillons.

III — 1.4.   Le réseau de résistances électriques équipant l'élément chauffant est constitué de deux zones distinctes, intéressant respectivement la zone centrale de mesures et la zone de réglage de surchauffe. *Deux variantes sont pratiquées* :
Dans la première, l'anneau de garde occupe entièrement la zone périphérique comprise entre la fente de discontinuité thermique et les bords de la plaque.

292

Dans la seconde, l'anneau de garde est limité à la zone externe de la plaque et il n'utilise qu'une fraction, de l'ordre du tiers, de la surface comprise entre le bord de la plaque et la zone centrale de mesure.

III — 2. Caractéristiques des cinq plaques expérimentées

Nous avons utilisé cinq plaques de forme carrée (50×50 cm) présentant les caractéristiques suivantes :

III — 2.1.

— *Plaques d'uniformisation* en duralumin, oxydées anodiquement sur leurs deux faces ;
  nuance : A G 3 — épaisseur 5 mm ;
— *Zone de mesure* délimitée par un entrefer d'une largeur de 1,2 mm bordant une zone centrale de 250×250 mm, la surface active étant délimité par le ½ entrefer, soit $251,2 \times 251,2$ mm.
— *Réalisation de l'entrefer* — deux variantes :
  — Pour *quatre* plaques (A, B, B′, C) la liaison entre la zone de mesure et l'anneau de garde est réalisée par 8 pontets (4 aux sommets du carré de mesure et 4 suivant les axes de symétrie de ce carré) ;
  — Section unitaire des pontets : 10 mm^2, soit au total 80 mm^2.
— *Une plaque de référence* (D) a été constituée en supprimant la totalité des pontets. Dans ce cas, la zone périphérique (anneau de garde) et la zone de mesure sont fixées directement sur l'élément chauffant.

III — 2.2. *Éléments chauffants*

III — 2.2.1. *Réseau chauffant*

Constitué en fil de nichrome (Nic 5) diamètre 0,8 mm, disposé en spirales carrées à sommets arrondis (rayon de courbure environ 2,5 mm) avec un pas de 10 mm. *Trois variantes* :

— *Variante A*, spirale simple, divisée en 2 zones, zone centrale débordant de 8 spires au-delà de l'entrefer — zone de surchauffe comportant 4 spires.
— *Variante B*, spirale double, divisée en 2 zones, zone centrale débordant de 4 doubles spires au-delà de l'entrefer — zone de surchauffe comportant 2 doubles spires.
— *Variante (C et D)*, identique à la variante B, à l'exception de la zone centrale qui occupe exclusivement la zone délimitée par l'entrefer, la zone de surchauffe occupant la totalité de l'anneau de garde.

III — 2.2.2. *Matériau d'enrobage*

— *Variante A*, carton d'amiante constitué de 3 feuilles d'une épaisseur de 1 mm — épaisseur totale 3 mm — liant de silicate.
— *Variante B, B′, C et D*, feutre réfractaire moulé sous pression — utilisable jusqu'à 1 100 °C — épaisseur totale 4 mm.

III — 2.3. *Thermopiles de réglage de surchauffe*

Le support des thermopiles est formé de feuillets en tissu de verre (épaisseur initiale 2/10 mm). Ces feuillets sont imprégnés de silicone et enrobent une thermopile de 20 thermocouples en chromel-alumel, en fil de 0,2 mm. Les soudures de ces thermocouples sont disposées alternativement suivant deux spires carrées de 220 et 280 mm, symétriquement de part et d'autre de l'entrefer. Chaque soudure est située sur une portion d'isotherme d'une longueur d'environ 50 mm.

Après enrobage, on réalise l'affleurement des soudures sur l'une des faces du feuillet. L'épaisseur d'un feuillet, après enrobage, atteint 4 à 5/10 mm.

Deux dispositions distinctes de thermopiles ont été utilisées :

— *Variantes A, B, C et D*

Deux feuillets de thermopiles montés en série, soit 40 thermocouples sont utilisés, ils sont placés respectivement entre la face externe de l'élément chauffant et la face interne de la plaque d'uniformisation correspondante.

— *Variante B'*

Dans cette variante, on a conservé la disposition précédente, mais on a ajouté un deuxième jeu de deux thermopiles placées respectivement entre la face chaude de chaque échantillon et la face externe de la plaque d'uniformisation. Les thermopiles externes sont utilisées pour le réglage de la surchauffe, tandis que les thermopiles internes sont affectées au contrôle du déséquilibre résiduel.

III — 3. MÉTHODE D'ESSAI

III — 3.1. *Dispositif d'essai*

On a utilisé systématiquement le même couple d'échantillons de laine de roche, épaisseur 25 mm définie au moyen de 4 cales de quartz placées dans des encoches périphériques.

Pour les quatre expérimentations, le dispositif d'essai et l'appareillage de mesure ont été maintenus rigoureusement identiques.

L'empilage était constitué, mis à part les plaques chauffantes et les échantillons, par deux plaques isothermes et quatre feuillets de mesure de température de surface d'échantillons.

La disposition des points de mesure de température de surface était conforme aux Règles RILEM (Un point de mesure au centre de chaque carré de mesure et quatre suivant les ½ diagonales.)

Ces quatre points de mesure formaient de part et d'autre de chaque échantillon une thermopile à 4 éléments.

La construction des feuillets de mesure est identique à celle des thermopiles.

Quant aux plaques d'uniformisation, elles sont constituées par des plaques de cuivre (épaisseur 5 mm) conditionnées par une circulation de liquide thermostatée utilisant un serpentin de tube de cuivre ($\varnothing$ 10/12 mm) disposé en spirale carrée double au pas de 25 mm. Ce serpentin est enrobé dans un liant graphité assurant la liaison thermique avec les plaques de cuivre.

L'empilage ainsi constitué est placé horizontalement dans un caisson à parois métalliques rempli d'isolant micacé (vermiculite). L'épaisseur latérale de l'isolation est de 28 cm, la conductivité moyenne du matériau utilisé étant de 0,04 kcal/h. m. °C.

III — 3.2. *Appareillage de mesure*

— *Mesure de la puissance* dissipée dans la zone centrale de mesure : méthode potentiométrique.

Température de surface des échantillons : potentiomètre de précision à décades.

Réglage de surchauffe : préamplification et détection sur appareil Leeds et Northrup.

Contrôle global de l'essai et du réglage de surchauffe par enregistreur potentiométrique.

— *Thermostatisation des plaques isothermes* : Ultra-thermostat Haake.

— *Alimentation des plaques chauffantes* en courant continu par alimentation stabilisée à $\pm 0,05 \%$.

III — 3.3. *Données expérimentales*

III — 3.3.1. Les cinq plaques ont été expérimentées suivant la méthode de Woodside [6]. Pour chaque niveau de température moyenne, on a fixé plusieurs valeurs du réglage de surchauffe correspondant à des valeurs positives et négatives du déséquilibre de température $\Delta\theta_E$ au niveau de l'entrefer. Les valeurs apparentes de la conductivité thermique λ ont été déterminées pour chacun des réglages utilisés.

La valeur vraie de λ soit λ_0 a été obtenue graphiquement à partir de la courbe $\lambda = f(\Delta\theta_E)$ en considérant le cas $\Delta\theta_E = 0$.

On a calculé ensuite l'écart $\Delta\lambda = |\lambda - \lambda_0|$ correspondant à chaque valeur ($\Delta\theta_E$) du déséquilibre, exprimé en grandeur et en signe.

Conformément à la méthode de Woodside, la courbe caractéristique d'erreur de chaque plaque a été définie en considérant l'écart relatif $\Delta\lambda/\lambda_0$ en fonction du déséquilibre relatif $\Delta\theta_E/(\theta_c-\theta_f)$.

III — 3.3.2. Les données expérimentales relatives à la valeur vraie de λ définie pour $\Delta\theta_E = 0$ figurent dans le tableau ci-dessous, la conductivité thermique λ_0 étant exprimée en fonction de la température moyenne dans l'échantillon, soit :
$\lambda_m = (\theta_c + \theta_f)/2$

Tableau A

VALEUR DE LA CONDUCTIVITÉ THERMIQUE λ_0 POUR $\Delta\theta_E = 0$

Plaque A		*Plaque B*	*Plaque B'*	*Plaque C*	*Plaque D*
m °C	λ_0 kcal/h . m . °C	λ_0 kcal/h . m . °C	λ_0 kcal/h . m . °C	λ_0 kcal/h . m . °C	λ_0 kcal/h . m . °C
56	0,0365	0,0375	0,0366	0,0355	0,0355
70	0,0399	0,0405	0,0399	0,0391	0,0381
85	0,0437	0,0439	0,04335	0,0430	0,0409

N. B. La conductivité du matériau étant pratiquement linéaire en fonction de la température, on peut utiliser en première approximation la relation $\lambda_0 = a\theta + b$ (λ_0 exprimé en kcal/h. m. °C et θ en °C).

Les valeurs de a et b correspondant à chaque type de plaque figurent dans le tableau ci-dessous.

Tableau B

Plaque utilisée	*Coefficient a*	*Constante b*
A	0,000245	0,02275
B	0,000225	0,02475
B'	0,000230	0,0240
C	0,0002575	0,02132
D	0,000187	0,0250

III — 3.3.3. *Équations caractéristiques des erreurs dues aux plaques*

Nous avons représenté graphiquement la relation $\Delta\lambda/\lambda_0$ en fonction du déséquilibre relatif $\Delta\theta_E/\theta_c - \theta_f)$ pour les quatre plaques A, B, B' et C.

Les graphiques des figures 2, 3 et 4 correspondent à des températures moyennes d'échantillons d'environ 56°, 70° et 86°C.

Ainsi que Woodside l'avait signalé, on a constaté que la relation entre $\Delta\lambda/\lambda_0$ et $\Delta\theta_E/(\theta_c - \theta_f)$ est pratiquement linéaire, même pour des valeurs déjà importantes du déséquilibre relatif de l'ordre de 0,02.

III — 3.4. *Analyse des résultats*

III — 3.4.1. Divergences sur les valeurs « vraies » de la conductivité thermique.

Ces divergences sont mises en valeur par la dispersion des coefficients caractérisant la relation entre la conductivité et la température.

D'après le tableau B, on observe que la plaque D présente le plus faible coefficient angulaire. Par contre, la constante *b* correspondant à titre comparatif à $\theta = 0$ présente la valeur la plus élevée.

La plaque C est caractérisée par la plus forte valeur du coefficient angulaire, ainsi que par la valeur la plus faible de la constante *b*. On notera que les plaques B et C ne diffèrent que par la disposition de la zone de surchauffe. Dans la plaque C, l'anneau de garde occupe la totalité de la surface comprise entre les bords de la plaque et l'entrefer, tandis qu'il n'occupe qu'une fraction de la même surface, dans le cas de la plaque B.

Le tableau ci-dessous fournit une comparaison des plaques A, B, B′ et D pour les trois niveaux de température utilisés, la plaque C étant prise comme référence.

Tableau C

Température moyenne d'échantillon θ_m, en °C	Valeurs du rapport $(\lambda_0)/(\lambda_0)c$			
	Plaque A	Plaque B	Plaque B′	Plaque D
56	1,03	1,05	1,03	1
70	1,01	1,03	1,01	0,97
86	1,01	1,015	1	0,95

On a observé ainsi que les écarts constatés varient entre -5 et $+5\%$ suivant le mode de construction des plaques.

III — 3.4.2. *Divergences en cas de réglage incorrect de la surchauffe* (fig. 2, 3 et 4).

On constate, en premier lieu, une influence apparemment faible de la température moyenne de l'échantillon sur les pentes des pseudo-droites traduisant pour chaque plaque la relation :

$$\frac{\Delta\lambda}{\lambda} = f\left(\frac{\Delta\theta_E}{\theta_c - \theta_f}\right).$$

De ce point de vue, il faut souligner que le domaine de variation des températures (50 à 90 °C) était limité et, d'autre part, que le mode de construction des plaques de répartition (nature du matériau, dimensions, entrefer et pontets) était identique pour 4 plaques. Il est donc difficile d'évaluer l'influence éventuelle qui résulterait d'une modification notable dans la construction des plaques de répartition.

En deuxième lieu, on constate que les divergences résultant d'un réglage incorrect du déséquilibre sont importantes suivant le type de plaques.

Les risques d'erreur de réglage sont les plus réduits avec la plaque D dont l'anneau de garde occupe la totalité de la zone comprise entre les bords de la plaque et l'entrefer et dont les pontets ont été supprimés.

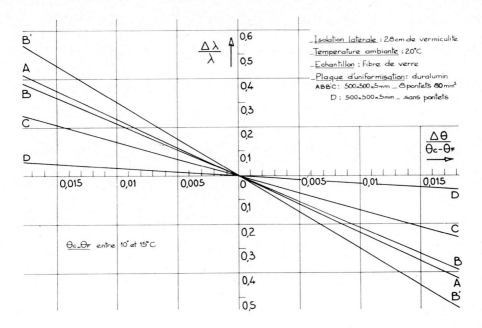

Fig. 2 — Erreur relative $\Delta\lambda/\lambda$ en fonction du déséquilibre relatif $\Delta\theta_E/\theta_c - \theta_f$ pour une température moyenne de 56 °C.

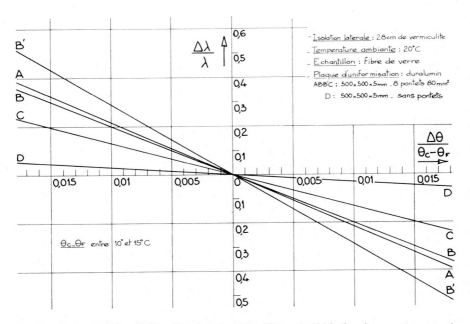

Fig. 3 — Erreur relative $\Delta\lambda/\lambda$ en fonction du déséquilibre relatif $\Delta\theta_E/\theta_c - \theta_f$ pour une température moyenne de 70 °C.

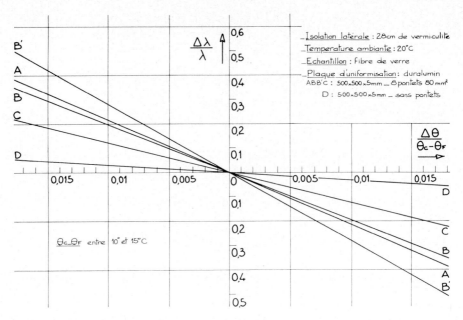

Fig. 4 — Erreur relative $\Delta\lambda/\lambda$ en fonction du déséquilibre relatif $\Delta\theta/\theta_c - \theta_f$ pour une température moyenne de 86 °C.

Le tableau D fournit une comparaison entre les erreurs relatives $\Delta\lambda/\lambda_0$ des 5 plaques pour la valeur $+0,010$ du déséquilibre relatif $\Delta\theta_E/(\theta_c - \theta_f)$ et une température moyenne de 70 °C.

Tableau D

Plaque	A	B	B′	C	D
$\Delta\lambda/\lambda$	0,215	0,200	0,290	0,130	0,03

Ainsi, l'erreur de réglage est 6,6 fois plus grande dans la plaque B que dans la plaque D.

On constate également que cette erreur est dans le rapport 1,45 pour le même type de plaque (B ou B′) suivant la position interne ou externe des thermopiles de réglage.

Nous avons également calculé le déséquilibre $(\Delta\theta_E)$ correspondant, dans les mêmes conditions, à une erreur relative de 1% pour un écart de température $\theta_c - \theta_f = 20\,°C$.

On obtient les valeurs ci-dessous :

Tableau E

Plaque	A	B	B′	C	D
$(\Delta\theta_E)$ en °C	0,01	0,01	0,007	0,016	0,07

On observe ainsi que la détection du déséquilibre donc, en définitive, le choix de l'appareillage est notablement influencé par la qualité de la plaque utilisée. Ces conditions peuvent devenir critiques, lorsque l'on s'oriente vers des appareillages automatiques de réglage de surchauffe, d'autant plus qu'elles interfèrent probablement avec les causes d'erreur dues à l'environnement.

IV — CONCLUSION GÉNÉRALE

En résumé, il apparaît que les deux causes principales d'erreur dans la méthode de la plaque chauffante à anneau de garde peuvent, actuellement, être analysées avec une certaine précision. Il nous paraît donc souhaitable que les prochaines confrontations entre laboratoires puissent, non seulement être étendues au domaine des basses températures, mais encore, qu'elles incorporent l'analyse des paramètres d'environnement et celle des causes d'erreur spécifique du type de plaque chauffante utilisée.

Il serait donc possible, par un effort collectif de comparaison, d'opérer une sélection des meilleurs appareillages et d'en définir avec précision le domaine d'application.

RÉFÉRENCES

[1] I. C. MARÉCHAL. *Matériaux et Constructions*, 1, No 5, p. 441. Mesure de la conductivité thermique par la méthode du champ thermique unidirectionnel.
[2] ASTM. « Method of Test for Thermal Conductivity of Materials by Means of the Guarded Hot-Plate » (C. 177-63) — 1966. Book of ASTM Standards Part — 14, p. 17.
[3] ASTM. Special Technical publication No. 217 (1957). Description de l'appareillage NBS.
[4] J. C. ROUSSELLE. « A Guarded Hot-Plate Apparatus for Measuring Thermal Conductivity from −80 +100 °C. » Proceedings of the Seventh Conference of Thermal Conductivity— Gaithersburg, Maryland (Nov. 13-16, 1967).
[5] H. W. ORR. « A Study of the Effects of Edge Insulation and Ambient Temperatures on Errors in Guarded Hot-Plate Measurements ». Proceedings of the Seventh Conference of Thermal Conductivity — Gaithersburg, Maryland (Nov. 13-16, 1967).
[6] W. WOODSIDE et A. G. WILSON. « Unbalance Errors in Guarded Hot-Plate Measurements », ASTM — Special Technical Publication, No 217 (1957), p. 32.

DISCUSSION pp. 317-319.

QUELQUES RÉFLEXIONS
SUR LA MÉTHODE NON STATIONNAIRE
POUR LA MESURE DE LA CONDUCTIVITÉ DES MATÉRIAUX

G. de LEPELEIRE

Laboratorium voor Koeltechniek en Klimaatregeling,
Kath. Universiteit, Leuven (Belgique)

Some remarks on a non-steady-state method for measuring the thermal conductivity of materials.

SUMMARY: *The method in question utilises temperature differences and variations in composite layers of the material under test. Measured values are interpreted according to a theoretical analysis of quasi-steady-state occurring in an homogeneous model under constant unidirectional heat flow.*

In fact, different errors occur; one of these errors in studied here.

As in most cases the thermocouple leads have thermal properties different from the tested material, a non steady state will occur which differs from the theoretical behaviour of the homogeneous model. It appears, from simplifying analysis, that in this context the diameter of the thermocouple leads is important. On the other hand, the thermal properties of the thermocouple material are almost without influence whether they be made from copper, nickel or any alloy. Practically, the error introduced can be kept at a reasonably low level.

La méthode stationnaire constituée essentiellement par l'application d'un flux thermique unidirectionnel constant et connu à une couche de matériau d'épaisseur connue, tout en mesurant les températures (constantes) de part et d'autre du matériau exige un temps de mise en régime assez important. On a élaboré une méthode plus rapide non stationnaire dans laquelle on réalise, dans un empilage judicieux de couches

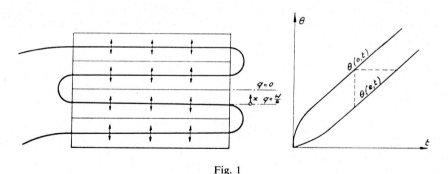

Fig. 1

de matériau sous essai, et de rubans chauffants, des zones en régime d'échauffement accessible au calcul (voir fig. 1). En effet, après une période de transition, on atteint rapidement un régime quasi stationnaire, dont l'équation différentielle s'écrit

$$a \frac{\partial^2 \theta}{\partial x^2} = \frac{\partial \theta}{\partial t} = K .$$

Par intégration on obtient

$$\theta_{(x,t)} = \frac{q}{2\lambda e} \cdot x^2 - \frac{q}{\lambda} \cdot x + K_t + c.$$

On voit que les écarts de température internes sont constants tandis que toutes les températures montent dans le temps. La répartition des températures est déterminée par la conductivité et le flux thermique, la variation par contre par le flux thermique et l'inertie.

La solution énoncée est valable pour le modèle théorique représenté en figure 1, constitué par des couches homogènes à contact thermique parfait. L'introduction de thermocouples pour la mesure des températures entraîne différentes modifications *.

1. Une résistance de transition entre le matériau de base et les thermocouples provoque une différence entre l'écart de température ($\Delta\theta$) théoriquement calculé et celui mesuré en réalité, même si ces résistances thermiques sont identiques des deux côtés. En effet les flux thermiques sont différents, ce qui entraîne des chutes de températures différentes sur des résistances identiques. On admet en général qu'un montage soigné permet de réduire cette erreur d'une façon efficace.

2. Admettons que les résistances de transition soient négligeables. Dès que les propriétés thermiques du thermocouple diffèrent de celles du matériau de base (ce qui est le cas très souvent) la situation réelle diffère du modèle homogène.

Il n'est donc pas certain, même improbable, que les thermocouples indiqueront l'écart de température tel que calculé vis-à-vis du modèle théorique.

Supposons que les thermocouples se présentent comme des fils rectilignes, parallèles entre eux, et considérons une coupe perpendiculaire à ces fils. Le flux thermique n'est plus unidirectionnel. Le manque d'homogénéité introduit par les thermocouples provoque une distorsion des plans adiabatiques et isothermiques (fig. 2). Le problème devrait donc être posé dans le plan (x, y) et la solution exacte sera de la forme

$$\theta_{(x,y,t)} = F_{(x,y,t)}.$$

Nous ne connaissons pas la solution exacte analytique de ce problème. L'étude numérique sur ordinateur d'un cas concret ne pose pas de problème spécial, cependant elle demande un temps de calcul assez important.

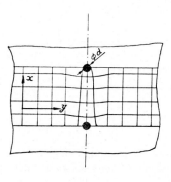

Fig. 2

(*) Nous supposons que les pertes latérales soient relativement négligeables moyennant des précautions adaptées.

Nous proposons ici une étude simplifiée, qui cependant mène à des conclusions utiles.

D'abord, on peut démontrer qu'il y a une différence entre l'écart de température mesuré, et celui calculé.

Moyennant analyse d'un modèle simplifié, des limites peuvent être indiquées pour l'erreur ainsi introduite.

L'hétérogénéité réelle, introduite par les thermocouples, se manifeste « en profondeur » (c'est-à-dire dans le plan des thermocouples, l'empilage n'est plus homogène) et « en surface » (c'est-à-dire: l'empilage n'est plus identique à l'endroit des thermocouples, et à côté de ces derniers). Considérons les deux aspects séparément :

a) L'hétérogénéité « en profondeur » n'offre pas de difficulté bien spéciale. Une situation comparable serait créée par l'empilage de couches homogènes telles que le ruban chauffant et l'échantillon par exemple. L'équation différentielle du flux unidirectionnel peut s'appliquer aux couches successives, quels que soient leur composition, leur nombre ou leur épaisseur. On obtiendra bien sûr pour chaque couche une expression $\theta(x,t)$ différente numériquement mais analogue à la formule déduite pour le modèle homogène.

b) La non-homogénéité « en surface » est bien plus difficile à manipuler, puisque le flux, strictement parlant, n'est plus unidirectionnel. On réalisera un régime quasi stationnaire, dans lequel cependant les plans adiabatiques et isothermiques présenteront une distorsion dans le voisinage des thermocouples, si ceux-ci ont des propriétés thermiques différentes de celles de l'échantillon.

Supposons ces plans connus. On peut diviser fictivement le système en « tranches » limitées par les plans adiabatiques parallèles aux fils de thermocouple, dont une comprenne les fils de thermocouple, les autres tranches se situant en dehors.

Ces tranches, de par la distorsion du réseau d'isothermes, n'auront pas une largeur dy uniforme.

L'analyse reste cependant possible en considérant ces tranches à flux uniquement « longitudinal » comme composées d'un grand nombre de couches, à caractéristiques thermiques adaptées.

Cette procédure assez encombrante, même sans être élaborée, laisse entrevoir qu'on arrivera à un régime quasi stationnaire légèrement différent de celui calculé pour le modèle homogène. Plus spécialement, l'écart de température entre les thermocouples (c'est-à-dire la mesure réelle) ne coïncidera pas avec l'écart théorique.

On peut démontrer l'existence d'une erreur de mesure d'une autre façon.

En effet: le gradient de température à l'interface du thermocouple et de l'échantillon peut être considéré comme associé à deux flux thermiques locaux: un flux q_e qui se consomme totalement dans la montée de la température locale. (Ce flux est nécessairement identique pour des thermocouples identiques). De plus, il y a un flux « transit » q_t qui est, comme le précédent, proportionnel à la vitesse de variation de température de l'ensemble, mais différent d'un point à l'autre. En milieu homogène, ce flux-transit est associé à un gradient de température proportionnel. Autour d'une discontinuité locale (comme par exemple les thermocouples) il y aura distorsion locale des gradients. Cette distorsion, en valeur relative, est fonction des répartitions de propriétés thermiques, en valeur absolue cependant cette distorsion est proportionnelle au flux local. Or, à la face chaude on applique un flux (q) tandis que la face « froide » est adiabatique. On en déduit que l'écart de température entre le thermocouple et l'échantillon (ou le ruban chauffant) dans le plan chaud sera différent de l'écart entre le thermocouple et l'échantillon dans le plan froid adiabatique. Les deux écarts ne peuvent donc se compenser; en d'autres termes, il y aura une erreur sur la mesure de la différence de température entre le plan chauffé et le plan adiabatique.

Ce n'est pas grave, si cette erreur reste limitée, et relativement négligeable vis-à-vis de la précision générale des mesures, ce qui reste à vérifier.

Il est certain que l'erreur introduite par la présence d'un fil de diamètre (d) serait inférieure à celle d'une feuille de même qualité thermique et d'une épaisseur (e) égale au diamètre du fil.

Nous pouvons donc dire, que l'erreur (calculable) due à la feuille, constitue une limite pour l'erreur réelle due au fil.

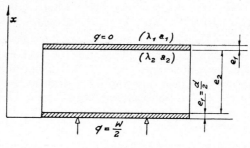

Fig. 3

Calculons l'erreur limite, due à des « feuilles », qui introduisent donc une hétérogénéité « en profondeur ».

La figure 3 montre la composition étudiée. Il s'agit donc d'une pile à trois couches : un échantillon d'épaisseur e_2, ayant les caractéristiques λ_2, a_2 serré entre deux couches minces d'épaisseur e_1, avec les caractéristiques λ_1, a_1. Sur l'une des faces, un flux uniforme et constant q est appliqué, l'autre face étant adiabatique par symétrie. Une fraction q_2 est absorbée par l'échantillon, une fraction q_1 par chacune des couches minces.

D'une façon générale, pour la période quasi stationnaire :

$$a_i \cdot \frac{\partial^2 \theta_i}{\partial x^2} = \frac{\partial \theta_i}{\partial t} = K = \frac{q}{\Sigma e_i \rho_i c_i}.$$

Cette équation est applicable aux différentes couches, tout en respectant les conditions de limite imposées. Nous en tirons les fonctions, $\theta_1(x,t)$, $\theta_2(x,t)$, $\theta_3(x,t)$.

Première couche mince ($0 < x < e_1$) :

$$x = 0 \rightarrow \frac{\partial \theta}{\partial x} = -\frac{q}{\lambda_1}$$

$$\frac{\partial \theta}{\partial x} = \frac{K}{a_1} x - \frac{q}{\lambda_1}$$

$$\theta = \frac{K}{2a_1} x^2 - \frac{q}{\lambda_1} x + f(t)$$

$$\frac{\partial \theta}{\partial t} = \frac{\partial f(t)}{\partial t} = K$$

$$f(t) = Kt + c$$

304

Finalement :

$$\theta_1 = \frac{K}{2a_1} x^2 - \frac{q}{\lambda_1} x + Kt + c .$$

Couche centrale $(e_1 < x < e_1 + e_2)$:

$$x = e_1 + e_2 \rightarrow \frac{\partial\theta}{\partial x} = \frac{-q_1}{\lambda_2}$$

$$x = e_1 \rightarrow \theta_1 - \theta_2 = \frac{q_1 + q_2}{\alpha}$$

$$\theta_2 = \frac{K}{2a_2} x^2 - \left[\frac{q_1}{\lambda_2} + \frac{K}{a_2}(e_1 + e_2) \right] x + Kt + c_3$$

$$c_3 = c + e_1^2 \left(\frac{K}{2a_1} - \frac{K}{2a_2} \right) + e_1 \left[\frac{K}{a_2}(e_1 + e_2) + \frac{q_1}{\lambda_2} - \frac{q}{\lambda_1} \right] - \frac{q_1 + q_2}{\alpha} .$$

Deuxième couche mince $(e_1 + e_2 < x < e_2 + 2e_1 = E)$:

$$x = E \qquad \frac{\partial\theta}{\partial x} = 0$$

$$x = e_1 + e_2 \qquad \theta_2 - \theta_3 = \frac{q_1}{\alpha}$$

$$\theta_3 = \frac{K}{2a_1} x^2 - \frac{KE}{a_1} x + Kt + c_5$$

$$c_5 = c_3 + (e_1 + e_2)^2 \left(\frac{K}{2a_2} - \frac{K}{2a_1} \right) + (e_1 + e_2) \left(\frac{KE}{a_1} - \frac{K}{a_2}(e_1 + e_2) - \frac{q_1}{\lambda_2} \right) - \frac{q_1}{\alpha} .$$

Pour le système homogène nous aurions obtenu :

$$\theta_H = \frac{K'}{2a_2} x^2 - \frac{q}{\lambda_2} x + K't + c .$$

Admettons $q' = q$ (c'est-à-dire le flux appliqué au modèle homogène est identique au flux réel).

Il s'ensuit que :

$$K' = \frac{Ke_2 + 2e_1 (c_1 \rho_1 / c_2 \rho_2)}{E} = A \cdot K$$

$$\theta_H = \frac{AK}{2a_2} x^2 - \frac{q}{\lambda_2} x + AK \cdot t + c .$$

Dans le système homogène, nous calculons une différence de température :

$$\Delta\theta_H = -\frac{K \cdot A \cdot E^2}{2a_2}.$$

Dans le système à trois couches, par contre :

$$\Delta\theta_K = KE^2 \left[-\frac{K}{2a_1} + (1-r)^2 \left(\frac{1}{-2a_2} - \frac{1}{2a_1} \right) + (1-r)\frac{1}{a_1} + r^2 \left(\frac{-3}{2a_1} - \frac{1}{2a_2} \right) + \right.$$

$$\left. + (1-r)\,r\,\frac{1}{\alpha^2} + r(1-2r)\left(-\frac{\lambda_1}{a_1\lambda_2} - \frac{\lambda_2}{a_2\lambda_1} \right) \right] - \frac{q}{\alpha}.$$

Dans cette formule :

$$r = \frac{e_1}{E}.$$

Calculons l'erreur relative :

$$(ER) = \frac{\Delta\theta_K - \Delta\theta_H}{\Delta\theta_H}.$$

En mettant :

$$m = \frac{a_1}{a_2}, \quad n = \frac{\lambda_1}{\lambda_2}, \quad \frac{q}{\alpha} = 0$$

et après quelques transformations, nous obtenons :

$$(ER) = \frac{r^2 \cdot 4[(m/n) + n - m - 1] + r \cdot 2 \cdot [m - (m/n)]}{-m + r \cdot 2(m-n)}. \qquad (1)$$

En pratique, $r \ll 1$, ce qui permet de réduire la formule précédente :

$$(ER) \simeq \frac{r \cdot 2 \cdot [m - (m/n)]}{-m} = 2r\left(\frac{1-n}{n} \right).$$

Si, de plus, $n \gg 1$, on obtient :

$$(ER) \simeq 2r \simeq 2\frac{d}{E}.$$

Voilà donc une conclusion remarquable. L'erreur introduite par les couches minces est déterminée surtout par l'épaisseur relative r, tandis que les propriétés thermiques n'interviennent que d'une façon très secondaire, tant que les couches sont vraiment minces.

Il paraît logique d'admettre que l'erreur introduite par des fils de thermocouples varie d'une façon semblable, tout en restant nettement en dessous de l'erreur calculée pour les feuilles minces.

(¹) Voir en annexe quelques résultats numériques, à partir de la formule exacte (1).

Tableau 1

CARACTÉRISTIQUES THERMIQUES MOYENNES DE QUELQUES MATÉRIAUX
(unités : kg, m, h, °C, kcal)

	a	λ	Comparé au fer			Comparé au cuivre		
			m	n	m/n	m	n	m/n
Cuivre	0,41	340	0,142	0,15	0,95	1	1	1
Fer	0,058	51	1	1	1	—	—	—
Plomb	0,087	30,2	0,66	1,69	0,39	—	—	—
Béton	0,0022	0,95	26,3	53,5	0,49	186	358	0,52
Bois	0,00044	0,12	132	425	0,31	930	2840	0,33
Mousse plastique (densité 25 kg/m³)	0,0039	0,031	14,9	1650	0,009	105	11000	0,0095

Tableau 2

ERREUR RELATIVE CALCULÉE D'APRÈS LA FORMULE COMPLÈTE,
POUR FEUILLE MINCES : (en %)

Echantillon	Feuilles en fer			Feuilles en cuivre		
	$r = 0,1$	0,01	0,001	$r = 0,1$	0,01	0,001
Cuivre	107,7	10,773	1,0773	0	0	0
Plomb	−8,1755	−0,81810	−0,0818	—	—	—
Béton	−19,627	−1,9627	−0,1962	−19,98	−1,994	−0,1994
Bois	−19,953	−1,995	−0,1995	−19,87	−1,999	−0,1999
Mousse	−19,988	−1,9988	−0,2385	−19,99	−1,999	−0,1999
Mauvais conducteurs	∼−20	−2	−0,2	∼−20	−2	−0,2

Ce n'est donc pas la qualité thermique du fil (cuivre, nickel ou autre), mais son diamètre qui est important.

D'autre part, l'erreur probable est parfaitement acceptable. En appliquant par exemple des thermocouples de ⌀0,1 mm, sur un échantillon de 20 mm d'épaisseur, l'erreur sera nettement inférieure à 0,5 %.

SYMBOLES

θ	température
λ	conductivité thermique
$1/\alpha$	résistance de transition
ρ	densité
A	coefficient numérique
a	conductivité « de température »
c	chaleur spécifique/constantes d'intégration
d	diamètre
	épaisseur de couche
E	épaisseur totale
f, F	fonction

K	variation de température
m	rapport des conductivités de température
n	rapport des conductivités thermiques
q	flux thermique (kcal/m^2·h)
r	épaisseur relative
t	temps
W	puissance dissipée par le ruban chauffant, par unité de surface
x, y	coordonnées

RÉFÉRENCES

[1] O. Krischer et H. Esdorn, *VDI, Forsch. Heft*, 450 (1955).
[2] H. Myncke, A. Van Itterbeek et J. Uyttenbroek, *Annexe 1964-2, Bull. I.I.F.*, Commission 2, Turin, pp. 61-69.
[3] P. Vernott, *C.R. Acad. Sci.*, Paris (1937), 204, 563.

DISCUSSION pp. 317-319.

MODIFIED GUARDED HOT PLATE APPARATUS FOR MEASURING THERMAL CONDUCTIVITY OF MOIST MATERIALS

Th. LUND MADSEN

The Thermal Insulation Laboratory, The Technical University
Lyngby (Denmark)

Méthode modifiée de la plaque chaude gardée pour la détermination de la conductivité thermique des matériaux humides

RÉSUMÉ : *La plupart des matériaux de construction contiennent un certain pourcentage d'eau, ce qui a pour résultat une conductivité thermique élevée par rapport à celle du matériau sec. On estime que les meilleures méthodes de détermination de la conductivité thermique de matériaux humides sont généralement les méthodes rapides utilisant l'établissement d'un flux de chaleur non permanent. Toutefois une méthode qui permettrait de mesurer la conductivité thermique sous des flux simultanés de chaleur et d'humidité serait plus en rapport avec les conditions pratiques.*
L'article décrit un dispositif modifié de la plaque chaude gardée où l'échantillon est séparé de la plaque chaude et de la plaque froide par des lames d'air dans lesquelles circule un air conditionné bien défini. Après l'établissement de régimes stationnaires de températures et d'humidité, le flux de chaleur et l'écart de température à travers l'échantillon sont mesurés et la conductivité est calculée à partir de ces valeurs.

INTRODUCTION

It is generally known that the determination of thermal conductivity presents problems because the change in temperature gradient, which the test involves, creates an unwanted vapor movement in the material. By measuring with an ordinary guarded hot-plate apparatus the vapor will move towards the cold surface, from where it has no possibility of escaping, as it normally can do in correctly made, practical constructions.

To get round this problem, a series of non-stationary methods has, in the course of time, been developed (O. Krischer and H. Esdorn, van der Held, Vos and others). The common factor in all these methods is that attempts are made to determine thermal conductivity before the vapor has managed to move appreciably. This does not correspond particularly well to practical circumstances where one usually has both a heat and a vapour transport through the building material from the hot to the cold side.

In an attempt to simulate the practical situation during laboratory experiments better than till now, the laboratory has constructed a modified guarded hot-plate apparatus where it is possible to impress a constant vapor gradient simultaneously on a specimen.

MODIFIED GUARDED HOT-PLATE APPARATUS

In principle, the apparatus is built like a classical double-sided guarded hot-plate apparatus. It is intended for specimen with a total area of 30×30 cm of which the actual measuring area amounts to 20×20 cm.

The construction of the guarded hot-plate apparatus is shown in figure 1. The heating element itself is made of 3 mm pertinax sheet, which is evenly winded with 1.5×0.1 mm kanthalband. On both sides of the heating element, and electrically

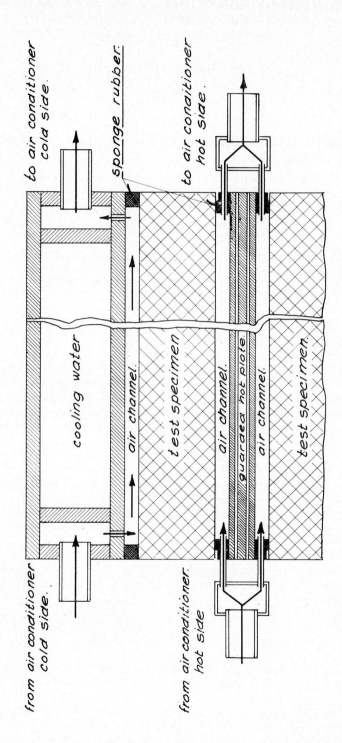

to air conditioner cold side.

sponge rubber.

to air conditioner hot side.

from air conditioner cold side.

from air conditioner hot side.

cooling water

air channel.

test specimen

air channel.

guarded hot plate.

air channel.

test specimen.

Fig. 1 — Cross section of guarded hot plate apparatus.

310

isolated from it, is a 2 mm thick aluminium sheet with a 2 mm gap which thermally divides the metering panel and the border panel. Thermal elements are moulded in the grooves in the aluminium sheets so that it is possible to determine the difference in temperature between the metering panel and the border panel with an accuracy of ± 0.01 deg C.

The cooling plates are constructed as flat aluminium boxes through which cooled water flows, like through a labyrinth. The temperature of the cooled water is regulated by help of a unit whose principle is shown in figure 2.

A volume of water of 90 litres circulates between the two cooling plates. These are fixed in series with a pump so as to yield about 100 litres per minute. After the passage through the cooling plates, the water is guided through a heat exchange where the water's temperature is reduced about 1/10 deg C, because the cooling effect is constant and about 500 Watt.

Regulation of the temperature is made with an electric heating element, which is governed with the help of a transistorised contact thermometer. The thermometer is placed in such a way that water passes it directly after it passes the heating element. This causes a quick regulation frequency, which combined with the large quantity of water between the heating element and the outlet to the cooling plate, gives the result that these surface temperatures do not measurably change.

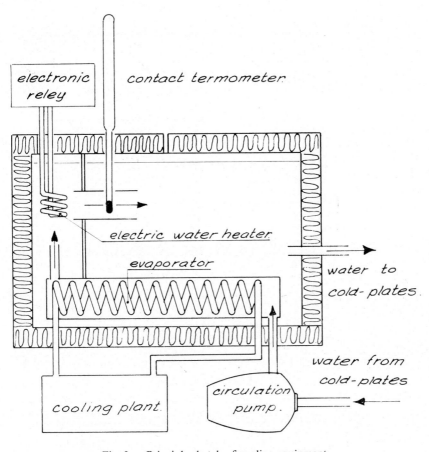

Fig. 2 — Principle sketch of cooling equipment.

With the intention of creating vapor transport through the specimen, both the cold plate and the hot plate are equipped with a row of sponge rubber lists; it can be seen as a cross section in figure 3. Conditioned air flows continuously through these channels.

On the cold side this ventilation does not present particular problems. The air is guided through a chamber to a series of holes in the side of the cold plate, from which it passes through the channels and is sucked up on the other side for re-conditioning (fig. 4).

On the other hand it is important that on the hot side conditioned air neither conveys nor removes heat from the test piece. This is achieved by measuring the temperature of air immediately before it flows into the channel and again immediately after it is sucked out, and next, to make sure, by help of a little heating element, that these two temperatures are identical. Air is drawn by two double-working piston pumps which work with a constant speed to provide a constant heat leakage resistance between the hot plate and the test piece. The air is conditioned by being guided through closed containers with saturated saline solution. By weighing these containers before and after measurement, it is possible partly to determine the material diffusion figure during the thermal conditions concerned, and partly to get control over the ventilation system's tightness, as the weight loss from the hot side must be the same as the weight increase on the cold side.

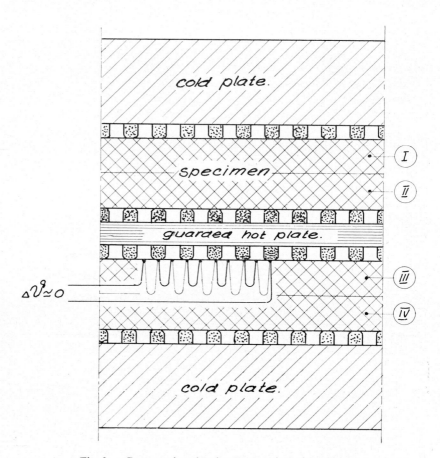

Fig. 3 — Cross section showing the location of specimens.

Until now, the following saline solutions have been used:

$MgCl_2$ = 33% RF
$Na_2Cr_2O_1$ = 52% RF
NaCl   = 75% RF
KNO_3 = 90% RF

MEASURING EQUIPMENT

The power which supplies the hot plate comes from a stabilised direct current distributor, which is itself supplied by a mains voltage stabiliser, and which also conveys current to the motor that drives the air pumps.

Supplied power is determined by measuring current and voltage by the hot plate terminals with a Hartmann and Brown measuring instrument, class 0.2.

The various temperature differences are measured with copper-constantan thermal elements, and are registered on a Philips 12 channel recorder.

The difference in temperature between the test piece hot and cold side is determined by five thermal elements in series, and glued to the test piece so that the wires go at right angles to the channels. This direction is chosen so that it is possible to even out differences in temperature which are a result of the heat resistance being different above sponge rubber and above air channel. The water content of conditioned air is determined by an Aqmel dewpoint meter, of the H.C.P.P type. This instrument measures the dewpoint directly, so that it can be established at which temperature the air in question causes dew formation on a mirror. This method has many advan-

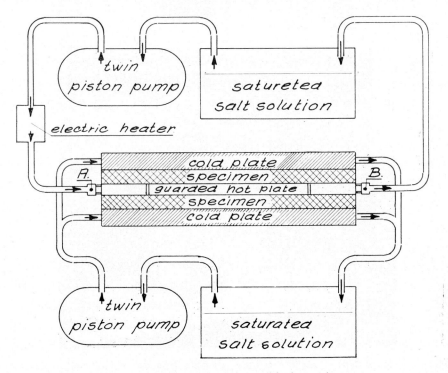

Fig. 4 — Principle sketch of air conditioning equipment.

tages for this purpose. First, only a very little volume of air is used to determine the dew point, secondly the measurement makes no appreciable change of the air condition, and thirdly the apparatus has a very good reproducibility.

It is clear that the given design of heating and cooling plates presents the risk of hot current in the specimens not being one-dimensional because the resistance varies from rubber list to air channel.

To be able to examine this case more closely, ten thermal elements in a series 5 mm apart, were attached so that all the "cold" joints met with a rubber list, and all the "hot" joints met with an air gap (see fig. 3). Measuring on this thermopile shows that the test piece surface temperature is, practically speaking, the same for rubber lists as for air gaps.

It was also expected that the thermal conductivity for a 5 mm wide air gap according to J.S. Cammerer is 0.038 kcal/h.m. °C, while the rubber list thermal conductivity is stated to be 0.040 kcal/h. m. °C.

Heat transport is carried out one-dimensionally through the whole test element,

Fig. 5 — General view of modified guarded hot plate apparatus. At top right is seen the dewpoint meter.

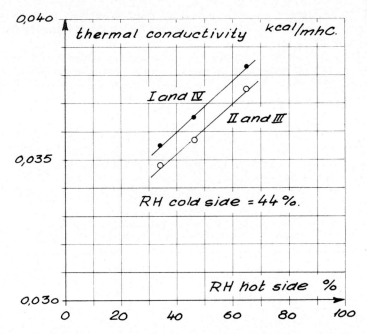

Fig. 6 — The thermal conductivity of soft fibre board in dependence on the relative humidity of the hot side.

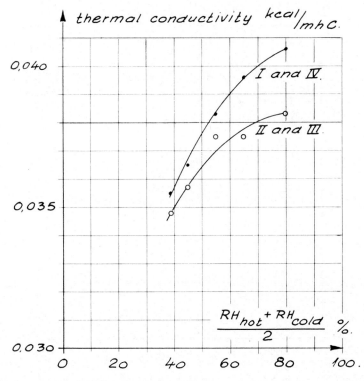

Fig. 7 — The thermal conductivity of soft fibre board in dependence on average relative humidity at cold and hot side.

whereas the transport of vapor is carried out two-dimensionally near the surface, because water vapor can only diffuse in and out of the test pieces where the specimens are limited by the air gaps.

During the measurement, the edges of test pieces are covered with vapor-tight tape, so that the water vapor and the outside space cannot be exchanged. The whole installation is as shown in figure 5 put in an insulated box whose interior is kept at the test piece average temperature by suitable exposure of the cold plates.

MEASURING RESULTS

Until now measurements have only been taken on 4 pieces of $\frac{1}{2}$ in. soft fibre board with a dry weight of 240 kg/m^3.

Table 1

THE THERMAL CONDUCTIVITY OF SOFT FIBRE BOARD IN DEPENDENCE
ON THE RELATIVE HUMIDITY AT THE COLD AND THE HOT SIDE

Relative air humidity %		Temperature °C		Vapor content weight percent		Thermal conductivity kcal/h.m. °C	
cold side	hot side	cold side	hot side	I and IV	II and III	I and IV	II and III
44	34	+18.3	23.6	5.7	5.3	0.0355	0.0348
44	46	+18.3	23.6			0.0365	0.0357
44	65	+18.8	23.8			0.0383	0.0375
57	74	+18.8	23.7			0.0396	0.0375
81	80	+18.8	23.7	16.0	11.4	0.0406	0.0383

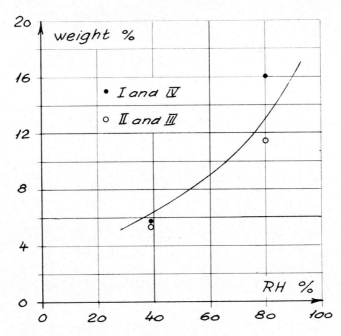

Fig. 8 — Vapor content in soft fibre board in dependence on relative humidity, compared with a sorption curve specified by P. Lund-Hansen, 1967.

The result of these measurements is shown in table 1 and figures 6 and 7. The position of the boards is shown in figure 3.

So as not to change the conditions of the experiment en route, the boards were only weighed immediately before the thermal measurement begun and again immediately after they were finished. Afterwards, the boards were dried at a temperature of $+55\,°C$ until they reached a constant weight. From this weight, the vapor content from the first and last measurements was calculated. The result is shown in figure 8 together with a sorption curve for soft fibre board as specified by P. Lund-Hansen in " Moisture Transfer in Building Materials", 1967.

DISCUSSION

B. H. VOS (The Netherlands) — In the first place I want to state that as far as I know, this is the first apparatus that can teach us more about the variation of thermal conductivity under realistic boundary conditions. We have to bear in mind that thermal conductivity is not at all a constant property:

$$\lambda = \lambda(\overline{T}, \text{grad. } T, \Psi, \text{grad. } \overline{\Psi}, \alpha)$$

thickness (α). This parameter enters into the function as a consequence of radiation, depending strongly upon temperature.

temperature (T). The influence of this parameter is well known.

moisture content (ψ). These vapour transfer in the pores — evaporation and condensation — plays an important part.

moisture and temp. gradient (grad. ψ, grad. T)

Both gradients together determine flow of vapour and water in liquid form in the material.

Of course one can give a definition of thermal conductivity by:

$$\lambda = \lim \lambda(\Psi, T, \text{grad. } \Psi, \text{grad. } T, \alpha)$$

$$\left.\begin{array}{l} \text{grad. } T \\ \text{grad. } \Psi \end{array}\right\} \to 0.$$

However a property is found that is only of academical value. We cannot work with it. I think the apparatus as it has been developed by the author can provide us with more information about what really happens than any other way of approach. Especially thermal conductivity of structures that are closed at one side and open at the other side can be compared with two-sided open structures. Unfortunately the author whill have to restrict himself to the hygroscopic moisture range. In practice also the higher region are of great interest, primarily around critical water content. Therefore I hope that also measuring devices can be suitably developed for measurement in this region.

D. FOURNIER (France) — 1. Je pense qu'il faut tenir compte dans les problèmes relatifs à la mesure de la conductivité thermique et en particulier dans les mesures comparatives telles que celles organisées par l'I.I.F., non seulement des paramètres déjà nombreux cités par M. Vos et qui caractérisent l'influence des conditions de mesure sur la conductivité thermique d'un échantillon donné, mais aussi de paramètres caractéristiques du matériau lorsque plusieurs échantillons sont en cause.

En effet un isolant industriel n'est pas un corps pur parfaitement homogène, loin de là. Pour un isolant fibreux par exemple il faut ajouter au paramètre « épaisseur » le paramètre « résistance spécifique au passage de l'air » qui est lui-même fonction de la masse volumique apparente du produit, de la finesse des fibres, de leur arrangement spatial, et dont la valeur varie d'un échantillon à l'autre, si soignée que soit même une fabrication spéciale. Il n'y a donc aucun intérêt à mesurer de façon parfaite la conductivité thermique d'un échantillon (sauf en tant qu'éprouvette de référence destinée à des étalonnages comparatifs) si l'on est pas à même d'effectuer la mesure sur de nombreux échantillons pour obtenir des résultats statistiques.

Cette notion prend tout son sens pour apprécier à leur juste valeur les résultats chiffrés de la confrontation internationale de l'I.I.F., effectués sur plusieurs échantillons d'un produit fibreux dont nous avons l'expérience et dont la conductivité thermique peut varier de $\pm 1\%$ d'un échantillon à l'autre.

On peut donc considérer que des écarts de 2 % entre laboratoires au cours de cette confrontation ne sont pas significatifs sur le plan métrologique, mais peuvent être dus uniquement aux inhomogénéités du produit isolant lui-même.

2. Le rapport présenté par M. DOUSSAIN (p. 289) met en évidence le fait que les résultats obtenus à l'aide d'une plaque chaude gardée possédant des circuits de chauffe «centre» et «garde» indépendants, sont meilleurs que ceux fournis par une plaque gardée surchauffée sur le pourtour de la garde, ceci à même déséquilibre de température centre-garde.

Nous pouvons confirmer ces conclusions à la suite d'expériences menées depuis 1962 sur ces deux technologies, la première solution, proposée d'ailleurs au Symposium ASTM de 1957, s'étant avérée meilleure à tous les points de vue.

3. En ce qui concerne la question (controversée) de « l'environnement » le plus adéquat à prévoir autour de l'empilement de mesure, nous pensons qu'il n'y a pas de solution idéale convenant à tous les cas. Notre expérience sur divers appareils fonctionnant à température assez proche de l'ambiante (-20 °C à $+90$ °C) a révélé que la solution d'un environnement gazeux, régulé à la température moyenne de mesure, est très satisfaisante.

Elle fait intervenir de plus, pour les températures inférieures à l'ambiante, un « point froid » qui a l'avantage de constituer un préventif contre la condensation d'humidité sur les plaques froides ou dans les zones froides des éprouvettes.

En ce qui concerne les essais à températures plus élevées, où le rayonnement joue un rôle important dans les transferts de chaleur, nous avons constaté que cette solution n'était pas satisfaisante et que l'emploi d'une poudre superisolante ajouté à celui d'une couronne de garde extérieure à l'empilement de mesure, tel que l'a réalisé Dynatech Corporation, est des plus judicieux.

A basse température, où c'est alors la convection naturelle qui peut provoquer des transferts de chaleur considérables, cette dernière solution semble également satisfaisante.

4. Enfin à notre avis il faut se méfier de l'utilisation d'un seul et même appareillage dans des conditions de température trop différentes, ainsi que pour des matériaux de conductivité thermique trop différentes (des superisolants, du béton ou du verre par exemple) ou ayant une anisotropie trop marquée (conductivités thermiques de rapport 1 à 10 suivant deux axes rectangulaires).

J.C. MARÉCHAL (France) — Les essais normalisés doivent rester une détermination bien précise des caractéristiques de base des matériaux servant autant à la normalisation des matériaux qu'au contrôle des fournitures ou aux expertises et également de références aux études des paramètres influençant les comportements de ces matériaux.

Mais de nombreux laboratoires ont une double mission qui tout en pratiquant les tests officiels de référence ont également des programmes de recherches sur l'étude des matériaux en fonction de divers paramètres. Les laboratoires bien équipés pratiquent en général plusieurs méthodes en régime stationnaire ou non stationnaire et il reste à leur charge de choisir les méthodes qui conviennent le mieux. J'en profiterai pour répondre à M. Vos qui craint que notre barbe soit longue avant d'en avoir terminé avec les mesures en régime permanent sur matériaux humides qu'il y a de nombreux cas ou les méthodes non stationnaires ne sont pas toujours applicables et qu'il n'y a pas d'autre issue que les systèmes stationnaires. Ces méthodes peuvent être assez rapides si l'on veut bien essayer les matériaux non pas en fonction d'une teneur en eau imposée, mais à une température imposée correspondant à l'état d'équilibre thermohygrométrique imposé par cette température et ceci évidemment sous faible gradient de température.

MESURES DE CONDUCTIVITÉ THERMIQUE DES MATÉRIAUX ISOLANTS PAR UN APPAREIL ORIENTABLE À PLAQUE CHAUDE BI-GARDÉE

D. FOURNIER et S. KLARSFELD

Centre de Recherches Industrielles, Compagnie de Saint-Gobain
Rantigny (France)

Revolving apparatus with a doubly-guarded hot plate for measuring the thermal conductivity of insulating materials

SUMMARY: *A guarded hot plate apparatus was developed and perfected in the latterly years. It now enables one to determine the thermal conductivity of insulants at temperatures between $-20\,°C$ and $+90\,°C$. Besides the conventional lateral guard-zone of the hot plate, the apparatus is provided with a second hot plate placed directly against the first and maintained at an identical temperature. This device enables unidirectional heat flow between the metering hot plate and a single cold plate: measurements are consequently performed on a single specimen of insulating material. Furthermore, the apparatus can be revolved around a horizontal axis. This feature reveals even the smallest thermal conductivity variations as a function of the orientation of the specimen temperature gradient in relation to the vertical. Examples are given to show the presence or absence of heat transfer in an insulating material by natural convection.*

The results obtained with this new apparatus are in agreement, to less than $\pm 1\%$, with determinations made under the same circumstances using an apparatus with a heat flowmeter (presented at Trondheim in 1966) calibrated against NBS standard specimens.

I — INTRODUCTION

La mesure de la conductivité thermique par la méthode dite de la plaque chaude gardée, ou encore de Poensgen, a fait l'objet de très nombreuses publications, travaux, normalisations et continue, à la fois sur le plan mathématique relativement au calcul des fuites latérales, et sur le plan technologique et expérimental, de faire l'objet de nombreux travaux.

Notre but essentiel n'a pas été de reprendre une fois de plus l'étude de la technologie de la plaque chaude gardée, mais de réaliser, en tenant compte des travaux effectués par les divers auteurs et en choisissant délibérément certaines options, un appareil permettant :

1° d'effectuer la mesure de la conductivité thermique d'un échantillon *unique* (et non d'un *couple* d'échantillons comme c'est pratiquement toujours le cas);

2° d'être en mesure de *modifier l'orientation des plaques*, l'échantillon étant positionné entre elles, de façon à pouvoir mettre en évidence l'existence éventuelle d'une contribution de la convection naturelle au transfert de chaleur à travers cet échantillon.

L'unicité de l'échantillon est précieuse pour deux raisons. La première tient au problème même de l'échantillonnage. Lors de la mise au point de nouveaux produits et également, lors d'une expertise, il est difficile parfois de se procurer une quantité notable de produits; d'autre part, il est impératif dans la méthode classique utilisant une paire d'échantillons que ces derniers soient aussi identiques que possible, ce qui est difficilement réalisable dans les deux cas précédents. La deuxième raison est que la mise en œuvre d'un flux thermique unidirectionnel donc d'un seul échantillon, est indispensable si l'on veut mettre en évidence une modification éventuelle de ce flux par la convection naturelle, suivant l'orientation du gradient de température vis-à-vis du champ de pesanteur.

L'ensemble de ces motifs nous a amenés à partir de 1965, après avoir antérieurement réalisé un appareil à plaque chaude gardée classique et à surchauffe de bord, à entreprendre la construction d'une nouvelle plaque chaude destinée à la réalisation d'un appareil non symétrique donnant les facilités indiquées précédemment.

II — Réalisation d'un appareil à plaque chaude bi-gardée, non symétrique, à un seul échantillon et à flux thermique unidirectionnel

Dans cet appareillage on peut distinguer 5 parties :

— La plaque chaude gardée;
— La plaque froide;
— Leur assemblage et leur positionnement relatif;
— L'environnement;
— Les organes de régulation et de mesure.

Aucun appareil de ce type n'a été décrit à ce jour à notre connaissance sauf par Achtziger [1], son principe étant simplement mentionné comme agréé dans la norme DIN 52612 [3].

Nous avons été néanmoins largement guidés par les principes de l'ASTM C. 177-63 [2] ainsi que par certaines données du Symposium 1957 de l'ASTM [4], en particulier pour la structure de la plaque chaude dans sa partie classique (avec par conséquent abandon de la surchauffe de bord).

Le but de ce rapport n'étant pas technologique, nous nous bornerons à donner les principes et les grandes lignes de la structure de l'appareil lui-même.

II.1 Plaque chaude bi-gardée

Dans un appareil à plaque chaude gardée classique deux échantillons du matériau à étudier sont disposés symétriquement de part et d'autre de cette plaque, entre elle et deux plaques froides. Le bilan thermique en régime permanent est alors effectué en considérant la somme des puissances thermiques W_1 et W_2 qui traversent les deux échantillons, cette somme étant prise égale à la puissance électrique dépensée dans le carré central de la plaque chaude dont les faces sont à températures θ_{c1} et θ_{c2} après établissement du régime permanent.

Si l'on supprime l'échantillon et la plaque froide situés du côté de la plaque chaude gardée correspondant à la puissance W_2, il suffit, à l'aide d'une seconde plaque chauffante maintenue à θC_2, de rendre nulle cette puissance W_2 pour retrouver les conditions de l'équilibre avec le seul échantillon traversé par W_1.

Ceci suppose évidemment de ne pas introduire ainsi de nouvelles fuites latérales ce qui est facilité en rapprochant au maximum la plaque chauffante de garde de la plaque chaude gardée. La plaque chaude bi-gardée comprendra donc (fig. 1) une plaque chaude I d'aspect classique, à laquelle sera adjointe une plaque chauffante III la recouvrant entièrement sur une face, séparée seulement d'elle par une tranche mince d'un matériau convenable.

Une thermopile positionnée entre les faces en regard de la plaque chaude gardée et de la plaque chauffante, réalise conjointement avec cette tranche de matériau le fluxmètre thermique de zéro II.

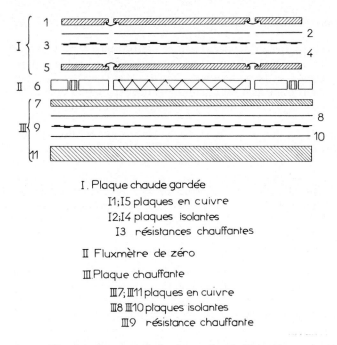

I. Plaque chaude gardée
    I1; I5 plaques en cuivre
    I2; I4 plaques isolantes
     I3 résistances chauffantes

II Fluxmètre de zéro

III. Plaque chauffante
    III7; III11 plaques en cuivre
    III8 III10 plaques isolantes
     III9 résistance chauffante

Fig. 1 — Structure de la plaque chaude bi-gardée.

On peut noter les points particuliers suivants :

— La plaque a été conçue pour travailler jusqu'à 100 °C ;

— L'ensemble des thermopiles a été réalisé par un procédé original de tissage de fils de verre et de fils de thermocouples permettant une grande régularité de structure, ainsi que l'emploi de fils très fins ($\varnothing = 0{,}1$ mm) en chromel et constantan, les soudures étant réalisées par points en grande série. La thermopile de garde latérale a pu être ainsi réalisée aisément avec 2×400 soudures, celle assurant l'annulation du flux arrière comportant 1000 soudures ;

— Il n'y a aucun pont métallique entre le carré central actif de la plaque chaude gardée et sa garde latérale. Le déjoint de 1 mm entre eux comporte simplement 4 ponts en résine époxy ;

— La plaque de cuivre rectifiée III 11, jointe à une technologie très soignée, assure à l'ensemble sa planéité ;

— Les structures des âmes chauffantes de la plaque chaude gardée et de la plaque chauffante de garde sont voisines de celles décrites par Gilbo [4].

— Les dimensions latérales hors-tout sont de 40 cm × 40 cm, pour permettre une comparaison avec un appareil précédemment décrit [5], celles du carré central de mesure utilisées dans le calcul du flux thermique étant de 23 cm × 23 cm.

— La plaque chauffante de garde est isolée thermiquement sur sa face libre.

II.2 PLAQUE FROIDE

C'est une plaque classique en cuivre, à double spirale carrée où la circulation du liquide se fait à contre-courant ; son épaisseur totale est 30 mm, les deux faces en cuivre étant de 5 mm, la face active étant rectifiée.

II.3 ASSEMBLAGE ET POSITIONNEMENT DES PLAQUES

Un bâti a été prévu pour supporter l'ensemble formé par la plaque chaude bi-gardée, l'échantillon et la plaque froide, et assurer en même temps à volonté sa rotation autour d'un axe horizontal, de façon continue. En particulier cette rotation peut être réalisée lorsque le régime permanent est atteint pour une inclinaison donnée, ce qui permet d'éliminer toutes causes de modification de ce régime autres qu'un phénomène dû à la convection naturelle.

La plaque de cuivre rigide (III.11, fig. 1) permet une fixation mécanique correcte de l'ensemble constituant la plaque chaude, avec un minimum de fuites thermiques parasites.

La plaque froide peut être déplacée en restant parallèle à la plaque chaude gardée, à l'aide de guides convenables. L'épaisseur de mesure de l'échantillon d'isolant fibreux ou alvéolaire étudié est déterminée à l'aide de 4 cales en silice usinées à 0,02 mm.

II.4 ENVIRONNEMENT

Comme il est précisé, entre autres, dans l'ASTM C 177-63, il a été prévu autour de l'ensemble précédent une enceinte étanche vis-à-vis de l'ambiance du laboratoire. Elle permet d'entourer l'appareil d'une atmosphère contrôlée en température et maintenue à un degré hygrométrique suffisamment bas pour empêcher toute condensation d'humidité sur la plaque froide ou dans l'échantillon. Ceci est réalisé par passage forcé de l'air de l'enceinte dans une chambre de refroidissement munie d'un radiateur maintenu à une température inférieure à celle de la plaque froide, puis dans une chambre de chauffage qui élève la température de cet air à la valeur désirée pour sa rentrée dans l'enceinte.

Cette température est choisie égale à la moyenne arithmétique entre les températures des faces chaudes et froides de l'échantillon. La précision de son réglage, indiquée plus loin, est très largement au-delà des conditions minimales précisées par Woodside [6].

II.5. ORGANES DE RÉGULATION ET DE MESURE

II.5.1. *Régulations*

La plaque froide est alimentée à partir d'un thermostat assurant la stabilité de sa température de surface et fonctionnant actuellement jusqu'à $-20\,°C$.

Le carré central de la plaque chaude est alimenté en courant continu stabilisé à 0,02 %. L'anneau de garde et la plaque chauffante de garde sont alimentés en courant alternatif régulé à 1 %. Ces deux éléments de garde comportent des circuits de chauffe dont l'alimentation est assurée par deux régulations indépendantes l'une de l'autre, commandées respectivement par les thermopiles latérales et par le fluxmètre de zéro II.

Ces régulations sont du type « tout ou peu » avec, comme détecteurs, des galvanomètres de zéro à cellules photoélectriques.

Entre carré central et anneau de garde les fluctuations de l'écart de température, calculé à partir d'enregistrements de la f.e.m. de la thermopile, sont normalement de l'ordre de 0,004 deg C.

Entre l'arrière du carré central de mesure et la plaque chauffante de garde, le chiffre correspondant est de 0,001 deg C.

La régulation de la température de l'enceinte s'effectue au niveau de la chambre de chauffe (voir II.4), par « tout ou rien ». Les enregistrements sur thermographe montrent une stabilité meilleure que 1 deg C dans tous les cas, le degré hygrométrique ne variant pas de plus de 2 %. Par exemple pour une température moyenne de mesure de 24 deg C le degré hygrométrique peut être maintenu à 14 % ±1 %.

II.5.2. *Dispositifs de mesure*

Outre l'épaisseur définie et mesurée comme indiqué au paragraphe II.3, les mesures de la puissance électrique fournie au carré central de la plaque chaude sont effectuées, ainsi que celles des f.e.m. des thermocouples, au potentiomètre de précision (1 μV, ou 0,1 μV dans les cas spéciaux).

II.5.2.1. *Mesures de puissance*

On mesure la f.e.m. aux bornes de l'enroulement de chauffe du carré central, les prises de tension étant faites au milieu du déjoint entre centre et garde. La mesure de l'intensité du courant est effectuée par l'intermédiaire d'une résistance étalon de précision. La précision obtenue ainsi sur la puissance est de l'ordre de 0,1 %.

II.5.2.2. *Mesures de température*

Les thermocouples utilisés actuellement sont des nickel chrome-nickel allié de 0,2 mm de diamètre, à demi-tolérances NBS; chaque thermocouple est de plus comparé individuellement à un thermomètre à résistance de platine étalon, ce qui permet de tracer sa courbe d'étalonnage à mieux que 0,5 μV près.

Conformément aux indications de la norme ASTM C 177-63 relative aux produits isolants non rigides, les prises de températures sont effectuées habituellement à la surface métallique des plaques chaude et froide. Outre des thermocouples de contrôle en nombre suffisant destinés à vérifier l'homogénéité de la température de surface des plaques chaude et froide, deux thermocouples de mesure sont positionnés au centre de ces plaques.

Les soudures froides de l'ensemble des thermocouples sont, suivant un procédé inspiré de Zabawsky [4], positionnées dans un bloc métallique thermostaté de référence, dont le niveau de température est lui-même mesuré par un couple dont la soudure froide se trouve dans une cellule à point triple de l'eau, stable à ±10^{-4}°C. La mesure des f.e.m. est effectuée normalement avec un potentiomètre donnant le microvolt mais peut être faite, si besoin est, à 0,1 μV près.

III — PERFORMANCES OBTENUES

III.1. OBTENTION ET STABILITÉ DU RÉGIME PERMANENT

Dans des conditions données de température, le régime permanent est obtenu en 12 heures environ et sa stabilité est vérifiée au cours des 6 heures suivantes.

L'expérience montre que le régime permanent est établi lorsque les indications des thermocouples de mesure sont stables à mieux qu'un microvolt, sans dérive, au cours de trois mesures séparées d'une heure.

Dans la plupart des cas la stabilité du régime établi est telle qu'aucune variation de température n'est décelable au potentiomètre de sensibilité 1 μV.

III.2. REPRODUCTIBILITÉ DES MESURES

Des mesures successives de la conductivité thermique d'un même échantillon d'isolant fibreux de propriétés stables (masse volumique 70 kg/m^3), ont montré à plusieurs mois d'intervalle une reproductibilité de l'ordre de 0,4 %. On peut considérer que dans tous les cas la précision en comparaison est meilleure que 1 %.

a) *Résultats obtenus sur échantillons étalons fournis par le National Bureau of Standards*

L'appareil décrit ci-dessus a fourni sur un échantillon de fibre de verre (162 kg/m³), mesuré au NBS, des résultats qui, en fonction de la température s'alignent de façon très satisfaisante autour de la droite fournie par les chiffres NBS (à mieux que 1 %) (fig. 2).

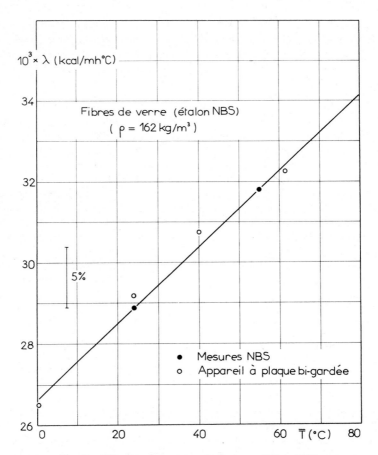

Fig. 2 — Comparaison entre résultats sur étalon NBS.

b) *Résultats obtenus sur un échantillon utilisé pour les mesures comparatives de l'I.I.F.*

Grâce à l'amabilité du Laboratoire National d'Essais, Paris, nous avons eu la possibilité de mesurer un échantillon fourni par Fibreglass Ltd. dans le cadre de la campagne récente d'essais comparatifs de l'I.I.F. Nous avons obtenu les résultats portés sur la figure 3 où nous avons reporté des éléments de comparaison figurant dans [7]. La droite obtenue est sensiblement parallèle à la droite moyenne du NPL et à 2,8 % au-dessus. Nous pensons d'après notre expérience sur ce type de matériau, qu'une bonne partie de cet écart peut être attribuée au fait qu'un échantillon individuel peut s'écarter normalement du produit-type « moyen », dans des proportions de cet ordre.

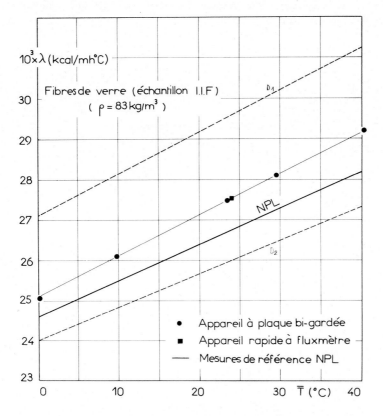

Fig. 3 — Mesures comparatives internationales I.I.F. [7]. D1-D2 zone de dispersion pour les mesures I.I.F. (cas des courbes parallèles à la courbe NPL).

c) *Comparaison de résultats obtenus à l'aide de l'appareil de mesure rapide* (5)

Le lambdamètre rapide décrit antérieurement à Trondheim (I.I.F. — 1966) a été comparé à l'appareil à plaque bi-gardée sur d'autres échantillons que ceux (NBS) servant à son étalonnage. On a reporté ainsi (fig. 3) les valeurs obtenues sur l'échantillon Fibreglass Ltd. On donne également, dans la figure 4, à titre d'exemple, un résultat obtenu sur polystyrène expansé (UNIMAT 12 kg/m^3) où l'accord est excellent entre les appareils.

D'une façon générale, dans tous les cas étudiés jusqu'à ce jour, l'accord obtenu a été nettement meilleur que 1 %.

d) *Comparaison de résultats obtenus sur l'appareil TCFG — N 18 de la Dynatech Corporation*

La figure 5 représente, dans leur plage commune de fonctionnement, les résultats obtenus par l'appareil à plaque bi-gardée et l'appareil Dynatech Corporation TCFG — N 18 à plaque chaude gardée conforme à l'ASTM C 177-63, sur des panneaux de fibres de verre à usage industriel. On voit que l'accord est de l'ordre de 1,5 %.

L'ensemble de ces résultats montre le comportement très satisfaisant de l'appareil à plaque bi-gardée vis-à-vis d'appareils de conception et de principe très différents, ainsi que vis-à-vis des résultats fournis par des laboratoires reconnus comme référence tels que le NBS et le NPL.

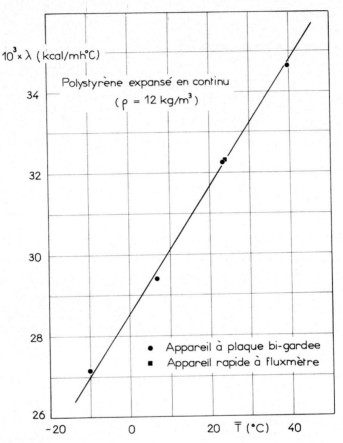

Fig. 4 — Comparaison appareil à plaque bi-gardée et appareil de mesure rapide (à fluxmètre).

IV — APPLICATION À LA MISE EN ÉVIDENCE
DE LA CONVECTION NATURELLE EN MILIEUX POREUX

IV.1. VÉRIFICATION EXPÉRIMENTALE DE L'ABSENCE DE TRANSFERT DE CHALEUR PAR CONVECTION DANS LES CONDITIONS NORMALISÉES DE MESURE DE CONDUCTIVITÉ THERMIQUE

Dans les conditions courantes de mesure du λ des isolants fibreux, où les gradients de température ne dépassent pas 10 °C/cm, on a vérifié en modifiant l'orientation des plaques par rapport à la verticale l'absence de toute convection décelable, cette absence se maintenant même pour des gradients de température plus élevés (20 °C/cm), et pour des isolants fibreux à faible masse volumique (8 kg/m³). Ceci pourrait à l'avenir supprimer toute contestation relative à la nécessité d'utiliser des appareils à faces verticales ou horizontales dans l'intervalle de température considéré ici. Il est à noter également que la modification du gradient de température n'influe pas sur les résultats obtenus.

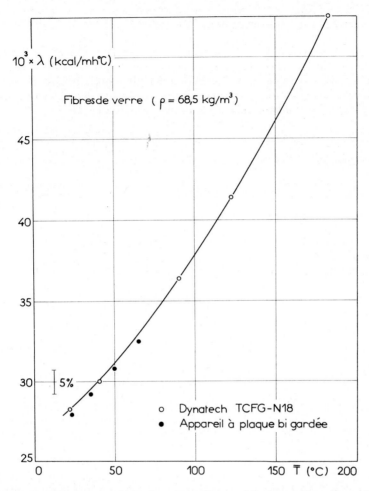

Fig. 5 — Comparaison des résultats obtenus sur l'appareil à plaque bi-gardée et l'appareil TCFG-N18 « DYNATECH ».

D'autre part il est intéressant de rapprocher ce résultat des conclusions obtenues récemment dans le cadre d'une recherche Euratom effectuée conjointement sous contrat par la Société Bertin et Cie et la Compagnie de Saint-Gobain, et relative au transfert de chaleur par convection naturelle dans les milieux fibreux, en espace confiné [8, 9]. Il a été montré par le calcul théorique et confirmé par l'expérience que le déclenchement de la convection est lié entre autre, à la valeur d'un nombre sans dimensions, le nombre de Rayleigh de filtration, qui dépend en particulier des caractéristiques du milieu poreux (nature du fluide, perméabilité), du gradient de température entre les faces chaude et froide, etc...

Dans les conditions de mesure précédentes, l'application de ces résultats montre que le déclenchement de la convection est impossible. On peut donc penser aller plus loin et réaliser des vérifications expérimentales dans des conditions simulées, à même nombre de Rayleigh, ce qui donne un intérêt particulier à l'expérimentation suivante avec l'appareil à plaque bi-gardée.

IV.2. Mise en évidence expérimentale du transfert de chaleur par convection dans un milieu fibreux immergé dans un liquide

La figure 6 représente la variation de la conductivité thermique en fonction de la différence de température (ou gradient) entre les plaques chaude et froide de l'appareil bi-gardé, maintenues horizontales, pour un échantillon constitué par de la fibre de verre ensachée dans un film plastique étanche rempli d'un liquide (eau dans le cas présent). La température moyenne a été maintenue constante (20 °C).

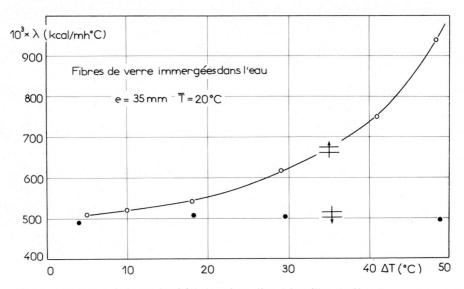

Fig. 6 — Variation de la conductivité thermique d'un échantillon de fibre de verre immergé dans l'eau en fonction de la différence de température entre ses faces et de l'orientation du gradient thermique par rapport au champ de pesanteur.

Dans le cas où la plaque chaude est au-dessus de la plaque froide on constate la constance de la conductivité avec la variation du gradient de température. Par contre l'inversion de position de deux plaques montre une augmentation rapide de la conductivité, donc un effet convectif, qui croît avec le gradient de température.

On peut espérer que l'exploitation de ce nouveau moyen d'investigation, combinée avec l'emploi des nombres sans dimensions, pourra permettre une exploration systématique de tels phénomènes dans une gamme étendue de conditions dimensionelles, de températures, etc... en particulier pour prévoir le comportement des isolants à basse température où les phénomènes de convection prennent toute leur importance.

V — Conclusion

On a décrit un appareil de mesure de conductivité thermique à plaque bi-gardée, dont la réalisation technologique très soignée présente certains points originaux. On a donné des résultats expérimentaux obtenus par son intermédiaire et qui s'alignent de façon très satisfaisante sur ceux donnés par d'autres appareils classiques et normalisés ainsi que par des laboratoires faisant autorité en la matière. On a montré enfin deux exemples d'application de cet appareil à l'étude de la convection naturelle en milieux fibreux, résultats qui ne sont que l'amorce d'un travail de recherche en cours de développement.

Les auteurs tiennent à exprimer leurs remerciements à deux collaborateurs du Laboratoire de Physique du Centre de Recherches Industrielles de SAINT-GOBAIN, MM. VIEILLERIBIÈRE et J. BOULANT, qui ont pris une part essentielle à la construction et à la mise en route de l'appareil à plaque bi-gardée entre 1965 et 1968.

RÉFÉRENCES

[1] J. ACHTZIGER. Wärmeleitfähigkeitsmessungen an Isolierstoffen mit dem Plattengerät bei tiefen Temperaturen. *Kältetechnik* (1960), 12, pp. 372-375.
[2] Norme ASTM—C 177-63, Standard method of test for thermalconductivity materials by means of the guarded hot plate.
[3] Norme DIN 52612—63, Bestimmung der Wärmeleitfähigkeit mit dem Plattengerät.
[4] 1957 Philadelphia Meeting. Symposium on thermal conductivity and applications of thermal insulations. *ASTM Special Technical Publications*, n° 217.
[5] D. FOURNIER, G. ANDRÉ et S. KLARSFELD. Études récentes sur la conductibilité thermique des isolants fibreux effectuées à l'aide d'un appareil de mesure rapide de la conductibilité thermique en régime permanent. *Annexe 1966-2 Bull. I.I.F.*, pp. 163-174, Commission 2, Trondheim.
[6] W. WOODSIDE. *Rev. Sci. Instr.* (1957), 28, p. 1933.
[7] Commission 2, I.I.F., Mesures comparatives internationales de conductivité thermique. *Bull. I.I.F.* (1968), n° 1, pp. 29-59.
[8] P. MICHEAU, R. GROSSIN et S. KLARSFELD. Convection naturelle en milieux fibreux. Réunion d'information sur les travaux relatifs aux cuves de réacteurs en béton pré-contraint et à leur isolation, Euratom, Bruxelles (7-8 nov. 1967), non publié.
[9] G. MORDCHELLES-RÉGNIER et al. Recherches récentes effectuées en France sur l'isolation thermique des réacteurs nucléaires. Colloque sur les réacteurs poussés et à haute température refroidis par un gaz, Jülich 21/25 Oct. 1968. Advanced high temperature gas-cooled reactors, International Atomic Energy Agency, Vienna (1969), pp. 529-544.

DISCUSSION

R. LOTZ (France) — 1. Dans la comparaison des résultats avec l'appareil à fluxmètre, comment l'étalonnage du fluxmètre avait-il été obtenu?

2. Les soudures des fluxmètres de fuites latérales sont noyées dans une résine. Un léger déplacement en hauteur de certaines soudures peut modifier l'indication de la thermopile puisque cette thermopile est traversée par le même flux de chaleur que l'échantillon, et donner un zéro indiqué différent du zéro réel. La qualité des mesures sur un montage similaire serait donc liée à la précision de la mise en place de ces soudures? (cette remarque est également valable pour un appareil à plaque chaude gardée classique).

3. Vous indiquez que la stabilité de l'état de température entre plaque chauffante centrale et plaque chauffante de garde reste meilleure que 1/1000 °C.

A quelle flux de chaleur supplémentaire vers le bas correspond une variation de cet écart de 1/1000 °C?

D. FOURNIER — 1. Nous n'étalonnons pas le fluxmètre, car ceci n'est pas la méthode la plus efficace. En effet, comme expliqué dans le document référencé [5], on détermine une « constante d'appareil » qui inclut la sensibilité du fluxmètre à la température moyenne de mesure considérée. Ceci est effectué à partir d'échantillons étalonnés au « National Bureau of Standards » sur appareil à plaque chaude gardée, la stabilité de cet étalonnage étant vérifiée fréquemment, ce qui permet de détecter tout anomalie du fluxmètre ou de l'appareillage.

2. Cette possibilité n'est pas absurde. Mais, justement, la technologie ici décrite permet d'éviter tout risque de cette nature.

3. Le calcul montre qu'une erreur *systématique* de 1/1000 °C sur le réglage de la consigne « flux nul » pour le fluxmètre de garde arrière, correspond à un flux parasite à 0,046 Watts/m² ; les flux utilisés pour la mesure étant en général au moins égaux à 100 Watts/m², l'erreur relative serait alors de 5.10^{-4}, chiffre qui parle de lui-même.

J.C. MARÉCHAL (France) — Je formulerai une remarque concernant la désignation « plaque chaude bigardée ». Le mot garde étant réservé au plan de l'isotherme. Lorsque les laboratoires ont à mesurer un seul échantillon ils utilisent sur l'autre face de la plaque chaude un échantillon auxiliaire maintenu sous gradient nul par une autre plaque chauffante—ou bien un fluxmètre auxiliaire. Vous avez réuni les deux. Néanmoins il conviendrait de l'appeler méthode à échantillon unique.

D. FOURNIER — Le terme « bigardée » n'est peut être pas parfait, mais utiliser l'expression « méthode à échantillon unique » ne fait pas référence à la plaque chaude gardée qu'utilise cette méthode, et donc la désigne de façon beaucoup trop vague.

A CALORIMETER FOR MEASURING
VERY SMALL HEAT FLUXES

H. GLASER

Institut für Technische Thermodynamik
Universität Stuttgart (W. Germany)

Un calorimètre pour la mesure de très petits flux de chaleur

RÉSUMÉ : *On a mis au point un calorimètre pour déterminer la conductivité thermique des isolants humides.*
Cet appareil est de forme cylindrique et se compose de 3 parties. Seule la partie centrale est utilisée pour la mesure des flux de chaleur entrant dans le calorimètre. Les deux autres parties protègent la partie centrale contre les effets limites indésirables. A l'intérieur de la partie centrale est montée une plaque perforée contenant un liquide. L'air sec traverse la plaque et forme des bulles dans le liquide, atteignant au cours de son passage un équilibre de température parfait avec le liquide. Lorsque l'air quitte le liquide il est complètement saturé. La chaleur nécessaire à l'évaporation de la portion correspondante de liquide est absorbée dans le calorimètre en traversant, par exemple, une isolation thermique de conductivité thermique inconnue. En régime stable le flux de chaleur traversant l'isolation thermique est donné par le débit d'air sec, la température de l'air sec et la température de l'air saturé.
On examine les principes thermodynamiques et hydrodynamiques de la méthode de fonctionnement du calorimètre. L'appareil décrit permet de mesurer des flux de chaleur de l'ordre de quelques watts par m².

It is well known that under certain conditions water may be precipitated in thermal insulations [1]. Consequently the thermal conductivity of the insulation as well as its resistance against moisture transmission will change their numerical values.

A prediction of the time dependent influence of water precipitation in an insulated wall on the thermal insulation resistance is possible in principle. However, for this prediction the relationship between thermal conductivity and resistance against moisture transmission of insulations and of building materials on one hand and the water content on the other hand has to be known.

Experimental investigations of these phenomena may be only considered to be reliable as long as the measurements do not disturb the thermal conduction and the moisture transmission. A safe way to obtain good results is to measure the heat flux which is flowing through the test material at the cold side. However, using a handy experimental apparatus the described investigation method requires measurements of heat fluxes in an order of magnitude of only a few watts. The operation of a calorimeter which complies with the requirements will be described as follows.

THE OPERATION OF THE CALORIMETER

A schematic representation of the calorimeter is shown in figure 1. The cylindrical shell is divided into three parts. Only the middle part (1) is used for measurements, whereas the purpose of the two other parts, (2) and (3), is to protect the middle part from unwanted boundary effects. At the bottom of the middle part a sieve tray (4) is placed which carries a liquid, for instance water. The amount of the liquid is regulated with the help of a tube (6). Dry air flows steadily through the tube (5), passes the sieve tray, bubbles through the liquid (7) producing a bubbling bed, and leaves the middle part of the calorimeter through the tube (8). Figure 2 shows such a bubbling bed. Because of the many small air bubbles which are mixing the liquid completely an excellent temperature equilibrium is produced within the bubbling bed. Therefore, the temperature at the inner wall of the middle part is constant and hence well defined.

Since the air which is flowing to the sieve tray is totally dry it becomes loaded with

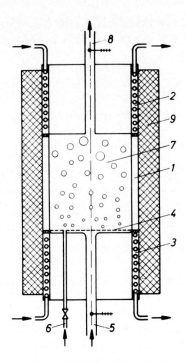

Fig. 1 — Schematic representation of the calorimeter :

1. test section
2. 3. protection from boundary effects
4. sieve tray
5. air inlet

6. liquid inlet
7. bubbling bed
8. air outlet
9. thermal insulation

vapour of the liquid when passing the bubbling bed. Because of the large bubble surface the air leaves the apparatus completely saturated.

According to this evaporation a certain amount of heat is needed. Heat transfer is also to be taken into account if the temperatures of the air flowing into the calorimeter and the air flowing out of the calorimeter are different.

The calorimeter is surrounded by the material whose thermal conductivity coefficient is to be measured. Provided that the temperature outside the calorimeter is higher than in the bubbling bed heat flows into the calorimeter. At steady state conditions this heat flux can easily be determined by means of the air flow rate and the temperatures at the air inlet and the air outlet. This will be discussed later.

The heat flux and the difference between the temperatures in the bubbling bed and at the outside of the tested material lead to the wanted thermal conductivity coefficient of the material. The results of the measurements obtained in the described way might be wrong because of axial heat fluxes in the calorimeter shell and in the insulating material. To prevent these heat fluxes part 2 and part 3 of the calorimeter are used. They contain channels through which a temperature controlled liquid streams so that they have the same temperature as the bubbling bed. By that means any unwanted boundary effects are completely eliminated.

Essential conditions for a well working calorimeter are that the bubbling bed occupies nearly the whole volume of the middle part and that the air bubbles are as small as possible. There are no difficulties to design a reasonable sieve tray since

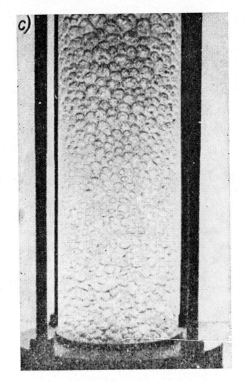

Fig. 2 — Bubbling bed.

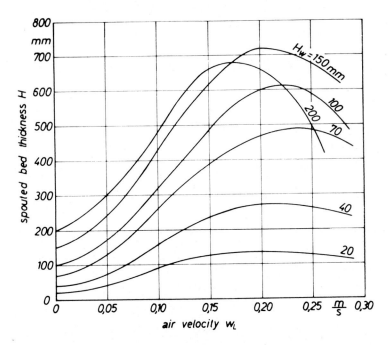

Fig. 3 — Bubbling bed thickness.

A. Mersmann [2] has carefully investigated bubbling beds such as they are to be used in the calorimeter. For example, figure 3 shows in which way the thickness H of the bubbling bed depends on the air velocity w_L. The air velocity is related to the entire flow cross-section. H_w means the thickness of the liquid on the sieve tray without any air bubbling through it.

H_w is used as a parameter in figure 3. The results shown in figure 3 are valid for water and for a sieve tray with 20 000 holes per m^2 with a diameter of 0.5 mm, According to this figure it is possible to produce a bubbling bed with a thickness of about 300 mm, provided the correct air flow rate and the correct amount of liquid on the sieve tray are used. By means of the informations given in the paper of A. Mersmann it is possible to determine the thickness of bubbling beds for arbitrary geometr'c shapes and for arbitrary working conditions. Therefore it is not difficult to design the calorimeter according to the particular requirements.

CALCULATION OF THE HEAT INFLOW

Now the calculation of the heat flux streaming into the calorimeter is discussed in detail. It can be shown theoretically that the air leaving the calorimeter has the same temperature t_f as the bubbling bed and that it is completely saturated. That means that the air has the saturation humidity ratio x'. This theoretical conclusion has been verified by accurate measurements. Furthermore, it is of advantage to fill the calorimeter with liquid which has the same temperature as the bubbling bed.

In figure 4 heat and mass balances are shown. A mass balance for the liquid is obtained as follows: the sieve tray carries the amount m_w of liquid. The air flow m_L streaming towards the bubbling bed has the humidity ratio x. That means, per unit of time the amount of liquid $m_L \cdot x$ is added to the liquid on the sieve tray. Further, the calorimeter is supplied with a steady liquid flow m_{we} in order to maintain a certain thickness of the bubbling bed. On the other hand, the leaving air carries the liquid amount $m_L \cdot x'$, where x' is the saturation humidity ratio according to t_f. Furthermore,

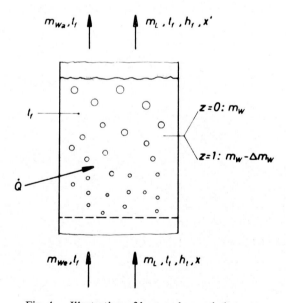

Fig. 4 — Illustration of heat and mass balances.

336

the leaving air may carry an amount m_{wa} of liquid out of the calorimeter—for instance as small drops. Δm_w is the diminution per unit of time of the liquid content of the calorimeter. Then the following balance holds:

$$m_w + m_L \cdot x + m_{we} = (m_w - \Delta m_w) + m_L x' + m_{wa} \tag{1}$$

or

$$m_{we} - m_{wa} = m_L(x' - x) - \Delta m_w . \tag{2}$$

In the same manner a heat balance can be obtained. However, additionally it has to be taken into account, that there is a heat flux $\dot{Q}$ flowing into the calorimeter. The specific heat of the liquid is c_w, then using the symbols for the temperatures t and the enthalpies h which are given in figure 4 the heat balance related to the unit of time is:

$$m_w c_w t_f + m_L h_t + m_{we} c_w t_f + \dot{Q} = (m_w - \Delta m_w) c_w t_f + m_L h_f + m_{wa} c_w t_f \tag{3}$$

or

$$\dot{Q} = m_L(h_f - h_t) + (m_{wa} - m_{we} - \Delta m_w) c_w t_f . \tag{4}$$

The combination of equations (4) and (2) yields

$$\dot{Q} = m_L(h_f - h_t) - m_L(x' - x) c_w t_f . \tag{5}$$

It is remarkable that it is entirely unimportant how much liquid is added to or withdrawn from the calorimeter, as long as the added liquid has the same temperature t_f as the bubbling bed.

The amount of heat which is taken up by the air flow per unit of mass is of practical interest.

From equation (5) it follows that:

$$q = \dot{Q}/m_L = h_f - h_t - (x' - x) c_w t_f . \tag{6}$$

Because the last term on the right side of equation (6) is usually very small, the heat taken up by the air is approximately equal to the increase of enthalpy of the air in the calorimeter.

Provided, the ideal-gas equation of state is valid for the air as well as for the water-vapour, the enthalpy of humid air may be expressed as

$$h = c_p t + x(r_0 + c_{pd} t). \tag{7}$$

In equation (7) c_p is the specific heat of dry air, t is the temperature of dry air, x is the humidity ratio, r_0 is the latent heat of water at $0\,°C$, and c_{pd} is the specific heat of water vapour.

With the use of equation (7), equation (6) gives

$$q = c_p(t_f - t_t) + (x' - x) r_0 + x' c_{pd} \cdot t_f - x \cdot c_{pd} \cdot t_t - (x' - x) c_w t_f . \tag{8}$$

In an experiment it is necessary to measure the temperatures t_f and t_t. By it the saturation humidity ratio x' is known. However, the humidity ratio x is still unknown. To determine it an additional temperature would have to be measured and, therefore, another error in measurements would have to be taken into account. A better way is to use completely dry air, which may be easily produced by adsorption of the humidity with silicagel. Then x is equal to zero and equation (8) may be rearranged to give

$$q = c_p(t_f - t_t) + x' \cdot [r_0 - (c_w - c_{pd}) t_f]. \tag{9}$$

The specific heat flux which is taken up by the calorimeter under the above mentioned assumptions is shown in figure 5 depending on the temperature t_t of the dry air and on the temperature t_f over the bubbling bed.

At a constant temperature t_f, q decreases with an increasing temperature of the dry air. At a constant temperature of the dry air, the heat flux increases with an increasing temperature of the bubbling bed.

It should be noticed that the calorimeter is a small refrigerating machine, because the heat removed is adequate to a refrigerating effect. This refrigerating effect is produced by the unsaturated air.

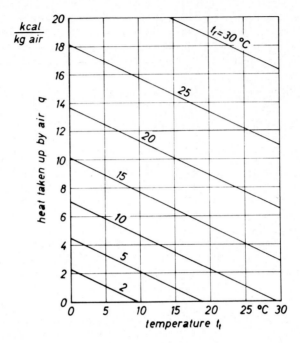

Fig. 5 — Heat taken up by air per mass unit.

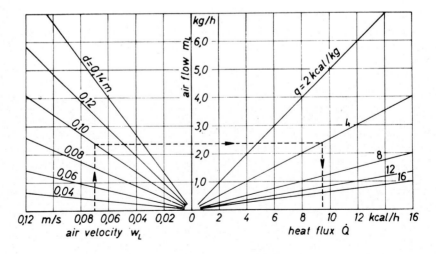

Fig. 6 — Determination of the heat flux.

Finally, the connection between the heat removed and the size of the calorimeter is discussed. If w_L represents the air velocity related to the entire flow cross-section in the calorimeter, ρL is the density of air and d is the diameter of the bubbling bed, then the heat removed by the air flow rate m_L may be expressed by

$$\dot{Q} = m_L \cdot q = \frac{\pi}{4} d^2 w_L \rho_L q .$$ (10)

This relationship is shown in figure 6 as a nomogram. From it the heat flux $\dot{Q}$, which is identical with the refrigeration effect of the calorimeter can be found depending on the air velocity w_L and the diameter d of the calorimeter for arbitrary specific amounts of heat. For instance, the refrigeration effect of a calorimeter with a diameter of 10 cm, for an air velocity $w_L = 0.07$ m/s and a specific refrigerating capacity $q = 4$ kcal/kg will be 9.4 kcal/hr.

Additionally, reasonable dimensions and proper working conditions for the calorimeter can be determined with the help of this monogram according to any demands. At last the used air flow rate can be found by it. The nomogram is valid for a density of air $\rho_L = 1.2$ kg/m^3. For other densities it should be corrected by means of equation (10).

EXPERIMENTAL PROCEDURE

The heat flux is determined by three different measurements. According to equation (9) the specific heat flux is obtained by measuring the temperature t_t of the dry air at the air inlet and the temperature t_f of the bubbling bed. The saturation humidity ratio is a function of t_f. To find the real heat flux $\dot{Q}$, the air flow rate m_L is needed due to equation (10). Clearly the accuracy of the results depends on carefully measured t_t, t_f and m_L.

More details are presented in the paper of H. Auracher, p. 341.

REFERENCES

[1] H. GLASER, "Wärmeleitung und Feuchtigkeitsdurchgang durch Kühlraumisolierungen", *Kältetechnik* **10** (1958), pp. 86-91.
[2] A. MERSMANN, "Druckverlust und Schaumhöhen von gasdurchströmten Flüssigkeits-schichten", *VDI-Forschungsheft*, 491, VDI-Verlag Düsseldorf (1962).

A BUBBLE-COLUMN-APPARATUS FOR MEASURING THE THERMAL CONDUCTIVITY AND THE DIFFUSION RESISTANCE FACTOR OF INSULATING MATERIALS

H. AURACHER

Institut für Technische Thermodynamik, Universität Stuttgart (W. Germany)

Un appareil à colonne à bulles pour la mesure de la conductivité thermique et le facteur de résistance à la diffusion des isolants

RÉSUMÉ : *On décrit un appareil destiné à la mesure de la conductivité et du facteur de résistance à la diffusion des isolants en régime stable. Le principe de cet appareil s'appuie sur la méthode de la colonne à bulles de Glaser. Il convient à des mesures au-dessus et au-dessous de 0 °C. Le domaine de température choisi dans la colonne à bulles détermine le fluide bouillonnant. On fait quelques observations sur l'appareil et son utilisation.*

On a suffisamment étudié la précision de l'appareil à l'aide d'une méthode expérimentale spéciale. Les mesures étudiées sont suffisamment précises même dans le cas de très faibles flux de chaleur.

On présente les premiers résultats de mesure de la conductivité thermique.

The measurement of the thermal conductivity of moist insulations is difficult. This is because usually the moisture content varies, in general, with respect to the location and to the time during the measurement. The impact of these variations on the resulting measurements has to be taken into account. Essential for the change of the moisture content due to time is, among others, the momentary value of the diffusion resistance factor μ, which in turn depends on the moisture content. As already reported by Glaser [1], it is therefore appropriate to measure simultaneously diffusion and conductivity, without essentially disturbing either one. This is possible by using the calorimeter described by him.

If, for instance, a temperature drop and a partial pressure drop of the water vapour are maintained across the insulation (1, fig. 1) to be tested, a heat and vapour flux enters into the insulation surrounding the calorimeter. The vapour is partly condensing in the insulation, so that it is accumulating moisture in the course of time. It is now possible to calculate the thermal conductivity and the diffusion resistance factor as a function of the moisture content. This requires the local moisture content, the temperature gradient, and the heat flux to be measured as functions of time. The working condition and the accuracy of the calorimeter with respect to the measurement of the heat flux will be investigated in this paper. The influence of the moisture may be disregarded for these investigations. Such experiments are being carried out at the present. Later on a separate report will be given on the results to be obtained.

In the following part, the construction of the calorimeter will be described as well as the apparatus which is necessary for its proper functioning. Later the accuracy of the calorimeter will be theoretically and experimentally tested. Finally, first results on the thermal conductivity measurement of an insulation made of Styropor will be reported.

THE CALORIMETER

The construction of the calorimeter is shown in figure 1. The test section consists of a hollow cylinder (2) made of brass, 300 mm long and 100 mm in diameter. The cylinder contains a liquid into which dry air is blown through a sieve tray (3). The air is saturated with the vapour of the liquid whereby the radially inflowing heat is removed. The temperature of the spouted bed is measured by thermocouples (4) at various points. In this fashion, the temperature equilibrium of the spouted bed can be checked. It is obtained with satisfactory accuracy under practically required working conditions.

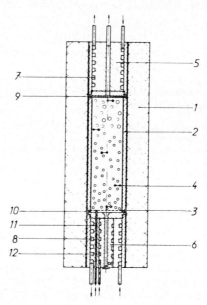

Fig. 1 — Cross section of the calorimeter:

1. thermal insulation
2. test section
3. sieve tray
4. thermocouple
5,6. protection cylinder

7, 8. channel for the cooling liquid
9, 10. Teflon ring
11. liquid inlet
12. channel for the dry air

Both protection cylinders (5, 6) are also made of brass. They are kept on a constant temperature by a cooling liquid flowing through bifilar-wound channels (7, 8). At a liquid flow rate of only a few litres per minute the temperature differences in the protection cylinders are negligible. If the temperatures of the protection cylinders and the test section are not identical, small axial heat fluxes will be generated. These are kept negligible by means of two insulating Teflon rings (9, 10), located between the protection cylinders and the test section. Through a hole (11) in the lower protection cylinder the liquid is always refilled in such a way that the thickness of the spouted bed remains constant. Thereby the liquid obtains the temperature of the spouted bed already in the protection cylinder. The air flows at first through a flat channel (12) in the lower protection part before entering the sieve tray. Thereby it obtains the temperature of the spouted bed within about 0.1 grade deviation. Consequently, the temperature measurement of the dry air is no longer necessary. This will be discussed later.

Glaser [1] has shown that because of measurement problems it is advisable that the air supplied to the calorimeter be dry. Based on this requirement the complete equipment of the apparatus shown in figure 2 was devised as follows:

THE BUBBLE-COLUMN-APPARATUS

A diaphragm compressor (1) delivers oil-free air from the environment into an air-mixing chamber (2) and from there to a heat exchanger (3), which is kept at constant temperature in a thermostat (5). Linked to the heat exchanger (3) is a pressure regulator (4) which is also enclosed by the thermostat (5). By that the pressure variations of the compressed air are eliminated and the mass flow rate is kept exactly constant. Next the air flows through a water filter (6), which separates entrained water drops. In two driers (7) arranged in parallel the compressed humid air is dried by means of silica gel. Thereby a dew point temperature of $-60\,°C$ is obtained. The dry state of the air is checked by a blue-gel indicator (8). The parallel arrangement of the driers permits for replacing the moist silica gel without interrupting the measurements. Before entering the gas flow meter (10) the air flows through a filter (9) in which dust and dirt particles are separated. The measurement of the air flow rate together with the measurement of the spouted bed temperature is essential for the accuracy of the procedure. Since, in extreme cases, very small air flow rates of about 0.05 kg/hr must be measured, a suitable gas flow meter has been constructed especially for this purpose. Essentially it consists of several capillaries, arranged in parallel, through which the compressed air flows. The pressure drop in the capillaries occurs in direct proportion to the flow rate. The gas flow meter has been calibrated with very good accuracy. The reproducibility was found to be better than $\pm0.5\%$ for individual measurements. Details on the construction and the method of calibration cannot be discussed here. The air mass flow can be accurately controlled by a needle valve (11).

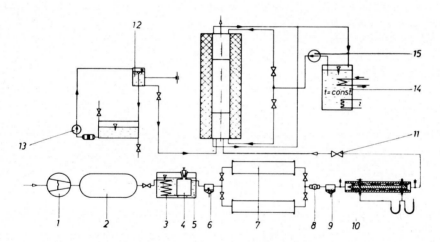

Fig. 2 — Schematic representation of the bubble-column-apparatus:

1. diaphragm compressor
2. air-mixing-chamber
3. heat exchanger
4. pressure regulator
5. thermostat
6. water filter
7. drier
8. blue-gel indicator
9. dust filter
10. gas flow meter
11. needle valve
12. storage tank
13. pump
14. thermostat
15. pump

The amount of liquid carried by the air out of the calorimeter is replaced from a storage tank (12) in which the liquid level is kept constant by a pump (13). The temperature of the cooling liquid for the two protection cylinders is kept constant by means of a thermostat (14). Its pump (15) supplies the cooling liquid through parallel leads to both the protection cylinders of the calorimeter.

Now, at first it is important to test how accurately the heat flux can be measured with this apparatus. A theoretical estimation of the accuracy is possible if the general equation given by Glaser [1] to determine the refrigeration effect of the calorimeter is considered.
It is:

$$\dot{Q} = \dot{m}_L \cdot c_{p_L} \cdot (t_f - t_t) + \dot{m}_L \cdot x' [r_0 - (c_f - c_{pa}) t_f]. \tag{1}$$

The refrigerating effect, in the state of equilibrium, is identical to the heat flux flowing through the insulation. This equation is valid assuming that the supply liquid enters the test section with the spouted bed temperature and the air passes the sieve tray in dry condition. According to the design of the apparatus both conditions are met. In equation (1) $\dot{m}_L$ is the dry air flow rate, c_{p_L}, c_f, and c_{pa} are the specific heats of the dry air, of the bubbling fluid and of the vapour, t_f and t_t are the temperatures of the spouted bed and of the dry air, r_0 is the latent heat of the liquid at $0°C$, and x' is the specific humidity of the saturated air.

Equation (1) can still be simplified. Considering that, due to the construction of the calorimeter, the temperature t_t of the dry air deviates only about 0.1 grade from the temperature t_f of the spouted bed, the left part of equation (1) becomes negligible as compared to the right part. A further simplification is obtained, if the enthalpy h' of the saturated liquid and the enthalpy h'' of the saturated vapour are known. And usually this is the case.
Through substituting:

$$h'' = r_0 + c_{pa} \cdot t_f \tag{2}$$

and:

$$h' = c_f \cdot t_f \tag{3}$$

as well as:

$$r = h'' - h' \tag{4}$$

and:

$$x' = \frac{R_L}{R_f} \cdot \frac{p'_d}{p - p'_d} \tag{5}$$

a very simple form of the equation is obtained for the refrigerating effect of the calorimeter:

$$\dot{Q} = \frac{R_L}{R_f} \cdot \frac{p'_d}{p - p'_d} \cdot r \cdot \dot{m}_L. \tag{6}$$

In this equation r is the latent heat of the liquid, R_L and R_f are the gas constants of air and liquid, p and p'_d are the absolute atmospheric pressure and the saturated vapour pressure of the liquid. p'_d and r are functions of the spouted bed temperature t_f.

Consequently, the selection of the liquid depends on one hand on the bubbling behaviour of the liquid but primarily on how accurate the coefficients used in equation (6) are known. At spouted bed temperatures above $0°C$ water will qualify best. In the temperature range from $0°C$ to $-40°C$, for instance, the refrigerant R-11 $(CFCl_3)$ is suitable.

Now, for R-11, the measuring error will be estimated according to Gauß based on equation (6). The errors of the factors of this equation are on the one hand accidental and on the other hand systematical. The latter are not considered because they can be eliminated by a special test measurement, as will be shown later.

The spouted bed temperature is assumed to be $t_f = -20\,°C$ and the atmospheric pressure $p = 1$ bar. The air flow rate $\dot{m}_L$ be measured with an error of $\pm 1\%$. The absolute error of the pressure measurement be $\Delta p = \pm 1$ Torr corresponding to a relative error of 0.13%. If the temperature of the spouted bed is measured with an error of $\Delta t_f = \pm 0.1$ grade, the relative error of the vapour pressure p'_d is $\pm 0.5\%$ and that of the latent heat $r \pm 0.014\%$. Consequently, the probable relative error of the refrigerating effect $\dot{Q}$ according to Gauß is

$$\frac{\Delta \dot{Q}}{\dot{Q}} \cdot 100\% = \pm 1.17\%. \tag{7}$$

Considering that not all of the measuring errors can be taken into account mathematically, the probable error range might be somewhat larger.

TEST MEASUREMENTS

Whether or not this assumption is realistic was proven by a test measurement. Figure 3 shows the arrangement of the test set-up. In the calorimeter, there is a

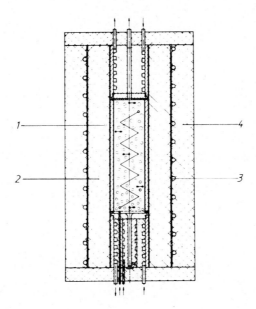

Fig. 3 — Experimental set-up for the test measurement :

1. heating coil
2. thermal insulation
3. copper cylinder
4. thermal insulation.

heating coil (1) made of thermocoax wire, fed with direct current. This heating coil supplies the bubbling liquid in the calorimeter with the heating power N, which has been measured with an error of $\pm 0.1\%$, approximately. The calorimeter is surrounded by an insulation (2) of 120 mm thickness. On the insulation there is a copper cylinder (3) the temperature of which is enforced by a thermostat to be equal to that of the spouted bed. The copper cylinder is also surrounded by an insulation (4).

Under steady state conditions the heat input N is to be equal to the refrigerating effect Q of the calorimeter because no heat flux can enter from outside. The results of the measurements are shown in figure 4. Here, the refrigerating effect $\dot{Q}$ of the

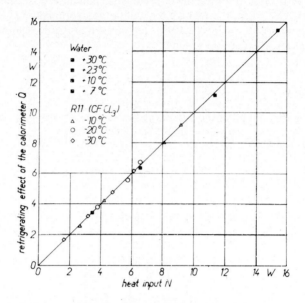

Fig. 4 — Test measurements.

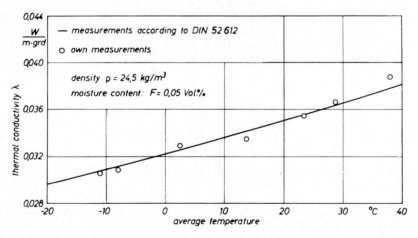

Fig. 5 — Thermal conductivity of Styropor.

calorimeter is plotted against the heat input N. Under ideal conditions, the measured points should result in a 45°–line in the diagram. The measurements were made in a temperature range from +30°C to −30°C and in a heat power range from 1.6 W to 15.4 W. At temperatures above 0°C, distilled water was used as bubbling fluid, and at temperatures below 0°C, R-11.

At first it can be stated that no systematic deviation from the nominal value (45°-line) occurs; this means that, in this case, the systematic errors of the measurements are obviously negligible or, at least, they eliminate each other. Besides, it may be particularly pointed out that systematic errors which may occur under other working conditions, can be eliminated by this test measurement. The largest deviation from the 45°-line is about 3.5%, however, most of the measurements are much more accurate. Thus the theoretically expected accuracy of the measuring method is confirmed by experiment.

THE THERMAL CONDUCTIVITY OF STYROPOR

First measurements of the thermal conductivity resulted also in a satisfactory accuracy. Figure 5 shows the thermal conductivity of Styropor as a function of its average temperature. An insulation shell with an inside diameter of 100 mm and a wall thickness of 40 mm was tested. The density was 24.5 kg/m^3. The moisture content of the test sample amounted to 0.05 Vol. % and remained constant, because it was surrounded by a moisture barrier. The temperature drop from the outside to the inside of the insulation amounted to 40 grade in average corresponding to a heat flux of about 4.5 W. The full line curve represents average values of numerous measurements obtained by the producer according to DIN 52 612. These data are in good accordance with the measurements obtained by the bubble-column-apparatus.

From the foregoing it can be concluded that the functional suitability of the measuring method has been proven. Its accuracy is satisfactory so that the application of the apparatus to measure the thermal conductivity of moist insulations is justified.

REFERENCE

[1] H. GLASER, "A calorimeter for measuring very small heat fluxes". *Annex 1969-7 Bull. I.I.R.*, pp. 333-339. Commissions II and VI, Liège.

MEASURING THE THERMAL DIFFUSIVITY OF INSULATING MATERIALS BY MEANS OF A PLATE APPARATUS USED FOR MEASUREMENTS OF THERMAL CONDUCTIVITY

W. GOGÓŁ

Institute of Heat Engineering, Warsaw (Poland)

Mesure de la diffusivité thermique des isolants à l'aide de l'appareil à plaques destiné à la mesure de la conductivité thermique

RÉSUMÉ : *On décrit une méthode de détermination de la diffusivité thermique à l'aide de l'appareil à plaques destiné à la mesure des isolants. L'échantillon est refroidi entre deux refroidisseurs d'eau à plaques; le débit d'eau à travers le refroidisseur doit être assez élevé pour permettre d'obtenir* Bi = ∞ *en régime instable de l'échantillon. On mesure la température en un point de l'échantillon à l'aide d'un thermocouple et on applique la théorie du régime thermique stable pour déterminer la diffusivité thermique.*

A method is presented for the measurement of the thermal diffusivity of insulating materials based on the theory of thermal regular regime [1]. The plate apparatus for measurements of thermal conductivity can be used as a calorimeter with metal walls. The method enables the determination of all the thermal properties in one system, using the same test sample.

Assessing the thermal properties of materials means the determination of two of the three properties sought: thermal conductivity λ, thermal diffusivity a and specific heat c. Inasmuch as the measurement of density ρ generally offers no difficulty, the third magnitude can be calculated from the relation $\lambda = ac\,\rho$.

Thermal conductivity is generally determined under steady state in a popular type plate apparatus (apparatus by Bock, Poensgen, etc.). A separate apparatus is indispensable for the measurement of thermal diffusivity or specific heat, another sample of the material tested being generally taken for such measurements: this sample should not influence the accuracy of the results obtained or create problems in obtaining the right representative sample of the material (granular, loose or fibrous bodies).

The method described in this paper can be applied only to certain plate apparatus (e.g., Poensgen apparatus).

After removing the heater from the apparatus (fig. 1), the tested sample is placed between two water coolers. An opening is made in one of the coolers in which one junction of the differential thermocouple is to be found. The other junction of the thermocouple is placed in the tested sample. The position of this junction, in conformity with the thermal regular regime, does not play any role; it is most convenient, for measurement purposes, to place the junction more or less in the middle of the tested sample. Placing the junction inside of the grain is recommended when testing granular material.

Water (or another liquid) flows from the ultrathermostat through the cooler. The measurement proper begins after the sample attains a temperature approximate to the temperature of the cooler. It is not necessary to attain complete thermodynamic equilibrium in the system prior to the measurements because the distribution of temperature during the initial stage has no effect on the regular regime.

At the time the measurement is begun the temperature of the water changes, being regulated by the ultrathermostats to a lower (or higher) value (several degrees) depending on the kind and magnitude of the test sample. The sample is cooled (or heated) by the metal surface of the coolers and, after a certain time lapse, it is in a thermal regular regime. The cooling rate

$$m = \frac{\partial N / \partial \tau}{N}$$

can be calculated from the graph $\ln N = f(\tau)$ by reading N off the galvanometer during time τ measured with a stop-watch. The duration of the measurement proper ranges from several minutes to a half hour, depending on the type and magnitude of the plate apparatus and the thickness of the sample. It is not necessary to scale the thermocouple.

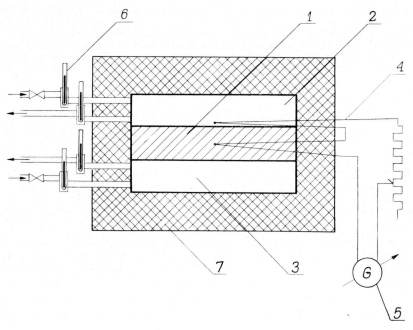

Fig. 1 — Lay-out of the Measurement Stand; 1) sample; 2) upper cooler; 3) lower cooler; 4) differential thermocouple; 5) galvanometer; 6) thermometers for measuring the temperature of water flowing through the coolers; 7) thermal insulation.

To achieve correct measurements, it is recommended to have the sample enter into regular regime and to achieve a thermal state of the sample corresponding to cooling in a calorimeter.

The first condition can easily be checked during measurement [graph $\ln N = f(\tau)$ accurately rectilinear] or a preliminary determination can be made by calculating the time necessary to enter regular regime from the Fourier number Fo = 0.5 assuming the probable value of thermal diffusivity of the body tested; when Fo > 0.5 the sample can be regarded as being in regular regime at negligible error [1].

The thermal state of the sample corresponding to cooling in the calorimeter is expressed by the condition Bi = ∞. The relation of the cooling rate to the Biot number is presented in figure 2.

The boundary value of the cooling rate m_∞ can be attained by increasing the heat transfer coefficient between the water flow and the surface of the cooler channels; in the method described, it means ensuring an appropriately high flow velocity through the cooler. When the channels are correctly designed, the pump of the ultrathermostat

350

ensures a sufficiently high flow. In order to test the condition Bi = ∞ several measurements can be made of the cooling rate by consecutively reducing the cooling water
discharge (throttling); within a certain range of the discharge the cooling rates will
be equal to m_∞ and with a sufficiently small water flow they will begin to diminish

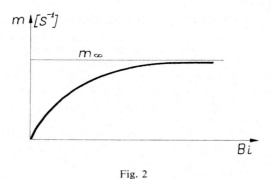

Fig. 2

(fig. 3). Such check measurements appear necessary in order to determine the limits
of the usability of the given plate apparatus for measurements of thermal diffusivity.
The surface of the cooler is isothermic and this is ensured, in principle, by the structure
of the apparatus used for measuring thermal conductivity.

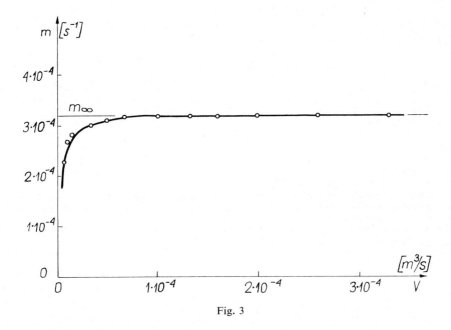

Fig. 3

Thermal diffusivity a of the material tested is calculated from the relation

$$a = m_\infty K,$$

where K is the coefficient of shape [1]. For an unlimited plate of δ thickness, the coefficient of shape is $K = (\delta/\pi)^2$.

The side losses from a sample with finite dimensions L (as compared to an unlimited plate) depend on the thickness of the sample, its thermal conductivity and the thermal insulation used in the apparatus. With good thermal insulation (e.g., Styropian) and a sample thickness of $\delta < (L/5)$ for insulating materials ($\lambda < 0.3$ W/mK) the effect of side losses on measurements is negligble; this effect can be estimated by means of Heissler's graph [3].

The method presented was used at the Research Center of Thermal Parameters of the Institute of Heat Engineering in a Poensgen plate apparatus with coolers (and sample) measuring 340×340 mm. Figure 3 presents one of the relations, experimentally determined, of the cooling rate m to the water rate flow V for this type of apparatus (tap water being used). The apparatus was insulated with Styropian. The thermal diffusivity was determined for various insulating materials (the method employed could be particularly useful in testing loose materials). The granular materials were ceramzite with a grain size $\varnothing$ 6.5 mm and a clinker size of about 12 mm down to a powder; measurements were carried out by means of samples (bed) of these materials of various thicknesses up to $\delta = 90$ mm. The properties of these materials were tested at the same time by other methods. Thermal conductivity was measured in a Poensgen and Bock apparatus and thermal diffusivity in a water calorimeter [1, 2]; the specific heat was determined by the calorimetric method and in regular regime via an air thermostat (in the latter case, ceramzite and clinker were crushed into a powder) [1, 2].

Satisfactory conformity of results was achieved. An error in one measurement did not exceed $\pm 6\%$ (the measurement error for thermal conductivity in a Poensgen apparatus averaged $\pm 5\%$).

The thermocouples were made from copper constantan wire 0.12 to 0.20 mm in diameter and insulated with lacquer. The mirror galvanometer had a sensitivity of 33.10^{-9} A/graduation, internal resistance 27 Ω and a fluctuation period of 2 seconds.

The method described will provide average accuracy in thermal diffusivity measurements without the need for additional apparatus (except a galvanometer). Other advantages of the method are: simplicity of measurements and calculations, short duration in making measurements and the possibilities of determining all the thermal properties unsig the same sample of the material tested.

REFERENCES

[1] W. Gogół, The Theory of the Thermal Regular Regime and Its Possible Applications, Theoretical Applied Mechanics (1966), 1.
[2] W. Gogół, An Unsteady State Method for the Measurement of Thermal Conductivity, Thermal Diffusivity and Specific Heat of Solids, International Institute of Refrigeration, Cambridge (1961).
[3] W. Gogół, Auxiliary Material to Heat Transfer Problems. Pub. Polytechnical University of Warsaw.

THE REPRODUCIBILITY OF
THERMAL DIFFUSIVITY MEASUREMENTS
IN QUASI-STEADY PERIODIC CONDITIONS

G. REALE and G. RUBATTO

Istituto di Fisica Tecnica, Università di Genova (Italy)

Reproductibilité des mesures de diffusivité thermique en régime périodique quasi-stabilisé

RÉSUMÉ : *L'utilisation du régime thermique variable (en particulier du périodique stabilisé) pour la détermination expérimentale de la diffusivité thermique (et donc de la conductivité) continue à intéresser les expérimentateurs parce qu'il permet, au moins en théorie, de réaliser des essais dont la durée est relativement brève.*

Les résultats expérimentaux que l'on peut ainsi obtenir ne sont pas trop en accord avec les résultats que l'on peut déduire, pour un matériel donné, des essais effectués selon les méthodes du régime thermique permanent.

Dans cette note, on rapporte des essais qui ont eu pour but de trouver les conditions à réaliser afin d'obtenir une approximation suffisante dans les déterminations de la diffusivité thermique effectuées en régime thermique périodique. Les expériences ont été effectuées sur un échantillon de polystyrol.

En raison du but de nos expériences, les résultats se rapportent à des essais qui ont été effectués avec continuité pendant six mois.

Les données expérimentales ont été représentées sur un diagramme; elles sont contenues dans une bande relativement large, qui précise, évidemment, les limites d'erreur. Ces résultats ont été obtenus en prenant la valeur moyenne de plusieurs déterminations expérimentales effectuées dans les conditions d'un régime qui peut être considéré comme quasi-permanent, dans des limites possibles en pratique.

1. This paper gives the results for a wide series of experiments directed towards the study of the reproducibility of a testing method for measuring the thermal diffusivity of insulating materials. The testing procedure is based on the variable heating method which, during latter years, has raised the interest of many researchers engaged in measurements on the thermal properties of materials.

The testing method and apparatus used in this investigation had already been built, erected and experimented upon by G. Rubatto some years ago at the Applied Physics Institute of the University of Genoa. Procedure and constructional details are given in refs. [1] and [2]; description in this paper is limited, therefore, only to features which are the most significant for the purposes of this present work.

The essence of this experimental method is based on the following considerations. Suppose a body C, bounded by plane faces of finite dimensions (fig. 1) and placed between two heat sources S_1 and S_2 in thermal contact with two opposite surfaces of the specimen at l (m) distance. Source S_2 is maintained at constant temperature T_0, while source S_1 is kept at a periodic sinusoidal temperature as expressed by:

$$T_1 = T_0' + \Delta T_0 e^{i\omega\tau}. \tag{0}$$

By assuming the constituent material of sample C as homogeneous, it is known that, under unidirectional heat flow conditions, the temperature T of any one section of the body, normal to the flow path, must satisfy as a function of time τ the well known Fourier differential equation.

If it is assumed that all the interface contact resistances may be neglected, the boundary conditions which enable the resolution of the problem are:

$$\Delta T = \Delta T_0 e^{i\omega\tau}, \quad \text{at} \quad x = 0$$

$$\Delta T = 0, \qquad \text{at} \quad x = l.$$

Under these conditions, the solution is found to be [3]:

$$\Delta T = \Delta T_0 \frac{\sinh\left\{(1+i)\dfrac{l-x}{l_0}\right\}}{\sinh\left\{(1+i)\dfrac{l}{l_0}\right\}} e^{i\omega\tau} \tag{1}$$

where the term $l_0 = \sqrt{2a/\omega}$ involves the thermal diffusivity a of the tested material and the angular velocity ω, being equal to 2π divided by the period (in hours) of the periodic surface temperature T_1 generated by source S_1. From solution (1), amplitude and time lag of the temperature wave are defined, for a given instant τ and for a given section $X-X$ of the specimen; this determines the thermal diffusivity value of the tested material.

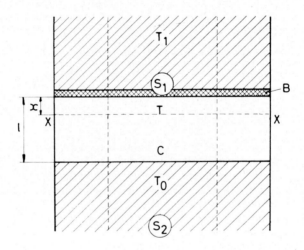

Fig. 1

2. Source S_1 used in our experiments is made up of 100×100 cm aluminium plate. Suitably distributed heating electric resistances and a copper refrigerating coil with constant temperature water recirculation through a feed thermostat, are placed on the upper face of the plate. Power is supplied to the electric heat source through a magnetic amplifier connected to a P.I.D. regulator. The regulator action is controlled by an error signal obtained by comparison of the actual temperature T_1 of source S_1 with the reference temperature t of a program unit, suited to vary the temperature t according to a preselected periodic time function. It appears, therefore, that the joint action of both the cooling water and the variable electrical supply to the heating resistances enables one to maintain the heat source S_1 at the required test temperature T_1 as given by function (0).

Notice that source S_1 has a built-in thermal filter (B) placed between the lower face of the hot plate and the upper face of the tested specimen in order to reduce the disturbing effect of harmonics still present in the thermal wave generated within the heat source.

Sample C is made up of six foamed polystyrene layers each one chosen among the 72 sheets, average 50×100 cm surface area and 5 cm thickness, obtained from a

354

single $100 \times 100 \times 200$ cm polystyrene slab stock and suitably cut with a sawing machine so that permanent deformations should not occur over the cut surface. Careful measurements of weight and thickness were carried out on all those slabs at an ambient temperature of about 22 °C. The six sheets found to be the most uniform in thickness (s) and specific weight (γ), were selected to make up the specimen C.

The average thickness and density values of the slabs used in our tests are given in the following table together with a progressive number corresponding to the sequence order of each layer along the heat flow path.

Layer no.	I	II	III	IV	V	VI
s (m)	0.0500	0.0499	0.0501	0.0505	0.0486	0.0510
γ (kg/m^3)	12.60	12.57	12.59	12.63	12.16	12.26

It was noted that the thickness of the sheets and number of the constituent layers of the tested specimen were chosen by taking various requirements, among which the thickness measurement error, into consideration.

Source S_2 is made up of an aluminium plate on the lower face of which a copper cooling coil is placed and maintained at constant temperature by recirculation of water fed through the same thermostat as source S_1. In order to make source S_2 independent of environmental thermal variations, the cold plate lies on a thick sand layer contained in a suitable wooden sandbox. A refrigerating coil embedded in the sand heat-sink is fed by constant temperature water recirculation. Water flow rate of source S_2 cooling coils is controlled, so that constancy of temperature T_0 (within ± 0.1 °C) is ensured over the upper surface of the aluminium plate adjacent to the lower face of the specimen.

It seems useful to point out that the unidirectional heat flow assumption is satisfied because two thermal independent regions are provided in each component (S_1, C and S_2) of the testing apparatus. One is the 50×50 cm measurement zone at the center of each component part, the other is the peripheral zone 24 cm thick and working, of course, as thermal guard ring; these regions are referred to in figure 1 as the $Y - Y$ axes.

3. Prior to taking observations the foamed polystyrene specimen C was allowed to undergo a natural aging process so that this phenomenon may have reduced effects on the experimental results [4]. For this purpose the sample C placed inside the measuring apparatus was maintained in the Test Laboratory several months in order to age. This room, where all the referred measurements were carried out has such thermal insulation characteristics that conditions of nearly constant temperature and relative humidity may be established in the ambient air for a long time.

Meanwhile, with an expressly made apparatus, tests were carried out on foamed polystyrene sheets cut out from the same slab stock as sample C layers in order to determine the temperature effects on the percentage elongation of the material under load.

Elongations were measured at 65 kg/m^2 pressure for a temperature range of 18 to 90 °C. Figure 2 shows the obtained linear relationship between percentage elongation and temperature, up to a temperature of 80 °C, at which the maximum percentage elongation is about 0.45%.

During the thermal diffusivity experiments, temperatures were measured in the testing apparatus by means of seven 28 A W G Chromel Alumel thermocouples having negligible thermal inertia and placed as follows: thermocouple No. 1 is set between the lower face of thermal filter B and the upper surface of layer I of specimen

C: by this thermocouple, therefore, the boundary temperature T_1 maintained by heat source S_1 is measured. Thermocouple No. 2, being placed on the interface surface between layer I and the underlying layer II, measures the temperature T_2 settled within the sample at $x = 5$ cm nominal distance from the source S_1, by the boundary

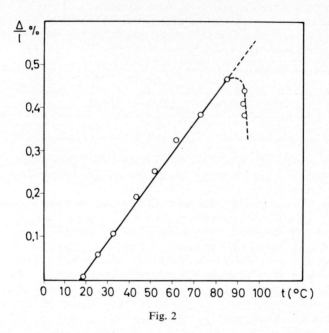

Fig. 2

temperature T_1. Thermocouples Nos. 3, 4, 5 and 6 are likewise placed at the interfaces between the next layer III, IV, V and VI. Finally thermocouple No. 7, being located between the lower surface of layer VI and the upper face of source S_2, is intended for the measurement of temperature T_o, which is to be maintained constant owing to the aforementioned assumptions. The thermocouple cold junctions were placed inside a Joseph Kaye Model 2110 Ice-Point Reference Standard, which maintains an immersion well at $0°C$ continuously. The leads of all couples were connected to a Leeds & Northrup Speedomax Type G "Azar" Recording Potentiometer which allows for 6 different temperatures to be continuously and simultaneously recorded.

For sake of convenience the whole research program was divided into three groups of experiments, each being different from one another according to both angular velocity ω and amplitude of the boundary temperature wave T_1.

In each group of tests, ω and T_1 were given the following values:

1st Group: $\omega = 0.66$ (rad. h^{-1}); $\Delta T_0 \cong 9°C$
2nd Group: $\omega = 3.14$ (rad. h^{-1}); $\Delta T_0 \cong 5°C$
3rd Group: $\omega = 15.70$ (rad. h^{-1}); $\Delta T_0 \cong 3°C$

The mean values $T'_0 = (70\pm2)$ °C of the applied thermal wave T_1 was maintained during all the tests. Each group of measurements was repeated several times; a total number of 12 series of tests were therefore carried out, each one lasting between 34 and 53 cycles.

At the end of each test series, provision was made to switch off source S_1 from the electrical power supply for about 10 days so that similar initial thermal conditions could be re-established in the specimen prior to the next series of experiments.

356

Finally it was noticed that, during each series of tests, the sinusoidal form of some of the measured temperature waves were verified by the Fourier expansion method: a few high-order harmonics were found which may be disregarded in view of their negligible amplitudes, the maximum being 6% of the amplitude of the fundamental wave.

4. It is recalled that equation (1) states that the thermal diffusivity of the material under test can be determined through amplitude measurements of both the boundary temperature wave T_1 supplied by source S_1 and the corresponding periodic tempe-rature T over a given inner surface of specimen C. Amplitudes were measured directly on the recorded temperature charts. As an example, a certain length of one such chart is reproduced in figure 3; numbers marking the different curves refer to the order number given to the thermocouples, as specified in Section 2.

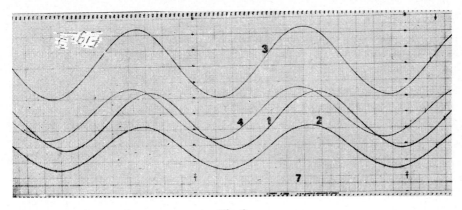

Fig. 3

Once the thermal wave amplitudes ΔT and ΔT_0 are determined, one can evaluate their ratio:

$$\theta = \frac{\Delta T}{\Delta T_0} \tag{2}$$

which was the particular object of our experiments *.

It was noted that the main advantage of the testing method was the possibility of obtaining the required values of θ as an average under conditions enabling the satisfactory reproducibility of measurements. It can be said that such conditions essentially depend on the features of the test system that enable a quasi-steady pe-riodic regime of heat flow to be established at the boundary of and inside the specimen: the experiments have shown that, in order to obtain the required steady periodic state conditions, each test needs to be continued long enough beyond the 15th-20th cycle at least, this last number being independent of the angular velocity of the applied boundary temperature wave.

(*) It is known that equation (1) enables one to determine the thermal diffusivity of the tested material also by measurements of the phase angle between the temperature T, in a given section inside the slab, and the imposed boundary temperature T_1. It seemed convenient, however, to carry out the diffusivity tests by considering amplitudes only, because of the higher accuracy attainable in such measurements.

The results of one series of tests carried out with $\omega = 0.66$ rad. h^{-1} and $\Delta T_0 \cong 9\,°C$ are plotted by way of example in figure 4, as ratio θ versus the order number of the temperature cycles following one another during the considered series of experiments.

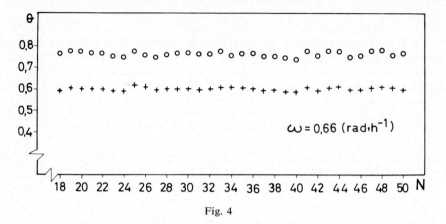

Fig. 4

The circles denote measurements pertaining to layer I ($l = 0.30$ m; $x = 0.05$ m; $\theta_{1 \cdot 0} = \Delta T_1 / \Delta T_0$), while the crosses refer to measurements taken across layers I and II considered as a whole ($l = 0.30$ m; $x = 0.10$ m; $\theta_{2 \cdot 0} = \Delta T_2 / \Delta T_0$). The results of one of the 2nd group series of tests ($\omega = 3.14$ rad.h^{-1}; $\Delta T_0 \cong 5\,°C$) and those of one of the 3rd group ($\omega = 15.70$ rad. h^{-1}; $\Delta T_0 \cong 3\,°C$) are likewise represented in figures 5 and 6 respectively.

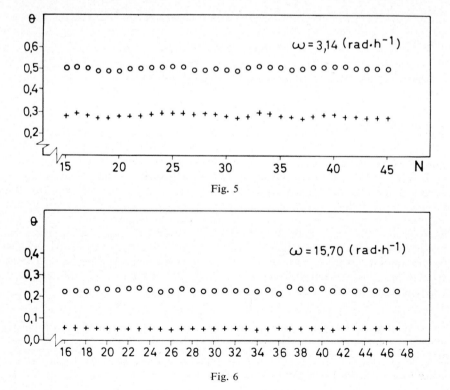

Fig. 5

Fig. 6

358

These figures show within what limits the required conditions for quasi-steady periodic flow had settled down in the system during the tests and to what extent the resulting reproducibility of measurement was attained. The fluctuations of ratio θ obviously indicate the range of reproducibility of the thermal diffusivity measurements which may be defined through the maximum θ_M and the minimum θ_m of the values of θ within a certain number of quasi-steady thermal cycles. In our investigation we took the seven last cycles for each test series into consideration.

All results obtained from our experiments are collated in Table B where, for each series of tests and layer of the specimen, we give: the date of the start and finish of the test; the angular velocity ω of the imposed temperature wave; the total number N of cycles carried out; the number n of quasi-steady cycles; the values θ_M and θ_m of ratio θ, defining the reproducibility range in the last seven cycles.

It was observed that the test series made with the same angular velocity did not follow one another continuously as it can be read in Table B, and this in order to make the tests independent of further aging effects. Consequently, a problem arises which concerns the repeatability of significant experimental data obtained from such discontinuously subsequent tests. Repeatability may be expressed by the mean value θ^* among the seven values of ratio θ evaluated in the aforementioned reproducibility range. The values of θ^* so obtained are also given in Table B thus completing the informations that may be acquired from the tests as far as the fundamental experimental result is concerned i.e., ratio θ.

TABLE B

Series No.	Date of		ω	N	n	Layer I			Layer (I + II)		
	Start on	End on	rad. h^{-1}			θ_M	θ_m	θ^*	θ_M	θ_m	θ^*
A	16/5/68	31/5/68	0.66	34	20	0.774	0.750	0.760	0.612	0.576	0.594
C	6/7/68	22/7/68	0.66	38	20	0.786	0.769	0.774	0.612	0.592	0.602
F	30/8/68	16/9/68	0.66	40	24	0.771	0.754	0.764	0.608	0.597	0.603
H	17/10/68	2/11/68	0.66	37	22	0.773	0.757	0.765	0.597	0.581	0.590
L	30/11/68	21/12/68	0.66	50	32	0.774	0.755	0.765	0.608	0.589	0.599
B	17/6/68	21/6/68	3.14	45	30	0.516	0.504	0.510	0.291	0.278	0.283
D	5/8/68	7/8/68	3.14	40	24	0.520	0.500	0.512	0.290	0.271	0.280
G	1/10/68	5/10/68	3.14	44	28	0.523	0.509	0.518	0.301	0.283	0.289
I	16/11/68	19/11/68	3.14	36	20	0.535	0.500	0.519	0.296	0.257	0.274
M	3/1/69	7/1/69	3.14	40	25	0.552	0.502	0.519	0.302	0.242	0.281
E	20/8/68	21/8/68	15.70	47	31	0.244	0.230	0.236	0.060	0.054	0.055
N	17/1/69	18/1/69	15.70	53	38	0.243	0.230	0.239	0.056	0.051	0.053

From Table B it may be inferred that our testing method enables fairly satisfactory results to be obtained. The possible variations of θ (hence the reproducibility range of the measurements) extend within rather restricted limits with deviations from the mean value θ^* that are less than 5%. Moreover, since the differences among the values of θ^* obtained from tests carried out with same angular velocity are rather small, we may conclude that a good repeatability of measurements has been achieved via the satisfactory pre-aging acquired by the material.

The values for the thermal diffusivity of the tested material, i.e., the ultimate results of the present research, are plotted in figures 7 and 8.

The maximum a_M and the minimum a_m of the values of thermal diffusivity evaluated from the measurements carried out across layer I of the specimen during each group of tests, are depicted in figure 7 as a function of the common logarithm of the respective angular velocity ω. The value of a_M pertaining to a given group was calculated by inserting equation (1) with the maximum θ^* obtained for the same group of tests at constant ω (see Table A); equation (1) was solved by the successive approximations method. The value of a_m was likewise evaluated for each group by

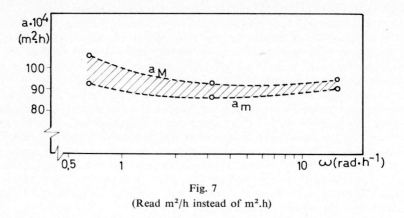

Fig. 7
(Read m²/h instead of m².h)

taking the minimum among the values for θ^* from the same group of tests. By following the same procedure the graph of figure 8 was drawn where the thermal diffusivity values concerning layers I and II, considered as a single layer 10 cm thick, are plotted.

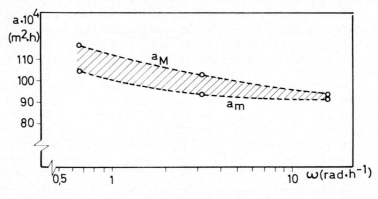

Fig. 8
(Read m²/h instead of m².h)

In virtue of the approach adopted for handling the experimental data the charts reproduced in figures 7 and 8 can therefore be considered as suitable to represent the reproducibility range of the method for measuring the thermal diffusivity of foamed polystyrene.

5.   This method, therefore, enables one to specify the reproducibility and repeatibility of the results as well as to better define the average value of the actual temperature of the material under test due to the possibility of keeping the fluctuations of such a temperature within rather restricted limits.

360

It should be observed, that the measurement sensitivity of the method depends upon three quantities, two of which are merely geometric, as the total thickness l of the 6-layer specimen and the thickness x of the layer across the boundary surfaces of which temperature measurements are taken; the third quantity $l_0 = \sqrt{2a/\omega}$ finally takes into account the physical properties, which pertain to the imposed thermal wave (as the angular velocity ω) and to the tested material (as the thermal diffusivity a). It follows, therefore, that to achieve good sensitivity of measurements, it is necessary that the total thickness l of the specimen and the angular velocity ω of the boundary temperature oscillation be suitably chosen in order to obtain just the proper values of the first derivative of function θ with respect to l_0. This choice, however, will be subjected to the other operative requirements that must be satisfied, i.e., to maintain a satisfactory constancy of the boundary temperature T_0.

REFERENCES

[1] G. Rubatto, « Misura della diffusività termica in regime periodico stabilizzato » *La Termotecnica*, no. 1 (1969).
[2] G. Rubatto, « Determinazione della diffusività termica del polistirolo col metodo a regime periodico stabilizzato », *Il Calore*, no. 2 (1968).
[3] R.A. Smith, F.R. Jones and R.P. Chasmar, "The detection and measurement of infra-red radiation", Oxford (Clarendon press) (1957), pp. 47-55.
[4] E. Bettanini, A. Cavallini e P. Di Filippo "Effetto dell'invecchiamento sulla condutti vità termica del poliuretano espanso", Relazione presentata al 14° Congresso del Freddo Padova (1965).

ÉCHANGES THERMIQUES
AVEC LES FLUIDES FRIGORIGÈNES BIPHASÉS

HEAT TRANSFER WITH TWO-PHASE REFRIGERANTS

LIQUID-VAPOUR PHASE EQUILIBRIUM
IN SOLUTIONS OF R-22 IN DIBUTYL PHTHALATE AND TETRAETHYLENE GLYCOL DIMETHYL ETHER

V.P. LATYSHEV

Scientific Research Institute of Refrigerating Industry of USSR, Moscow (U.S.S.R.)

Equilibre liquide-vapeur de solutions de R-22 dans du phthalate de butyle et dans de l'éther méthylique de tétraéthylène glycol

RÉSUMÉ : *Ce rapport traite des résultats d'une étude expérimentale sur l'équilibre liquide-vapeur de solutions de R-22 dans du phthalate de butyle et dans de l'éther méthylique de tétra-éthylène glycol; ces résultats sont obtenus en utilisant un instrument employé initialement pour la détermination de l'équilibre de phase de mélanges suivant la méthode piézométrique à volume constant. Cet instrument permet d'effectuer des mesures sur les conduites à concentration constante avec une grande précision sans rechargement et n'exige pas de connaissance des propriétés de la phase gazeuse.*

Les recherches portent sur une gamme de températures comprise entre 0 et 180°C et sur 9 concentrations réparties sur toute la série de concentrations de chaque solution.

L'utilisation d'équations empiriques pour déterminer le coefficient d'activité en fonction de la concentration et de la température, ainsi que l'emploi de méthodes numériques de calcul sur ordinateur, permettent de caractériser les données expérimentales par une équation pour chaque solution avec un grand degré de précision. En s'appuyant sur les calculs de la pression du mélange à l'aide d'équations d'interpolation, on a déduit, pour chaque solution, les diagrammes d'influence de la pression du mélange sur la concentration et la température.

On a effectué les recherches de l'équilibre de phase des solutions indiquées ci-dessus, en fonction d'un nombre de paramètres beaucoup plus étendu que celui des données connues jusqu'à maintenant.

NOMENCLATURE

$P \cdot 10^5$ N/m^2	bubble-point vapour pressure;
$P_1^0 \cdot 10^5$ N/m^2	saturation pressure of pure R-22;
$\Delta P \cdot 10^5$ N/m^2	pressure difference;
$T^1 = 0.01$ T°K	temperature/T°C = T°K − 273.15°K;
A, B, C, D, J, K, б, в, г, д, е, ж	coefficients;
ψ	mole concentration of R-22 in the liquid phase;
γ	activity coefficient of R-22;
$i - 1, 2, 3, \dots n$	number of experimental points;
R-22	Refrigerant 22;
DBPh	dibutyl phthalate;
E-181	tetraethylene glycol dimethyl ether.

Solutions of R-22 in DBPh [1] and E-181 [2] are the most promising of new working substances for Freon absorption refrigerating machines for whose calculation it is necessary to have accurate data on liquid-vapour phase equilibrium.

APPARATUS

During the experiments, a piezometer and zero-indicator were used (fig. 1). Constructional features of the piezometer are a removable wall (5) in plastic material fixed between the flanges of the piezometer body (1) and its cover (6).

The thickness of the removable wall (5) is defined by the permissible elastic deformation caused by maximum pressure drop at the wall. On the cover (6) of the piezometer, a bolt (8) with a semisphere (7) produces deformation of the removable wall and thus slowly decreases the inner volume of the device. It is possible to increase the volume with the help of an additional vessel.

The piezometer contains an agitator (3) driven by a solenoid (4) mounted outside the body.

The capacity membrane zero-indicator (10) reads the equilibrium of pressures inside the piezometer and the measuring system.

The membrane can travel between two limiting surfaces located a little distance from each other. Therefore a change of volume, due to movement of the membrane, is small and can be neglected.

For operation in the temperature range 373-453 °K, the TC-24 thermostat with oil "Banop-T" as a heat transfer medium was used.

In order to achieve a temperature precision of 0.01 to 0.05 °K the temperature control scheme had to be altered.

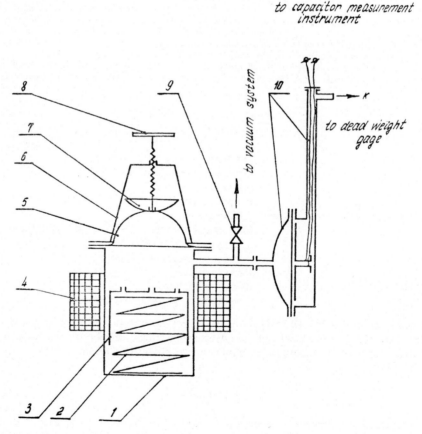

Fig. 1 — Instrument for investigating the phase equilibrium of mixtures:
   1) piezometer body; 2) spring; 3) agitator; 4) solenoid; 5) removable wall; 6) piezometer cover; 7) semisphere; 8) pressure bolt; 9) stop valve; 10) membrane capacity zero-indicator.

The mixture was compounded by weight: the non-volatile component was poured into the piezometer and the refrigerant was condensed into the piezometer by cooling.

Both components were carefully purged of air and moisture by vacuum. The charge was accurate to 0.01 g.

The investigated components had the following characteristics:

DBPh corresponded to ГОСТ-8728-58(GOST-8728-58).

E-181 corresponded to MPTУ 6-09 3132-66 "4".

During investigation of the R-22/DBPh mixture up to 373°K, impurities in R-22 were no more than 0.055% [3]. At temperatures more than 373°K for mixture mentioned and for R-22/E-181 mixture over the whole range of temperatures, impurities in R-22 were no more than 0.018%. A test on the saturation curve for R-22 gave good correlation with existing data [4] (divergence was not more than 0.24%).

The pressure in the piezometer was recorded after setting the temperature in the thermostat and stabilizing the pressure of the mixture. These temperatures points

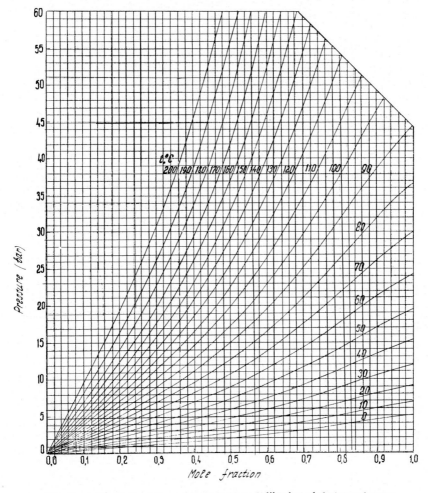

Fig. 2 — Pψ-diagram for refrigerant-22 and dibuthyl phthalate mixtures

were used for drawing the curve $P = P(t)$. Where there is an abrupt change in $(\partial P/\partial T)_v$, the concentration of the liquid phase solution is determined from quantities of the components introduced.

A further pressure check at the same concentration is obtained by decreasing the inner volume of the piezometer via deformation of the removable wall.

This procedure is repeated till one obtains the necessary number of pressure points for the constant concentration line.

A further plot for the constant concentration line was obtained after recharging piezometer. Each recharge produces a constant concentrations line.

TREATMENT OF EXPERIMENTAL RESULTS

Exchange between two phases of a binary system has been described in equations by Van-der-Waals. A special case of these equations (i.e., at constant concentration of the vapour and liquid phases) is the Clapeyron-Clausius equation from which the interpolations for a saturation vapour pressure curve are mainly founded.

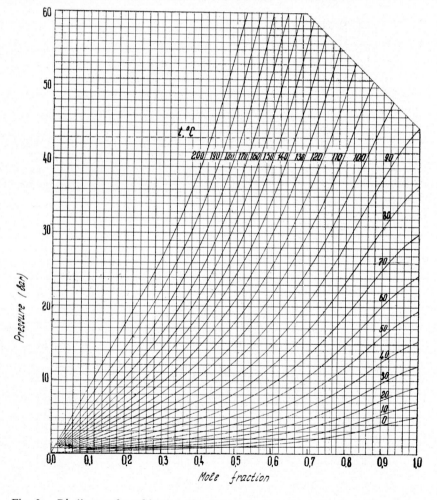

Fig. 3 — Pψ-diagram for refrigerant-22 and tetraethylene glycol dimethyl ether mixture.

In both the investigated solutions, one of the components possesses a very low saturated vapour pressure in the temperature range studied.

Thus, the superheated vapour of the first pure component (R-22) was present in the gas phase, i.e., concentration of the vapour phase was maintained constant automatically. Concentration of the liquid phase was maintained constant by a special procedure.

The various interpolations produced the following equation:

$$\ln P = A + \frac{B}{T'} + C \ln {}_iT' + DT'^6 \quad \text{at} \quad \psi = \text{constant}. \tag{1}$$

Scaling the temperature was necessary since the coefficients of the interpolation equation (1) had to be found with the help of a special computor programme.

The computed coefficients of equation (1) are given in table 1, where temperatures T_{min} and T_{max} determine the range of experimental investigations.

Table 1

No.	Ψ	A	B	C	D	T_{min}	T_{max}
SOLUTION OF REFRIGERANT-22 IN DIBUTYL PHTHALATE							
1	0.7877	15.303841	-31.409544	-2.5260146	$0.11517359 \cdot 10^{-4}$	272.0	374.0
2	0.7732	-3.5098569	-5.1907998	6.7220968	$-0.14272950 \cdot 10^{-3}$	272.0	374.0
3	0.7727	13.271920	-28.441470	-1.5940741	$0.58940931 \cdot 10^{-5}$	272.0	374.0
4	0.6193	5.5051622	-18.046047	2.0654526	$-0.69264891 \cdot 10^{-4}$	272.0	374.0
5	0.4094	9.1448752	-24.452559	-47.032730	$-0.30679870 \cdot 10^{-4}$	382.0	444.0
6	0.4025	8.1517376	-22.823979	0.42168301	$-0.33685359 \cdot 10^{-4}$	382.0	444.0
7	0.3209	-3.4940542	-6.0405720	5.9075608	$-0.12933930 \cdot 10^{-3}$	272.0	374.0
8	0.2386	235.05853	-377.37238	-101.29120	$0.47396983 \cdot 10^{-3}$	382.0	444.0
9	0.2210	-11.282880	5.3595095	9.2578758	$-0.15187777 \cdot 10^{-3}$	272.0	374.0
10	1.0000	16.832016	-32.653609	-3.2917567	$0.86835072 \cdot 10^{-4}$	272.0	444.0
SOLUTION OF REFRIGERANT-22 IN TETRAETHYLENE GLYCOL DIMETHYL ETHER							
1	0.9376	14.112766	-29.369963	-1.8530306	$0.88519168 \cdot 10^{-5}$	282.0	364.0
2	0.7346	17.907031	-37.269053	-3.5657314	$0.22349683 \cdot 10^{-4}$	282.0	374.0
3	0.7265	25.720596	-48.928785	-7.1927677	$0.33823833 \cdot 10^{-4}$	302.0	374.0
4	0.6050	118.22950	-192.63878	-48.963611	$0.25914165 \cdot 10^{-3}$	332.0	364.0
5	0.4530	16.992199	-38.759462	-3.1857352	$-0.14848887 \cdot 10^{-4}$	302.0	444.0
6	0.4044	51.836689	-93.820756	-18.712548	$0.49589798 \cdot 10^{-4}$	372.0	474.0
7	0.3843	-58.572963	74.241015	31.513636	$-0.26625783 \cdot 10^{-3}$	332.0	424.0
8	0.3515	-98.557352	127.62604	51.685018	$-0.63102324 \cdot 10^{-3}$	282.0	404.0
9	0.1804	18.866837	-43.017269	-4.7372121	$0.12106555 \cdot 10^{-4}$	312.0	454.0

Isothermal data on the saturated pressure for a liquid mixture, depending on the concentration, may be described by the polynomials of different types.

While computing equilibrium between liquid and vapour it is better to deal with the mixture pressure, using the activity coefficient, a change in which depends upon the concentration; this is described by the different polynomials [5].

The activity coefficient is:

$$\gamma = \frac{P}{P_1^0 \cdot \psi} \quad \text{at} \quad T = \text{constant}. \tag{2}$$

369

Similar assumptions have been made by Mastrangelo during the computation of the liquid-vapour equilibrium of mixtures [6].

The most well known Van-Laar and Margules equations describe the activity coefficients as dependent on concentration.

From preliminary computations it was found that our assumptions gave the best description of the experimental data when we used the empirical equation:

$$\ln \gamma = j(1-\psi)^2 \exp[K(1-\psi)] \quad \text{at} \quad _iT = \text{constant}. \tag{3}$$

The factor $[K(1-\psi)]$ takes into account the deviation in the true properties of the solution from the usual one.

The present interpolation equation is useful in computing the heat of mixing from data on phase equilibrium. Activity coefficients of the absorbent may be found according to Gibbs-Duhem equations by substituting equation (3) for the isotherm into them (less the influence of pressure)

$$\ln \gamma_2 = j\left(\psi^2 - \psi - \frac{1}{K}\right) \exp[K(1-\psi)] + \frac{j}{K} \exp K. \tag{4}$$

Equation coefficients (3) are found from the method of the least squares when assuming the equal accuracy of initial data. Values for K are found from:

$$\frac{\sum_{i=1}^{n} (1-\psi_i)^2 \ln \gamma_i \exp[K(1-\psi_i)]}{\sum_{i=1}^{n} (1-\psi_i)^4 \exp[2K(1-\psi_i)]} - \frac{\sum_{i=1}^{n} (1-\psi_i)^3 \ln \gamma_i \exp[K(1-\psi_i)]}{\sum_{i=1}^{n} (1-\psi_i)^5 \exp[2K(1-\psi_i)]} = 0, \tag{5}$$

j is derived from:

$$j = \frac{\sum_{i=1}^{n} (1-\psi_i)^2 \ln \gamma_i \exp[K(1-\psi_i)]}{\sum_{i=1}^{n} (1-\psi_i)^4 \exp[2K(1-\psi_i)]}, \tag{6}$$

a definite value for the root of the equation (5) is found by Vegstein 's method [7]. In this case we took a zero approximation for K_0, from the expression:

$$K_0 = \frac{1}{2(\psi_2 - \psi_1)} \ln\left[\frac{\ln \gamma_1}{\ln \gamma_2}\left(\frac{1-\psi_2}{1-\psi_1}\right)^2\right]^2, \tag{7}$$

where the subscripts 1 and 2 give values taken from the investigated range of concentrations, but $\psi = 0$ and $\psi = 1$.

According to the known value K, j can be accurately found from expression (6).

It was found from preliminary computations that the dependence of coefficients j and K from temperature, may be expressed by

$$j = \sigma T^2 + в T = г \tag{8}$$

$$K = д T^2 + e T + ж \tag{9}$$

370

The constant values for д, e, ж were found by the method of the least squares according to known values for K. For prefixed temperatures, the values K were calculated from equation (9) and substituted into expression (6) for the calculation of j.

According to the values obtained, we calculated the coefficients σ, в, г, in equation (8) the calculations were the subject of a computor programme. The equation coefficients (8) and (9) are given in table 2.

Table 2

д	e	ж	δ	в	г
Solution of refrigerant-22 in dibutyl phthalate					
$-0.53801782 \cdot 10^{-4}$	$0.42541615 \cdot 10^{-1}$	-10.294762	$-0.28018082 \cdot 10^{-3}$	0.17184295	-29.670242
Solution of refrigerant-22 in tetraethylene glycol dimethyl ether					
$-0.89461751 \cdot 10^{-4}$	$0.69533651 \cdot 10^{-1}$	-15.363185	$-0.92294045 \cdot 10^{-3}$	0.70731661	-143.13739

The bubble-point pressure of the mixture is calculated from equation (2) taking into account expressions (1), (3), (8), (9). The results of calculations for the mixture pressure from equation (2) were used to draw the P, ψ-diagrams given in figures 2 and 3.

DISCUSSION OF RESULTS

The maximum relative error in the experimental data was estimated as 2.5% according to [8].

Comparison of pressure values calculated from equations (1) and (2), for mixtures of R-22 in DBPh and E-181 shows that for 77% and 66% of the experimental points, the divergence is not more than 2.5%. The mean error in equation (2) therefore does not exceed the experimental error.

In the range of the parameters, important in the calculation of absorption refrigerating machines, the maximum divergence between the experimental and calculated values for pressure did not exceed 2.5%.

Table 3 gives the pressure differences and the pressures which determined the maximum percentage divergence in the pressure of R-22 and E-181 on results of the different experiments.

Table 3

Source	Computation [6]	Computation [9]	Computation [10]	[9]-[10]	[9]-[6
ΔP	0.21	0.33	0.14	-0.17	0.44
P	0.56	0.66	0.57	0.69	4.0

The divergence in the values of the mixture pressure may be explained by the different impurities of the mixture components especially on tests at low pressures

At high pressures the computed data agreed with data in [6] within the error of the experiment.

The author expresses his thanks to Professor I.S. Badylkes for his help in carrying out this work.

REFERENCES

[1] V.M. SELIVERSTOV, Use of exhausted gas heat from the main engines of refrigerated river ships. Dissertation, Leningrad (1966).
[2] B.J. EISEMAN, *ASHRAE journal*, **1** (1959), 12.
[3] V.P. LATYSHEV, *Kholodilnaya tekhnika* (1968), 4.
[4] A.V. KLETSKY, *Kholodilnaya tekhnika* (1964), 6.
[5] R. RID and T. SHERWOOD, Gas and liquid properties. *Gostoptekhizdat* (1964).
[6] S.V.R. MASTRANGELO, *ASHRAE journal*, **1** (1959), 10.
[7] J.N. LANS, Numeral methods for digital computor. *Izdateltvo inostrannoi literatury* (1962).
[8] B.M. CSHIGOLEV, Mathematical treatment of data, *Phizmatgiz* (1960).
[9] M. KRIEBEL and H.J. LÖFFER, *Kältetechnik* (1965), 17, 9.
[10] L.F. ALBRIGHT, P.T. SHANNON, F. TERIER and PING LIN CHUEH, *A.I.Ch.E. Journal*, **8** (1962), 5.

TRANSMISSION DE CHALEUR DANS LES TUBES DES ÉVAPORATEURS NOYÉS : INFLUENCE MUTUELLE DES TUBES

M. GÜTTINGER et R. WALLNER

Institut für Wärmetechnik, Universität Stuttgart (Allemagne)

Heat transfer in flooded evaporator tubes: mutual influence between tubes

SUMMARY: *In flooded shell and tube evaporators nucleate boiling does not occur easily. High hysteresis is observed as compared to separate tubes. Thus nucleate boiling is uncertain, especially in the case of thermal flux related to refrigeration.*

The bubbles are activated only by high thermal flux; but this is rarely found in practice. It was therefore studied whether an improvement in heat transfer could be obtained via a suitable geometry of tubes since the action of vapour bubbles forming on the surface of those tubes located below the test tube showed a higher heat flux.

Tests carried out on several tubes placed one above the other showed that there is a flow of vapour bubbles provided the lower tube has a high heat flux and there is evidence that bubble formation has commenced. This bubble flow, depending on the heat flux in the lower tube, has a favourable effect on heat transfer in tubes located above it even if there is no activation of bubbles.

Surface heat transfer coefficients are higher than those observed in separate tubes.

Dans la technique du froid on utilise en substance deux types d'évaporateurs, l'évaporateur sec et l'évaporateur à calandre multitubulaire noyé. A puissance élevée et surtout avec les compresseurs centrifuges, l'évaporateur à calandre multitubulaire noyé, avec évaporation du fluide frigorigène à l'extérieur des tubes, se montre le plus avantageux.

Le calcul de la transmission de chaleur du côté évaporation, dans un tel appareil, pose la question de savoir si les lois applicables aux tubes séparés sont valables pour les tubes à l'intérieur d'un faisceau de tubes. Les raisons des modifications peuvent être :

— le mouvement de la charge liquide entre tubes et calandre;

— l'influence mutuelle du processus d'ébullition sur les tubes d'un faisceau.

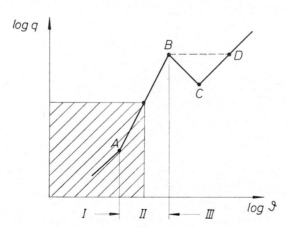

Fig. 1 — Courbe d'ébullition selon Nukiyama [1].

I   : ébullition en surface en convection naturelle.
II : ébullition nucléée.
III : ébullition par film.

Le mouvement de la charge liquide résulte de la recharge de liquide en compensation de la vapeur enlevée de l'évaporateur et d'une circulation provoquée par les bulles de vapeur. Ce mouvement peut être influencé par des mesures constructives. En général, la transmission de chaleur s'améliore par l'installation de chicanes guidant l'écoulement.

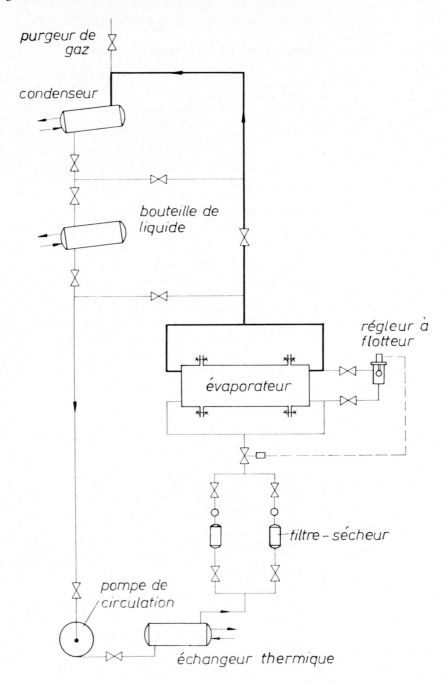

Fig. 2 — Schéma de l'installation d'essai

Jusqu'à présent l'influence mutuelle des tubes fut à peine étudiée et les recherches sur l'ébullition se limitaient à l'étude des phénomènes se produisant dans des tubes séparés. On peut rappeler la courbe d'ébullition établie pour la première fois par Nukiyama en 1934, une courbe qui indique les différents modes d'ébullition [1].

Le flux thermique apparaissant à l'ébullition est rapporté à l'écart de température entre surface de chauffe et liquide. Dans la technique du froid, on ne s'intéresse pratiquement qu'à la partie hachurée de l'abaque. C'est la région où l'ébullition passe de l'ébullition en surface à l'ébullition nucléée et où l'on trouve aussi la partie inférieure de l'ébullition nucléée, où l'écart de température n'est pas trop grand. On a eu raison de supposer que c'est justement dans cette région que l'écoulement du liquide provoque des modifications considérables.

Pour l'étude expérimentale, on a construit une installation d'essai dont le schéma simplifié est représenté dans la figure 2. La section de mesure se compose d'un évaporateur en verre de 150 mm de diamètre qui permet l'observation du processus d'ébullition. Comme la pression admissible est limitée, on a choisi comme fluide frigorigène le R 11 (CFCl$_3$), dont le point d'ébullition se trouve à 23,77°C (760 Torr). Cet évaporateur permet l'étude d'un tube séparé ainsi que l'étude de séries et de faisceaux de tubes, afin de rendre possible une comparaison directe des expériences. La circulation du fluide frigorigène se fait par différence de densités du liquide et de la vapeur. Pour l'étalonnage, on a aussi prévu une circulation forcée.

Les tubes d'évaporation sont des tubes en porcelaine de 10 mm de diamètre qui sont chauffés électriquement. Ils sont couverts à l'extérieur d'une couche très mince mais uniforme de Ni-Co, dont la résistance électrique permet très exactement la détermination de la température moyenne de surface. La rugosité de surface correspond à celle des tubes employés en pratique. La température du liquide se détermine à l'aide de plusieurs thermocouples.

LE TUBE SÉPARÉ

Les phénomènes se produisant lors de l'ébullition, à l'extérieur d'un tube séparé, furent bien recherchés expérimentalement, mais on n'a pas encore réussi à les concevoir théoriquement de façon exacte.

Nos résultats (fig. 3) correspondent à ceux des autres auteurs. Nos points de mesure montrent des déviations de ±5% au maximum dans le cas le moins favorable.

Le processus d'ébullition est représenté dans la figure 4 où on voit d'abord qu'on obtient un flux thermique croissant, selon la ligne droite A-B-C pour le coefficient de transmission superficiel dans l'abaque logarithmique. Cela désigne la région de l'ébullition en surface, en convection naturelle, qui peut être bien décrite par les formules connues pour la transmission de chaleur en convection libre, par exemple celle de Michejew [6]:

$$Nu = 0,54 \cdot (Gr\,Pr)^{0,25}. \tag{1}$$

En augmentant le flux thermique, la formation de bulles de vapeur se produit assez vite dans un des points C et à travers toute la surface chauffée du tube. Le coefficient de transmission superficiel atteint des valeurs beaucoup plus élevées et il obéit à une loi de puissance

$$\alpha = C \cdot q^m \tag{2}$$

pour laquelle nous trouvons à partir de nos expériences

$$\alpha = 1,32 \cdot q^{0,72}. \tag{3}$$

375

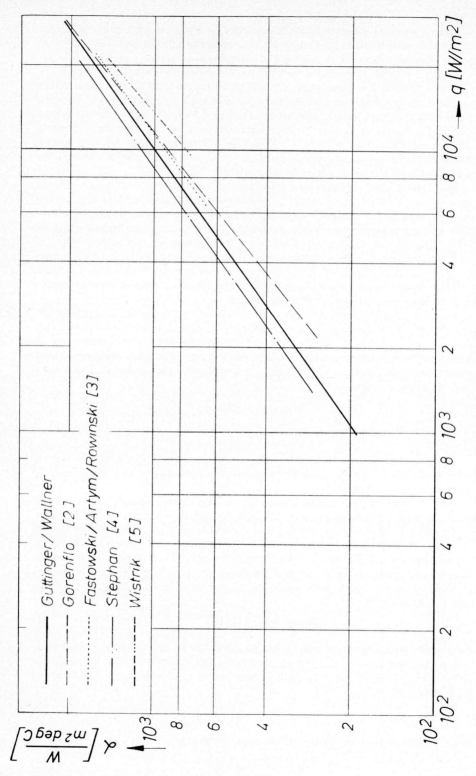

Fig. 3 — Résultats expérimentaux de différents auteurs.

Güttinger/Wallner
Gorenflo [2]
Fastowski/Artym/Rowinski [3]
Stephan [4]
Wistrik [5]

$q \; [W/m^2]$

$\alpha \; \left[\dfrac{W}{m^2 \, deg \, C} \right]$

376

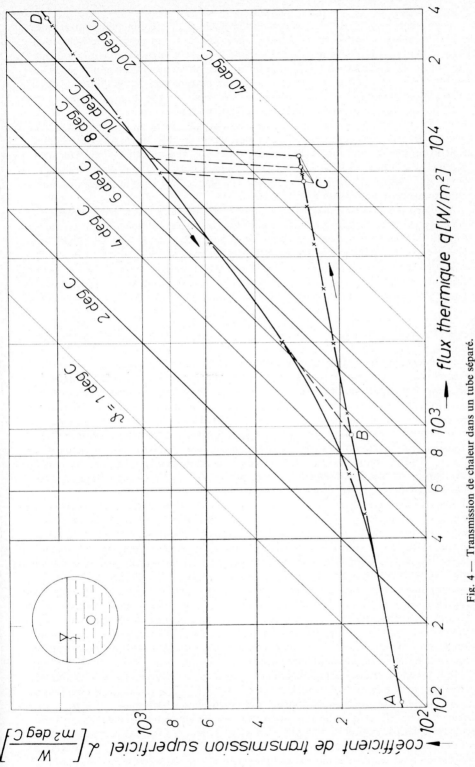

Fig. 4 — Transmission de chaleur dans un tube séparé.
A-B : ébullition en surface, en convection naturelle;
B-D : ébullition nucléée;
B-C : région de l'hystérésis.

Si on abaisse alors, après que l'ébullition nucléée ait été complètement établie, le flux thermique — par exemple à partir du point D — on descend le long de la droite d'ébullition D-B. Le point B représente le point d'intersection entre la droite d'ébullition et la droite A-C. L'ébullition nucléée se maintient au moins jusqu'au point B. Comme les points de mesure le montrent, la transition de l'ébullition nucléée à l'ébullition en surface, en convection naturelle, ne se fait que successivement.

Le fait que la formation de bulles, pour un flux thermique croissant, ne se produise pas au point B mais seulement pour des écarts de températures beaucoup plus grands, dans un des points C, alors qu'elle continue, pour un flux thermique décroissant, jusqu'au point B, est appelé hystérésis. L'hystérésis se produit non seulement avec un tube séparé mais aussi dans des évaporateurs à calandre multitubulaire comme des essais industriels l'ont montré. Cependant, l'influence mutuelle des tubes entraîne une certaine réduction de l'hystérésis.

Au cours des essais avec le R 11, le point B se trouvait à 5 à 6 deg C d'écart de températures entre surface et liquide, tandis que cet écart montait jusqu'à 35 deg C au point C. Seulement, avec des écarts de températures de 35 deg C, l'ébullition nucléée est assurée dans toutes les circonstances, même dans les conditions les moins favorables. Toutefois, de tels écarts ne sont pas possibles dans une installation frigorifique en général. Ainsi est-on forcé de prendre pour base les relations valables pour l'ébullition en surface en convection naturelle, quand on désire établir des calculs sûrs.

SÉRIE DE TUBES

L'influence mutuelle des tubes fut d'abord étudiée avec une série de tubes et en commençant par deux tubes placés l'un au-dessus de l'autre. En régime d'ébullition en surface en convection naturelle, on n'a trouvé que des déviations insignifiantes dans les coefficients de transmission superficiels en comparaison avec ceux du tube séparé. De même, la formation retardée de bulles — l'hystérésis — se fait aux mêmes écarts de températures que dans le tube séparé (point C). De l'autre côté, il est important de savoir si la formation de bulles se fait d'abord dans le tube supérieur ou dans le tube inférieur. Comme ceci est lié directement à l'influence mutuelle du processus d'ébullition, on décrira d'abord les expériences effectuées en ébullition nucléée établie et avec un flux thermique décroissant.

Deux tubes placés l'un au-dessus de l'autre et chauffés avec un flux thermique identique nous donnent les deux courbes représentées dans la figure 5. Le tube inférieur (1) a les mêmes valeurs du coefficient de transmission superficiel et la même caractéristique que le tube séparé, tandis que le tube supérieur atteint des valeurs plus élevées. En augmentant le flux thermique, l'écart des courbes diminue et disparaît enfin. Ainsi, l'ébullition du tube supérieur est influencée par le tube placé en dessous.

Une étude systématique de cette influence nous mène enfin aux courbes représentées dans la figure 6. Ici furent portés les coefficients de transmission superficiels du tube supérieur (2), déterminés à la suite des flux thermiques constants différents du tube inférieur (1). Tant que le tube inférieur fonctionne en ébullition nucléée établie, les expériences ont montré les mêmes valeurs pour le tube supérieur, soit en flux thermique croissant, soit en flux thermique décroissant.

Si le tube inférieur n'est pas chauffé, on obtient pour le tube supérieur les valeurs valables pour le tube séparé. D'autre part, si le tube inférieur est chauffé en l'ébullition nucléée établie, les expériences donnent des coefficients de transmission superficiels beaucoup plus élevés pour le tube supérieur, des valeurs qu'autrement on observe seulement dans la région de l'ébullition nucléée établie et avec des écarts de températures plus grands. En augmentant le flux thermique du tube supérieur, la courbe passe enfin à la courbe d'ébullition du tube séparé. Cette transition se fait, comme on le

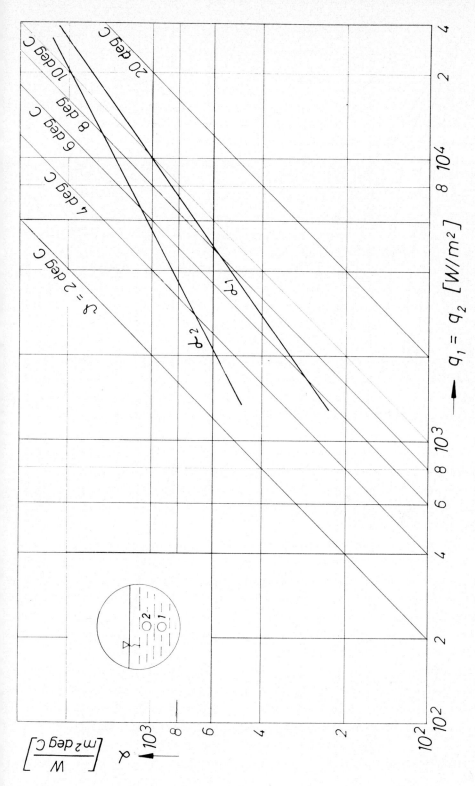

Fig. 5 — Deux tubes placés l'un au-dessus de l'autre, avec un flux thermique identique $(q_1 = q_2)$.

379

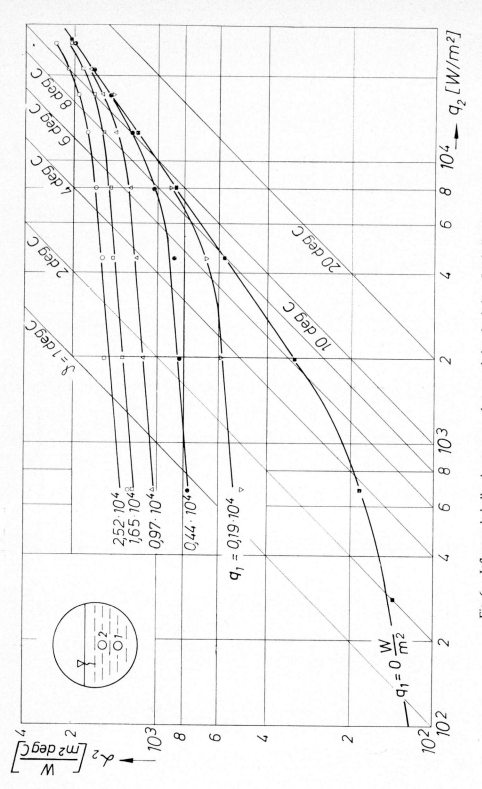

Fig. 6 — Influence de bulles de vapeur sur la transmission de chaleur. Les bulles de vapeur, formées dans le tube inférieur, influencent favorablement la transmission de chaleur dans le tube supérieur.

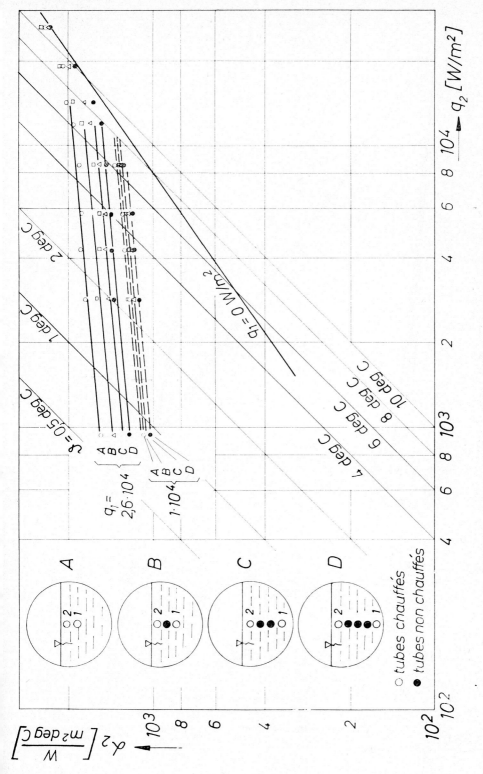

Fig. 7 — Influence de tubes non chauffés placés entre deux tubes chauffés.

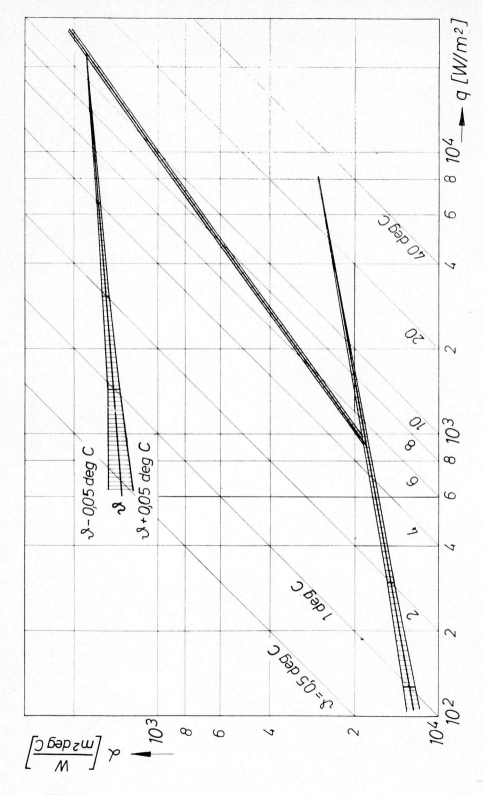

Fig. 8 — Effet d'une erreur de mesure sur le résultat, en supposant que l'écart de températures est lié à une erreur de ±0,05 deg C.

voit dans la figure 6, avec un écart de températures de 5 deg C d'environ, correspondant à l'écart de températures d'un tube séparé, c'est-à-dire correspondant au point B de la figure 4. Dans ce cas, on n'observe pas du tout d'hystérésis. L'amélioration du coefficient de transmission superficiel, pour des flux thermiques faibles, avant la transition à la courbe d'ébullition, est causée par des bulles de vapeur produites dans le tube inférieur. Avec des écarts de températures de 5 deg C environ, la formation de bulles se produit au tube supérieur sans hystérésis, l'impact de bulles de vapeur ne le permettant pas. Les bulles de vapeur produites au tube supérieur déplacent les bulles de vapeur arrivant de dessous, donc l'influence sur le coefficient de transmission superficiel disparaît elle aussi.

Dans plusieurs tubes placés les uns sur les autres on observe le même processus. Dans la région où l'impact de bulles est efficace, tous les tubes, excepté le tube le plus bas, ne forment aucune bulle. Ainsi, l'amélioration du coefficient de transmission superficiel dépend directement du comportement du tube le plus bas. L'effet sur les tubes supérieurs diminue un peu vers le haut avec l'augmentation du nombre de tubes, mais il est tout de même assez grand. En augmentant le flux thermique, l'écoulement de bulles devient inefficace aussitôt que la formation de bulles commence aux tubes supérieurs et tous les tubes se comportent comme le tube séparé.

La diminution de l'effet d'écoulement avec le nombre de tubes placés en dessus devient nette dans une expérience où des tubes non chauffés agissent comme des obstacles d'écoulement. La figure 7 démontre le résultat, où la diminution causée par 1, 2 et 3 tubes non chauffés est représentée.

Ces expériences ont posé des problèmes considérables concernant l'établissement des écarts de températures, qui sont du même ordre que l'erreur de mesure. Des erreurs très faibles modifient déjà considérablement les résultats, comme il est démontré dans la figure 8. Là, les déviations sont inscrites pour une erreur de ± 0.05 deg C.

C'est le processus d'ébullition dans plusieurs tubes placés les uns au-dessus des autres qui fut étudié d'abord. L'étude d'un faisceau de tubes n'est pas encore terminée. En résumant, on peut dire que l'étude a montré ce qui suit. Avec plusieurs tubes placés les uns sur les autres, le processus d'ébullition diffère considérablement de celui du tube séparé, dans la région des flux thermiques faibles. Dans le cas où les tubes reçoivent des bulles de vapeur provenant des tubes inférieurs, le coefficient de transmission superficiel atteint des valeurs considérablement plus élevées, avant le commencement de l'ébullition nucléée. L'ébullition nucléée ne se produit que plus tard, avec des flux thermiques plus élevés et sans aucune hystérésis. Avec de tels flux thermiques les tubes supérieurs se comportent comme le tube séparé. Les lois de transmission de chaleur constatées pour le tube séparé ne valent que pour le tube le plus bas.

RÉFÉRENCES

[1] V. NUKIYAMA, Maximum and Minimum Volumes of Heat Transfer from Metal to Boiling Water under Atmospheric Pressure. *J. Soc. Mech. Engrs.*, Jap., 37 (1934), n° 206, pp. 367/374.
[2] D. GORENFLO, Zur Druckabhängigkeit des Wärmeübergangs an siedende Kältemittel bei freier Konvektion. *Chemie-Ingenieur-Technik*, 40 (1968), n° 15, pp. 757/762.
[3] W.G. FASTOWSKI, R.J. ARTYM et A.E. ROWINSKI, La transmission de chaleur à l'évaporation de R11, chlorure de méthyle et benzène. (en russe) *Teploenergetika*, 5 (1958), n° 2, pp. 77/80.
[4] K. STEPHAN, Beitrag zur Thermodynamik der Wärmeübergangs beim Sieden. Abhandlungen des Deutschen Kältetechnischen Vereins (DKV) n° 18 (1964), Verlag C.F. Müller, Karlsruhe.
[5] W. WISTRIK, Untersuchung des Wärmeübergangs von der Aussenseite horizontaler Rohre an verdampfendes R11. Informations d'étude du Kältetechnisches Institut, Technische Hochschule Karlsruhe 1959, (non publié).
[6] MICHEJEW, Grundlagen der Wärmeübertragung. *VEB Verlag Technik*, Berlin.

DISCUSSION

J. MULDER (The Netherlands) — What was the influence of the vertical distance of the tubes?

R. WALLNER — The vertical distance of the tubes was varied from 0.5 to 5 tube diameters and we were not able to detect any influence on heat transmission in our experiments.

G. DE LEPELEIRE (Belgique) — M. WALLNER a exposé les variations des coefficients de transfert sur les tubes noyés *superposés*.

En pratique, on rencontre souvent des *faisceaux* tubulaires c'est-à-dire la juxtaposition de différents ensembles tels qu'examinés par M. WALLNER.

M. EMERSON est étonné de ne pas retrouver dans les résultats présentés une *diminution* du coefficient de transfert pour les tubes groupés. N'est-il pas probable que cet effet se produira effectivement en répétant les essais pour les faisceaux?

En effet, l'écoulement ascendant pourrait être modifié, l'accès du liquide gêné. Dans ce cas, une influence notable de la distance (horizontale) entre tubes paraît probable.

R. WALLNER — La quantité de vapeur en rapport avec la charge liquide de l'évaporateur noyé à calandre multitubulaire étant beaucoup plus élevée qu'à l'expérience décrite, il est bien possible que l'échange de liquide dans la couche limite — qui sert d'explication au mécanisme de la transmission de chaleur améliorée — soit défavorisé. Toutefois, il n'est pas évident en quoi cela amène une diminution du coefficient de transmission de chaleur, ce que M. EMERSON aurait attendu.

La présomption de M. de LEPELEIRE, que la distance horizontale des séries de tubes ait une influence, est certainement justifiée. L'influence de toute la géométrie de l'évaporateur sur la transmission de chaleur représente l'objet propre de notre recherche, dont nous n'avons rapporté ici que des résultats partiels concernant le problème spécial d'une série verticale de tubes.

L. DELVAUX (Belgium) — Can you give the physical explanation of your results?

R. WALLNER — The improvement of heat transmission on the upper tubes, as it is shown in figures 6 and 7, results from rising vapour bubbles produced on the outside of the lowest tube. These vapour bubbles and their liquid wake get in interaction with the boundary layer of the upper tubes thus destroying the boundary layer locally and causing instationary heat transmission.

With increasing heat flux of the upper tubes the wall temperature increases too and enables the upper tubes to produce vapour bubbles of their own which are displacing gradually the vapour bubbles arriving from below.

HIGH VELOCITY CONDENSATION OF R-11 VAPOR INSIDE VERTICAL TUBES

A. CAVALLINI and R. ZECCHIN

Istituto di Fisica Tecnica dell'Università
e Laboratorio per la Tecnica del Freddo, Padova (Italy)

Condensation de vapeur de R-11 à l'intérieur de tubes verticaux

RÉSUMÉ : *On décrit un appareil pour la mesure du coefficient moyen de transmission de la chaleur et de la perte de charge de vapeurs qui se condensent à l'intérieur de tubes verticaux dans le cas de haute vitesse de la vapeur. On donne les résultats expérimentaux relatifs au R-11 se condensant à l'intérieur d'un tube long de 170 cm avec un diamètre de 2 cm, pour des vitesses d'entrée jusqu'à 50 m/s. Les résultats expérimentaux sont comparés avec les données calculées au moyen des corrélations théoriques ou semi-empiriques disponibles.*

INTRODUCTION

Film condensation of a pure saturated vapor on vertical surface was originally dealt with by Nusselt in 1916 [1]. Nusselt studied this phenomenon assuming laminar flow of liquid film and that there was no shear stress at the liquid-vapor interface. Under these conditions Nusselt's theory leads to the following relationship for the mean heat transfer coefficient:

$$\bar{\alpha} = 0.943 \left[\frac{\bar{\lambda}_l^3 \bar{\rho}_l^2 g \, r}{\bar{\mu}_l \, L \Delta t} \right]^{0.25} \tag{1}$$

This same relationship can be written in the following dimensionless form:

$$\bar{\alpha} \left[\frac{\bar{\mu}_l^2}{\bar{\lambda}_l^3 \bar{\rho}_l^2 g} \right]^{1/3} = 1.47 \left[\frac{4 \Gamma_u}{\mu_{lu}} \right]^{-1/3} \tag{2}$$

The term $[4\Gamma/\mu_l]$ appearing in equation (2) represents the Reynolds number of the condensate.

The major portion of the experimental work on laminar film condensation has demonstrated that equations (1) and (2) are generally acceptable for describing the transfer coefficient. However the actual experimental values of the heat transfer coefficients have been found to be on average of 20 percent higher than those predicted by Nusselt's theory. This difference can be attributed to the formation of ripples in the liquid film which results in lower resistance to heat transfer [2].

The assumption of laminar flow of the condensate film is valid up to condensate Reynolds number, Re_l, of about 1800. When $Re_l > 1800$, flow of the condensate film becomes turbulent with a resulting increase in the local heat transfer coefficient. An empirical relationship for evaluating the mean heat transfer coefficient when both laminar and turbulent flow are present is given by Kirkbride [3]:

$$\bar{\alpha} \left[\frac{\bar{\mu}_l^2}{\bar{\lambda}_l^3 \bar{\rho}_l^2 g} \right]^{1/3} = 0.0077 \, Re_{lu}^{0.4} \tag{3}$$

Equation (3) is applicable when the flow of the condensate film is laminar on the upper part of the condensing surface but becomes turbulent on the lower part of the surface.

In high velocity condensation, when shear stress is present at liquid-vapor interface, this results in a decrease in the condensate film thickness and consequently an increase in the heat transfer coefficient.

Nusselt in dealing with the high velocity condensation of saturated vapor, assumed that the entire condensate film was in laminar flow and that its thickness was affected by both gravitational forces and shear stress at the liquid-vapor interface.

Carpenter and Colburn [4] determined from experimental data on steam, methanol, ethanol, toluene and trichloroethylene that the condensate film flow becomes turbulent at lower values of the condensate Reynolds number, Re_l, in the presence of vapor velocity. These workers studied condensation in vertical tubes with downward flow of the condensate. Inlet vapor velocities of up to 150 m/s were utilized. Carpenter and Colburn assumed that the entire resistance to heat transfer was localized in the laminar sublayer. They estimated the thickness of laminar sublayer using the equations for the single-phase fluid flow in pipes. The local heat transfer coefficient could then be calculated from the following relationship:

$$\frac{\alpha \mu_l}{\lambda_l \rho_l^{\frac{1}{2}}} = 0.043 \, \mathrm{Pr}_l^{\frac{1}{2}} \, \tau_w^{\frac{1}{2}}. \tag{4}$$

The wall shear stress, τ_w, is the sum of the shear stress due to gravity, τ_g, the shear stress due to friction at the liquid-vapor interface, τ_f, and the shear stress due to the change in momentum of the condensing vapor, τ_m:

$$\tau_w = \tau_f + \tau_g + \tau_m. \tag{5}$$

Soliman et al. [5] made the same initial assumption as Carpenter and Colburn. However, these workers utilized the equations for two-phase flow for estimating the value of the wall shear stress, τ_w. The local heat transfer coefficient was correlated using an equation similar to equation (4). However different values were used for the constant and exponent on the Prandtl number, Pr_l.

Different theoretical treatments of high velocity condensation in pipes have been given by Rohsenow et al. [6], Dukler [7], Akers and Rosson [8], Shekriladze and Gomelauri [9], Hilding and coworkers [10].

Semiempirical relationships for computing the mean heat transfer coefficient in high velocity condensation are of great practical importance in the design of heat transfer equipment. One such relationship is that of Colburn [4]:

$$\bar{\alpha} = 0.065 \left[\frac{\bar{c}_l \, \bar{\rho}_l \, \bar{\lambda}_l}{2 \, \bar{\mu}_l \, \bar{\rho}_v} \right]^{0.5} \overline{G} f^{*0.5} \tag{6}$$

where $\overline{G}$ is an average value of the vapor mass velocity in the tube, given by:

$$\overline{G} = \left[\frac{G_i^2 + G_i G_u + G_u^2}{3} \right]^{0.5}$$

The friction factor f^* can be calculated from the usual charts for single-phase flow. In these computations a mean value of the mass velocity $\overline{G}$ is used and the presence of the condensate is neglected.

Another semiempirical relationship is that of Akers et al. [11] for the mean heat transfer coefficient:

$$\overline{\mathrm{Nu}} = a\overline{\mathrm{Pr}}_l^{1/3} \left[\overline{\mathrm{Re}}_v^* \left(\frac{\bar{\mu}_v}{\bar{\mu}_l} \right) \left(\frac{\bar{\rho}_l}{\bar{\rho}_v} \right)^{0.5} + \overline{\mathrm{Re}}_l^* \right]^n. \tag{7}$$

These workers utilized the concept of an equivalent liquid mass velocity in equation (7).

While there have been a number of experimental studies on high velocity condensation [12, 13], the experimental data are quite limited and it has not been possible to obtain complete verification of either the theoretical or semiempirical relationships. Consequently an extensive experimental study of this subject has been initiated at "Laboratorio per la Tecnica del Freddo" at the University of Padua. This paper describes our initial experimental results. Mean heat transfer coefficients were determined with condensing R-11 in a vertical tube which had a diameter of 2 cm (I.D.) and was 170 cm in length. Experimental work was performed at inlet vapor Reynolds number up to 600 000.

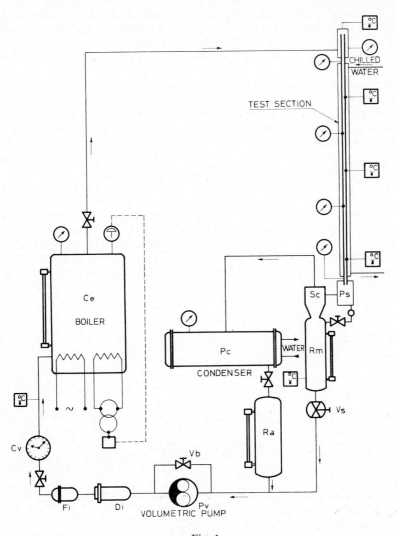

Fig. 1

Figure 1 presents a diagram of experimental apparatus. The R-11 vapor was generated in an electrically heated boiler (Ce). Entrained liquid was removed from the R-11
vapor by means of a separator which was located immediately before the test section.
The condensate leaving the test section was separated from the uncondensed vapor by
two traps (Ps and Sc) which were located in series. After being separated the condensate
was collected in a calibrated liquid receiver (Rm). The uncondensed vapor leaving the
test section was condensed in a shell and tube after-condenser (Pc) and collected in a
receiver (Ra). A rotary volumetric pump (Pv) was used to recirculate liquid R-11 from
the receiver to the boiler. A drier (Di) and a filter (Fi) were located between the pump
and the boiler to insure the purity of the liquid entering the boiler. A displacement
type flow-meter (Cv) was used to measure the inlet flow rate. The test section was
fabricated in the form of two concentric tubes. Chilled water was circulated in the
annular space between the tubes in order to effect condensation of the R-11 vapor. The
inlet temperature of the chilled water was controlled by means of a thermostat. An
externally finned copper tube was used as the inner tube in the test section in order to
obtain very high rates of heat transfer. This fact, coupled with very high flow rates of
chilled water, resulted in, essentially, a constant temperature along the length of the
test section.

The temperature of the inside tube of the test section was measured by means of
copper-constantan thermocouples which were inserted in thermocouple wells. A copper sheat was used to protect the thermocouple leads which passed through the annular
space of the test section. A mercury manometer was utilized to measure static pressure
at four points along the length of tube. The pressure taps had a diameter of 1 mm.
Copper tubing was used to connect the manometer to the pressure taps. This tubing
was electrically heated to prevent condensation of the R-11 vapor.

The flow rate of condensate from the condenser was determined by measuring the
time required to collect a given amount of liquid in the calibrated liquid receiver. The
inlet flow rate was measured by means of the flowmeter, Cv, after steady-state conditions had been reached. It was assumed that steady-state had been reached after the
temperature of the inlet vapor, the temperature of the liquid in the calibrated receiver,
the temperature of the liquid entering the boiler, the temperature of tube wall and the
temperature of the chilled water reached constant values. A recorder was used to
monitor these temperatures in order to determine when steady-state conditions had
been reached. In addition, the after-condenser (Pc) was purged periodically to remove
non-condensable gases from the R-11.

EXPERIMENTAL RESULTS

a) Heat Transfer Coefficient

The mean heat transfer coefficient was evaluated from the following relationship:

$$\bar{\alpha} = \frac{q}{S \, \overline{\Delta t}} \tag{8}$$

where $\overline{\Delta t}$ is the difference between the mean temperature of the vapor and the mean
temperature of the tube wall:

$$\overline{\Delta t} = \bar{t}_{vs} - \bar{t}_{w}. \tag{9}$$

The mean temperature of the vapor $\bar{t}_{vs}$ was the arithmetic average of the saturation temperatures at the inlet and outlet of the test section. The mean temperature of the tube wall, $\bar{t}_w$, was the average of the thermocouple readings for the tube wall. It should be noted that the change in saturation temperature between the inlet and outlet of the test section was small relative to $\overline{\Delta t}$. In addition, the temperature of the tube wall remained essentially constant between the inlet and the outlet.

The physical properties of the condensate were evaluated at a mean film temperature $\bar{t}_f$, which was defined as follows [15]:

$$\bar{t}_f = \bar{t}_{vs} - 0.75\overline{\Delta t}$$

The first series of tests were conducted at conditions under which essentially complete condensation of the vapor occurred. In addition, heat flow rates were quite low. This was done in order to have a negligible vapor velocity. The data obtained under these conditions are presented in figure 2 and are labeled "Re_{vi} negligible". These data are in agreement with the work of McAdams [16] who found that the actual heat transfer coefficients are approximately 20% higher than those predicted from Nusselt's theory. This first series of tests was limited to condensate Reynolds number of less than 700 since above this level vapor velocity could not be considered negligible.

It was impossible to operate with truly complete condensation of the vapor due to the accumulation of non-condensable gas in the test section under these conditions.

As already indicated, the high velocity condensation results in a substantial increase in the mean heat transfer coefficient. Data under conditions of high vapor velocity are reported in figure 2 for purpose of comparison. In this presentation inlet vapor Reynolds number, Re_{vi}, is used as a parameter.

A correlation similar to equation (7) of Akers *et al.* was employed for the interpolation of the data obtained at high vapor velocities. Under the conditions studied

$$Re_l \gg Re_v \left(\frac{\mu_v}{\mu_l}\right)\left(\frac{\rho_l}{\rho_v}\right)^{0.5}$$

therefore:

$$\overline{Nu} = a\,\overline{Pr}_l^{1/3}\left[\overline{Re}_v\left(\frac{\overline{\mu}_v}{\overline{\mu}_l}\right)\left(\frac{\overline{\rho}_l}{\overline{\rho}_v}\right)^{0.5}\right]^n \tag{10}$$

where the mean value of the vapor Reynolds number is defined as:

$$\overline{Re}_v = \left[\frac{Re_{vi}^2 + Re_{vi}\,Re_{vu} + Re_{vu}^2}{3}\right]^{0.5} \tag{11}$$

The constants a and n appearing in equation (10) were evaluated by regression analysis. The resulting correlation coefficient was 0.992 and the values of the constant were $a = 0.0724$ and $n = 0.764$. Therefore, equation (10) becomes:

$$\overline{Nu} = 0.0724\,\overline{Pr}_l^{0.33}\left[\overline{Re}_v\left(\frac{\overline{\mu}_v}{\overline{\mu}_l}\right)\left(\frac{\overline{\rho}_l}{\overline{\rho}_v}\right)^{0.5}\right]^{0.764}. \tag{12}$$

A comparison of the experimental data with equation (12) can be found in figure 3. Representative experimental data are reported in table I.

The experimental data can be compared with the semiempirical correlation of Akers *et al.* [11], and of Carpenter and Colburn [4].

389

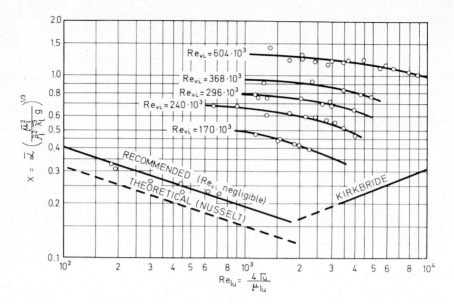

Fig. 2

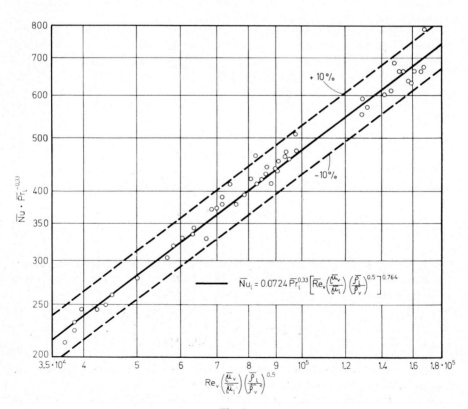

Fig. 3

Akers relationship is the following:

$$\overline{Nu}\,\overline{Pr_l}^{-0.33} = 0.0265\left[\overline{Re}_v\left(\frac{\overline{\mu}_v}{\overline{\mu}_l}\right)\left(\frac{\overline{\rho}_l}{\overline{\rho}_v}\right)^{0.5}\right]^{0.8}.\tag{13}$$

when

$$\left[\overline{Re}_v\left(\frac{\overline{\mu}_v}{\overline{\mu}_l}\right)\left(\frac{\overline{\rho}_l}{\overline{\rho}_v}\right)^{0.5}\right] > 5\cdot 10^4$$

Equation (11) has been used to define the mean vapor Reynolds number $\overline{Re}_v$ and the condensate Reynolds number has been ignored.

The comparison of the experimental data with equation (13) showed that the calculated values were 40 to 45 percent lower than the experimental data. This observation was also made by Soliman *et al.* [5] who found that the correlation of Akers predicted local heat transfer coefficients which were lower than the reported experimental values of other workers.

The correlation of Carpenter and Colburn can be written in the following form which involves some minor approximations:

$$\overline{Nu} = 0.046\,\overline{Pr_l}^{0.5}\left(\frac{\overline{\mu}_v}{\overline{\mu}_l}\right)\left(\frac{\overline{\rho}_l}{\overline{\rho}_v}\right)^{0.5}\overline{Re}_v\,f^{*0.5},\tag{14}$$

where f^* is the Fanning friction factor for single-phase flow which can be expressed in the following form within a limited range of Reynolds numbers for smooth tubes:

$$f^* = b\,\overline{Re}_v^{-m}\tag{15}$$

Combining equation (14) and (15) the following result is obtained:

$$\overline{Nu} = k\,\overline{Pr_l}^{0.5}\left(\frac{\overline{\mu}_v}{\overline{\mu}_l}\right)\left(\frac{\overline{\rho}_l}{\overline{\rho}_v}\right)^{0.5}\overline{Re}_v^{[1-(m/2)]}.\tag{16}$$

It should be noted that this equation has the same form as equation (12). The exponent in equation (15) has a value of approximately 0.2 for Reynolds numbers between 10^5 and 5.10^5. Consequently, equation (16) indicates that the Nusselt number is proportional to the $\overline{Re}_v^{0.9}$ However, the experimental data reported in figure 3 indicate that the Nusselt number is proportional to the $\overline{Re}_v^{0.764}$. Therefore, the correlation of Carpenter and Colburn would appear to be inappropriate for describing the experimental data shown in figure 3.

b) Friction Losses

The application of the equation of momentum to the fluid between inlet and outlet of the test section results in the following expression for mean wall shear stress:

$$\overline{\tau}_w = \overline{\tau}_m + \overline{\tau}_p + \overline{\tau}_g\tag{17}$$

where

$$\overline{\tau}_m = [(W_t - W_c)u_{vu} + W_c u_{lu} - W_t u_{vi}]\frac{1}{\pi\,dL},\tag{18}$$

Table I

Run No.	W_t kg/h	t_{si} °C	$\overline{\Delta t}$ °C	$p_l - p_u$ N/m²	W_c kg/h	$\overline{Re}_{vi} \cdot 10^{-3}$	$\overline{\alpha}$ kcal/m²·h·°C	$\overline{Re}_{vr} \cdot 10^{-3}$	$|\overline{\tau}_m|$ N/m²	$|\overline{\tau}_p|$ N/m²	$|\overline{\tau}_q|$ N/m²	$|\overline{\tau}_w|$ N/m²
29	192.0	26.0	9.20	3458	63.5	340	2868	286	6.62	9.83	0.76	17.21
34	194.3	26.4	8.95	3990	62.8	344	2910	290	6.65	11.34	0.74	18.73
46	216.0	30.5	4.40	5054	33.3	381	3230	352	3.44	13.99	0.53	17.96
47	209.8	29.9	8.19	4522	62.7	369	3159	316	6.42	12.85	0.76	20.03
48	213.5	30.5	13.27	3857	92.9	376	2849	298	8.97	7.18	1.12	17.27
51	206.1	31.1	20.65	2527	136.9	362	2673	252	11.52	7.18	2.30	21.00
57	341.1	35.4	11.51	9310	114.8	596	4207	500	14.98	26.47	0.95	42.40
58	347.5	35.8	15.43	7980	138.9	607	3776	491	18.15	22.69	1.13	41.97
60	341.7	36.2	21.79	3990	199.6	597	3715	435	23.35	11.34	1.94	36.63
65	342.2	36.7	8.72	10306	79.0	597	3983	531	9.80	29.30	0.75	39.85
66	343.8	35.5	7.15	10640	62.4	601	3976	548	7.90	30.25	0.64	38.79
68	343.3	35.7	5.25	10640	44.2	600	4026	563	5.06	30.25	0.56	35.87
71	97.4	29.4	7.59	665	30.4	172	1601	146	1.48	1.89	0.80	4.17
72	94.0	29.6	10.89	531	41.6	166	1527	131	1.81	1.51	1.13	4.45
73	98.5	29.5	10.77	798	41.0	174	1523	139	1.85	2.27	1.06	5.18
74	95.5	29.1	13.90	267	50.8	168	1462	126	2.19	0.76	1.45	4.30
82	137.6	27.7	9.05	1731	47.0	241	2128	202	3.37	4.92	0.82	9.11
83	137.4	27.3	11.25	1330	57.4	241	2078	193	4.03	3.78	1.00	8.81
84	137.6	27.2	13.28	865	76.0	242	2033	186	5.14	2.46	1.44	9.04
87	137.9	28.6	21.27	397	95.5	243	1812	166	5.45	1.13	2.44	9.02
90	221.0	30.1	11.11	4520	79.9	390	2966	322	8.34	12.85	0.90	22.09
91	219.3	29.2	13.24	3855	93.2	387	2889	308	9.54	10.96	1.05	21.55
93	226.6	29.0	18.21	2659	126.7	399	2834	295	12.67	7.56	1.56	21.79
96	170.4	29.3	19.59	999	107.9	300	2226	213	7.72	2.84	2.00	12.56
97	168.6	29.0	13.92	1994	81.2	297	3512	229	6.33	5.67	1.24	13.24
98	166.4	29.3	12.05	2262	72.0	293	2394	234	5.56	6.43	1.06	13.05
99	168.8	29.1	9.13	2459	54.6	298	2454	251	4.58	6.99	0.81	12.38
100	167.8	29.5	5.73	2927	34.8	296	2534	265	2.93	8.32	0.61	11.86
101	168.4	21.9	13.69	2128	82.5	300	2481	231	8.15	6.05	1.06	15.26
105	334.9	27.1	12.05	10771	111.4	592	4039	498	18.35	30.62	0.74	49.72
106	249.4	24.2	8.37	7317	67.9	443	3527	385	10.29	20.80	0.60	31.69

is the contribution to the wall shear stress resulting from the change in momentum of the fluid;

$$\bar{\tau}_p = (p_u - p_i)\frac{d}{4L} \tag{19}$$

is the contribution to the wall shear stress resulting from the change in static pressure;

$$\bar{\tau}_g = -\bar{\rho}g\frac{d}{4} \tag{20}$$

is the contribution to the wall shear stress due to gravitational forces.

The vapor velocities u_{vu} and u_{vi} appearing in equation (18) were evaluated by assuming that the liquid film was of negligible thickness. Consequently the vapor velocities could be calculated from the internal diameter of the tube and the saturated vapor properties.

The mean liquid velocity u_{lu} was evaluated by means of the following expression:

$$u_{lu} = \frac{(1 - x_u)W_t}{\rho_{lsu}(1 - \vartheta_u)A} \tag{21}$$

The void fraction ϑ_u was calculated using the relationship of Zivi [17] wich is valid for annular two-phase flow:

$$\vartheta_u = \frac{1}{1 + \left(\dfrac{1 - x_u}{x_u}\right)\left(\dfrac{\rho_{vsu}}{\rho_{lsu}}\right)^{2/3}} \tag{22}$$

The mean density of the fluid in the test section $\bar{\rho}$, appearing in equation (20), was calculated from the following approximate relationship:

$$\bar{\rho} = \bar{\rho}_{vs}\bar{\vartheta} + \bar{\rho}_{ls}(1 - \bar{\vartheta}) \tag{23}$$

where:

$$\bar{\vartheta} = \frac{\vartheta_u + 1}{2}$$

In the tests involving incomplete condensation the term $\bar{\tau}_g$ is small relative to $\bar{\tau}_w$ and $\bar{\tau}_p$. Consequently, this approximation does not have any serious effect on the results.

The values of $\bar{\tau}_w$ can be found in Table I. The mean value of the friction factor can be defined by the following expression:

$$f = \frac{2\bar{\tau}_w}{\rho_{vi}u_{vi}^2} \tag{24}$$

The calculated values of the friction factor f are shown in figure 4 as a function of $\overline{Re}_v$. These data can be described in an approximate manner by the following empirical expression:

$$f_i = 0.0383\,\overline{Re}_v^{-0.128} \tag{25}$$

393

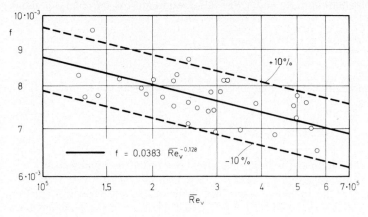

Fig. 4

Conclusion

The mean heat transfer coefficients for high velocity condensation in a vertical tube have been determined using R-11 as the condensing vapor. The vertical test section used in this work was 2 cm in diameter and 170 cm in length. Inlet vapor Reynolds numbers varying from 150 000 to 600 000 were investigated.

The experimental results were correlated using an expression of the following form:

$$\overline{Nu} = 0.0724 \overline{Pr_l}^{0.33} \left[\overline{Re}_v \left(\frac{\overline{\mu_v}}{\overline{\mu_l}} \right) \left(\frac{\overline{\rho_l}}{\overline{\rho_v}} \right)^{0.5} \right]^{0.764} \tag{12}$$

The available semiempirical correlations for predicting mean heat transfer coefficients were not appropriate for describing the present data.

A friction factor was defined in order to describe the friction losses and this friction factor was correlated using the following empirical relationship:

$$f = 0.0383 \overline{Re}_v^{-0.128} \tag{25}$$

The lack of agreement between the experimental data and existing correlations indicates the need for additional work. Different fluids and various geometries should be studied in order to obtain more general conclusions.

Acknowledgements

Special thanks are due to Professor L. Mattarolo, Director of the "Laboratorio per la Tecnica del Freddo" for his continuing interest and counsel during the course of this work.

Nomenclature

A   cross sectional area;
c specific heat;
d inside diameter of test section;

f, f^* friction factors;
g acceleration due to gravity;
G   mass velocity;
L   heat transfer surface length;
Nu Nusselt number (characteristic length: d);
p static pressure;
Pr Prandtl number;
q heat flow rate;
r latent heat of condensation;
Re Reynolds number (characteristic length: d);
S   heat transfer surface;
t temperature;
u velocity;
W   mass flow rate;
x vapor quality.

Greek

α heat transfer coefficient;
Γ condensate rate per unit periphery;
Δt temperature difference;
ϑ void fraction;
λ thermal conductivity;
μ dynamic viscosity;
ρ density;
τ shear stress.

Subscripts

c refers to condensate;
g due to gravitational forces;
i refers to test section inlet;
l refers to liquid;
m due to momentum change;
p due to static pressure change;
s saturation;
t total;
u refers to test section outlet;
v refers to vapor;
w wall.

$^-$ (bar) indicates mean value.

REFERENCES

[1] W. NUSSELT. *Z. Ver. Deut. Ing.*, **60**, 541 (1916).
[2] S.J. FRIEDMAN and C.O. MILLER. *Ind. Eng. Chem.* **33**, 885 (1941).
[3] C.G. KIRKBRIDE. *Trans. AIChE*, **30**, (1933-1934) pp. 170-186; *Ind. Eng. Chem.* **26** (1934), pp. 425-428.
[4] F.G. CARPENTER and A.P. COLBURN. Proceedings of the General Discussion of Heat Transfer, The Institute of Mechanical Engineers and the ASME (July 1951), pp. 20-26.
[5] M. SOLIMAN, F.R. SCHUSTER and P.J. BERENSON. *Trans. ASME* (May 1968), pp. 267-276.
[6] W.M. ROHSENOW, J.H. WEBBER and A.T. LING. *Trans. ASME*, **78** (1956), pp. 1637-1643.
[7] A.E. DUKLER. *Chem. Eng. Prog.*, **55** (1959), pp. 62-67.

[8] W.W. AKERS and H.F. ROSSON. *Chem. Eng. Prog.*, Symposium Ser. No. 30, **56**, (1960), pp. 145-149.

[9] I.G. SHERKRILADZE and V.I. GOMELAURI. *Int. J. Heat Mass Transfer*, **9** (1966) pp. 581-591.

[10] W.E. HILDING and C.H. COOGAN *et al.* Report submitted to NASA, Code SC-NsG-204.62 S1 (1965).

[11] W.W. AKERS, H.A. DEANS and O.K. CROSSER. *Chem. Eng. Prog.*, Symposium Series No. 29, **55** (1959), pp. 171-176.

[12] J. BORCHMANN. *Annexe* 1966-2 *Bull I.I.F.*, pp. 299-315. Commission 2, Trondheim.

[13] J.H. GOODYKOONTZ and W.F. BROWN. NASA TN D-3952 (1967).

[14] Anon. Thermodynamic properties of Freon-11 Refrigerant; Bul. No. T-11, E.I. Du Pont de Nemours Company.

[15] W.H. McADAMS. "Heat Transmission"; third ed., McGraw Hill Book Co., Inc. (1954), p. 330.

[16] ibid., p. 335.

[17] S.M. ZIVI. *Trans. ASME*, Series C, **68**, No. 2 (May 1964), pp. 247-252.

DISCUSSION

L. DELVAUX (Belgium) — I would like to have some details on the nature of the finned tube and on the way used for measuring the temperature of the tube.

A. CAVALLINI — The longitudinal fins on the outer surface of the test tube were obtained by grooving a copper tube 5 mm in wall thickness. The grooves were 3 mm in depth and 2.5 mm in thickness, with a pitch of 5 mm (measured at the outer surface).

The grooves were slightly interrupted at the temperature measuring points, where the termocouple wells were drilled. These wells were 11 mm in length, and 1 mm in diameter. The thermocouple wires were soldered at the bottom of the well and insulated with glass fiber.

The well was then filled with soft soldering.

FREEZING OF LIQUIDS IN PIPES WITH TURBULENT FLOW

K. GENTHNER

Institut für Thermodynamik, Technische Universität, Berlin (W. Germany)

Congélation des liquides dans les tuyaux à écoulement turbulent

RÉSUMÉ : *La solidification ou la congélation dans les tuyaux ne provoquent pas forcément une obstruction complète. La solidification initiale le long de la surface interne s'arrêtera si la forme de la couche solide est telle que le taux de transfert de chaleur par convection à l'interface solide-liquide est juste égal au taux de transfert de chaleur à l'agent de refroidissement.*

On a calculé la forme de la couche solide stable obtenue pour le cas où le liquide est de l'eau ou de la saumure et où l'écoulement dans les zones d'étranglement est turbulent. En appliquant les travaux récents de Stephan, on a considéré un transfert de chaleur arbitraire à travers la paroi du tuyau. On a fait varier d'autres paramètres : chute de pression à travers l'ensemble de la tuyauterie, température d'entrée du liquide, température de refroidissement ambiante, longueur de la zone de refroidissement. On examine l'influence de ces paramètres et l'influence de l'isolation thermique.

Les résultats peuvent avoir un intérêt pratique dans certaines applications de la congélation normale, telles que l'extraction de solvants, la purification, la concentration, le dessalement de l'eau ou dans les cas où la solidification est nuisible comme dans les canalisations d'eau, les échangeurs de chaleur métalliques pour les liquides, les refroidisseurs de saumure.

INTRODUCTION

A liquid flowing through a tube will solidify along the wall of this tube if it is cooled by a surrounding which is below freezing temperature and if convective heat transfer between fluid and tube is sufficiently low. Starting at the section of the tube where external cooling begins (see fig. 1), the temperature along the inner surface o the tube will decrease in flow-direction until, at a certain point, solidification temperature is reached. Downstream of this point, liquid will solidify and the solid-liquid interface maintains solidification temperature. Solidification proceeds as long as the heat flux transmitted to the coolant is greater than the convective heat transfer in the tube, thus permitting the developement of latent heat of fusion. Solidification will come to a stop when the shape of the solid layer is such that convective and transmitted heat flux become equal. A tube will freeze up if anywhere within the given lenght of tube such conditions are not attainable.

The object of this paper is to calculate the resulting stationary-state-profiles and the corresponding distribution of temperature, heat transfer, and pressure drop, to determine the conditions necessary to prevent or to cause blockage of the pipe, and to discuss the effective parameters. These parameters are, on one side, the flow rate and the temperature of the liquid at the entrance to the cooling zone which govern the convective heat transfer and, on the other side, the ambient cooling temperature and thermal resistance between solid phase and coolant which govern the transmitted heat flux. The theory is applied to water as the liquid solidifying in the tube, and restriction is made to turbulent flow.

The subject treated here may be of interest in such practical application as ice production, normal freezing applied to purification or to separation of organic or inorganic substances or the concentration of impurities from a liquid [1], the concentration of juices, desalination of sea-water [2], [3]. In other cases one is anxious to avoid solidification or blockage, for instance in water mains, brine coolers, liquid metal heat exchangers.

Stephan [4] developed a theory to calculate the instationary liquid-solid change of phase, which allows for variable heat transfer from the liquid phase and arbitrary thermal resistance between solid phase and coolant. Some of the results Stephan obtained were the impetus for this work. Hirschberg [5], assuming uniform thickness of the solid phase and a short cooling zone, derived a relation between solid phase thickness and the effective parameters. Zerkle [6] recently presented a theory to determine the solid layer profile, heat transfer, and pressure distribution for laminar flow and uniform wall temperature, i.e. neglecting thermal resistance between solid phase and coolant.

Mathematical formulation

Figure 1 shows the geometrical configuration of the problem. A tube of constant diameter 2R is cooled between the axial coordinates $z = 0$ and $z = 1$ by a surrounding medium of which the temperature T_u is below solidification temperature T_s of the liquid flowing within the tube. The thermal resistance between the inner surface of the tube and the surrounding medium is specified by the coefficient of heat transmission k. The liquid in the tube enters the cooling zone at $z = 0$ with fully developed turbulent flow, average velocity w, and a temperature T_0 higher than T_s. Downstream of a point $z = z_e$ a stationary solid phase layer has developed.

The geometry of the problem suggests the use of cylindrical coordinates. For sake of briefness, dimensionless variables as defined in the list of symbols shall be applied. The steady state condition in the tube is mathematically described by the following set of equations and boundary conditions.

The Navier-Stokes equations, the continuity equation, and the boundary condition at the wall for the radial and axial velocity component, which formally belong here, are not stated, because it is intended not to apply them for the solution.

A boundary condition defining the rate of flow can be either

$$\mathrm{Eu}_v = \Delta \mathrm{P_D}/(\rho_1 \bar{w}_v^2/2) = \text{const. (i.e. given a total pressure drop } \Delta p_\mathrm{D}$$
$$\text{across a fixed length of tube} \tag{1}$$

or

$$\mathrm{Re}_0 = (\bar{w}_0 \cdot 2\,\mathrm{R})/\mathrm{v} = \quad \text{const. (i.e. given a certain input velocity)} \tag{2}$$

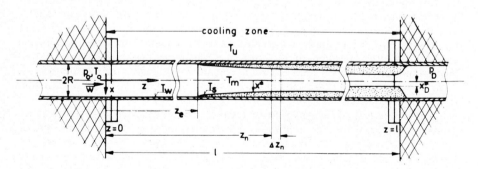

Fig. 1 — Geometrical configuration of the problem.

The energy equation for the liquid phase is:

$$W \frac{\partial \vartheta}{\partial Z} + U \frac{\partial \vartheta}{\partial X} = \frac{1}{Re_0} \frac{\partial}{\partial X} \left(\frac{X}{Pr} \frac{\partial \vartheta}{\partial X} \right) + \frac{1}{Re_0} \left(\frac{\partial W}{\partial X} \right)^2 \tag{3}$$

$\vartheta = (T - T_s)/(T_0 - T_s)$ is the dimensionless temperature
$W = w/w_v$, $U = u/w_v$ are the dimensionless axial and radial velocity components.
The boundary condition for convective heat transfer to the tube wall and to the solid phase are, respectively:

$$Z < Z_e: \quad -\left(\frac{\partial \vartheta}{\partial X} \right)_{X=1} = \frac{Bi}{2} \frac{\lambda_s}{\lambda_L} (\vartheta(X=1) - \vartheta_u) \tag{4a}$$

$$Z \geqslant Z_e: \quad \vartheta(Z, X^*) = \vartheta_s = 0 \tag{4b}$$

ϑ_u is the dimensionless temperature of the coolant,
ϑ_s is the dimensionless solidification temperature.
Bi is $k \cdot 2 R/\lambda_s$ defines the thermal transmissivity to the surrounding.
The boundary condition specifying the dimensionless temperature ϑ_0 at the thermal entrance to the cooling zone is:

$$\vartheta(Z = 0, X) = \vartheta_0 = 1 \tag{5}$$

The energy equation for the solid phase is:

$$\frac{\partial^2 \vartheta}{\partial X^2} + \frac{1}{X} \frac{\partial \vartheta}{\partial X} = 0. \tag{6}$$

Conservation of energy along the inner surface of the tube demands:

$$Z > Z_e: \quad -\left(\frac{\partial \vartheta}{\partial X} \right)_{X=1} = \frac{Bi}{2} (\vartheta(X=1) - \vartheta_u). \tag{7}$$

Neglecting axial conduction of heat in equations (6) and (7) causes practically no loss of accuracy [7].

SOLUTION

An accurate numerical treatment of equations (1) to (7) and of the flow equations by the method of finite differences is feasible. Such a solution seems commendable for the case of laminar flow. In order to treat turbulent flow, one would have to adopt transport coefficients on empirical basis. It therefore appeared reasonable to dispense with an accurate solution of the flow equations and the energy equation for the liquid phase and instead to make use of the vast amount of empirical work and boundary layer theory existing on the flow in tubes and convergent ducts.

The relation between flow rate and pressure drop is determined by energy considerations. The total energy loss $(p_0 + (\rho/2)w_0^2) - p(z) + (\rho/2) w (z)^2$ between a point z and the beginning of the tube, which is assumed to be at $z = -(A-1)l$ (A is a constant), can be equated with the sum of
frictional losses between $z = -(A-1)l$ and z_e:

$$\psi_0 \frac{\rho_L}{2} \bar{w}_0^2 \frac{z + (A-1)l}{2 R},$$

frictional losses between z_e and z:

$$\frac{\rho_L}{2} \int_{z_e}^{z} \psi(z) \cdot \bar{w}(z)^2 \frac{dz}{2x^*},$$

possible energy loss caused by rapid swelling of the solid phase at $z = z_e$, with $w_e = w(z = z_e + 2\,R)$:

$$\zeta_e \cdot \frac{\rho_L}{2} \bar{w}_e^2.$$

The resulting pressure drop then is

$$p_0 - p(z) = \frac{\rho_L}{z} (\bar{w}(z)^2 - \bar{w}_0^2) + \frac{\rho_L}{2} \psi_0 \bar{w}_0^2 \frac{z_e + (A-1)l}{2\,R} +$$

$$+\zeta_e \frac{\rho_L}{2} \bar{w}_e^2 + \frac{\rho_L}{2} \int_{z_e}^{z} \psi(z) \bar{w}(z)^2 \frac{dz}{2x^*}. \tag{8}$$

ξ_e is an empirical factor becoming $\zeta_e = 0.04$, if the slope of the solid layer at z_e is less than 20° [9]. $\psi(z)$ is the local friction coefficient which can be determined with any desired accuracy.

As long as $dx^*/dz < 0.01$, Prandtl's resistance law for smooth tubes

$$\frac{1}{\sqrt{\psi}} = 2.0 \log (\mathrm{Re}\,\sqrt{\psi}) - 0.8 \tag{9}$$

is considered valid here. By introducing the continuity condition

$$\bar{w}(z) = \bar{w}_0 \frac{R^2}{x^{*2}} \tag{10}$$

and the dimensionless parameters already defined, equation (8) becomes

$$\mathrm{Eu}(Z) \cdot \mathrm{Re}_v^2 =$$

$$= \mathrm{Re}_0^2 \left[\frac{1}{X^*(Z)^4} - 1 + \psi_0(Z_e + AL - L) + \frac{\zeta_e}{X^{*4}} + \int_{Z_e}^{Z} \psi(Z) \frac{dZ}{X_e^{*5}} \right]. \tag{11}$$

That part of the tube which has a solid phase profile, $Z \geqslant Z_e$, is now subdivided into n segments ΔZ_n. By introducing average values of $\psi(Z)$, defined by

$$\psi_n \int_{z_n}^{z_{n+1}} \bar{w}(z)^2 \frac{dz}{x^*(z)} = \int_{z_n}^{z_{n+1}} \psi(z) \bar{w}(z)^2 \frac{dz}{x^*(z)},$$

and changing the integration variable to

$$X^* = X_n^* + \frac{\Delta X_n^*}{\Delta Z_n} (Z - Z_n),$$

equation (11) becomes

$$\text{Eu}(Z) \cdot \text{Re}_v^2 = \text{Re}_0^2 \left[\psi_0(Z_e + AL - L) + \frac{\zeta_e}{X_e^{*4}} + \right.$$

$$\left. + \sum_{i=1}^{i=n-1} \left(1 + \frac{\psi_i}{4} \frac{\Delta Z_i}{\Delta X_i^*}\right)\left(\frac{1}{X_{i+1}^{*4}} - \frac{1}{X_i^{*4}}\right) \right]. \qquad (12)$$

Similarly, one obtains for the pressure drop across the complete length of the cooling zone

$$\text{Eu}_v \cdot \text{Re}_v^2 = \text{Re}_0^2 \left[\psi_0(Z_e + AL - L) + \frac{\zeta_e}{X_e^{*4}} + \zeta_D \left(\frac{1}{X_D^{*2}} - 1\right)^2 + \right.$$

$$\left. + \sum_{i=1}^{i=n_D-1} \left(\frac{\psi_i}{4} \frac{\Delta Z_i}{\Delta X_i^*}\right) \cdot \left(\frac{1}{X_{i+1}^{*4}} - \frac{1}{X_i^{*4}}\right) \right]. \qquad (13)$$

$$\text{Re}_0^2 \cdot \zeta_D \left(\frac{1}{X_D^{*2}} - 1\right)^2$$

is the energy loss due to sudden expansion at the end of the cooling zone.
The energy equation for the liquid shall be replaced by an equation adopting cross-sectional averages,

$$\vartheta_m(Z) = (T_m(Z) - T_s)/(T_0 - T_s),$$

of the dimensionless liquid temperature:

$$X^* \frac{\text{Re}(Z) \text{ Pr}}{4} \frac{\partial \vartheta_m}{\partial Z} = -\text{Nu}^*(\vartheta_m - \vartheta_w) + Q_R. \qquad (14)$$

ϑw is the dimensionless local temperature of the tube wall or of the solid liquid interface.
N_u^* is $\alpha \cdot 2R/\lambda_L$, the local Nusselt-number, is calculated by a relation which assumes a direct coupling between shear stress and convective heat transfer, [8]

$$N_u^* = \frac{\psi/8 \text{ Re}(Z) \cdot \text{Pr}}{1.02 + 13 (\psi/8)^{1/2} (\text{Pr}^{2/3} - 1)}. \qquad (15)$$

Equation (4a) now becomes

$$\text{Bi} \frac{\lambda_s}{\lambda_L} (\vartheta_w - \vartheta_u) = \text{Nu}^*(\vartheta_m - \vartheta_w). \qquad (16)$$

Boundary conditions (4b) and (5) remain unaltered.
The point Z_e, where $\vartheta_w = 0$, is obtained by integration of equation (14) with regard to boundary conditions (5) and (16), yielding

$$Z_e = \frac{\text{Re}_0 \text{ Pr}}{4 \text{ Bi}} \left(\frac{\text{Bi}}{\text{Nu}^*} + \frac{\lambda_L}{\lambda_s}\right) \cdot \ln\left(\frac{1 - 1/\vartheta_u}{(\lambda_s/\lambda_L \text{ Bi}/\text{Nu}^*) + 1}\right), \qquad (17)$$

401

and the temperature at $Z = Z_e$ is

$$\vartheta_m(Z = Z_e) = -\frac{\text{Bi}}{\text{Nu}^*}\frac{\lambda_s}{\lambda_L}\vartheta_u. \tag{18}$$

The mean liquid temperature downstream of the coordinate Z_e is determined by integrating equation (14) with regard to boundary conditions (4b) and (16):

$$\ln\left[\frac{\vartheta_m(Z_n)}{\vartheta_m(Z_e)}\right] = -\frac{1}{2}\sum_{i=0}^{i=n}\frac{\Delta Z_n \psi_i}{1.02 + 13\,(\psi_i/8)^{1/2}\,(\text{Pr}^{2/3}-1)}\cdot\frac{\ln(X_{i+1}^*/X_i^*)}{X_{i+1}^* - X_i^*}. \tag{19}$$

The temperature drop within a segment ΔZ_n is given by:

$$\ln\left[\frac{\vartheta_m(Z_{n+1})}{\vartheta_m(Z_n)}\right] = -\frac{\Delta Z_n}{2}\frac{\psi_n}{1.02 + 13\,(\psi_n/8)^{1/2}\,(\text{Pr}^{2/3}-1)}\cdot\frac{\ln(X_{n+1}^*/X_n^*)}{X_{n+1}^* - X_n^*}. \tag{20}$$

The heat conduction equation (6), combined with equation (7), can be accurately solved if the interface heat flux $q = \alpha(T_m - T_s)$ between liquid and solid phase is known. If the dimensionless heat flux is given by

$$\frac{q(Z)\cdot 2\,\text{R}}{ha_s\rho_s} = q^*(Z) = -\left(\frac{1}{\text{Ph}}\frac{\lambda_L}{\vartheta_u\,\lambda_s}\right)\text{Nu}^*(z)\cdot\vartheta_m(Z) \tag{21}$$

the exact solution, see [4], is

$$q^*(Z)\cdot\text{Ph} = -\left(\frac{1}{\vartheta_u}\frac{\lambda_L}{\lambda_s}\right)\text{Nu}^*\vartheta_m = \frac{2\,\text{Bi}}{X^*(2 - \text{Bi}\ln X^*)}. \tag{22}$$

The expression on the right hand side of equation (22) was plotted in figure 2 for various Biot-numbers. It shows that the interface heat flux does not necessarily decrease with increasing thickness of the solid layer. The diagram is helpful for predictions about the resulting solid state layer.
Substitution of equation (20) for $\vartheta_m(Z_{n+1})$ in (22) yields:

$$\Delta Z_n = \frac{\ln \text{B}_n}{\text{A}_n}, \qquad \text{with} \tag{23}$$

$$\text{A}_n = \frac{1}{2}\frac{\psi_n}{1.02 + 13\,(\psi_n/8)^{1/2}\,(\text{Pr}^{2/3}-1)}\cdot\frac{\ln(X_{n+1}^*/X_n^*)}{\Delta X_n^*} \tag{24}$$

$$\text{B}_n = -\frac{\vartheta_u}{\text{Nu}_{n+1}^*\,\vartheta_m(Z_n)}\frac{\lambda_s}{\lambda_L}\frac{2\,\text{Bi}}{X_{n+1}^*(2 - \text{Bi}\cdot\ln X_{n+1}^*)}. \tag{25}$$

Equation (23) enables us to calculate the length of a tube segment ΔZ_n corresponding to a finite small increase of solid phase thickness ΔX_n^*. (The converse problem would have to be iteratively solved.)
Summarizing, the equations to be solved are now:

$$Z_e = \frac{\text{Re}_0\cdot\text{Pr}}{4\,\text{Bi}}\left(\frac{\text{Bi}}{\text{Nu}^*} + \frac{\lambda_L}{\lambda_s}\right)\cdot\ln\left[\frac{1 - 1/\vartheta_u}{(\lambda_s/\lambda_L\,\text{Bi}/\text{Nu}^*)+1}\right] \tag{17}$$

$$\vartheta_m(Z_e) = -\frac{\text{Bi}}{\text{Nu}^*}\frac{\lambda_s}{\lambda_L}\vartheta_u \tag{18}$$

$$\Delta Z_n = \frac{\ln B_n}{A_n},$$

A_n and B_n being defined by equations (24) and (25), respectively.

$$\text{Eu}_n =$$

$$= \left(\frac{\text{Re}_0}{\text{Re}_v}\right)^2\left[\psi_0(Z_e+AL-L) + \frac{\zeta_e}{X_e^{*4}} + \sum_{i=1}^{i=n-1}\left(1+\frac{\psi_i}{4}\frac{\Delta Z_i}{\Delta X_i^*}\right)\left(\frac{1}{X_{i+1}^{*4}} - \frac{1}{X_i^{*4}}\right)\right] \tag{12}$$

$$\ln\left(\frac{\vartheta_m^{(Z_n)}}{\vartheta_m^{(Z_e)}}\right) = -\frac{1}{2}\sum_{i=0}^{i=n}\frac{\Delta Z_n\,\psi_i}{1.02+13\,(\psi_i/8)^{1/2}\,(\text{Pr}^{2/3}-1)}\,\frac{\ln\,(X_{n+1}^*/X_n^*)}{X_{n+1}^* - X_n^*}. \tag{19}$$

For a given flow rate these equations can be explicitly solved in steps for finite segment-lengths ΔZ_n. If the total pressure drop over a fixed length of cooling zone is the given parameter, equation (13) has to be included, i.e.:

$$\text{Eu}_v \cdot \text{Re}_v^2 = \text{Re}_0^2\left[\psi_0(Z_e+AL-L) + \frac{\zeta_e}{X_e^{*4}} + \zeta_D\left(\frac{1}{X_D^{*2}}-1\right)^2 +\right.$$

$$\left. + \sum_{i=1}^{i=n_D-1}\left(\frac{\psi_i}{4}\frac{\Delta Z_i}{\Delta X_i^*}\right)\left(\frac{1}{X_{i+1}^{*4}} - \frac{1}{X_i^{*4}}\right)\right] \tag{13}$$

This makes an iterative solution necessary.

Results

Equation (18) renders a value of the reduced dimensionless liquid temperature $\vartheta_m(Z^2)/\vartheta_u$ at $Z = Z_e$ wich depends only on the Biot-number and the flow rate. Bearing this in mind, inspection of equations (20) and (23) shows that the form of the resulting solid phase layer is independent of ϑ_u, i.e. independent of the cooling temperature, solidification temperature, and the temperature at the entrance section of the cooling zone. This not immediately obvious conclusion implies that the results discussed below are valid for arbitrary values of ϑ_u and ϑ_0, as far as the interface profiles, reduced heat transfer rates, and pressure distribution are concerned. Also, the resulting interface profiles may be considered independent of the solidifying liquid if it produces the same heat transfer rates. Liquids having the same Prandtl-numbers and thermal conductivities as water, which, for instance, can be assumed of brines up to certain concentrations, will produce identical interface profiles. Of course, the coordinate Z_e of the solid phase front does depend on the given temperatures.

In figure 2 the reduced interface heat flux $q^* \cdot \text{Ph}/2$ is plotted as a function of the solid phase thickness for various Biot-numbers. If $\text{Bi} \leqslant 2$, the interface profile must have an inclination which causes a continuous downstream increase of interface heat transfer. If $\text{Bi} > 2$, the inclination must be such that heat transfer initially decreases. This is demonstrated by the solid phase profiles plotted in figure 3a for various Biot-numbers. For $\text{Bi} \leqslant 2$ the profiles are short and slightly convex (very nearly linear), whereas for large Biot-numbers the profile is concave and can be remarkably long without resulting in freeze-up. For $\text{Bi} < 0.1$ the profiles are practically independent of the Biot-number.

Figure 3b shows, for a wide range of Biot-numbers, the variation with solid phase thickness $(1-X^*)$, of the reduced liquid temperature ϑ_m/ϑ_u, the local heat transfer rates, and the dimensionless local pressures. The distributions of pressure vary only little with the Biot-number. This is to be expected, because the Bernoulli-change-of-pressure, varying with the fourth power of X^*, dominates against the frictional losses.

Figures 4a and 5a show the interface profiles for a wide range of flow rates (Re_0), when $Bi = 10$ and $Bi = 0.1$ respectively. The effect of an alteration of flow velocity is moderate.

In figures 4b and 5b, the position of the solid phase front Z_e is plotted as a function of the dimensionless cooling temperature ϑ_u. If the point z_e reaches the entrance section of the cooling zone, $z = 0$, the solid phase profile is no longer independent of ϑ_u. The dashed curves in figures 3a and 4a are examples of solid phase shapes which develop in such a case.

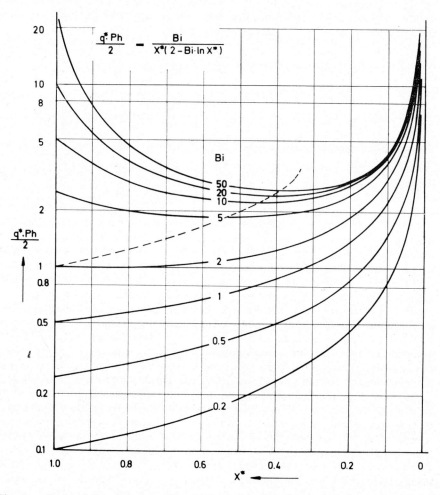

Fig. 2 — Reduced interface heat flux $q^*.Ph/2$ as a function of solid phase thickness for various Biot-numbers.

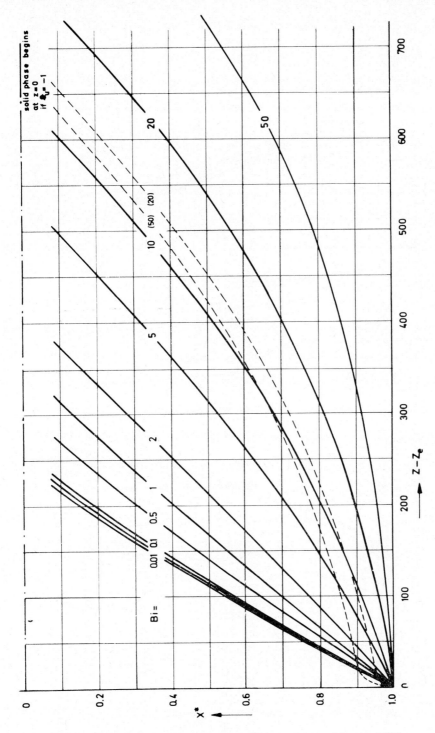

Fig. 3a — Solid phase profile for various Biot-numbers and $Re_0 = 5,000$.

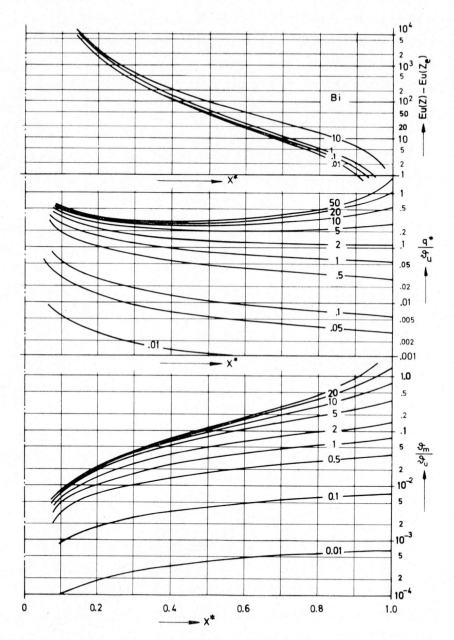

Fig. 3b — Liquid temperature, local heat transfer and pressure drop versus solid phase thickness for various Biot-numbers and $Re_0 = 5,000$.

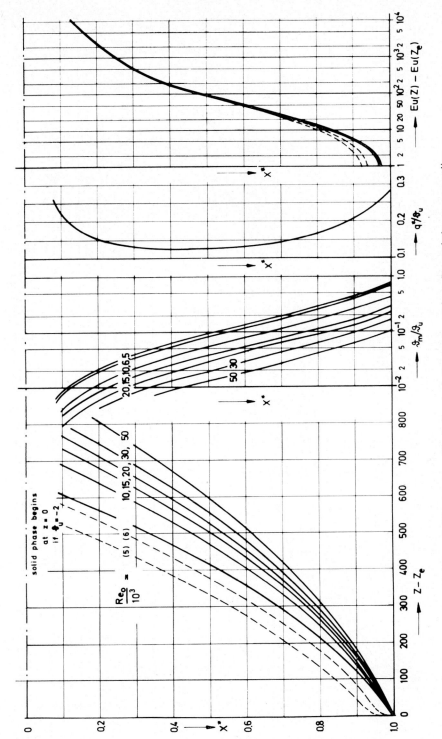

Fig. 4a — Solid phase profile for various Reynolds-numbers Re and the corresponding variation of the dimensionless local liquid temperature, heat transfer and pressure drop. Bi = 10.

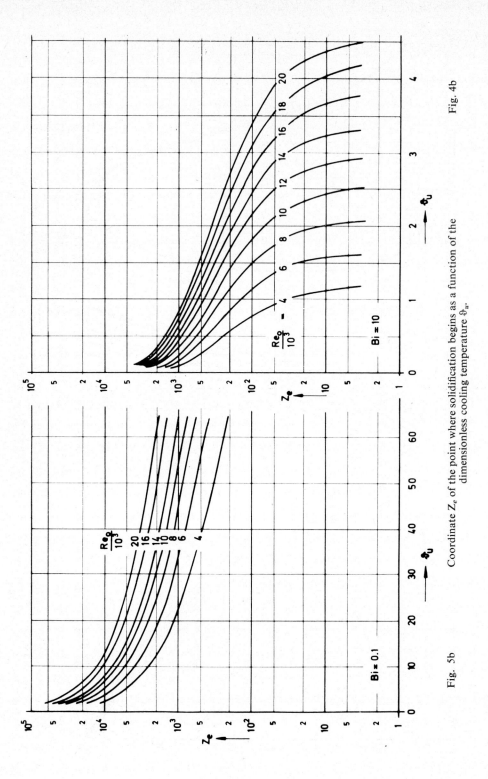

Fig. 4b

Fig. 5b

Coordinate Z_e of the point where solidification begins as a function of the dimensionless cooling temperature ϑ_u.

408

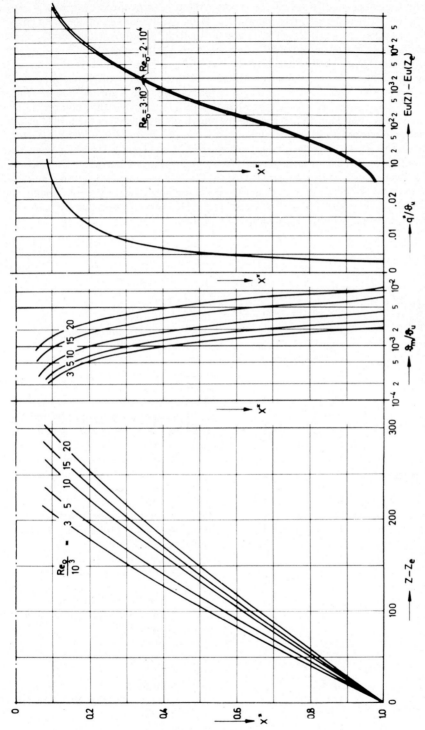

Fig. 5a — Solid phase for various Reynolds-numbers Re and the corresponding variation of dimensionless local liquid temperature, heat transfer and pressure drop. Bi = 0.1.

409

The following predictions can be made if the total pressure drop across a given length, Al, of tube is the given parameter. If the coordinate z_e of the solid phase front, calculated by equation (17), is greater than the length l of the cooling zone, naturally no solidification is possible. If z_e lies within l, the constriction caused by the growing solid phase shell will reduce the rate of flow. This, in turn, causes the solid phase front z_e to move upstream until the entrance section of the cooling zone, $z = 0$, is reached. Therefore, if the length of the cooling zone is greater than the length of the complete interface profile corresponding to the particular Biot-number, a tube will inevitably freeze up. In cases where the length l is smaller than the profile length, further predictions can be made: careful inspection of equation (13) shows that, if the tube length $(A-1)l$ outside of the cooling zone is sufficiently large, an increasing constriction by solidification near $z = 0$ is accompanied by an increase of flow velocity and of heat transfer in the constricted zone. This can eventually produce conditions permitting the existence of a stationary solid phase layer. If, on the other hand, the additional tube length $(A-1)l$ is small, flow velocity and heat transfer can only decrease, thus again implying blockage of the tube.

LIST OF SYMBOLS

Symbols with dimension

a_L, a_s	(m^2/s)	thermal diffusivities
c_s	(J/m^3K)	specific heat of solid phase
h	(J/kg)	latent heat of fusion
k	(J/m^2Ks)	coefficient of heat transmission
l	(m)	length of cooling zone
p	(kg/ms^2)	local pressure
p_0	(kg/ms^2)	pressure at tube entrance
Δp_D	(kg/ms^2)	total pressure drop across tube length
q	(J/m^2s)	local heat flux from liquid to solid
R	(m)	radius of tube
T_0	(K)	temperature at entrance to cooling zone
T_m	(K)	cross-sectional average of liquid temperature
T_s	(K)	solidification temperature
T_u	(K)	ambient cooling temperature
T_w	(K)	inside wall temperature of tube
u	(m/s)	radial velocity component
w	(m/s)	axial velocity component
$\bar{w}$	(m/s)	cross-sectional average of flow-velocity
$\bar{w}_0$	(m/s)	w at entrance to cooling zone
$\bar{w}_v$	(m/s)	w for given pressure drop if no solid phase existed
x	(m)	radial coordinate
x^*	(m)	radial coordinate of solid-liquid interface
z	(m)	axial coordinate
α	(J/m^2sK)	coefficient of convective heat transfer
λ_L, λ_s	(J/msK)	thermal conductivity
ρ_L, ρ_s	(kg/m^3)	density
v	(m^2/s)	kinematic viscosity

410

Dimensionless parameters

Bi	$=$	$k\, 2R/\lambda_s$	Biot-number
Eu(z)	$=$	$p(z) - p_0/(\rho_L \bar{w}_0^2/2)$	local Euler-number
Eu_v	$=$	$\Delta p_D/(\rho_L w_v^2/2) = \psi_v \mathrm{Re}_v$	Euler-number for total pressure drop across cooling zone
L	$=$	$\frac{1}{2}\,R$	dimensionless length of cooling zone
Nu*	$=$	$\alpha\, 2R/\lambda_L$	Nusselt-number
Ph	$=$	$h/c_s(T_s - T_u)$	phase conversion parameter
Pr	$=$	ν/a_L	Prandtl-number
q*	$=$	$q\, 2R/h \cdot a_s \rho_s$	dimensionless heat flux at solid-liquid interface
Re	$=$	$w(z) \cdot 2R/\nu$	local Reynolds-number
Re_0	$=$	$w_0\, 2R/\nu$	Reynold-number at entrance to cooling zone
Re_v	$=$	Eu_v/ψ_v	
U	$=$	$u/\bar{w}_v$	
W	$=$	$w/\bar{w}_v$	
X	$=$	x/R	
X*	$=$	x^*/R	dimensionless radial coordinate of solid-liquid interface
Z	$=$	$z/2R$	
Z_e	$=$	$z_e/2R$	dimensionless coordinate of point where solid phase begins
ΔZ_n			dimensionless length of tube-segment downstream of Z_n
ζ_e, ζ_D			energy-loss coefficients
ϑ	$=$	$(T - T_s)/(T_0 - T_s)$	dimensionless temperature
$\vartheta_m, \vartheta_0, \vartheta_u, \vartheta_w$			compare T_m, T_0, T_u, T_w
$\psi\,(Z)$			local friction coefficient
ψ_n			average friction coefficient of segment Z_n

Indices

D	end of cooling zone
L	liquid
n	beginning of *n*-th segment
0	at entrance to cooling zone
s	solid
u	ambient conditions

REFERENCES

[1] T.H. Gouw, Normal Freezing. Progr. Separation Purification, 1 (1968), 57/82.
[2] I.L. Dschu, Ausfrieren von Eis aus einer strömenden Salzlösung an gekühlten Oberflächen. *Kältetechnik-Klimatisierung*, 19 (1967), 9, 278/283, and 21 (1969), 2, 43/45.
[3] W. Schneider and K. Fischbeck, Über die Kristallisation von Eis aus Salzlösungen. *Desalination*, 5 (1968), 217/230.
[4] K. Stephan, Influence of Heat Transfer on Melting and Solidification in Forced Flow. *Int. J. Heat and Mass Transfer*, 12 (1969), 2, 199/214.
[5] H.G. Hirschberg, Das Einfrieren von Rohrleitungen. *Kältetechnik*, 14 (1962), 10, 314/321.

[6] R. D. ZERKLE and J. E. SUNDERLAND, The Effect of Liquid Solidification in a Tube Upon Laminar-Flow Heat Transfer and Pressure Drop. *J. Heat Transfer, Trans. ASME*, C 90 (1968), 2, 183/190.

[7] G. S. SPRINGER, The Effect of Axial Heat Conduction on the Freezing or Melting in Cylinders. *Int. J. Heat and Mass Transfer*, 12 (1969), 521/524.

[8] H. BARTHELS, Dissertation Kernforschungszentrum Jülich, Jül-506-KB.

[9] VDI-Wärmeatlas (1963), p. Lc1.

412

MODÈLE À GLISSEMENT VARIABLE
POUR L'ÉTUDE DES DÉBITS CRITIQUES
EN DOUBLE PHASE

D. MEUNIER et A. FRITTE

Université de Louvain (Belgique)

Model with variable slip for the study of two-phase critical flows

SUMMARY: *The study of two-phase critical flows is important in the sizing of pipes and exchangers for refrigerating plant. The first theoretical approaches to this problem considered a model of homogenous two-phase flow. But the flows calculated from this model were too low as compared to experimentally recorded flows. The phase separating model providing slip between the phases gave better numerical results, except at low levels. The conventional theories for this model however neglect the variations in slip at the critical section and calculate a maximum flow value independent of the slip law. These numerical values obtained are usually higher than the test values. Furthermore, the system of equations used in this case is inconsistent. A more complete description of the slip model rectifies these drawbacks. It also enables the calculation of the characteristics for flow above the critical section.*

LISTE DES SYMBOLES

D	m	diamètre de conduite;
E	J/kg	énergie massique;
G	kg/m^2.s	débit massique;
I	N/m^2	quantité de mouvement;
L	m	longueur de canalisation;
S	J/kg.°K	entropie massique;
T	°K	température;
h	J/kg	enthalpie massique;
k	–	coefficient de glissement u_2/u_1;
p	N/m^2	pression;
q	J/kg.m	source de chaleur;
u	m/s	vitesse;
v	m^3/kg	volume massique;
x	–	titre thermodynamique;
z	m	coordonnée;
α	–	fraction de vide;
ρ	kg/m^3	masse volumique;
τ	N/m^2	frottement;
1		indice phase liquide;
2		indice phase vapeur.

1. INTRODUCTION

L'étude des débits critiques en écoulement biphasé est importante pour le dimensionnement des conduites et des échangeurs des installations frigorifiques. Des débits critiques peuvent être atteints à la vanne de régulation thermostatique et dans la conduite d'amenée à l'évaporateur.

Deux types d'études théoriques des débits critiques ont donné des résultats numériquement satisfaisants :

— Les théories de Fauske [6] et de Moody [14] pour les très longues canalisations (rapport longueur/diamètre L/D > 25);

— Les théories sur la métastabilité de Silver et Mitchell [17] et de Bailey [1] pour les étranglements de faible longueur (L/D < 3).

Pour une canalisation (3 < L/D < 25) située après un étranglement et avant un vase d'expansion, il n'existe pas de modèle satisfaisant. En effet d'une part le déséquilibre de phase s'est atténué de sorte que la théorie de métastabilité ne donne plus satisfaction [16]. D'autre part le glissement entre phases n'a pas atteint une valeur suffisante pour que les théories de Fauske ou de Moody soient applicables.

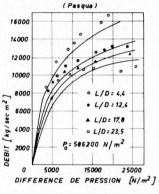

Fig. 1

La faillite de ces modèles est particulièrement mise en évidence aux figures 1 et 2. La figure 1 résulte des expériences de Pasqua [16] sur le fréon 12; la figure 2 est extraite des travaux de Fauske et Min [7] sur le fréon 11.

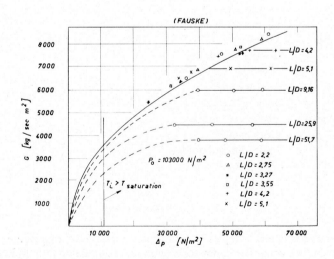

Fig. 2

414

On y remarque que pour des rapports L/D décroissants, les débits expérimentaux obtenus croissent constamment. Cela est contraire à ces théories qui donnent une diminution des débits critiques lorsque le glissement entre phases diminue.

Nous nous proposons de considérer dans l'étude du débit critique une théorie plus générale du glissement qui rende compte de ce phénomène.

2. Modèle à glissement et équations constitutives

Pour les débits élevés, l'expérience montre qu'un écoulement biphasé est à phases séparées dès que le titre atteint une valeur ...0, 1 ... La figure 4 montre une répartition de débit typique en canal vertical. On observe un noyau central de vapeur entouré d'un anneau de liquide séparé par un interface fortement turbulent.

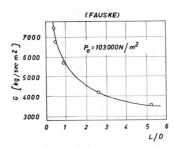

Fig. 3
(Erratum: en abcisse lire 10 L/D)

On schématisera un tel écoulement par les hypothèses suivantes :

1) La stabilité moyenne de l'interface nous oblige à admettre l'égalité de pression dans les deux phases;

2) Compte tenu du développement de l'interface et de sa forte turbulence, on peut supposer l'équilibre thermodynamique réalisé;

3) Les profils de vitesse à l'intérieur de chaque phase ne sont pas connus. On admettra, en première schématisation, que ces vitesses sont uniformes, c'est-à-dire que les gradients de vitesse à la paroi et à l'interface sont supérieurs aux gradients de vitesse à l'intérieur de chacune des phases. Ce faisant, nous admettrons également que les contraintes de cisaillement sont localisées d'une part, le long de la paroi et d'autre part, à l'interface. La figure 5 schématise l'écoulement théorique considéré.

Les équations constitutives de ce modèle sont :

— équation de continuité :

$$dG = d\left[\alpha \rho_2 u_2 + (1-\alpha)\, \rho_1 u_1\right] = 0 \tag{1}$$

— équation dynamique :

$$dI = d\left[p + \alpha \rho_2 u_2^2 + (1-\alpha)\, \rho_1 u_1^2\right] = -\tau_e \frac{4}{D}\, dz - \left[\alpha \rho_2 + (1-\alpha)\rho_1\right] g \cos\theta\, dz \tag{2}$$

— équation d'énergie :

$$d(GE) = d\left[\alpha \rho_2 u_2\left(h_2 + \frac{u_2^2}{2}\right) + (1-\alpha)\, \rho_1 u_1\left(h_1 + \frac{u_1^2}{2}\right)\right] = q_e\, dz. \tag{3}$$

415

En simple phase, ce système suffit à décrire l'écoulement. En effet, compte tenu de l'équation d'état, il n'existe alors que trois paramètres indépendants : p, ρ et u.

En double phase, les paramètres sont au nombre de quatre : la pression p à laquelle est liée la température, le taux de vide α ou le titre x, les vitesses u_2 et u_1 des deux phases. La loi d'évolution différente des vitesses doit être recherchée [11] en écrivant l'équation dynamique des deux phases.

On a pour la phase liquide :

$$\frac{\mathrm{d}}{\mathrm{d}z}\left[(1-\alpha)\,\rho_1\,u_1^2\right]+(1-\alpha)\,\frac{\mathrm{d}p}{\mathrm{d}z} = -\frac{4}{D}\,\tau_e + \frac{4\sqrt{\alpha}\,\tau_i}{D} - (1-\alpha)\,\rho_1\,g\,\cos\theta. \quad (4)$$

On a de même pour la phase vapeur :

$$\frac{\mathrm{d}}{\mathrm{d}z}\left[\alpha\rho_2\,u_2^2\right]+\alpha\,\frac{\mathrm{d}p}{\mathrm{d}z} = -\frac{4\sqrt{\alpha}\,\tau_i}{D} - \alpha\rho_2\,g\,\cos\theta. \quad (5)$$

Mais on a déjà tenu compte de la loi globale de quantité de mouvement (2); il suffit donc d'écrire la loi d'échange entre phases qui s'obtient en éliminant $\mathrm{d}p/\mathrm{d}z$ des équations (4) et (5).

Fig. 4   Fig. 5

On a :

$$\frac{1}{\alpha}\frac{\mathrm{d}}{\mathrm{d}z}\left[\alpha\rho_2\,u_2^2\right] - \frac{1}{1-\alpha}\frac{\mathrm{d}}{\mathrm{d}z}\left[(1-\alpha)\,\rho_1\,u_1^2\right]$$

$$= -\frac{4}{D}\frac{\tau_e}{1-\alpha} - \frac{4\sqrt{\alpha}\,\tau_i}{D\alpha(1-\alpha)} + (\rho_1 - \rho_2)g\,\cos\theta. \quad (6)$$

L'évolution des quatre caractéristiques principales est donc décrite par le système du tableau 1.

Les équations de continuité et d'énergie n'ont pas été écrites pour chacune des deux phases par suite de l'hypothèse d'équilibre thermodynamique. Cet équilibre suppose un transfert continu du liquide au gaz et une homogénéisation permanente de la température. D'ailleurs l'équation de continuité d'une phase par rapport à l'autre

$$k = \frac{u_2}{u_1} = \frac{1-\alpha}{\alpha}\cdot\frac{x}{1-x}\cdot\frac{\rho_1}{\rho_2} \quad (7)$$

ne fait que relier le taux de vide au titre et introduit une variable auxiliaire, le glissement.

3. Loi de débits critiques

Le système de quatre équations différentielles précédent est soluble à condition que son déterminant soit non nul. Lorsque ce déterminant tend vers zéro, les gradients dp/dz, $d\alpha/dz$, du_2/dz et du_1/dz tendent vers l'infini et un état critique est obtenu à l'annulation du déterminant. Cette condition donne une relation d'état critique de la forme :

$$f(p, \alpha, u_2, u_1) = 0.$$

Dans le but de simplifier les écritures, on éliminera la variable u_1 au profit du glissement k par

$$\frac{du_1}{u_1} = \frac{du_2}{u_2} - \frac{dk}{k} \tag{8}$$

ainsi que le taux de vide au profit du titre par :

$$\frac{d\alpha}{\alpha(1-\alpha)} = \left(\frac{d\rho_1}{\rho_1} - \frac{d\rho_2}{\rho_2}\right) + \frac{dx}{x(1-x)} - \frac{dk}{k}. \tag{9}$$

On obtient ainsi le système du tableau 2, et la condition d'état critique

$$\begin{vmatrix} xv_2' + k(1-x)v_1' & v_2 - kv_1 & -\dfrac{1}{G} & (1-x)v_1 \\[2mm] \dfrac{1}{G} & \left(1-\dfrac{1}{k}\right)u_2 & \left(x + \dfrac{1-x}{k}\right) & -\dfrac{(1-x)u_2}{k_2} \\[2mm] xh_2' + (1-x)h_1' & \dfrac{u_2^2}{2}\left(1-\dfrac{1}{k^2}\right) + (h_2 - h_1) & \left(x + \dfrac{1-x}{k^2}\right)u_2 & -\dfrac{(1-x)u_2^2}{k^3} \\[2mm] 0 & \dfrac{k(1-x)u_2}{v_2} + \dfrac{xu_2}{kv_1} & \dfrac{kx(1-x)}{v_2} - \dfrac{x(1-x)}{v_1 k} & \dfrac{x(1-x)u_2}{v_1 k^2} \end{vmatrix} = 0.$$

Après simplification et élimination de u_2 au profit de G, on obtient une équation bicarrée pour le débit critique :

$$A_1 G^4 + A_2 G^2 + A_3 = 0$$

avec :

$$A_1 = \frac{1}{2}\frac{xv_2}{\alpha}\left(1 - \frac{1}{k^2}\right)\left[xv_2' + k(1-x)v_1'\right]$$

$$A_2 = 2(v_2 - kv_1)\left[-1 + \frac{\alpha}{xv_2}\left[xh_2' + (1-x)h_1'\right] + (1-\alpha)\left(1 - \frac{1}{k}\right)\right]$$

$$-\frac{\alpha}{xv_2}(h_2 - h_1)\left[xv_2' + k(1-x)v_1'\right] + \frac{1}{2}\left(1 - \frac{1}{k^2}\right)\left[k^2 v_1 + \alpha(v_2 - k^2 v_1)\right]$$

$$A_3 = -\frac{\alpha^2}{x^2 v_2^2}(h_2 - h_1)\left[k^2 v_1 + \alpha(v_2 - k^2 v_1)\right].$$

417

Système 1

$$\alpha u_2 \rho_2' + (1-\alpha) u_1 \rho_1' \qquad\qquad \rho_2 u_2 - \rho_1 u_1 \qquad\qquad \alpha \rho_2$$

$$1 + \alpha u_2^2 \rho_2' + (1-\alpha) u_1^2 \rho_1' \qquad\qquad \rho_2 u_2^2 - \rho_1 u_1^2 \qquad\qquad 2\alpha\rho_2 u_2$$

$$\alpha u_2 (h_2 \rho_2)' + \tfrac{1}{2}\alpha u_2^3 \rho_2' +$$
$$+ (1-\alpha) u_1 (h_1 \rho_1)' + \tfrac{1}{2}(1-\alpha) u_1^3 \rho_1' \qquad \rho_2 u_2 h_2 + \tfrac{1}{2}\rho_2 u_2^3 - \rho_1 u_1 h_1 - \tfrac{1}{2}\rho_1 u_1^3 \qquad \alpha\rho_2 h_2 + \tfrac{3}{2}\alpha\rho_2 u$$

$$u_2^2 \rho_2' - u_1^2 \rho_1' \qquad\qquad \frac{\rho_2 u_2^2}{\alpha} + \frac{\rho_1 u_1^2}{1-\alpha} \qquad\qquad 2\rho_2 u_2$$

Système II

$$xv_2' + k(1-x)v_1' \qquad\qquad v_2 - kv_1 \qquad\qquad -\frac{1}{G}$$

$$\frac{1}{G} \qquad\qquad \left(1 - \frac{1}{k}\right)u_2 \qquad\qquad x + \frac{1-x}{k}$$

$$xh_2' + (1-x)h_1' \qquad\qquad \frac{u_2^2}{2}\left(1 - \frac{1}{k^2}\right) + h_2 - h_1 \qquad\qquad \left(x + \frac{1-x}{k^2}\right)u_2$$

$$0 \qquad\qquad \frac{k(1-x)u_2 v_1}{xv_2} + \frac{u_2}{k} \qquad\qquad \frac{k(1-x)v_1}{v_2} - \frac{1-}{k}$$

$$
\begin{bmatrix}
(1-\alpha)\rho_1 \\[4pt]
2(1-\alpha)\rho_1 u_1 \\[4pt]
(1-\alpha)\rho_1\left(h_1+\tfrac{3}{2}u_1^2\right) \\[4pt]
-2\rho_1 u_1
\end{bmatrix}
\begin{bmatrix}
\dfrac{dp}{dz} \\[8pt]
\dfrac{d\alpha}{dz} \\[8pt]
\dfrac{du_2}{dz} \\[8pt]
\dfrac{du_1}{dz}
\end{bmatrix}
=
\begin{bmatrix}
0 \\[4pt]
-\tau_e \dfrac{4}{D} - \big[(1-\alpha)\rho_1+\alpha\rho_2\big]g\cos\theta \\[8pt]
q_e \\[8pt]
\dfrac{\tau_e}{(1-\alpha)}\dfrac{4}{D} - \dfrac{4\sqrt{\alpha}\,\tau_i}{D\alpha(1-\alpha)} + (\rho_1-\rho_2)g\cos\theta
\end{bmatrix}
$$

$$
\begin{bmatrix}
(1-x)v_1 \\[6pt]
-\dfrac{(1-x)u_2}{k^2} \\[8pt]
-\dfrac{(1-x)u_2^2}{k^3} \\[8pt]
\dfrac{(1-x)u_2}{k^2}
\end{bmatrix}
\begin{bmatrix}
\dfrac{dp}{dz} \\[8pt]
\dfrac{dx}{dz} \\[8pt]
\dfrac{du_2}{dz} \\[8pt]
\dfrac{dk}{dz}
\end{bmatrix}
=
\begin{bmatrix}
0 \\[6pt]
-\dfrac{1}{G}\dfrac{1}{k(1-x)v_1+xv_2}\big[k(1-x)+x\big]g\cos\theta - \\
\qquad\qquad\qquad -\dfrac{4}{GD}\tau_e \\[8pt]
\dfrac{q_e}{G} \\[8pt]
\dfrac{1}{G}\dfrac{4}{D}\tau_e - \dfrac{1}{G}\dfrac{4\tau_i}{D}\sqrt{1+\dfrac{k(1-x)v_1}{xv_2}} + \\
\qquad +\dfrac{1}{Gv_2}\dfrac{k(1-x)}{k(1-x)v_1+xv_2}(v_2-v_1)g\cos\theta
\end{bmatrix}
$$

Cette équation se réduit pratiquement, compte tenu de la faible valeur numérique de A_1, à :

$$G^2 = - \frac{A_3}{A_2}.$$ (11)

Les courbes 1 de la figure 6 montrent l'évolution en fonction du glissement, du débit critique calculé par cette formule. Elles ont été tracées pour le fréon 12 à la pression de 98 100 N/m² et au titre $x = 0{,}1$, ainsi que pour le fréon 11 à la pression de 62 200 N/m² et au titre $x = 0{,}1$. On remarque que le débit critique est pratiquement constant aux glissements élevés mais se relève fortement lorsque le glissement tend vers zéro.

4. MÉTHODES APPROCHÉES DE DÉTERMINATION DE DÉBITS CRITIQUES

En simple phase, la conservation de la masse, de la quantité de mouvement et de l'énergie entraîne la constance de l'entropie. Il est alors possible de se servir de deux lois de conservation et de la loi d'entropie pour caractériser un état critique. Cela donne lieu à une interprétation physique évidente : sous les conditions de conservation considérées, l'entropie est maximale à la section critique.

En double phase, on a montré [12] que les lois de conservation précédentes entraînent la loi de variation d'entropie :

$$T \frac{ds}{dz} - \frac{dx}{dz} \frac{(u_2^2 - u_1^2)}{2} = \frac{1}{G} \left[q_e + \tau_e u_1 \frac{4}{D} - \tau_i (u_2 - u_1) \frac{4\sqrt{\alpha}}{D} \right].$$ (12)

On y voit apparaître au membre de droite une production d'entropie due à l'apport calorifique et au cisaillement de paroi et d'interface, mais également au membre de gauche un terme de production d'entropie dû à la vaporisation entre des phases animées de vitesses différentes.

Si on néglige ce terme, on peut à nouveau, comme en simple phase, éliminer une équation de conservation au profit de celle :

$$\frac{ds}{dz} = \left[xs_2' + (1-x)s_1' \right] \frac{dp}{dz} + (s_2 - s_1) \frac{dx}{dz} = 0$$ (13)

maximant le flux d'entropie.

Les courbes 2 et 3 de la figure 6 ont été calculées avec cette approximation. Les courbes 2 ont été obtenues en remplaçant l'équation d'énergie par celle d'entropie; elles répondent à la loi de débit critique :

$$-\frac{1}{G^2} = \frac{\left[xv_2' + k(1-x)v_1' \right] + 2(v_2 - kv_1) \left(\dfrac{dx}{dp} \right)_s}{\dfrac{\alpha k}{xv_2} \left[\dfrac{v_2}{k} - (1-\alpha) \left(\dfrac{v_2}{k} - kv_1 \right) \right]}.$$ (14)

Les courbes 3 ont été obtenues en remplaçant l'équation de quantité de mouvement par celle d'entropie; la loi de débit critique est alors

$$\left\{\left[xv_2' + k(1-x)v_1'\right] + 2(v_2 - kv_1)\left(\frac{dx}{dp}\right)_s\right\}\left(\frac{1}{k^2\alpha} + \frac{1}{k(1-\alpha)}\right) -$$

$$-\frac{1}{G^2} = \frac{-\frac{1}{2}\left(1 - \frac{1}{k^2}\right)\left(\frac{dx}{dp}\right)_s\left(\frac{kv_1}{\alpha} + \frac{v_2}{k(1-\alpha)}\right)}{\frac{\alpha^2 v_2}{x^2 v_2}\left[\frac{kv_1}{\alpha} + \frac{v_2}{k(1-\alpha)}\right]} \qquad (15)$$

La comparaison des courbes 1, 2 et 3 montre que le fait de négliger la source d'entropie due à la vaporisation modifie légèrement les valeurs de débits critiques

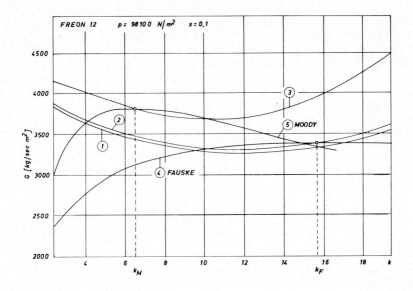

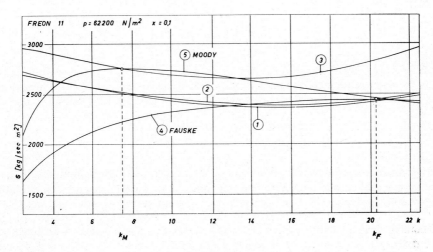

Fig. 6

mais ne change pas l'allure du phénomène. En particulier, les débits critiques augmentent aux faibles glissements. Le remplacement de l'équation énergétique ou dynamique par celle d'entropie offre par ailleurs l'avantage de donner des formules explicites du débit critique.

5. Comparaison avec les théories de Fauske et de Moody

Ces auteurs utilisent la condition d'entropie maximum [13] pour décrire les débits critiques. Ils font donc l'approximation de négliger la production d'entropie due à la vaporisation :

$$\frac{\mathrm{d}x}{\mathrm{d}z} \frac{(u_2^2 - u_1^2)}{2}.$$

Fauske utilise cette condition pour un écoulement à débit et impulsion donnés, tandis que Moody considère un écoulement à débit et énergie donnés. Leurs résultats sont donc à comparer respectivement aux formules (14) et (15) et aux courbes 2 et 3.

Par ailleurs, ces auteurs considèrent une loi de glissement de la forme $k = f(p)$. Ils ne considèrent donc pas le glissement comme une variable évolutive semblable à la pression, au titre ou à la vitesse mais comme une « caractéristique d'état ». Cette façon de voir donne de bons résultats en longue canalisation où l'on peut supposer un glissement « établi », mais ne convient pas pour un étranglement de faible longueur.

Les formules de ces auteurs sont ainsi :

$$-\frac{1}{G_F^2} = \left(x + \frac{1-x}{k}\right)\left[x v_2' + k(1-x) v_1'\right] - x(1-x)\left(\frac{v_2}{k^2} - v_1\right)\frac{\mathrm{d}k}{\mathrm{d}p} +$$

$$+ \left[\left[x v_2 + k(1-x) v_1\right]\left(1 - \frac{1}{k}\right) + (v_2 - k v_1)\left(x + \frac{1-x}{k}\right)\right]\left(\frac{\mathrm{d}x}{\mathrm{d}p}\right)_s \quad (16)$$

$$-\frac{v}{v_k}\frac{1}{G_M^2} = \left(x + \frac{1-x}{k^2}\right)\left[x v_2' + k(1-x) v_1'\right] - x(1-x)\left(\frac{v_2}{k^3} - v_1\right)\frac{\mathrm{d}k}{\mathrm{d}p} +$$

$$+ \left[\left(x + \frac{1-x}{k^2}\right)(v_2 - k v_1) + \frac{1}{2}\left(1 - \frac{1}{k^2}\right)v_k\right]\left(\frac{\mathrm{d}x}{\mathrm{d}p}\right)_s \quad (17)$$

avec $v_k = x v_2 + k(1-x) v_1$.

Ces formules sont indépendantes de la forme de la loi de glissement (annulation du terme $\mathrm{d}k/\mathrm{d}p$) pour les glissements particuliers de Fauske et de Moody :

$$k_F = \left(\frac{v_2}{v_1}\right)^{1/2} \quad \text{et} \quad k_M = \left(\frac{v_2}{v_1}\right)^{1/3}.$$

Ces glissements correspondent d'ailleurs [3] aux écoulements avec minimum d'impulsion d'une part et minimum d'énergie cinétique d'autre part. Ils donnent de plus aux formules précédentes une valeur maximale en fonction du glissement.

Les courbes 4 et 5 de la figure 6 tracées à partir de ces formules montrent que ce maximum est fort aplati. Ces formules sont donc relativement indépendantes du glissement dès que celui-ci atteint une valeur suffisante. L'expérience le confirme d'ailleurs pour de longues canalisations.

Si par ailleurs, nous comparons les courbes 2 et 3 à celles de Fauske et de Moody, nous voyons que pour les glissements respectifs k_F et k_M, nous obtenons les mêmes débits critiques; ce qui est naturel puisque, pour ces valeurs de k, la loi de glissement n'intervient pas.

De plus, les courbes 2 et 3 présentent, comme celles de ces auteurs, des variations lentes, pour les glissements élevés. Toutefois, nous obtenons un relèvement de ces courbes lorsque k diminue. Ce qui est conforme aux résultats expérimentaux (fig. 1 et 2) suivant lesquels le débit critique augmente pour des conduites de plus en plus courtes, correspondant à des glissements de plus en plus petits.

6. Conclusions

1. Le modèle à glissement à phases séparées et à l'équilibre thermodynamique doit être décrit en considérant les équations globales de conservation et l'échange de quantité de mouvement entre phases. Dans ces conditions, le glissement est une variable évolutive dépendant de la configuration de toute la conduite. Ce n'est que pour les conduites longues qu'on peut admettre une « relation d'état » de glissement.

2. Avec ce modèle, nous obtenons une loi de débit critique donnant des valeurs peu variables aux glissements élevés mais qui augmentent rapidement aux faibles glissements. Cette allure de phénomène est confirmée par l'expérience en longue canalisation d'une part et en étranglement de faible longueur d'autre part.

3. Cette théorie jointe aux expériences de Fauske et de Pasqua montre qu'il est peu vraisemblable que le débit critique survienne à l'étranglement même d'une vanne à liquide saturé. Par suite de la métastabilité, l'écoulement à cet endroit est de type liquide à débit critique élevé. Il est plus vraisemblable au contraire que l'état critique soit obtenu à la sortie de la vanne ou dans la tuyauterie de l'évaporateur lorsque le glissement entre phases s'établit, ce qui réduit fortement les valeurs de débits critiques.

4. Nous espérons, grâce à la littérature récente sur les contraintes τ_e et τ_i, pouvoir lier la loi de débit critique non plus au glissement, variable difficilement mesurable, mais au rapport L/D de la canalisation.

RÉFÉRENCES

[1] J. F. Bailey, Metastable flow of saturated water, vol. 73, 1109. *Trans A.S.M.E.* (1951).
[2] J. Boure, Propagation des petites perturbations dans les écoulements diphasés avec transfert de masse et glissement. *S.H.F., VIIIᵉ Journées de l'hydraulique* (1964), Question Vb, Rapport n° 4, 462/469.
[3] J.E. Cruver and R.W. Moulton, Critical flow of liquid vapor mixtures. *A.I.Ch.E. Journal* (1967), 13, 1, 52/60.
[4] D.W. Faletti and R.W. Moulton, Two-phase critical flow of steamwater mixtures. *A.I.Ch.E. Journal* (1963), 9, 2, 247/253.
[5] H.K. Fauske, Some ideas about the mechanism causing two-phase critical flow. *Appl. Sci. Res.* (1963), A 13, 149/160.
[6] H.K. Fauske, A theory for predicting pressure gradients for two-phase critical flow. *Nuclear Science and Engineering* (1963), 17, 1/7.
[7] H.K. Fauske and T.C. Min, A study of the Flow of Saturated Freon-11 through Apertures and Short Tubes. ANL 6667 (Jan. 1963).
[8] M. Giot et D. Meunier, Méthodes de détermination du débit critique en écoulements monophasiques et diphasiques à un constituant. E.P.E. (1968), IV, 1-2, 1/23.
[9] H.S. Isbin, Critical Two-Phase Flow (March 1965).
[10] H.S. Isbin, J.E. Moy, and J.R. Da Cruz, Two-Phase, Steam-Water Critical Flow. *A.I.Ch.E. Journal* (1957), 3, 3, 361/365.

[11] S. Levy, Steam slip. Theoretical prediction from momentum model. *Journal of Heat Transfer* (1960), 5, 113/124.
[12] D. Meunier et M. Giot, Equations de conservation du modèle à glissement. Séminaire de Thermodynamique. Rapport n° 403 UCL (1969).
[13] F.J. Moody, Maximum flow rate of a single component, two-phase mixture. *Journal of Heat Transfer* (1965), 2, 134/142.
[14] F.J. Moody, Maximum two-phase vessel blowdown from pipes. *Journal of Heat Transfer* (1966), 8, 285/293.
[15] H. Ogasawara, A Theorical Prediction of Two-Phase Critical Flow. *Bulletin of J.S.M.E.* (1967), 10, 33, 279/290.
[16] P.F. Pasqua, Metastable Flow of Freon-12. Refrigerating Engineering 61, 1084A-1088, 1131 (Oct. 1953).
[17] R.S. Silver and J.A. Mitchell, The discharge of saturated water trough nozzles. *Trans North East Coast, Inst. Engineers and Shipbuilders* (1945), Vol. 62, 51.
[18] S.M. Zivi, Estimation of steady-state steam void fraction by means of the principle of minimum entropy production. *Journal of Heat Transfer* (1964), 5, 247/252.

TEMPERATURE DISTRIBUTION IN A FROST LAYER FORMED ON A CYLINDRICAL TUBE IN CROSS FLOW

H. W. SCHNEIDER

Institut für Technische Thermodynamik, Universität Stuttgart (W. Germany)

Répartition des températures dans une couche de givre se formant sur un tube cylindrique, dans le cas du courant transversal

RÉSUMÉ : *Une recherche sur la transmission de chaleur, pour des surfaces refroidissantes givrées, a conduit à des mesures de la répartition des températures dans une couche de givre sur un tube cylindrique horizontal, dans le cas du courant transversal. L'appareil utilisé pour la mesure est une grlile de 20 thermocouples fins, placés perpendiculairement à l'axe du tube, à 1 mm les uns des autres. On peut faire varier l'angle de la grille par rapport à la direction du courant d'air. La surface soumise à l'essai est placée dans un tunnel où l'air, qui circule en circuit fermé, est refroidi par un refroidisseur d'air du type humide. C'est pourquoi les conditions de transfert de chaleur et de masse ne varient pas et permettent de stabiliser la température de l'air et le point de rosée, à moins de 0.05 °C, pendant une longue durée.*

La forme circulaire des isothermes obtenus dans la couche de givre conduit à conclure que la température est presque uniforme sur toute la surface. À partir de la répartition des températures étudiée en fonction du temps, on calcule la répartition du coefficient de conductivité thermique du givre.

A survey of literature on heat transfer under frosting conditions shows that further work is needed to enable prediction of heat and mass transfer rates through a frost layer for given environmental conditions.

In most of the experimental investigations it has not been possible to control the temperature and humidity of the air stream within the desired accuracy. Thus additional processes take place rendering more difficult a fundamental insight into the mechanism of frost formation. The later presented test unit has been designed and constructed especially with regard to this aspect.

An investigation on frosting surfaces usually requires to know the surface temperature of the frost layer. One of the procedures consists of the measurement of the temperature distribution in the frost layer normal to its surface. The surface temperature is then extrapolated [1].

This procedure leads to the question, at which point around the tube the average surface temperature may be obtained that controls the transfer processes through the frost layer. Results on this subject obtained for a cylindric tube in cross flow will be presented.

EXPERIMENTAL APPARATUS

A schematic plan of the test unit is given in figure 1. It consists of a closed-loop wind tunnel a, where air is circulated by a blower b. The air stream is cooled in a wet-type air cooler g, which consists of a packed section g_1 of Raschig rings. The air passes the packing from the bottom while the coolant is distributed on the top. After leaving the cooler the air is nearly saturated. Therefore the psychrometric conditions can be controlled only by means of the coolant temperature. Beyond this it is a remarkable advantage of this type of heat exchanger, that the conditions of heat and mass transfer do not vary with time. The air temperature and the dew-point can be held to less than $\pm 0.05\,^{\circ}\text{C}$ of the desired value during 24 hours. The upstream positioned electric heater h allows action upon the air temperature independently of the temperature of the coolant.

425

The test cylinder d is placed at the beginning of the test section c. It measures 47.5 mm OD and 380 mm in length. A methyl alcohol coolant is circulated through the test tube. Its rise in temperature is used in calculating the total heat flux. The alcohol is cooled by means of a thermostat l. It is possible to control the alcohol temperature to within ± 0.05 °C.

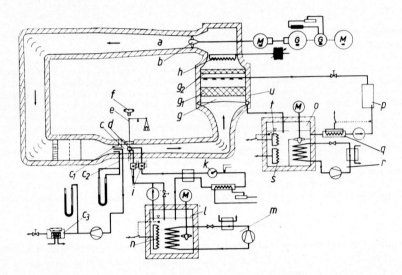

Fig. 1 — Schematic diagram of experimental apparatus.
  a wind tunnel; b turbo-blower; c test section; c_1 thermocouple measuring air temperature; c_2 pitot-static tube; c_3 dew-point hygrometer; d test cylinder; e balance; f camera; g wet-type air cooler; g_1, g_2 packed sections; h electric heater; i thermocouples measuring coolant temperatures; k flowmeter; l thermostat; m refrigerating unit; n temperature guided electric heater; o reservoir; p mixing vessel; q temperature guided electric heater; r refrigeration plant; s electric heater; t temperature guided electric heater; u insulation

The psychrometric conditions of the air are determined at the beginning of the test section c. The dry-bulb temperature is measured by a copper-constantan thermocouple c_1, the wet-bulb temperature is determined by a dew-point hygrometer c_3 operating on the principle of Glaser [2]. The air velocity is read from an inclined micromanometer connected to the pitot-static tube c_2.

Since it is necessary to measure the surface temperature of the test cylinder, eleven copper-constantan thermocouples have been inserted into the surface. The instrumentation for the measurement of the temperature distribution perpendicular to the axis of the tube consists of a grid of 20 thin (0.1 mm OD) thermocouples, mounted at equal distances of 1 mm apart. It is possible to vary the angle of the grid relative to the direction of the air flow. Arrangements similar to the described one have been used earlier by K. Schropp [3] and E. Hofmann [4].

DISCUSSION OF RESULTS

In order to verify the accuracy of the equipment, the total heat transfer rates have been measured. Figure 2 shows the results obtained for the bare cylinder. They can be expressed by the equation $\mathrm{Nu} = 0.245\,\mathrm{Re}^{0.589}$ which satisfies the measured values with an error of $\pm 2\%$. On the same figure have been plotted for comparison the resulting curves found by Hilpert [5] and Reiher [6]. It can be seen that the measured straight line falls within the two curves. From the descriptions of the test apparatus of the two authors it can be concluded that the turbulence level of Hilpert's test set-up surely has been lower while that of Reiher's unit propably has been higher than the turbulence level of the present wind tunnel. Therefore Reiher's curve is above and Hilpert's curve below the measured one.

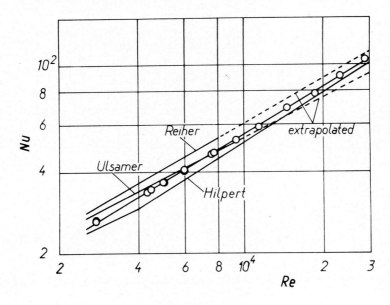

Fig. 2 — Nusselt numbers obtained for bare cylinder.

The temperature distribution in the frost layer normal to the test surface is shown in figure 3 as a function of time. The angle between the grid composed of 20 evenly-spaced thermocouples and the flow direction is 22.5°. The average Reynolds number for the frosted cylinder is 11 200, the air temperature 3.5 °C, the humidity 99.3% and the average temperature of the test surface -17.2 °C. The dotted lines refer to the temperatures in the air stream close to the frost-air interface.

The temperature distribution obtained for the early stages of deposition are nearly straight lines. It must be noted that there seems to be no break within the line at the frost-air interface. Results obtained by the grid positioned at several angles relative to the flow direction confirm this statement. It may be explained by the rough and porous structure of the frost surface creating a boundary layer quite different from that observed in the case of a solid wall.

The frost-air interface reaches 0 °C within 2½ hours after starting the test. At this temperature the condensation of water vapour directly occurs on the frost surface causing a progressively increasing density of the top layer. This explains the corresponding increase in thermal conductivity. Due to this fact the rising curves of temperature slightly level off to the actual value of the surface.

Close to the tube surface a quite similar process takes place. The water vapour molecules, after passing through the frost layer by diffusion, directly condense at the surface of the cooled tube, causing the frost density to increase. Due to the increasing density the thermal conductivity increases, too, causing the temperature curves to level off to the present wall temperature.

As the frost continuously becomes more dense at the top layer the porous structure of the surface slowly transforms into smoother formation more similar to that of a solid wall. This produces a break at the frost-air interface, which can be seen from the curves taken 5, 10 and 20 hours after starting the test.

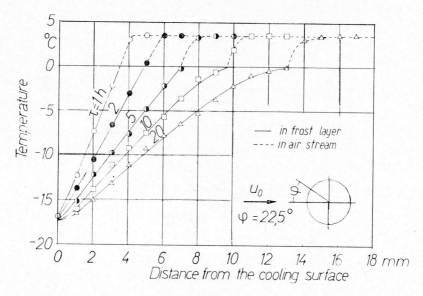

Fig. 3 — Temperature distribution in frost layer normal to the cooling cylinder as a function of time.

Isotherms obtained in the frost layer of a cylindrical tube are presented in figure 4. They have been taken 1, 5 and 20 hours after the beginning of the test run, the physical conditions being the same as in figure 3. Changing the angle of the grid relative to the

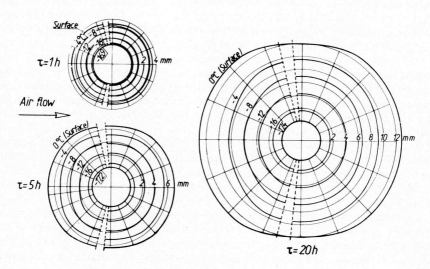

Fig. 4 — Isotherms in frost layer of a cylindrical tube.

air flow by steps of 22.5°, several runs have been carried out in order to obtain the distribution along the circumference.

The distance of the isotherms initially being very small, increases during the run. Therefore the temperature gradient decreases, causing a continual decrease of the diffusion rate of water vapour through the permeable structure. This may serve as a partial explanation why water vapour progressively condenses at the frost surface.

Initially the shape of the isotherms is similar to that of the outer edge of the frost layer. Both resemble concentric circles except near the sides of the cylinder ($\varphi = 90°$). This is due to the air speed which is highest in that region. It has been observed that some of the loose dendritic frost has continuously blown off the surface as the frost layer thickened.

Some hours after starting the test the isotherms differ from the outline of the frost layer near the stagnation point and at the rear of the tube. However, it was thought that the high transfer coefficient is not the only reason. As the pressure is highest the air may get through the permeable structure in this region, producing an additional increase of the density. This creates an increase in the thermal conductivity causing the temperature gradient to lessen.

A definite discontinuity in the thickness and density of frost has been observed in the range of $\varphi = \pm 80°$ from the stagnation point. Looking at figure 4 this is indicated by the dotted lines. At this point the boundary layer evidently becomes turbulent producing a gap that nearly reaches down to the cooled surface.

From the circular shape of the isotherms it may be concluded that the temperature of the frost-air interface takes an approximately equal value over the entire surface which is independent of the heat flow at any given point. Therefore it seems to be of less importance in what position relative to the air flow the temperature distribution has been measured, from which the surface temperature is extrapolated. It may be assumed, too, that any frost layer formed on a cooled solid has a nearly equal temperature over the entire surface.

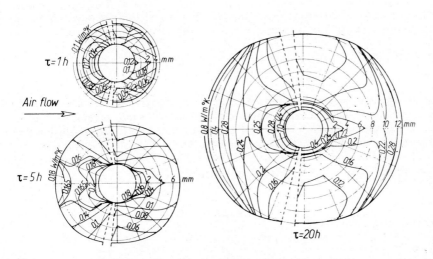

Fig. 5 — Local distribution of thermal conductivity of frost.

In figure 5 the local distribution of thermal conductivity is shown 1, 5 and 20 hours after starting the test. It has been calculated from the local temperature distribution. The dotted lines indicate the discontinuity in the frost layer. The thermal conductivity has been computed by the following equation

$$K_f(r, \varphi) = \frac{q(r, \varphi)}{\partial t/\partial r} - L \frac{\delta}{\mu R T} \frac{\partial P}{\partial t}$$
(1)

where:

$K_f(r,\varphi)$ represents the local thermal conductivity of frost at the polar coordinates r, φ;

$q(r,\varphi)$ the total heat flux at a given point;

$\partial t/\partial r$ the temperature gradient;

L   the latent heat of sublimation;

δ the mass diffusivity;

μ the diffusional resistance factor;

R   the gas constant of water vapour;

T   the absolute temperature in Kelvin degrees and

$\partial P/\partial t$ the pressure gradient of saturated water vapour at the local temperature t.

For the calculation it has been assumed that the distribution of transfer along the circumference of the rough frost layer is similar to that of a bare cylinder. These transfer coefficients have been measured e.g. by Schmidt and Wenner [7], Winding and Chenney [8] and Krückels [9]. The diffusional resistance factor has been adopted as being unit, thus causing an error of less than $\pm 3.6\%$ in case of the present problem.

The relationship between the thermal conductivity of frost and its density is usually expressed by a square function. With this in mind the following can be deduced from figure 5. In the first hour after starting the run the density of frost reaches a maximum at the stagnation point and at the rear of the tube while it is minimal at the sides However, the density decreases progressively from the cooled wall to the frost-air interface.

When the surface temperature is at 0 °C, the top layer of the frost becomes more dense. This process starting in the regions of maximal transfer rates, that is at the stagnation point and at the rear of the tube, extends over the entire surface.

CONCLUSIONS

A test unit has been described capable to control the temperature and the dewpoint of the circulated air stream to within ± 0.05 °C for long periods of time.

The isotherms in the frost layer of a cylindric tube lead to the conclusion that the temperature is almost equal at every point on the surface of the frost layer.

The distribution of the density based on the calculated values of the thermal conductivity has been discussed in a qualitative manner.

REFERENCES

[1] L. PRINS, Wärme- und Stoffübertragung in einem quer angeströmten, bereifenden Luftkühler, *Kältetechnik* **8** (1956), pp. 160-164 and pp. 182/187.
[2] H. GLASER, Feuchtigkeitsmessung mit Hilfe des Taupunktspiegels, *Kältetechnik* **2** (1950), pp. 126-131.
[3] K. SCHROPP, Untersuchungen über die Tau- und Reifbildung an Kühlrohren in ruhender Luft und ihr Einfluß auf die Kälteübertragung, *Z. ges. Kälte-Ind.*, **42** (1935), pp. 81-85, pp. 126-131 and pp. 151-154.
[4] E. HOFMANN, Wärmeübergangsversuche an einem Plattenkühler unter besonderer Berücksichtigung der Reifschicht, *Kälte*, **1** (1948), pp. 25-31.

[5] R. HILPERT, Wärmeabgabe von geheizten Drähten und Rohren im Luftstrom, *Forschung*, **4** (1933), pp. 215-224.

[6] H. REIHER, Wärmeübergang von strömender Luft an Rohre und Rohrbündel im Kreuzstrom. *VDI-Forschungsheft*, 269 (1925).

[7] E. SCHMIDT und K. WENNER, Wärmeabgabe über den Umfang eines angeblasenen, geheizten Zylinders. *Forschung*, **12** (1941), pp. 65-73.

[8] C.C. WINDING and A.J. CHENEY, Mass and Heat Transfer in Tube Banks, *Ind. Engng. Chem.*, **40** (1948), pp. 1087-1093.

[9] W. KRÜCKELS, Eine Methode zur photometrischen Bestimmung örtlicher Stoffübergangszahlen mit Hilfe chemischer Nachweisreaktionen, Dissertation Stuttgart (1968).

DISCUSSION

E. EMBLIK (Suisse) — Mr. Feniger has shown in a publication made in 1949 that frost formation takes place even in the porous frost layer itself. Therefore it is difficult to calculate the frost formation per unit of time. Do your investigations enable us to calculate, theoretically, the speed of frost formation?

H.W. SCHNEIDER — The results I have spoken of are only the results of the first stages of an investigation which will be continued in the future. Therefore I hope that one day I will be able to answer your question.

POSSIBILITIES FOR INTENSIFYING THE HEAT EXCHANGE IN "DRY" EVAPORATORS THROUGH INFLUENCE ON THE DYNAMICS OF TWO PHASE FLOW

Z.R. HUELLE

Danfoss A/S, Nordborg (Denmark)

Possibilités d'intensifier la transmission de chaleur dans les évaporateurs secs, par action sur la dynamique de l'écoulement biphasique

RÉSUMÉ : *Les évaporateurs, alimentés par des détendeurs thermostatiques, sont largement employés dans les équipements frigorifiques à cause de leurs avantages pratiques particuliers. Il semble qu'on n'a pas encore complètement étudié certaines possibilités d'améliorer la transmission de chaleur dans les évaporateurs secs.*

Le rapport présente les résultats d'essais de laboratoire sur des évaporateurs secs, pour lesquels la transmission de chaleur a été intensifiée par action sur la dynamique de l'écoulement biphasique.

INTRODUCTION

During the extension of investigations concerning evaporators fed through thermostatic expansion valves [1, 2, 3] numerous observations were made regarding flow pattern and heat exchange in two phase flow. The characteristic flow oscillations, observed in the bare tube, are correspondingly referred to in the previously published literature [4, 5, 6, 7, 8, 9].

As is well known, the dynamics of two phase flow, during the evaporation period, is very complicated in character [11] and therefore, only a few partial descriptions of the problem, limited to specific conditions, are known and of real value to the industry.

A study of the results from our own research on boiling heat transfer, leads the author to the opinion that the proper changes in flow dynamics may bring about an increase in heat transfer and the damping of oscillations. This last aspect is of importance to automatic controls [3]. The idea was realised, simply by fitting a tight wire mesh screen to the internal surface of the tube, where the evaporating process takes place.

It was expected that the narrow space between screen and tube wall would influence the liquid transport on the tube surface in the proper way, and that the increase in pressure drop would be tolerable – since the main cross force channel in the tube would remain open to flow.

The results of the first tests carried out on the glass tube have established the expected changes in flow dynamics and heat transfer, this instigated the commencement of the extended research programme. This paper discloses the results of experiments made with a double tube evaporator (water chiller), specially designed for this purpose. The object of this paper is to discuss the interesting results obtained from the test. The test results on hand at present are only explored dependencies in their dimensional form, which are corresponding to earlier findings in this sphere. The remaining results are to be elaborated upon for future discussion.

EXPERIMENTAL APPARATUS

The schematic diagram of experimental apparatus appears in figure 1. The refrigerant and water circuits are distinguishable as two separate circuits. The refrigerant used was R 12 ($CF_2 Cl_2$).

The refrigerant flow was measured by two flowmeters, mounted in series. One meter was indicating the instantaneous flow, the other was metering the liquid.

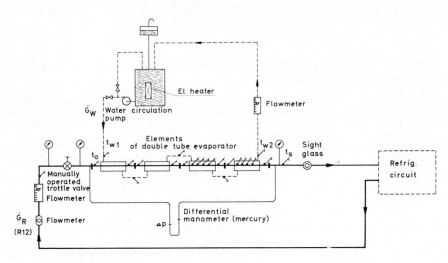

Fig. 1 — Schematic diagram of experimental apparatus.

The refrigerant flow was regulated by a manually operated throttle valve. The valve outlet was mounted directly into the inlet of the experimental double tube evaporator; this evaporator consists of four identical elements, each 1 m in length. The outlet of the evaporator was mounted onto the suction pipe, where the sight glass was also located.

For the remainder of the refrigerant circuit a conventional piston compressor and water cooled condenser was used. The pressures in the refrigerant circuit were measured with the use of a precision dial manometer and a differential mercury manometer. The location of pressure measuring points is shown in figure 1.

For the water circuit, the water elements of the experimental evaporator; flowmeter, heating element and water circulation pump—were mounted in series. The rate of water flow was established by proper bypassing of the main circuit. A schematic drawing of one of the elements for the experimental evaporator is shown in figure 2; the internal and external dimensions of the tube are also given.

The screen was made of brass wire mesh, no. 80.

The location of temperature measurement points for both circuits, is shown in figures 1 and 2. The temperatures recorded at these points were measured with the use of thermocouples, soldered onto the tube wall. Several selected temperature measuring elements were joined to the 12 point recorder, the remainder to the 48 point switch—which was connected to the two stripe recorder to make simultaneous continuous observations with greater accuracy. The accuracy for temperature measurement was 0.2 and 0.1 °C respectively.

A number of the temperatures in the water circuit were additionally controlled by mercury thermometers, with an accuracy of 0.1 °C. As shown in figures 1 and 2, two of the elements for the experimental evaporator were fitted with additional temperature measuring elements which were soldered onto the refrigerant tube on the water flow side. The distance between these elements was 100 mm. The intention here was to form a picture of the movement of the transient point, but not to measure the temperature of the tube wall with absolute accuracy. This arrangement was very successful in operation.

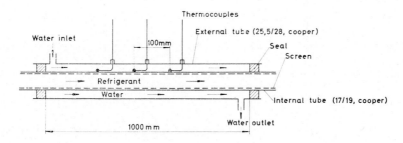

Fig. 2 — Schematic drawing of the element for experimental evaporator.

The oil concentration in the refrigerant was not estimated. It may be assumed that this was within the conventional limits for the size of refrigeration machinery employed. The temporary observations on refrigerant flow, made through the sight glass in the suction line, may support this assumption.

TEST PROCEDURE

In pursuance of the technique used by the author under previous investigations [1, 2, 3] the temperature t_S (see fig.1) was chosen as a signal. With given evaporating temperating t_0, water flow rate $\dot{G}_W$ and heat load $\dot{Q}_E$ (kept constant), the refrigerant flow rate $\dot{G}_R$ was changed by carefully stepping up the manually operated throttle valve, until the temperature t_S was balancing with MSS (minimum stable signal) [see 1, 2, 3]. Then the temperatures necessary for temperature difference calculations were measured and the value of $\dot{G}_R$ and pressure drop Δp were recorded. This procedure was repeated, but with a different heat load for the bare tube evaporator. The same programme was carried out again after the installation of the screen (see fig. 2), into all four elements of the experimental evaporator. In this way the function $\dot{Q} = f(\Delta t_m)$ for the evaporator—with screen and without (bare tube)—was estimated. Other dependencies can be derived from this basic function.

TEST RESULTS

The experimental estimated function $\dot{Q} = f(\Delta t_m)$ for the double tube evaporator, both with bare tube and tube with screen, is illustrated in figure 3. The observed increase in capacity with a given temperature difference is considerable.

The size of this capacity increase may be measured with the use of the heat intensification factor.

$$H_i = \frac{\dot{Q} \text{ (tube with screen)}}{\dot{Q} \text{ (bare tube)}} \qquad (1)$$

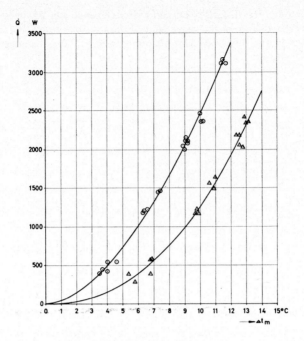

Fig. 3 — Test results. Capacity versus Δt_m
  △ bare tube
  ⊙ tube with screen.

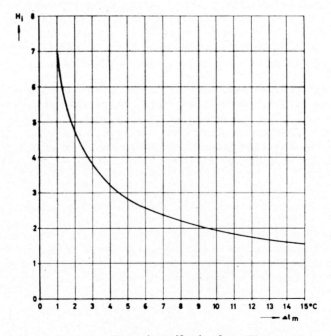

Fig. 4 — The heat exchange intensification factor H_i versus Δt_m.

This factor versus Δt_m is shown in figure 4. In the following, the percentage increase in capacity is given by equation:

$$P = (H_i - 1) \cdot 100 \qquad\qquad (2)$$

(see fig. 5)

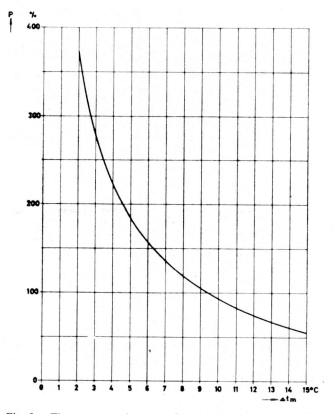

Fig. 5 — The percentage increase of the heat exchange versus Δt_m.

As is shown, the heat intensification factor H_i decreases with the increase of Δt_m, and an Δt_m figure of approx. 9.5 °C is equal to 2. This means a capacity increase o 100 per cent.

The estimated values for overall heat transfer coefficient k and heat flux $\dot q$ versus Δt_m are given in figure 6. Correspondingly, the mean logarithmic temperature difference Δt_m and overall heat transfer coefficient k versus $\dot q$ are given in figure 7. As indicated in figure 6, with $\Delta t_m \sim 8.5$ °C, the overall heat transfer coefficient k for the S-tube (tube with screen), has reached the considerably high value of 1000 W/m². °C. The measured pressure drop Δp versus $\dot q$ is shown in figure 8. It can be seen that the increase in pressure drop remains within the tolerable limit.

The above test results are estimated for parallel flow, with evaporating temperature $t_0 = +5$ °C and water velocity in the ring space of the double tube evaporator $w_V \cong 0.5$ m/s. The values of the heat flux $\dot q$ and of the overall heat transfer coefficient k, are found with the use of the internal area of the refrigerant tube (diameter 17 mm).

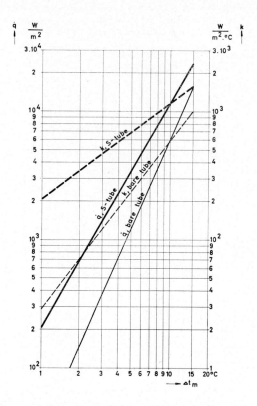

Fig. 6 — Test results. Heat flux $\dot{q}$ and overall heat transfer coefficient k versus Δt_m.

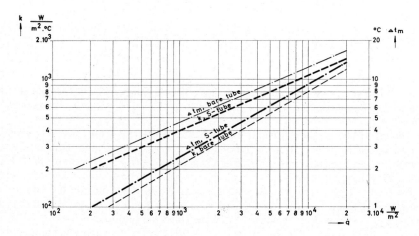

Fig. 7 — Test results. Logarithmic mean temperature difference Δt_m and overall heat transfer coefficient k versus $\dot{q}$.

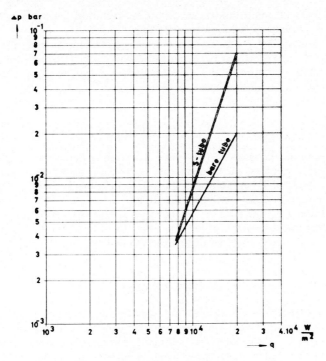

Fig. 8 — Test results. Pressure drop Δp versus $\dot{q}$.

DISCUSSION OF TEST RESULTS

It must be noted here that the primary purpose of these investigations was to provide an answer to the question: "How big is the increase in heat exchange and pressure drop for a given industrial evaporator, with the use of the previously mentioned idea on influencing the dynamics of two phase flow?"

Therefore, the estimated values for the total capacity of the evaporator and the overall heat coefficient, were of interest and are included in this paper. Of course, some local values were estimated simultaneously during the test, but they are not of general validity and are, therefore, not presented here.

The dependency $\dot{Q} = f(\Delta t_m)$ for the bare tube evaporator and the S-tube evaporator, are approximated from the measurements by means of the least square method. The mathematical form for this dependency was found to be as follows:

$$\dot{Q} = B \cdot (\Delta t_m)^n \qquad W \qquad (3)$$

The values for B and n are: $B_B = 6.27$; $n_B = 2.31$; $B_S = 43.61$; $n_S = 1.75$; (see full lines in figure 3).

Furthermore the function $k = f(\dot{q})$ was estimated as follows:

$$k = C \cdot \dot{q}^m \qquad W/m^2 \cdot {}^{\circ}C \qquad (4)$$

The values for C and m are: $C_B = 3.94$; $m_B = 0.58$; $C_S = 19.5$; $m_S = 0.44$; (see fig. 7).

All of these dependencies (see figs. 5, 6 and 7 and also figs. 4 and 5), point to the fact that the greatest increase in heat exchange is caused by lower temperature dif-

439

ferences. This is of great importance to the refrigeration industry, and the S-tube may be a very acceptable solution for the future.

The dependency $\Delta p = f(\dot{q})$, was also approximated from the measurements by use of the least square method. As is pointed out in figure 8, the values below 3.10^{-3} bar, are excluded because in this case the measurement accuracy was not better than 1 mmHg.

Conclusion

Changes in the dynamics of two phase flow during the evaporating of the tube are effected as a result of the screen (wire mesh), fitted tightly to the tube wall, as observed in the test. Here there was a favourable influence on the heat exchange due to a tolerable increase in the pressure drop.

It seems that this innovation will be of great importance to the refrigeration industry, since in many cases, it will represent the optimal solution to forthcoming problems. The S-tube evaporator idea is patent pending.

Final remarks

The tests carried out on various industrial evaporators where the S-tube was introduced, correspond well with the results presented in this paper. The S-tube evaporator is especially suitable for automatic feeding with thermostatic expansion valves. It may be noted here that the research as reported in this paper was correlated with our thermostatic expansion valve development programme.

It may be expected that the S-tube evaporator system, fed by modern expansion valves, will in a short space of time replace the systems used today. The advantages of this modern system cannot be overlooked.

Notation

$\dot{Q}$	Capacity	W
Δt_m	Logarithmic mean temperature difference	°C
H_i	Heat intensification factor	
k	Overall heat transfer coefficient	$W/m^2.°C$
$\dot{q}$	Heat flux	W/m^2
Δp	Pressure drop	bar
B	Index, bare tube	
S	Index, tube with screen	
mmHg	Millimeter of mercury rise	

REFERENCES

[1] Z. R. HUELLE, "Heat load influences on evaporator parameters" I. I. R. XIIth Int. Cong. Refrign, Madrid (1967), II, pp. 985-1000.
[2] Z. R. HUELLE, "Thermal balance of evaporators fed through thermostatic expansion valves", I. I. R.-XIIth Int. Cong. Refrign, Madrid (1967), II, pp. 1001-1010.
[3] Z. R. HUELLE, "New points of view on evaporator liquid supply control, by thermostatic expansion valves", *DANFOSS Journal*, no. 1/68; 3/68 (english edition).
[4] G. L. WEDEKIND and W. F. STOECKER, "Transient response of the mixture-vapor transition point in horizontal evaporator flow", Paper presented at the ASHRAE 73rd annual meeting in Toronto, Ont. Canada, (27-29, June 1966).

[5] J. M. Chawla, "Wärmeübergang und Druckabfall in waagerechten Rohren bei der Strömung von verdampfenden Kältemitteln", *VDI-Forschungsheft* 523 (1967).
[6] P. Worsøe-schmidt, "Some characteristics of flow pattern and heat transfer of Freon 12 evaporating, in horizontal tubes", *Ingeniøren, Int. Ed.* 3 (1959), no. 3, pp. 98/104.
[7] W. R. Zahn, "A visual study of two phase flow whilst evaporating in horizontal tubes", *ASME* paper no. 65-WA-166 (1965).
[8] S. W. Gouse jr., "Heat transfer and fluid flow inside a horizontal evaporator tube", I. I. R.-XIIth Int. Cong. Refrign, Madrid (1967), II, pp. 485-497.
[9] S. W Gouse jr. and K. G. Coumou, "Heat transfer and fluid flow inside a horizontal tube evaporator. Phase 1", *M. I. T. eng. proj. lab. report.* DSR 9649-1 (June 1964).
[10] E. V. Schlünder and J. M. Chawla, "Local heat transfer and pressure drop for refrigerants, evaporating in horizontal, internally finned tubes", I. I. R.-XIIth Int. Cong. Refrign, Madrid (1967), II, pp. 601-612.
[11] P. Grassmann, "Zweiphasenströmungen in Rohrleitungen", *Kältetechnik*, 17, (1965), no. 2, pp. 42-46.

DISCUSSION

G. de LEPELEIRE (Belgium) — 1. Was the (Δt_m) figure calculated on the basis of suction saturation temperature, or was the influence of pressure drop along the tube included?

2. I suppose that tests are carried out with complete evaporation. In this case I wonder what the result would be for an evaporator with an higher L/D ratio, where heat transfer is fixed to a great part by mass velocity, more than surface loading.

Z. R. HUELLE — 1. The pressure drop was included, but in the conditions presented it would be neglected because the pressure drop was very small.

2. Yes, the tests are carried out with complete evaporation. During the tests it was also observed, that with the usual high L/D ratio for the industrial evaporators the heat transfer was a function of the heat flux.

H. JUNGNICKEL (East Germany) — Have you made experiments on the influence of length?

Z. R. HUELLE — Not directly in the present tests. It was observed that the length normally used in industry is acceptable by using of the S-tube evaporator.

L. DELVAUX (Belgium) — What is the physical explanation of the effect of the wire screen?

Z. R. HUELLE — This is given briefly in the introduction to the paper. The narrow space between the screen and tube wall influences the liquid transport on the tube surface in the proper way.

SECTION 5

ESSAIS DES ÉQUIPEMENTS DE TRAITEMENT D'AIR

TESTING AIR TREATMENT EQUIPMENT

SOME EXPERIENCE WITH A NEW AIR CONDITIONING TEST ROOM

H.B. BOUWMAN

Bronswerk, Amersfoort (The Netherlands)

Expérience d'une nouvelle chambre d'essai de conditionnement d'air

RÉSUMÉ : *L'une des tâches les plus importantes d'une société de construction et d'installation d'appareils de conditionnement d'air est de montrer à l'utilisateur futur que le système choisi pourra assurer le niveau désiré de confort, une fois le bâtiment construit et équipé.*

Bronswerk a donc construit une chambre d'essai de conditionnement d'air transformable présentant les caractéristiques suivantes :

1. *Dimensions variables (maximum : 6 × 10 × 3,5 m);*
2. *Une ou deux parois vitrées, avec conditions climatiques extérieures réglées indépendamment;*
3. *Production d'entrées ou de pertes de chaleur réglables à travers les parois;*
4. *Technique de visualisation et de photographie des déplacements d'air, sans nécessité d'utiliser des couleurs sombres dans la chambre;*
5. *Appareils automatiques pour la mesure de la vitesse de l'air, de sa température et de l'écoulement de chaleur à des points choisis à l'avance dans la chambre;*
6. *Traitement des données permettant d'obtenir les résultats sous forme de diagrammes.*

On présente les expériences réalisées au moyen de ces appareils.

1 — THE HEAT BALANCE

For the steady state situation the heat balance of the test room was considered. The large glass windows were insulated with 50 mm rockwool covered with paper. The k value was estimated to be 1.16 W/m².°C. The temperature in the weather cases was held near the mean air temperature in the test room (ca. 42°C).

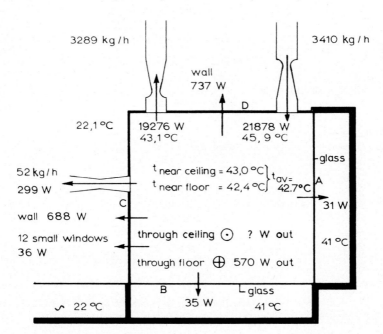

Fig. 1 — Scheme of heat balances test. Top view of test room 6 × 5.6 × 3.0 m. Glass windows insulated. Circulation rate: 30 times per hour.

Heat flow was measured on the floor, on the sandwich walls and on the observation windows in the walls. Heat loss through the ceiling was estimated to be small due to the fact that the air was exhausted through the ceiling. Heat flow of the airstreams as calculated with regard to the mean temperature in the hall at mean test room height (ca. 22 °C).

The measurements were made at about 8, 12 and 30 circulations per hour. The last situation is shown in figure 1.

It will be clear that when Δt is small, a small error in Δt will cause a considerable difference in the heat balance. Up to now the results appear useful.

Circulation: 8 times per hour.

Mass flow

In:	Out:
934 kg/h;	920/h
	Difference 1,5%

Heat flow

In: air 5,290 W; $\Delta t = 4.9$ deg C;

Out:

outlets	4,014 W
leak venturi	0
large window A	33
large window B	50
floor $k = 0.92 \dfrac{W}{m^2.°C}$	283
wall C $\}$ wall D $\}$ $k = 1.79 \dfrac{W}{m^2.°C} \rightarrow$	$\begin{cases}436\\467\end{cases}$
small windows extra $k = 3.65$ W/m^2.°C	23
ceiling	?

Difference ~ 0.4% ± ?   5,310 ± ?W

Circulation: 12 times per hour.

Mass flow

In:	Out:
1,411 kg/h	1,389 + 8 = 1,397
	Difference 1%

Heat flow

In: air 8,702 W; $\Delta t = 4.4$ deg C

Out:

air outlets	6,860 W
leak venturi	40
large window A	73
large window B	99
floor $k = 1.02$ W/m^2.°C	363
wall C $\}$ wall D $\}$ $K = 1.83$ W/m^2.°C $\rightarrow$	$\begin{cases}515\\551\end{cases}$
12 small windows $k = 3.85$ W/m^2.°C extra	27
ceiling	?

8,528 ± ?W

Difference ~ 2% ± ?

Circulation: 30 times per hour.

Mass flow

In:
3,410 kg/h

Out:
3,289 + 52 = 3,341 kg/h
Difference ~ 0,9%

Heat flow

In:
air 21,878 W; $\Delta t = 1.99$ deg C

Out:

air outlets	19,276 W
leak venturi	299
large window A	31
large window B	35
floor $k = 1.21$ W/m².°C	570
wall C $\}$ K = 2.01 W/m².°C → wall D	$\begin{cases}688\\737\end{cases}$
12 small windows $k = 4.34$ W/m².°C extra	36
ceiling	?

$$21,672 \pm \ ?\text{W}$$

Difference ~ 1% ± ?

Theses differences are smaller than expected.

2. For making airsteam pattern photographs a number of projectors (fig. 2) give parallel light beams which form together a "light wall" of about 25 cm thickness. White aldehyd flakes brought into the airstream are clearly visible in the light (fig. 3).

A rule of known length (20 cm) in the light beam gives a reference length making the study of air movement from the photograph possible. Exposure time should be known. Exposure time may be 3 to 6 seconds. To get more information about the air movement and air velocities the exposure time is interrupted periodically for a short time. In most cases four interruptions are chosen.

Flakes which have a velocity component perpendicular to the "light wall" will leave this lighted space, making part of the tracks invisible, which may suggest too low air velocity indications. With the short interruptions during exposure time the flakes make a number of marked tracks thus giving more chance that at least one full track will lie entirely in the "light wall".

By making the interruption pattern of the exposure time asymmetrical the direction of the flow may be determined if necessary. By taking a number of photographs of the same situation in the test room one gets an indication of velocity variations at certain points of the room. To check the exposure time pattern an exposure time control device is photographed at the same time (fig. 3). This device gives a number of light points on a circle, every light point corresponding with a certain interval (mostly 0.06 sec.). So the light points and its pattern give the indication of exposure time and exposure pattern on the photograph itself.

To give the camera free sight of the airstream the device giving optical time signals is photographed via a plane piece of glass making an angle of 45° with the optical axis of the camera (fig. 4). The device does not cut any part of the air stream picture and cannot disturb the air stream pattern. Nevertheless one gets large time marks on the photograph.

Of course it is possible to make these photographs with the camera either in or outside the testroom taking the shots through a glass window in the test room wall.

Fig. 2 — Stack of light beam projectors. At the right hand side heat flowmeters attached to the wall and peep window.

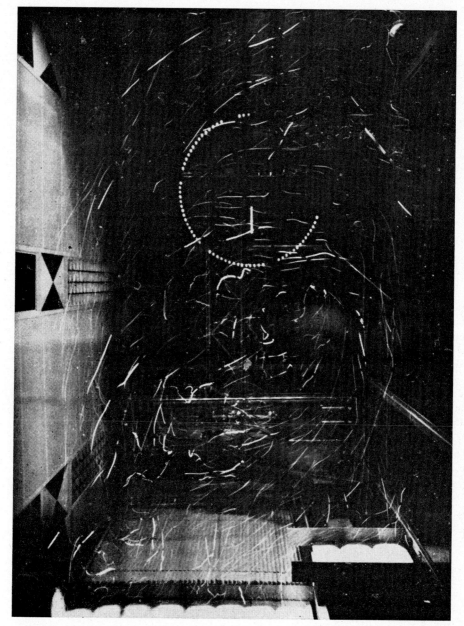

Fig. 3 — Air flow pattern photograph with horizontal rule of 20 cm and circular time marks.

449

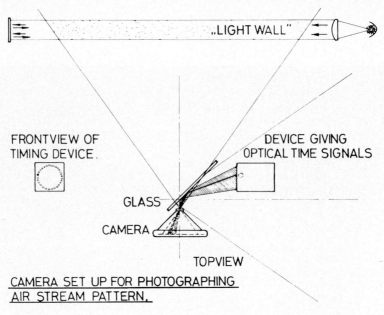

FRONTVIEW OF
TIMING DEVICE.

DEVICE GIVING
OPTICAL TIME SIGNALS

GLASS

CAMERA

TOPVIEW

CAMERA SET UP FOR PHOTOGRAPHING
AIR STREAM PATTERN.

Fig. 4 — Scheme of camera set up for air stream pattern photographs.

3. With the automatically movable measuring stand, temperatures and air velocities are measured on a large number of pre-selected points in the test room. Per hour,
air temperature and air velocity are measured on 500 points in the room. These data
are listed on a printer and also punched in a paper tape. After a test run in the test
room the punched papertape is fed into a computer which prints the measured points
and values in a section of the test room on a sheet.

An example of such a diagram for air velocities is given in figure 5. The curves are
drawn in by hand. At present trials are made to plot the lines of constant temperature
and velocity from the punched paper tape directly by the computer (off-line). One of
the first results is shown in figure 6 (p. 452). The hatched fields are below 23.5°C
or beyond 25.0°C. So the unhatched field has constant temperature within 1.5°C.

CONCLUSION

By using modern and automatised measuring apparatus and data processing in the
air conditioning test room it is possible to get easy readable results within an acceptable time.

REMARKS

The automatised measuring apparatus was developed and delivered by the Research
Institute for Public Health Engineering TNO, Indoor Climate Div., Delft. The beam
projectors were developed by the Institute of Applied Physics TNO-TH, Optical
Dept., Delft.

REFERENCES

[1] *Schip en Werf*, (25 Jul. 1969), pp. 327-330.
[2] *Holland Schipbuilding*, **18** (Jul. 1969), 5, pp. 57-59.

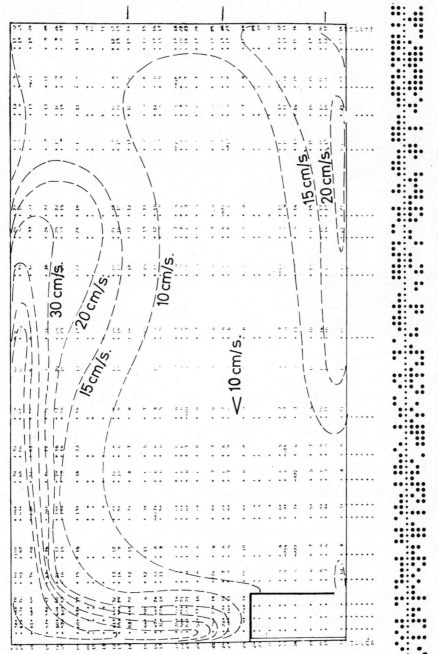

Fig. 5 — Printed diagram of air velocities in a vertical section of the test room. The curves are drawn by hand.

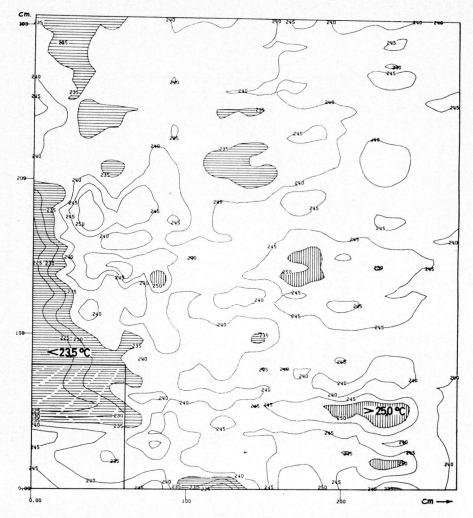

Fig. 6 — Isotherms for one half of a vertical section of the test room. Curves plotted by computer. The temperature is given in 0.1 °C, so 245 means 24.5 °C. Difference between the isotherms is one half degree centigrade.

DISCUSSION

A. JUDET DE LA COMBE (France) — La Sté Bronswerk a-t-elle déjà pu étudier l'influence d'obstacles (meubles) ou de personnes se déplaçant sur les champs de température ou de vitesse?

H. B. BOUWMAN — Certainly furniture or people in the test room will modify the air stream pattern. With furniture in the room the measuring program has to be split into parts to prevent interference between furniture and measuring facilities.

R. LANDSBERG (Israel) — What are the capacities of equipment which can be tested in the room described?

452

H.B. BOUWMANN — For instance maximum capacity for the long weather case will be concluded from the fact that 4,000 m³/h can be cooled to $-30°C$ or heated to $+$ ca 70°C.

In the heat balance experiments airflow could be maintained at a circulation factor of 30 times an hour at a temperature of 23°C over the temperature of the environment of the test room. Higher temperatures are not wanted for reasons of danger for deterioration of the construction.

V. KORSGAARD (Denmark) — Have you decided how to combine the measured values of air temperature and velocity and surface temperature to an index which can be related to the degree of thermal comfort?

H.B. BOUWMANN — This is an important remark. Of course the aim is to get comfort figures. But at this moment it is necessary to combine air temperature and air velocity figures with wall temperatures, and other data afterwards. But to have the apparatus modified in such a way that we get the comfort figures directly, we think this would be a large task.

J. LEBRUN (Belgique) — 1. Les observations sont-elles acceptables lorsque les miroirs sont disposés devant une bouche de pulsion (fig. 3)?

2. Quelle est la constante de temps des anémomètres?

H.B. BOUWMANN — 1. If the mirrors were to interfere with air flow they can be removed.

2. The time constant of response of the anemometers are the same as developed by Mr. Evan GUNST of the Institute of Public Health Engineering TNO, Delft, Inner Climate Department.

TESTS ON INDUCTION AIR CONDITIONERS

K.G. LIEM

Bronswerk-Refac, Amersfoort (The Netherlands)

Essais sur les conditionneurs d'air à induction

RÉSUMÉ : *Ce rapport présente les recherches concernant les mesures d'un appareil à induction, en ce qui concerne le taux d'induction et les problèmes de bruit.*

NOMENCLATURE:

q	= air quantity	m^3/hr
t	= temperature	°C
γ	= specific gravity	kg/m^3
c	= specific heat	kcal/kg. °C
Index p =	primary air	—
Index s =	secondary air	—
Index t =	total air	—
φ	= induction ratio	—
a	= turbulance factor	—
s	= induction length	$10^{-3}m$
b_0	= width of slot	$10^{-3}m$
l	= length of induction unit	$10^{-3}m$
n	= number of nozzles	—
v_x	= sound velocity	m/s
h	= nozzle length	m
f	= frequency	cycles ps

The induction unit in general has some characteristics which differ from other air distribution devices, e.g. duct mixing boxes, fan coil units, etc. In the following topic two important features of the induction units shall be discussed namely:

1. The induction ratio.
2. Some noise aspects.

THE INDUCTION RATIO

Due to animated discussions about air movement patterns in rooms, regarding optional comfort conditions, determination of the induction ratio will be discussed in this part of the topic. It is supposed to be known that the ratio between either total-, or secondary air quantity and primary air quantity can be called the induction ratio.

As concluded by Baturin (Lüftungsanlagen für Industriebauten) the induction ratio of a free slot-jet depends among others on the width of the slot, the turbulance factor of the air, induction length and certain constants, represented by the following relationship:

$$\varphi = \frac{q_t}{q_p} = 1.2 \sqrt{\frac{2 \cdot a \cdot s}{b_0} + 0.41}. \tag{1}$$

455

Above mentioned equation (1) is valid only for a free jet. In general, induction units have lower induction ratios as might be expected from Baturin's equation, due to the construction built-up which deviates from the theoretical synthesis. However, the characteristic of induction is the same as concluded by Baturin.

Due to the various construction-executions of induction units no theoretical calculations can be made beforehand and therefore measurements have to be made to know a certain correction factor on the theoretical value.

For induction units this correction factor turns out to be 0.82. After substitution of the constants valid for the test unit, equation 1 can be written as

$$\varphi = \frac{q_t}{q_p} = 0.74 \sqrt{\frac{l}{n}}. \tag{2}$$

In figure 1, the Baturin line is represented by the dotted line calculated for the Inductair. The drawn line represents the average measured induction ratio for the test unit.

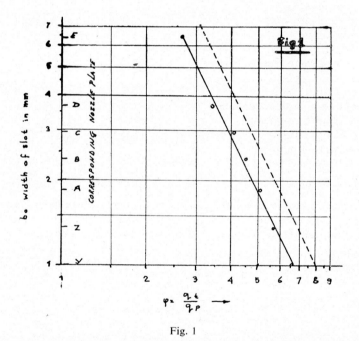

Fig. 1

It can be concluded that for a complicated aerodynamic system, as represented by induction units, the theoretical approach to the determination of induction ratios is being confirmed by the measurement values.

We now will continue with discussing the way in which the induction ratios can be measured.

Generally two possibilities are available: one based on air velocity measurements the other based on temperature measurements for heat balance calculation.

A. Velocity measurement

The most simple air measurement is to determine the primary air quantity with a (calibrated) orifice. A certain pressure loss along the orifice indicates a certain air quantity.

a) *Total air quantity*

Measuring the velocity is difficult due to the effects of primary air, because of the influence of the streams ejected by the nozzles.

This can be compensated by placing an enlarger at the outlet opening of the unit and measuring the air velocity at the opening of the enlarger. The influence of the enlarger on the induction ratio has to be known.

With the determination of the ultimate induction ratio one has to take into account the effect of the enlarger.

b) *Secondary air quantity*

Velocity measurements can be made before the heat-exchangers in a similar manner as mentioned above.

Above mentioned velocity measurements can be done by the normal instruments like the thermal anemometer or vane type anemometers.

B. Heat-balance

The second possibility to determine the induction ratio is by means of temperature measurements required for a heat-balance.

The equation of the heat-balance can be written as:

$$q_p \times \gamma_p \times c_p \times t_p + q_s \times \gamma_s \times c_s \times t_s = q_t \times \gamma_t \times c_t \times t_t. \tag{3}$$

When analysing this equation (3) of the heat-balance it turns out that for our purpose the equation can be reduced to the following form:

$$\frac{q_t}{q_p} = \frac{(t_p - t_s) \cdot \gamma_p}{(t_t - t_s) \cdot \gamma_t} \tag{4}$$

under the assumption that the specific heat is constant.

C. Comparison between the measuring systems

The accuracy of the results depends on the precision of the instruments used. To have a comparison between the two above mentioned measuring systems, the following accuracy calculation can be drawn up.

Velocity measurement

Assume that: $q_p = 100 \text{ m}^3/\text{hr}$ precision $\pm 5\% \rightarrow \pm 5 \text{ m}^3/\text{hr}$
$q_s = 200 \text{ m}^3/\text{hr}$ precision $\pm 10\% \rightarrow \pm 20 \text{ m}^3/\text{hr}$
$q_t = 300 \text{ m}^3/\text{hr}$ precision $\pm 10\% \rightarrow \pm 30 \text{ m}^3/\text{hr}$

The induction ratio is

$$\varphi = \frac{q_t}{q_p} = \frac{300}{100} = 3 \,.$$

a) *Total air quantity measurement*

The largest deviation becomes:

$$\varphi = \frac{q_t}{q_p} = \frac{300+30}{100-5} = \frac{330}{95} = 3.47 \,.$$

With regard to the real value the deviation is 15.7%

b) *Secondary air quantity measurement*

The largest deviation becomes:

$$\varphi = \frac{q_t}{q_p} = \frac{q_s+q_p}{q_p} = \frac{(200+20)+(100+5)}{100-5} = \frac{325}{95} = 3.42 \,.$$

With regard to the real value the deviation is 14%.

Heat-balance calculation

Assume that: $t_p = 30\,°C$ precision $\pm\, 0.1\,°C$

$t_s = 20\,°C$ precision $\pm\, 0.1\,°C$

$t_t = 23\,°C$ precision $\pm\, 0.1\,°C$

$\gamma_p = 1.16 \text{ kg/m}^3$ precision $\pm\, 0.01 \text{ kg/m}^3$

$\gamma_t = 1.19 \text{ kg/m}^3$ precision $\pm\, 0.01 \text{ kg/m}^3$

By means of equation (4) the induction ratio is:

$$\varphi = \frac{(30-20)\,1.16}{(23-20)\,1.19} = 3.25 \,.$$

The largest deviation is

$$\frac{(30.1-19.9)\,1.17}{(22.9-20.1)\,1.18} = 3.61 \,.$$

And in percentages it becomes 11.1%

Evidently the heat-balance determination is preferable due its accuracy.

It is stressed that the induction ratio as calculated for a free standing induction unit deviates considerably in practice as a result of:

a) Window sill construction;
b) Heat-exchanger in operation;
c) Outlet-inlet grilles;
d) Primary air condition.

458

The following example (fig. 2) illustrates the influence on the induction ratio by the application of outlet grilles.

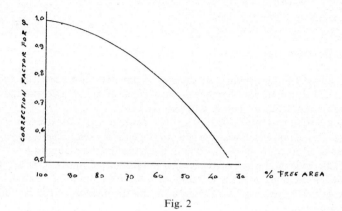

Fig. 2

As the air movement patterns in the air conditioned rooms are determined to a a great content by the air velocities at the outlet opening of the induction units and therefore by the induction ratio, for this reason more data must be gathered on the above mentioned effects on induction ratios. It is hoped, that this topic might be a contribution to the start of a further development in this field.

NOISE ASPECTS

All noise measurements for the test unit were done in a slightly echoing room by means of a sound level analyser with the microphone at a distance of 2 m and a height of 0.75 m from the unit, the sound pressure level is recorded in octave band measurements. The highest average level determines the ISO NR. value.
The test stand in which the sound measurements are performed consists of 4 rooms as shown in figure 3.

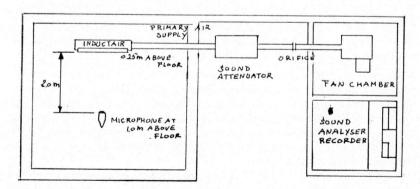

Fig. 3

Noise disturbance level of the induction unit

Two main sources contribute to the own sound production of the unit.

1. The noise produced by the primary air damper.
2. The noise produced by the nozzle.

Primary air damper noise

Sound caused by the primary air damper is normally attenuated in the primary air-chamber. The primary air-chamber is constructed in such a way that the air has to flow through lined "canals" before entering the nozzles.

The induction unit is provided with a single blade primary air damper. Depending on the place of the unit whether at the start or the end of a duct, the damper will have to be adjusted in order to allow a properly predetermined air quantity.

A noise increase might be expected when the primary damper is moved towards a closed position assuming that the primary air quantity remains constant.

Obviously this noise increase is caused by the turbulence of the air immediately after the damper.

In this aspect it is of great importance to analyse the noise increase as a function of the Δp which has to be throttled. When comparing a multitude of measurements with throttling dampers following general conclusions can be made (fig. 4).

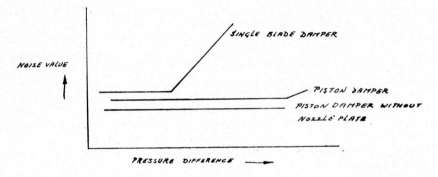

Fig. 4

1. The noise level of the unit itself is always given for a fully opened damper. Depending on the unit size and nozzle configuration this noise level remains practically constant over a range of 20-40 mm WG Δp.

2. When more throttling is required the noise level increases considerable, the angle of increase depending on unit size and nozzle configuration.

   Measurements were also executed with another primary air damper construction consisting of a perforated cylinder and piston (fig. 4).

   It turned out that the same conclusions were valid in general, however, these piston dampers have a lower noise level and also the range during which the noise levels are not increased is greater.

   It might be of interest to state that measurements on units executed with piston dampers and without nozzles plates resulted in a noise level reduction of appr. 3 ISO NR points.

Noise produced by the nozzles

Until now no general formula can be developed by means of which one can predict in advance the resulting room noise level with a certain nozzle arrangement.

Henceforth all data must be obtained in an empirical way. The survey hereafter may illustrate the development on this field:

Primary air quantity m³/hr	120	150	190	230	
a) Sheet metal perforated in one row, height 5 mm	49	56	58	61	
b) Oval nozzle length 30 mm	34	39,5	45	49	noise level in
c) Nozzle length 50 mm	27	34.5	40	45	ISO NR values
d) Nozzle length 70 mm	26	31	37.5	42.5	
e) Standard nozzle 120 mm	24	28	35	41	

The induction ratio is the same for all tests.

It is worthwhile to draw attention to the considerable reduction in noise production when going from execution *a*, *b* to *c*, particularly because this illustrates the necessity of using nozzles with a greater length. Analysis of the noise measurement data obtained with nozzle forms as indicated in *c*, *d* and *e* produced a relationship between the nozzle length and critical frequency band which determines the overall ISO NR value of the unit.

It turned out that this critical frequency band correlates with the resonance frequency of the nozzle.

Figure 5 shows the theoretical resonance frequency line as a function of nozzle length in accordance with the formula:

$$f = \frac{v_x}{2 \cdot h}$$

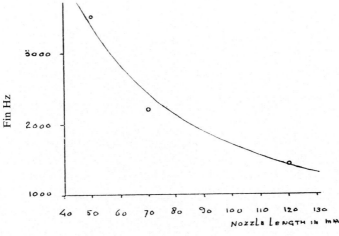

Fig. 5

461

In this figure also is indicated the critical frequency band for the three nozzle lengths found from noise measurements.

Figure 6 gives an arbitrary example of a sound pressure level measurement for an induction unit (INDUCTAIR) with 120 mm nozzle. Clearly can be seen that the critical frequency band corresponds with the theoretical noise frequency belonging to the 120 mm nozzle.

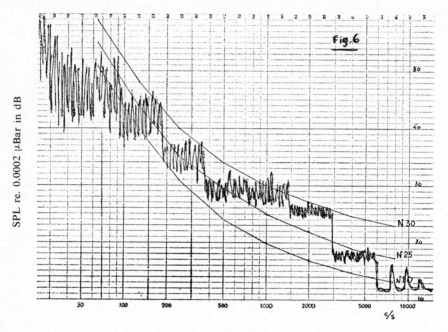

Fig. 6

LA TRANSMISSION DE LA CHALEUR DANS LES VENTILO-CONVECTEURS POUR LE CONDITIONNEMENT D'AIR

A. RASI et E. BETTANINI

Istituto di Fisica Tecnica dell'Università
Laboratorio per la Tecnica del Freddo del C.N.R., Padova (Italie)

Heat transfer in fan-coils for air conditioning

SUMMARY: *In the closed thermostatic chamber of the Technical Physics Institute of Padua University some heat transfer tests on fan-coil apparatus have been carried out. Experimental data are obtained both for heating and cooling conditions.*
The data obtained are reported and a simplified heat transfer correlation useful for most fan-coil applications is given.

Dans le rapport que nous présentons, nous avons employé la chambre thermostatique de l'Institut de Physique Technique de l'Université de Padoue, qui a été construite expressément pour les essais sur les éléments chauffants de différents modèles, alimentés par eau chaude, et dans laquelle on réalise la méthode d'essai bien connue de la «chambre fermée». Nous nous sommes proposés de voir le comportement de la chambre pour des essais sur des appareils ventilo-convecteurs avec fonctionnement intégral, et de fixer les caractéristiques d'échange thermique des appareils.

Ces appareils sont essentiellement constitués par une batterie à ailettes et par un ventilateur à vitesse variable; les batteries sont alimentées par eau chaude ou froide selon qu'on veut faires les essais dans l'hypothèse de la saison d'hiver ou d'été.

Il va sans dire qu'il y a une cuvette de récolte de l'eau condensée, et que les appareils sont munis d'une vanne thermostatique.

Les résultats que nous avons obtenus confirment l'emploi de la «chambre fermée» pour des essais qui ne sont pas de routine. Cela avait déjà été observé [1], mais le but principal était de trouver une valeur simplifiée du coefficient d'échange global et une relation entre le coefficient extérieur et le débit de l'air, en ce qui concerne l'échange de la chaleur sensible, et de l'ensemble chaleur sensible et latente.

La «chambre fermée» $(3,8 \times 3,70 \times 3 \text{ m})$ est entourée d'une autre chambre, thermiquement isolée, afin de former une couche entre les deux chambres, d'une épaisseur d'environ 60 cm et où l'air, chaud ou froid, peut circuler sous l'action du ventilateur.

La circulation de l'air peut être limitée à quelques couches d'air.

La quantité totale de chaleur qui est en jeu dans l'appareil est donnée par la relation :

$$q_T = c_A \cdot G_A \ (t_e - t_u) \text{ (kcal/h)} \tag{1}$$

avec t_e et t_u les températures d'entrée et sortie de l'eau dans la batterie; c_A la chaleur spécifique de l'eau (1 kcal/kg.°C); G_A le débit de l'eau en kg/h.

La figure 1 montre le schéma de la chambre, avec les circuits d'alimentation de l'eau chaude et froide, et les appareils de contrôle des températures dans les différentes sections; tout ceci fait comprendre le fonctionnement de la chambre quand on veut l'utiliser avec eau chaude ou froide.

La chaleur latente est donnée par la relation :

$$q_L = G_c \cdot r \ \text{(kcal/kg)} \tag{2}$$

où G_c est la quantité d'eau condensée par heure sur la batterie et récoltée dans la cuvette de l'appareil, et $r = 600$ kcal/kg, la chaleur de vaporisation.

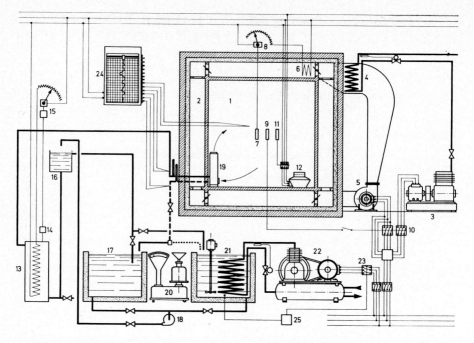

Fig. 1

1 Chambre thermostatique.
2 Couche d'air.
3 Groupe frigorifique.
4 Évaporateur.
5 Électroventilateur.
6 Chauffage électrique de l'air.
7 Thermostat d'ambiance.
8 Régulateur proportionnel.
9 Thermostat d'ambiance.
10 Interrupteur.
11 Humidostat d'ambiance.
12 Humidificateur.
13 Chaudière électrique.

14 Thermostat modulateur.
15 Régulateur proportionnel.
16 Vase d'expansion.
17 Bassin pour l'eau chaude.
18 Pompe de circulation.
19 Ventilo-convecteur.
20 Balance.
21 Bassin pour l'eau froide.
22 Groupe frigorifique.
23 Télérupteur contrôlé par le thermostat du bassin froid.
24 Enregistreur de températures.
25 Thermostat.

La chaleur sensible est :

$$q_S = q_T - q_L \quad (kcal/h) \qquad (3)$$

Dans les essais de chauffage les conditions observées étaient les suivantes :
— température ambiante : $+20°C = t_a$;
— température d'entrée eau chaude : 90, 70, 50°C $= t_e$;
— débit de l'eau = constante et égale à celle nécessaire pour obtenir $\Delta t_A = 20°C$ entre 90 et 70°C;
— vitesse de l'air dans l'appareil = minimum et maximum.

Dans les essais de réfrigération, les conditions observées étaient les suivantes :
— température ambiante = $+26°C$;
— humidité relative ambiante = 50%;
— température d'entrée de l'eau = $+5, +7,5°C$;
— débit d'eau égal à celui nécessaire pour avoir $\Delta t_A = 5$ deg C et $\Delta t_A = 10$ deg C;
— vitesse de l'air = minimum et maximum.

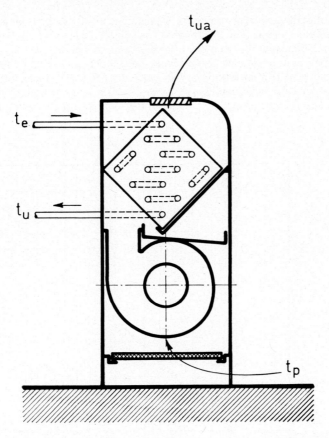

Fig. 2

Pendant les essais nous avons disposé dans le conduit d'air à la sortie de l'appareil, 3 thermocouples, protégés des radiations, pour la mesure de la température de l'air (t_{ua}). La valeur de cette température, employée dans les calculs, est la moyenne entre les trois valeurs des thermocouples.

Le débit de l'air G_a est donné par la relation :

$$G_a = \frac{q_S}{0,24\,(t_p - t_{ua})} \quad \text{(kg/h)} \tag{4}$$

où : t_p = température de l'air à l'entrée de l'appareil.

Table 1

Nombre d'ailettes	.205
Pas des ailettes	.2 mm
Dimensions des ailettes	.160 × 160 mm
Nombre de tubes	.16
Diamètre extérieur des tubes (en cuivre)	.(d_e) 16 mm
Diamètre intérieur des tubes	.(d_i) 14 mm
Surface intérieure des tubes	.(S_i) 0,36 m²
Surface extérieure des tubes	.(S_t) 0,345 m²
Surface ailettes des tubes	.(S_a) 9,274 m²
Surface extérieure totale	.(S) 9,62 m²

465

Le flux thermique q_S dans les essais de chauffage est le flux total en jeu.

Les résultats obtenus sur un appareil (fig. 2) dont les caractéristiques sont les suivantes (table 1), sont représentés sur les diagrammes 3 et 4.

La figure 3 concerne les essais en régime de chauffage, et la figure 4 concerne la réfrigération et la déshumidification. Les essais de la figure 3 ont été obtenus avec un débit d'eau constant (G_A), tandis que les essais de la figure 4 correspondent à des débits d'eau pour avoir un Δt_A de 5 ou de 10 deg pour chaque température d'entrée, t_e.

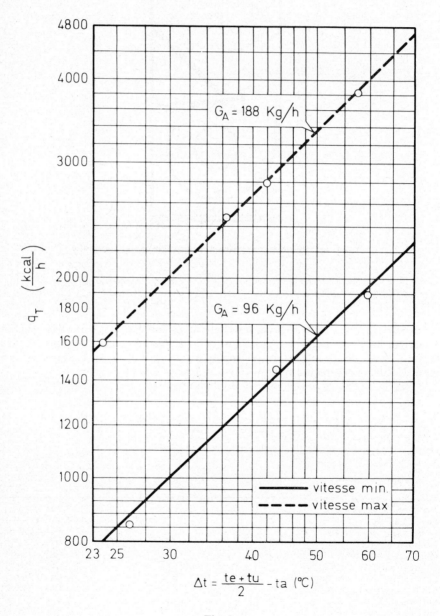

Fig. 3

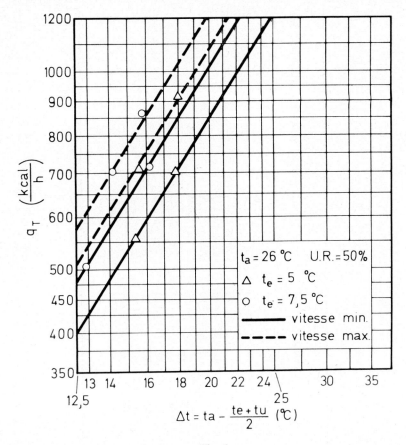

Fig. 4

À ce point nous avons vérifié, pour les essais complets, contenus dans la table 2, s'il était possible de donner une expression du flux thermique total échangé, avec un coefficient de transmission global H, lié à la chute de température moyenne pour les deux fluides, Δt_m, dans le cas de deux fluides en direction croisée, sans mélange entre eux [2]. On a donc utilisé l'expression suivante :

$$q_T = H.S. \, \Delta t_m \quad (kcal/h) \tag{5}$$

où H est donné par la relation [3] :

$$H = \frac{1}{(S/S_i)/\alpha_i + 1/\alpha} \quad (kcal/m^2.h.°C) \tag{6}$$

dans laquelle : α_i est le coefficient de convection intérieur dans les tubes et α le coefficient extérieur qui est fonction du coefficient air-ailette α_e, et du rendement de l'ailette η_a. L'expression de α est la suivante :

$$\alpha = \alpha_e (\eta_a \, S_a/S + S_t/S). \quad (kcal/m^2.h.°C) \tag{7}$$

Dans chaque essai on a cherché à mettre en relation les valeurs de α avec le débit d'air G_a, pour vérifier s'il était possible d'obtenir une expression simple unique, en tenant compte qu'il dépend de α_e et de η_a.

467

Table 2

N°	G_A (kg/h)	t_e (°C)	Δt_A (°C)	t_a (°C)	q_T (kcal/h)	q_L (kcal/h)	q_S (kcal/h)	Δt_a (°C)	G_a (kg/h)	Δt_m (°C)	H (kcal/m².h.°C)	α (kcal/m².h.°C)
3	95	70,60	15,35	19,60	1461	—	—	48,45	126	15,60	9,75	12,3
4	96	89,75	19,75	20,25	1892	—	—	60,75	130	23,40	8,38	9,94
6	96	50,40	8,90	19,90	851	—	—	28,25	124	9,30	9,55	12,6
7	188	88,20	20,45	20,50	3840	—	—	55,75	287	27,10	14,75	17,6
8	189	69,80	14,80	20,50	2797	—	—	41,75	279	18,90	15,45	19,3
10	188	49,50	8,50	21,40	1594	—	—	23,90	278	11,10	14,95	19,2
11	188	71,40	13,15	28,25	2470	—	—	37,60	274	16,10	16,00	19,8
17	137	5,40	5,10	25,75	701	226	435	13,95	130	7,45	9,78	13,5
18	180	5,50	5,10	26,00	919	219	700	12,25	240	9,25	10,35	13,6
19	74	5,75	9,65	26,25	712	72	640	10,50	254	8,62	8,60	14,1
20	62	6,50	8,90	26,25	552	157	395	11,75	157	7,50	7,65	12,5
21	52	7,60	9,65	26,50	505	160	345	11,90	122	6,60	7,98	14,5
22	70	7,90	10,00	27,75	702	10	692	10,65	270	7,92	9,20	15,5
23	177	7,50	4,90	25,75	869	189	680	10,70	265	8,50	10,70	14,1
24	134	7,25	5,35	26,00	717	267	450	12,40	152	6,70	11,10	14,0

Dans ce but, on a déterminé Δt_m en utilisant les températures d'entrée et de sortie d'air et d'eau dans la batterie; et la surface totale S étant connue, on a déterminé la valeur H avec la formule (5).

En employant la formule (6), nous avons déterminé le coefficient α, en adoptant la relation :

$$Nu = \alpha_i d_i/\lambda = 0,023 \cdot Re^{0,8} \cdot Pr^{0,33} \qquad (8)$$

avec laquelle nous avons déterminé α_i.

Le diagramme de la figure 5 a été construit avec les valeurs de α dans les cas du chauffage et de la réfrigération, en fonction du débit d'air G_a. Dans ce diagramme la ligne continue concerne le chauffage, et la ligne pointillée la réfrigération et la déshumidification.

Les pentes sont très différentes, comme nous le voyons dans les expressions suivantes :

$$\alpha = 1,10\ G_a^{0,5}\ (\text{kcal/m}^2.\text{h.}°\text{C}) \text{ pour le chauffage} \qquad (9)$$

$$\alpha = 8,50\ G_a^{0,1}\ (\text{kcal/m}^2.\text{h.}°\text{C}) \text{ pour la réfrigération} \qquad (10)$$

Cette conclusion peut être expliquée par la relation (7) et en rappelant les résultats obtenus, relativement à la valeur de α_e et du rendement η_a de l'ailette dans des travaux précédents, qui concernent l'échange thermique des surfaces avec ailettes où il y a échange de chaleur et de masse [4, 5, 6]. De ces travaux, il résulte que le rendement de l'ailette (avec la même valeur de α_e) descend notablement en présence d'échange de chaleur et de masse qui ont lieu dans le même temps. En outre on indique que la valeur de α_e augmente avec la même vitesse que l'air, s'il y a condensation sur la surface. On peut donc dire que la diminution de η_a n'est pas complètement balancée par l'accroissement de α_e, pour les débits d'air élevés.

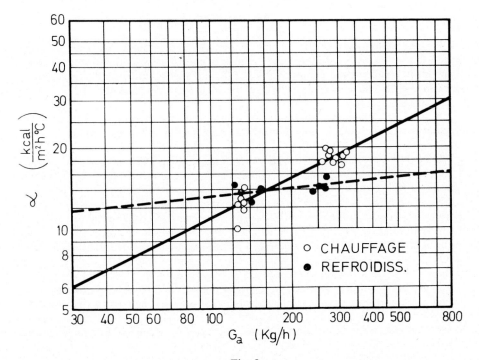

Fig. 5

Les résultats que nous avons obtenus sont, evidemment, valables pour l'appareil que nous avons éprouvé, tandis que la valeur des constantes qui sont comprises dans les relations (9) et (10) reste indéterminée pour des appareils de géométrie différente. Cependant, on peut toujours représenter la possibilité de l'échange thermique des batteries avec ailettes de ce modèle, avec l'expression du coefficient H, relativement à la chute de température pour fluides en directions croisées, comme nous l'avons montré.

Le travail de recherche peut être maintenant orienté vers la détermination des coefficients de convection qui permettent d'employer l'expression de α, à laquelle nous sommes parvenus, par tous les modèles d'appareils installés, quelles que soient leurs caractéristiques géométriques, constructives et de fonctionnement.

RÉFÉRENCES

[1] A. RASI. Sulla Misura della Trasmissione Termica dei Condizionatori Fan-Coil. *Atti del 6° Convegno Nazionale del Riscaldamento e della Ventilazione*, Padova (Juin 1966), pp. 99-111.
[2] R. PLANK. *Handbuch der Kältetechnik*, III Band, Springer-Verlag, Berlin (1959), pp. 208-209.
[3] G. BROWN. Theory of Moist Air Heat Exchangers. *Transactions of the Royal Institute of Technology*, Stockholm, no. 77 (1954).
[4] G. TRAPANESE. Calcolo del Rendimento Termico delle Superficie Alettate durante il Raffreddamento e simultanea Deumidificazione dell'Aria. *La Termotecnica*, 12, no. 6 (Juin 1968), pp. 242-250
[5] N. G. O'BRIEN et R. L. TURNER. Fin Thermal Efficiency during Simultaneous Heat and Mass Transfer. *A. I. Ch. E. Journal* (Mai 1965), pp. 546-548.
[6] G. TRAPANESE, P. DI FILIPPO et E. BETTANINI. Influence de la géométrie et des conditions de fonctionnement sur le comportement thermique des serpentins à ailettes. *I.I.F. — 12e Cong. Int. Froid*, Madrid (1967), II, pp. 525-552.

DISCUSSION

R. LANDSBERG (Israel) — I ask for details of the tubing: were there sharp-edged fins on expanded tubing, or sleeved fins covering part of the tube surface? Experiments made at the Israel Institute of Technology point to a noticeable difference in the fin efficiencies according to the manner of fin assembly.

A. RASI — The fan-coil apparatus that we have used in our test was a commercial apparatus and the coil, I think, was made with fins on expanded tubing.

J. HANNAY (Belgique) — Pour déterminer le débit d'air « G_a » vous placez 3 couples à la bouche de pulsion de l'appareil et vous déterminez « G_a » en comparant aux mesures de la puissance sur le débit d'eau.

Nous pensons qu'il serait préférable de mesurer également les vitesses et de pouvoir de la sorte obtenir le débit « G_a » par seules mesures sur le circuit d'air de l'appareil. La comparaison avec les mesures plus précises effectuées sur le circuit d'eau fournit alors un contrôle qui permet de juger de la validité de la mesure du débit d'air.

A. RASI — During our tests we have measured also the air flow rate and the outlet air temperature in many other points in the air flow. The results of these comparative tests were in good agreement with the results given with the three thermocouples as said in the report and with the balance of equation (4). For these reasons we take in our calculations the results thus obtained.

The approximation in the results was about 5 % in air flow rate.

J. LEBRUN (Belgique) — La méthode utilisée pour l'estimation du débit (p. 465 formule (4)) me paraît être la plus imprécise : il faudrait au moins effectuer une pondération du t° en fonction des vitesses, et encore ne peut-on espérer dépasser de cette façon une précision de l'orde de 10 %. M. HANNAY connait fort bien cette question. Pour obtenir une précision vraiment satisfaisante, il faudrait effectuer un *second essai* avec mesure du débit à l'aspiration au moyen d'un diaphragme et équilibrage de la pression; la conservation de *l'émision* est assurée si le débit l'est également (on est sûr ainsi de l'identité des conditions à l'aspiration dans les deux essais).

Si l'on ne désire pas effectuer 2 essais, il reste encore la possibilité d'utiliser un débitmètre thermique à l'aspiration.

A. RASI — During our tests we made both air flow measurements and other measures of temperature (in 6 other points) in the outlet air flow. The results obtained were in good approximation (5-6 %) with the data deduced with the means described in our report. With regard to more accurate results, I am sure that the advice given is exact.

LES PERTES DE CHALEUR DES ÉJECTO-CONVECTEURS DANS LES INSTALLATIONS À QUATRE TUYAUX AVEC RÉGULATION CÔTÉ AIR

G. SAGGESE

Politecnico di Torino, Istituto di Fisica Tecnica
Turin (Italie)

Heat losses from jet-convectors in 4-pipe installations having air-side regulation

SUMMARY: *Calorimeter tests on a jet-convector in a 4-pipe installation having air-side regulation have shown the energy losses —even during summer—from hot water batteries. These losses appear to be caused by convective air movement rather than air through the closed damper.*

CONSIDÉRATIONS GÉNÉRALES

Les installations de conditionnement d'air à quatre tuyaux ont permis de résoudre très efficacement le problème du réchauffement et du refroidissement simultanés des pièces d'un même édifice, exposées d'une manière différente ou bien des pièces ayant la même exposition mais ombragées à des heures différentes.

Chaque conditionneur local dispose toujours d'un fluide chaud et froid, ce qui rend inutile la commutation selon la saison et permet d'envoyer l'air de rechange à la température du point de rosée correspondant aux conditions intérieures qui doivent être maintenues. En ce cas l'air de rechange n'a aucune fonction thermique : il n'a comme but que de maintenir la pureté de l'air et son degré hygrométrique. Ce système n'a pas encore eu une diffusion rapide à cause de la difficulté de réaliser une vanne unique pour la régulation des deux circuits hydrauliques de chaque conditionneur local. On connaît déjà les solutions qui utilisent deux vannes agissant successivement (fig. 1/*a*) et deux batteries différentes pour les deux circuits chaud et froid et celles qui emploient deux vannes en phase et une batterie unique (fig. 1/*b*).

Par contre on vient de réaliser, avec un corps unique de vannes permettant une régulation de type proportionnel, un débit d'eau toujours constant aux pompes de circulation, soit avec une seule batterie (fig. 2/*a*), soit avec deux (fig. 2/*b*).

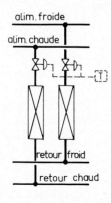

Fig. 1/*a*.

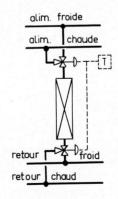

Fig. 1/*b*.

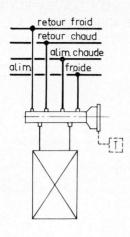
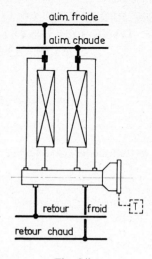

Fig. 2/a.   Fig. 2/b.

La difficulté de réaliser une régulation efficace et le prix de revient considérable ont poussé les techniciens à chercher d'autres solutions. On a donc construit des appareils où les variations de puissance thermique sont déterminées par le débit d'air aux batteries au lieu du débit d'eau.

La figure 3 donne une représentation schématique de cette solution : dans ce cas un seul servo-moteur, en général de type pneumatique, est à même d'effectuer la régulation complète en mettant en action des registres. Cet exposé a pour but de montrer les pertes d'énergie thermique subies par l'appareil en été, pertes qui viennent d'être relevées au cours d'épreuves faites sur un éjecto-convecteur pour une installation à 4 tuyaux à régulation sur le débit d'air.

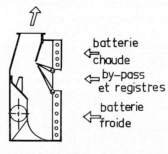

batterie
chaude

by—pass
et registres

batterie
froide

Fig. 3.

APPAREILLAGE D'ESSAIS

La figure 4 montre le schéma fonctionnel des circuits hydrauliques du calorimètre employé. L'eau, chaude et froide, est envoyée à deux vases séparés à niveau constant au moyen de déversoirs : les deux vases assurent un débit constant dans l'échangeur de l'appareil. La température de l'eau chaude (30-45 °C) et froide (8-12 °C) qui alimente les échangeurs de l'éjecto-convecteur est conservée à la valeur désirée au moyen d'un

474

réchauffeur final avec des résistances électriques dans chaque circuit. Un régulateur de type proportionnel, qui agit sur un circuit de puissance à diodes contrôlés, règle la tension des résistances électriques.

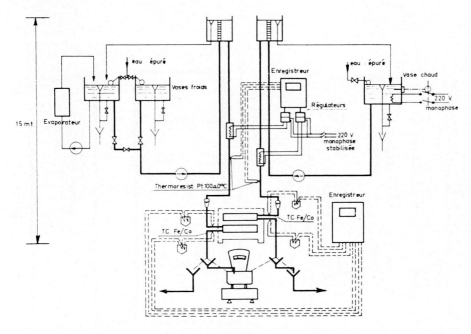

Fig. 4.

On mesure le débit d'eau chaude et froide par un fluxmètre et on le contrôle par une pesée.

Le calorimètre employé est construit selon les prescriptions de l'ISO dans sa Recommandation nᵒ 1150 : « Épreuves et déterminations des caractéristiques des conditionneurs d'air assemblés à l'usine » TC 86/5C 6. Il a été aussi complété par des dispositifs nécessaires pour l'essai d'appareils avec des échangeurs à eau

Les plans et les sections du calorimètre avec ses deux chambres de simulation sont représentés à la figure 5 : pendant les essais décrits dans cet exposé on n'a utilisé que le simulateur interne.

Les essais ont été faits sur un prototype d'éjecto-convecteur à double batterie du type représenté à la figure 3. Dans la chambre d'essais, on obtient des conditions d'ambiance constantes en introduisant l'énergie thermique nécessaire pour conserver la même température et pour faire évaporer l'eau destinée à maintenir une humidité relative invariée. Dans ce cas, on a maintenu une température de 26 °C et une humidité relative d'environ 45%. On obtient l'énergie thermique au moyen de résistances électriques : il est donc très facile, disposant d'un compteur électrique de précision, d'évaluer l'énergie introduite au cours de la période d'essai fixée.

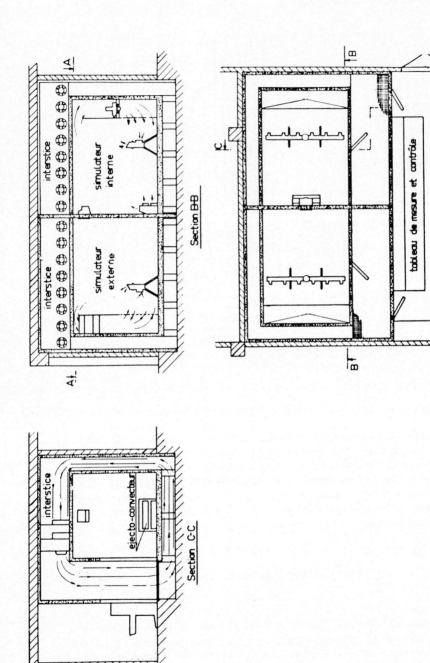

interstice

Section B-B

simulateur interne

simulateur externe

interstice

interstice

A

A

B

C

C

tableau de mesure et contrôle

Section A-A

B

interstice

ejecto-convecteur

Section C-C

Fig. 5.

476

On peut écrire ainsi l'équation qui exprime les conditions d'équilibre :

$$E_a = E_{ar} + q_l \, (i_e - i_u)$$

où :

E_a énergie thermique que l'éjecto-convecteur emprunte à l'ambiance ;

E_{ar} énergie thermique introduite dans l'ambiance au moyen des réchauffeurs électriques de l'air et de l'eau évaporée pour l'humidification et au moyen d'autres sources d'énergie éventuelles ;

q_l quantité d'eau introduite pour compenser la déshumidification causée par l'appareil ;

i_e enthalpie de l'eau pénétrant dans l'humidificateur ;

i_u enthalpie de l'eau sortant de la cuve qui recueille l'eau condensée par l'appareil.

Il faut encore considérer qu'il n'y a pas de flux thermique à travers les parois, car on maintient les interstices et le simulateur externe aux mêmes conditions que la chambre d'épreuve.

Dans le cas tout particulier des essais sur l'éjecto-convecteur, nous comprenons aussi dans le terme E_{ar} l'énergie nécessaire pour le moteur du ventilateur : celui-ci est placé dans la chambre d'essais et sert à alimenter les buses de l'éjecto-convecteur. On prend l'air primaire dans l'ambiance même, le ventilateur le comprime et le remet en circulation à travers les buses.

L'énergie électrique que l'on emploie pour alimenter le moteur, se retrouve dans l'ambiance sous forme d'énergie thermique soit pour le réchauffage du moteur soit pour le réchauffage causé par la compression de l'air qui est pratiquement un gaz parfait.

Dans des conditions d'équilibre on mesure aussi l'énergie thermique en la déduisant du produit :

$$q_a \Delta t C_p = E_a$$

où

E_a voir plus haut ;

q_a quantité d'eau traversant la batterie, dans le temps où l'on mesure E_a, déterminée par une pesée ;

Δt différence de température entre entrée et sortie de l'échangeur de l'appareil ;

C_p chaleur spécifique de l'eau à pression constante et à la température de l'épreuve.

Les différences entre les mesures d'énergie côté air et côté eau n'ont jamais dépassé 4 %. La température de l'eau froide, pendant les épreuves, était de 12 °C et la température de l'eau chaude de 42 °C.

On a conduit de la façon suivante les deux séries d'épreuves visant à déterminer les pertes d'énergie thermique, dues au passages d'eau chaude, même pendant le fonctionnement en été :

1° L'échangeur froid est parcouru par de l'eau froide et l'échangeur chaud n'est pas traversé par de l'eau chaude ;

2° L'échangeur froid est parcouru par de l'eau froide et l'échangeur chaud par de l'eau chaude.

Au cours des deux épreuves le servomoteur commandant les registres maintenait complètement ouvert le registre sur l'échangeur froid, et complètement fermé le registre sur l'échangeur chaud.

On peut voir les résultats des essais sur le diagramme de la figure 6 : il représente la puissance thermique de l'appareil en watts en fonction du débit d'eau chaude pour deux valeurs de la pression aux buses.

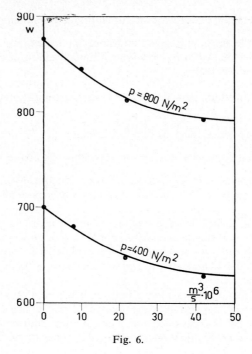

Fig. 6.

On a relevé les données quand on avait atteint les conditions de régime depuis une heure au moins, et on a continué de les relever pendant une heure au moins pour chaque essai.

CONCLUSIONS

La puissance thermique de l'appareil diminue quand la pression aux buses est constante et le débit d'eau chaude augmente : cette diminution de puissance est d'environ 7,5 %, la pression aux buses restant constante, et avec un débit d'eau chaude de $27,5 \times 10^{-6}$ m³/s (100 l/h) que l'on peut considérer comme le débit nominal de l'appareil. La diminution de puissance est plus lente ensuite, à mesure que le débit d'eau augmente, atteignant 9 % à peu près avec un débit de 42×10^{-6} m³/s environ (150 l/h).

On peut encore observer que l'augmentation de la puissance thermique de l'appareil, passant d'une pression aux buses de 400 N/m² à une autre de 800 N/m², reste presque constante.

On pense que les pertes de puissance thermique sont dues surtout à deux phénomènes : à un passage d'air à travers le registre fermé sur la batterie chaude et à des phénomènes convectifs intéressant la batterie chaude.

Les essais n'ayant pas été nombreux, on ne peut pas tirer des conclusions définitives sur la répartition des pertes entre ces deux phénomènes.

D'autre part la construction du registre est très soignée et l'augmentation de la puissance thermique est presque constante, lorsque la variation de la pression aux buses et le débit d'eau chaude augmentent. Tout cela engage à penser que la perte d'énergie thermique est due surtout à des phénomènes convectifs favorisés par l'air sortant de la partie supérieure de l'appareil.

La diminution de puissance est supérieure à celle qui se produit dans des appareils analogues pour les installations à 4 tuyaux avec réglage sur l'eau. Dans ce dernier cas, on a relevé une diminution de puissance thermique de 4 % environ qui pratiquement ne dépend pas du débit. On va entreprendre des essais analogues sur un prototype avec la batterie chaude placée sur la partie inférieure de l'éjecto-convecteur, pour constater si les phénomènes de convection changent d'importance. Enfin on peut encore faire une considération sur la bonne conservation des dispositifs de réglage dans le temps : d'après l'expérience on peut dire que la tenue d'une vanne se maintient dans le temps bien mieux que la tenue d'un registre, mais on n'a pas fait une série d'essais systématiques pour bien déterminer l'usure des organes de réglage dans les deux cas.

DISCUSSION

J.N. FAFCHAMPS (Belgique) — Il me paraît illusoire d'affirmer que les pertes aux parois sont nulles lorsque les conditions dans l'enceinte extérieure et dans la chambre sont les mêmes. La réalité est sans doute plus complexe car les conditions aux différentes parois peuvent être très différenciées par le type même d'appareil et surtout par son mode de pulsion de l'air (Cf. communication de M. HANNAY).

Même si les parois sont fort isolantes, il me semble utile d'effectuer un tarage préalable des échanges au travers des parois et ce pour chaque type d'appareil soumis à l'essai.

G. SAGGESE — Premièrement, le calorimètre employé, du type adiabatique, est construit selon les prescriptions ISO et représente le seul système actuellement connu pour ce genre d'épreuves.

Deuxièmement, la distribution de températures, que M. HANNAY a remarquée, correspond à des épreuves faites avec un appareil placé dans une chambre avec transmission de chaleur aux parois. Cette condition permet de prévoir de notables différences de température le long des parois.

Nous souhaitons bien que M. FAFCHAMPS (qui a posé la question) puisse proposer un calorimètre meilleur.

M. VAN BEVEREN (Belgique) — Tenue dans le temps du registre.

Il serait intéressant de refaire les essais sur le même appareil après par exemple un an de fonctionnement continu de l'appareil.

G. SAGGESE — Il est sans doute très intéressant d'exécuter des épreuves d'étanchéité des registres dans le temps. D'après l'expérience j'ai fait, dans mon rapport, une affirmation qui a aussi une signification évidente. En effet, la longueur de l'élément d'étanchéité est bien plus importante dans un registre que dans une vanne. De plus, dans une vanne la pression entre l'obturateur et le siège peut être plus intense et bien mieux répartie que dans le cas d'un registre.

H. MARCQ (Belgique) — Quelles sont les températures d'entrée et de sortie de l'eau chaude et de l'eau froide lors des essais dont les résultats sont représentés figure 6 ?

G. SAGGESE — Pendant mon exposé j'ai déjà dit quelles étaient les températures pendant les épreuves. Ces valeurs n'étaient pas indiquées dans mon rapport : je les répète donc :

— Température du bulbe sec de la chambre d'épreuves : 26°C;
— Humidité relative : 45% environ;
— Température de l'eau froide entrant à la batterie : 12°C
— Différence de température entre l'entrée et la sortie de la batterie froide : 2 à 2,5 deg C environ;
— Température de l'eau chaude entrant à la batterie : 42°C;
— Différence de température entre l'entrée et la sortie de la batterie chaude comprise entre 0,4 et 0,7 deg C environ.

J. HANNAY (Belgique) — Quelle est en fonction de la puissance de l'appareil, ou en fonction des dimensions de la chambre, le débit d'air dans le système de compensation placé à l'*intérieur* de la chambre?
— Quelles sont les vitesses d'air aux grilles de pulsion de ce système de compensation; peut-on être certain que ces jets d'air à température différente de celle du local, n'ont aucun effet sur les caractéristiques de l'appareil?

G. SAGGESE — Je ne me rappelle pas le débit d'air du simulateur interne. Cependant, je suis à même de dire que la vitesse avec laquelle l'air du simulateur entre dans la chambre, est très basse : la valeur maximale que les prescriptions ISO permettent est de 0.50 m/s. On n'a pas relevé de perturbations notables pendant les épreuves (dans ma réponse j'avais parlé de vitesse de 1 m/s qui est la vitesse dans les interstices, tandis que la vitesse de sortie du simulateur dans la chambre est de 0,5 m/s).

OBSERVATIONS WITH RESPECT TO DAMPER
AND WATER CONTROLLED INDUCTION UNITS

T. RÁKÓCZY

V.D.I., Köln (W. Germany)

Considérations économiques sur le réglage par robinet ou par volet d'air dans les appareils de conditionnement d'air à induction

RÉSUMÉ : *On a mis au point un nouveau mode de régulation de la température dans les systèmes de conditionnement d'air à haute pression et induction. Pour essayer de simplifier le réglage assez compliqué par l'eau, on est passé au réglage par volet d'air.*

En fonction de la température requise, on envoie l'air secondaire, par un système de volets, à travers un refroidisseur ou un réchauffeur ou sur un by-pass. L'A. compare économiquement le système normal à eau et le système à volet d'air. Les calculs sont faits sur des immeubles fictifs, en considérant les coûts d'installation et de fonctionnement ainsi que l'usure et l'encombrement des appareils.

With the growing number of various multi-storey buildings, materials such as sealed glass constructions, plastics, lightweight metals etc. are being used increasingly.

Buildings with closed facades have to be air conditioned or at least ventilated for reasons of temperature control and oxygen supply.

Depending upon the quantity of energy-exchange with heat losses and heat gains of the building during extreme outside air conditions in summer and winter, there are various forms and applications of air conditioning installations (fig. 1).

The question of whether one should apply the so-called high-pressure induction system for a multi-storey office building is determined by the physical properties of the building itself. There might be circumstances where if too much "heating" or "cooling" energy is transported to a room, one is forced to use not only air for heat transport (the case with low pressure air conditioning systems), but also water.

Apart from extremes in weather conditions there are other factors such as solar heat gains which, by nature, can vary their influence on the building in short periods of time. Because variable solar heat gains represent a considerable part of the total heat gain, it is necessary to develop a combined air-water system by means of which the largest variable, e.g. solar heat gain, can be controlled on the water-side of induction units.

For a room situated in the perimeter-zone and separated on one side from the outside air, heat transport media should preferably be positioned under the window-sill so that sudden variations in the solar load can be compensated for immediately; via this way, an energy balance with constant as possible temperatures can be obtained in the room.

Special air distribution devices have been developed for heat transport media (air and water) and which are suitable for installation in windowsills or false ceilings; they are called induction units after their operating principle. These induction units are applied mostly in combination with high pressure or high velocity air conditioning installations.

These induction units are, more or less, air distribution devices using high air induction via which the induced, so-called secondary air quantity is forced to flow through an heat exchanger backed up by a filter. With the secondary air passing through the heat exchanger, the air can be warmed or cooled depending upon requirements. By means of valve or damper controls, rapid room heating or cooling can be obtained; e.g. the valves mounted in the water circuit can provide, in a very short period, the necessary amount of energy to enable the induction unit to counteract the solar heat gain. In the operating of these installations, a large number of systems (regarding water circuiting, supply temperature diagrams and control systems) have been developed over the past 15 years, and much attention has been foccused on operation costs and economy during two years.

During the design of a high pressure installation, not only should the building construction or the installation of induction units be taken into account but also other factors such as exterior sun shading, building characteristics, building applica-

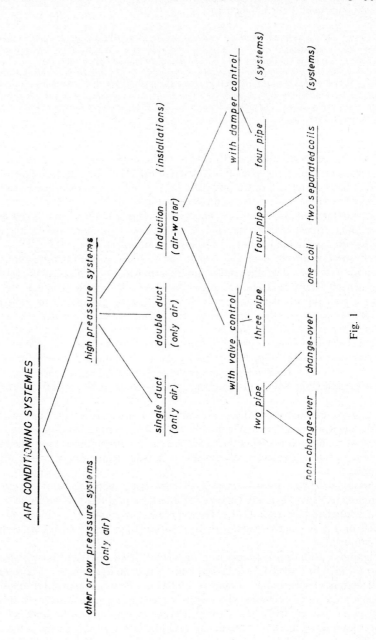

Fig. 1

tion, comfort requirements and operation economy; these factors will determine air conditioning system itself and the supply air temperature.

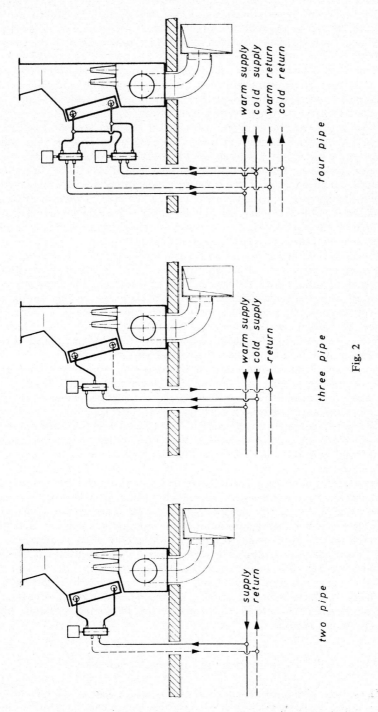

warm supply
cold supply
warm return
cold return

four pipe

warm supply
cold supply
return

three pipe

Fig. 2

supply
return

two pipe

The biggest unsolved problem today in air conditioning is to find one solution for a specific building that will meet requirements regarding investment costs and functioning as well as the most economic operation.

The water connections of a secondary heat exchanger can be hooked up to so-called two-pipe, three-pipe and four-pipe systems (fig. 2). Depending upon the shape of the building, a two-pipe system can also be divided in several zones of the water side.

With two-pipe induction systems, the divisioning into zones on the water side can be done according to facades and building orientation. However, for more than four facades of different orientations a two-pipe multi-zone induction installation is unacceptable, due partly to the relatively high installation costs and partly to complicated and vulnerable change-over control devices.

Regarding supply temperature diagrams for two-pipe systems, one can basically distinguish between change-over and non-change-over systems. The change-over relates to the water side and means that the water circuit of the units can supply, in sequence, chilled or warm water or, if during the year satisfactory operation can be obtained, supplying water at one constant temperature level. One should realise that for non-change-over systems economy can be achieved in permanently circulating chilled water in the secondary water circuit but for overnight winter heating there are certain problems with respect to operation costs since, heating can only be effected on the air side by means of expensive electrical apparatus.

In regions with higher outside air temperatures where, due to the building insulation, room temperatures will not drop below $+12\,°C$ during the year without night-heating a non-change-over system is best applied from the economy viewpoint However, with weather conditions in Western Germany, heating during the night normally cannot be omitted. One must, if one selects a two-pipe system, choose a change-over system but there is, however, an exception, i.e. if the solar heat gains are relatively very small. In this case the cooling energy can only be supplied by air and the application of an induction system then comes into discussion.

This article therefore goes further into the two-pipe change-over induction system. With the two-pipe system, water at only one temperature is available at any one time, the water circuit containing either chilled or warm water. Depending on the outside air conditions the water must be changed with these systems.

The three-pipe system is characterised by the fact that both chilled and warm water are available for the coils of the induction units. With these systems, a common return line is installed in the water circuit so that chilled or warm water can be mixed and returned to the plant-room.

With this four-pipe valve controlled system, two separated water returns are used in the chilled and warm water circuits thus excessive heat losses will occur in mixing the return water. With four-pipe damper-controlled systems one tries to avoid the relatively complicated controls and their resulting heat losses on the water side. Here the chilled or warm water is not quantity-controlled, depending upon requirements but the induced secondary air is discharged by means of a relatively complicated damper mechanism either through the cooling coil or the heating coil; the two coils are, naturally, fed continuously with water. It was found logical to include a by-pass possibility with these induction units (fig. 3) thus, if neither cooling or heating is required, the secondary air stream can, due to a specific damper position, flow to the primary air and be mixed.

HEAT LOSSES OF THE AFOREMENTIONED SYSTEMS

The various high pressure induction systems which have to maintain proper room air conditions operate with so-called energy losses. Depending on where these heat losses take place, the energy destruction can be divided in two different groups. Heat losses with valve-controlled systems take place in the coils, the water circuit and centrally in the plant room.

With damper controlled induction units these heat losses take place more or less in the air conditioned rooms because the dominating source of the heat losses is represented by the heat radiation of those coils which are by-passed by the secondary air. The energy losses in two-pipe change-over systems are due to the switching over in the water circuit only.

Depending on outside air conditions and the way in which the installation was designed with respect to control and energy consumption, the changing over from chilled to warm water and vice versa may be executed daily for outside air temperatures varying between $+5\,°C$ and $+15\,°C$. During the change over process heat losses are caused by the available water volume and temperature.

With the three-pipe system, mixing losses occur continuously because of a common return.

Although with the four-pipe valve-controlled system small heat losses take place, these losses can be calculated and measured; they are caused by the valves which supply, arbitrarily, either chilled or warm water to the coils of the induction unit. Due to changing-over of the valve, water in the coil is returned to the system which is working in opposition (heating or cooling) e.g. water which was originally used for heating the coil is returned, after change-over by the valve, to the chilled water circuit.

As mentioned before, the heat losses of damper-controlled four-pipe systems are

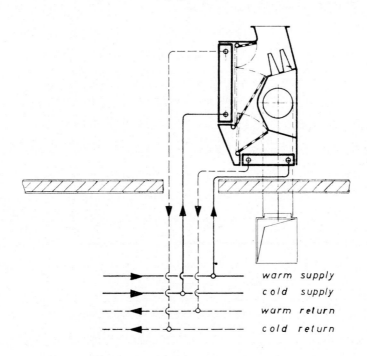

warm supply
cold supply
warm return
cold return

four pipe

Fig. 3

caused mainly by the fact that chilled or warm water is flowing continuously through the coil with a possibility that neither the cooling or heating coil is in operation on the secondary air side. Here, energy is supplied to the room continuously via conduction and radiation.

With these systems the heat losses are dominant due to more favourable radiation conditions. Added to this is the fact that heating and cooling energy is supplied at the same time to the rooms where the induction units are installed in order to maintain room air conditions.

In the case where room cooling takes place on the secondary side, the induction heat-exchangers will radiate and conduct heating energy into the same room. Therefore, on the cooling side, increased cooling capacity must be taken into account.

ECONOMY CALCULATION

In order to calculate the aforementioned heat losses for a fictive multi-storey building, an operation-economy calculation was executed for the various high pressure induction systems described. With the calculation, only the varying factors were taken into account and it was therefore assumed that the building construction, orientation, outside air temperatures, required interior air temperatures and minimal number of fresh air changes remained constant. It was assumed that in a building containing about 360 induction units, either a two-pipe change-over system, a three- or a four-pipe system could be installed; the latter is possible with either damper or valve controlled induction units.

Table 1

INVESTMENT COSTS

| | Valve control | | | Damper control | |
	two pipe	three pipe	four pipe	four pipe	
Unit including mounting	187.00	187.00	187.00	371.00	DM per unit
Valve incl. mounting	171.00	288.00	504.00		DM per unit
Pneum. damper motor incl. mounting				298.00	DM per unit
Cost for water connection	82.00	108.00	135.00	135.00	DM per unit
Cost of control-system	105.00	115.00	135.00	105.00	DM per unit
Total	545.00	698.00	961.00	909.00	DM per unit

Table 2

YEARLY COSTS FOR HEAT LOSSES

| | Valve control | | | Damper control | |
	two pipe	three pipe	four pipe	four pipe	
Heating	1.46	3.20	0.05	4.80	DM per unit, year
Cooling	2.20	5.35	0.54	2.75	DM per unit, year
Energy removal in the room				7.55	DM per unit, year
Total losses	3.66	8.55	0.59	15.10	DM per unit, year

Table 3

YEARLY DIFFERENCE-COSTS
(Indicated are only the variable costs for the various systems)

| | Valve control | | | Damper control | |
	two pipe	three pipe	four pipe	four pipe	
Investment	545.00	698.00	961.00	909.00	DM
Amortization	82.00	104.00	144.00	136.00	DM per unit, year
Heat losses	3.66	8.55	0.59	15.00	DM per unit, year
Total	85.66	112.55	144.59	151.00	DM per unit, year

Yearly constant energy costs are not taken into account because they are the same for the various systems.

Table 4

VALUATION OF DAMPER AND VALVE CONTROLLED INDUCTION UNITS

| Valve control | | Damper control | |
advantages	disadvantages	advantages	disadvantages
1. No variation in total supplied air quantity during temperature control.	1. Vulnerable valve construction.	1. No complicated and, vulnerable valves.	1. Bad control characteristics of the dampers.
2. Good control characteristics.	2. Considerable number of solder-points in the installation with increased possibility of water leakage and damage.	2. Smaller risk of water damage.	2. Temperature control is combined with variations in noise production.
3. Smallest overall dimensions possible.	3. Increased corrosion possibility due to water temperature fluctuations.	3. Simpler lay-out of the central water circuit.	3. The supplied total air quantity varies with repositioning of the damper blades.
4. Many years of experience.	4. Considerable expansion as a result of water temperature variations.	4. Lower corrosion danger.	4. Twisting of the damper blades.
		5. Effects of expansion in the water system are lesser.	5. Bigger overall dimensions.
			6. Continuous radiation losses in the rooms.
			7. Increased investment and operation costs.

Since the determination of the difference-values between the various systems was of interest, the comparison was concentrated on investment and operation costs.

The costs for other factors, e.g. heat losses or gains for the building, primary air production etc. remain the same and therefore need not to be taken into account in the comparative calculations.

The investment costs for induction units using damper or valve control and the relevant costs for the control system are indicated in table 1. This survey concludes that investment costs for damper or valve control are approximately the same. The yearly heat losses for various systems are indicated in table 2 on the basis of one unit per module.

In order to calculate the exact comparison between the various systems, the investment costs were amortized over a period of 10 years including an interest of 15%.

Added to these yearly amortization costs are the increased costs for heat losses in in the various systems.

By this the differences in total costs are obtained for the various systems on the basis per unit (see table 3).

Table 3 leads us to the conclusion that the most expansive possibility is represented by the newly developed damper controlled induction units for four-pipe systems.

The difference in value between the simplest and most complicated possibilities with various high pressure induction installations amounts to DM 20 000/per unit, based on an installation with about 360 induction units. If this figure is compared to the yearly operational costs of the complete installation, one can conclude that this figure is about 20-30% of the total yearly operation and maintenance costs.

The features of valve and damper-controlled units along with their advantages and disadvantages are indicated in table 4. Apart from this comparison-table it should be noted that experiences with damper controlled induction units are based on a much shorter period as compared to valve controlled induction units which have been well known since about 10 years. Whether this new trend will continue depends on further developments and improvements in these units.

LE BRUIT DES ÉJECTOCONVECTEURS

F. CLAIN

Co.S.T.I.C., Paris (France)

Noise in jet-convectors

SUMMARY: *The purpose of the study was to determine sound problems created by the operation of air-conditioning equipment located directly in air-conditioned rooms.*

The noise level of different jet-convectors were measured in a reverberating room by the Co.S.T.I.C. Discharge noise, at the level of the nozzles, is predominant and the results show how these levels evolve in terms of the static pressure in the container carrying the nozzles.

1. GÉNÉRALITÉS

Les éjectoconvecteurs produisent un sifflement dû au passage de l'air primaire à forte vitesse dans les buses de soufflage. De fortes vitesses sont nécessaires afin d'obtenir des taux d'induction d'air secondaire élevés, donc des émissions calorifiques ou frigorifiques suffisantes. Celles-ci croissent donc lorsque la pression d'air primaire dans le caisson portant les buses s'élève. Malheureusement, parallèlement à cet accroissement d'émission de l'appareil, la puissance acoustique s'élève également.

2. PUISSANCE ACOUSTIQUE DES ÉJECTOCONVECTEURS

Nous allons par la suite donner des résultats des mesures de puissance acoustique de quelques éjectoconvecteurs disponibles sur le marché français. Ces mesures ont été effectuées dans la salle réverbérante du Co.S.T.I.C. par différents expérimentateurs. Les relevés ont été effectués par tiers d'octave. Dans un premier stade [1], le bruit de fond provenant du ventilateur de soufflage était gênant aux basses fréquences.

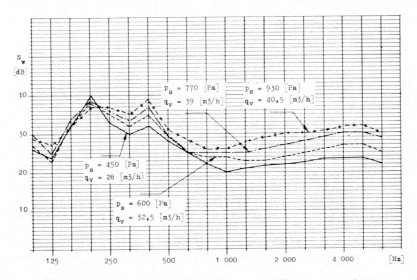

Fig. 1 — Puissance acoustique par tiers d'octave (re : 10^{-12} watt) d'un éjectoconvecteur (fabricant n° 1 — éjecteurs de type A) [1].

C'est le cas des appareils des fabricants n° 1 et n° 2 présentés aux figures 1 à 5 bis. Par la suite, le bruit de fond a été réduit : ceci est relatif aux essais sur les appareils des fabricants n° 3 et n° 4 (fig. 6). Pour terminer, nous présentons des mesures sur un aéroréchauffeur (fabricant n° 5, figure 7).

2.1 Des mesures de puissance acoustique en chambre réverbérante ont été effectuées sur un appareil d'un fabricant n° 1 [1]. Il s'agit d'un éjectoconvecteur d'allège de dimensions extérieures 900 × 620 × 215.

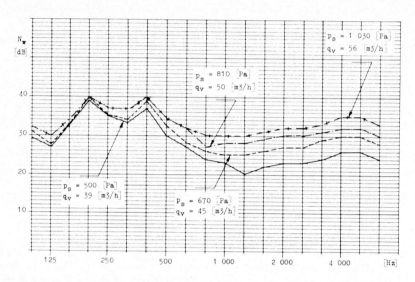

Fig. 2 — Puissance acoustique par tiers d'octave (re : 10^{-12} watt) d'un éjectoconvecteur (fabricant n° 1 — éjecteurs de type B) [1].

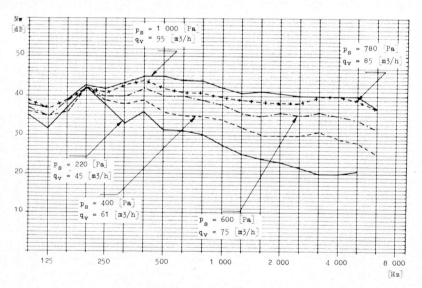

Fig. 3 — Puissance acoustique par tiers d'octave (re : 10^{-12} watt) d'un éjectoconvecteur (fabricant n° 1 — éjecteurs de type D) [1].

490

Les éjecteurs d'air construits en matière plastique sont de 5 types interchangeables : A, B, C, D et E. La forme de l'éjecteur reste inchangée; seule la dimension augmente du type A au type E.

Les spectres de puissance par tiers d'octave sont représentés aux figures 1 à 4. Sur une même figure, on a porté les spectres mesurés pour différentes pressions statiques, donc différents débits d'air primaire pour un même type de buses.

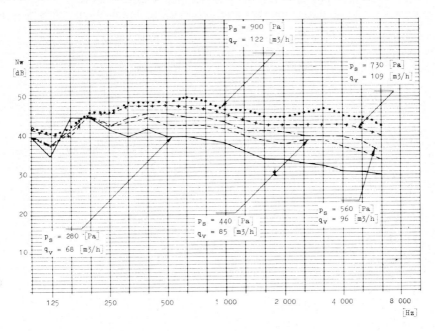

Fig. 4 — Puissance acoustique par tiers d'octave (re : 10^{-12} watt) d'un éjectoconvecteur (fabricant n° 1 — éjecteurs de type E) [1].

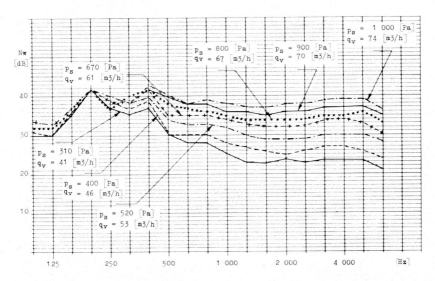

Fig. 5 — Puissance acoustique par tiers d'octave (re : 10^{-12} watt) d'un éjectoconvecteur (fabricant n° 1 — éjecteurs de type C) [1].

Le type de buses (A, B, C, D et E) est différent d'une figure à l'autre.

Les spectres sont plats, mais assez superposés dans les basses fréquences, ceci pour deux raisons :

a) le sifflement comporte peu de basses fréquences;

b) le bruit de fond provenant du ventilateur de soufflage apparaît néanmoins et s'ajoute au bruit des buses, surtout pour les petites buses (A et B).

Par contre, à 250 [Hz] et au-dessus, les spectres sont dus, quel que soit le type de buses, uniquement au bruit de soufflage sur celles-ci.

Ainsi, la puissance acoustique globale en fonction des paramètres pression d'air

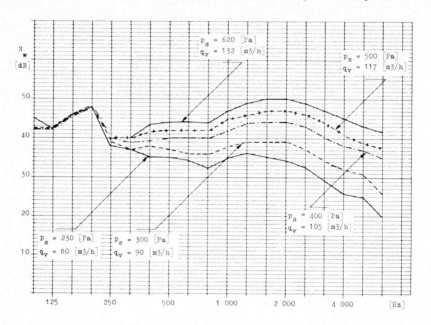

Fig. 5bis — Puissance acoustique par tiers d'octave (re : 10^{-12} watt) d'un éjectoconvecteur (fabricant n° 2) [1].

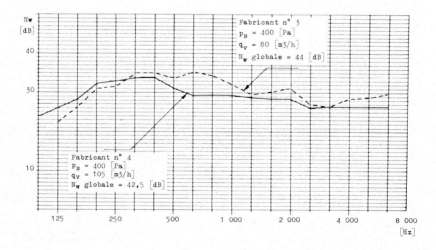

Fig. 6 — Puissance acoustique de 2 éjectoconvecteurs (fabricants n° 3 et n° 4).

492

p_s ou types de buses n'est pas absolument représentative du bruit produit par les éjecteurs. Nous les donnerons néanmoins par la suite à titre indicatif.

Nous constatons au vu des courbes les points suivants :

a) *Influence de la pression p_s pour un même type d'éjecteurs*

Lorsque la pression p_s croît, la puissance acoustique augmente aux moyennes et hautes fréquences, car relativement le débit croît selon $\sqrt{p_s}$. Nous donnons au tableau 1, à titre indicatif, les puissances acoustiques globales pour chaque essai, qui sont ici par excès (bruit de fond aux basses fréquences dû au ventilateur).

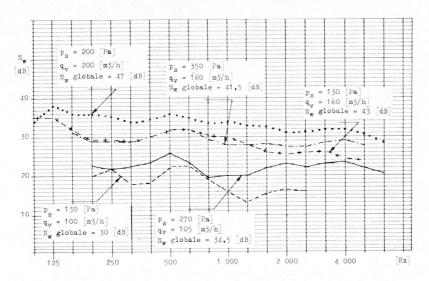

Fig. 7 — Puissance acoustique N_w par tiers d'octave (re : 10^{-12} watt) d'un aéroréchauffeur (fabricant n° 5).

Sur le plan pratique, il est plus judicieux d'examiner l'influence de la pression p_s sur le niveau acoustique exprimé en critère ISO. Ceci est possible en voyant de quelle façon les puissances à 2000 et 4000 [Hz] sont relevées en fonction des pressions p_s. En effet, les niveaux à 2000 et 4000 [Hz] sont ceux qui déterminent le critère ISO du fait de la forme plate du spectre. En moyenne, les niveaux ISO sont relevés suivant la pression p_s selon la relation :

$$30 \lg \left[\frac{p_{s2}}{p_{s1}} \right]$$

Ainsi, l'augmentation est de 8 à 9 indices ISO lorsque la pression p_s double. Les différents fabricants donnent d'ailleurs dans leur catalogue une augmentation de 6 à 10 dB (A) (ou une augmentation de l'indice ISO de 6 à 10 points) lorsque la pression double.

On peut aussi dire que l'augmentation en indices ISO est égale à :

$$15 \lg \frac{Q_2}{Q_1}$$

b) *Influence du type de buses* (même débit à pression variable)

Pour obtenir un même débit d'air aux buses, en changeant celles-ci, il faut alimenter

Tableau 1

PUISSANCES ACOUSTIQUES D'UN ÉJECTOCONVECTEUR (FABRICANT N° 1) SELON LA PRESSION p_s ET LE TYPE DE BUSES.

Type d'éjecteurs	Débit [m³/h]	Pression p_s [Pa]	Puissance acoustique globale (re 10^{-12} W) [dB]
A	28	450	43
A	32,5	600	43,5
A	39	770	44,5
A	40,5	930	45,5
B	39	500	44
B	45	670	45,5
B	50	810	47
B	56	1030	48
C	41	310	46
C	46	400	47
C	53	520	48
C	61	670	49,5
C	67	800	49,5
C	70	900	51
C	74	1000	51,5
D	45	220	46,5
D	61	400	48,5
D	75	600	51
D	85	780	53
D	95	1000	55
E	68	280	52,5
E	85	440	54
E	96	560	55,5
E	109	730	58
E	122	900	59,5

Nota : Ces puissances comportaient un bruit de fond de basse fréquence dû au ventilateur

l'appareil à des pressions p_s différentes, d'autant plus faibles que la section cumulée des buses est plus forte. Le niveau acoustique qui en résulte diminue lorsque les buses choisies sont plus grandes selon la loi :

$$15 \text{ a } 20 \lg \left[\frac{p_{s2}}{p_{s1}} \right]$$

Les catalogues des fabricants indiquent une variation égale à

$$15 \lg \left[\frac{p_{s2}}{p_{s1}} \right]$$

c) *Influence de la longueur de l'appareil*

Pour une même fabrication et un même type de buses, on dispose en général chez un fabricant d'appareils de différentes longueurs L. Le débit pour une même pression p_s varie alors proportionnellement à la longueur de l'appareil. Les puissances acoustiques, donc le niveau acoustique qui en résulte, augmente de :

$$10 \lg \left[\frac{Q_2}{Q_1} \right] = 10 \lg \left[\frac{L_2}{L_1} \right]$$

d) *Influence de la conception de l'appareil*

La forme des buses et la matière qui les constitue influent sur la puissance acoustique. L'emplacement des buses dans l'appareil, le capotage de celui-ci, l'environnement immédiat des buses, jouent également un rôle et il est impossible de donner une loi générale donnant la puissance acoustique selon le débit et la pression p_s. Quant à la forme du spectre, les figures 1 à 5 montrent que ce spectre se relève vers les hautes fréquences pour des débits croissants. La réduction spectrale n'est donc pas constante.

2.2 Fabrication n° 2

La puissance acoustique d'un éjectoconvecteur d'un second fabricant (fabricant n° 2) mesurée avec le même montage d'essai que précédemment [1] est représentée à la figure 5bis. Un seul type de buses a été essayé (buse type n° 11). L'augmentation d'indice ISO est environ égale à :

$$37 \lg \left[\frac{p_{s2}}{p_{s1}} \right]$$

Les puissances acoustiques globales sont données au tableau 2. Le bruit de fond dû au ventilateur, même aux basses fréquences, est négligeable.

Tableau 2

Puissances acoustiques globales d'un éjectoconvecteur (fabricant n° 2) selon la pression p_s.

Type d'éjecteurs	Débit [m³/h]	Pression p_s [Pa]	Puissance acoustique globale (re 10^{-12} W) [dB]
11	80	230	53
11	90	300	53,5
11	105	400	55,5
11	117	500	57
11	132	620	59

2.3 Le montage expérimental précédent a été amélioré dans le sens d'une diminution du bruit dû au ventilateur. La figure 6 représente les spectres de puissance par tiers d'octave de *deux autres* fabricants (n° 3 et n° 4). Les essais ont été effectués pour une seule pression p_s. On voit nettement mieux que précédemment la forme plate du spectre. Les puissances acoustiques globales indiquées sur les courbes sont réellement celles du bruit des éjecteurs, le bruit de fond étant très réduit.

3. Conclusion

La puissance est très variable d'une fabrication à une autre. Seule une mesure de puissance en laboratoire permet de connaître celle-ci.

RÉFÉRENCE

[1] R. Micard, Bruit des ventilo et éjectoconvecteurs. Thèse expérimentale effectuée au Co.S.T.I.C. (1965).

LE BRUIT DES VENTILOCONVECTEURS

F. CLAIN

Co.S.T.I.C., Paris (France)

Noise in fan-convectors

SUMMARY: *The purpose of the study was to determine sound problems created by the operation of air-conditioning equipment located directly in air-conditioned rooms.*

With fan-convectors, an initial series of noise level measurements in a reverberating room was made by the CoSTIC laboratory in 1965. The equipment used has since been developed. In order to localize the troublesome sound levels obtained, recordings were made in a typical office room using different units.

1. Ces appareils sont équipés d'un moteur à deux ou trois vitesses, ou plus rarement d'un moteur à variation continue de vitesses. Les modèles existent en grand nombre sur le marché.

Nous présentons ici deux catégories de mesures:

a) des mesures de puissances acoustiques effectuées sur 5 appareils en chambre réverbérante [1];

b) des mesures de niveaux sonores effectuées à 1,60 [m] de la grille de soufflage dans un bureau meublé de 45 [m³].

2. PUISSANCES ACOUSTIQUES MESURÉES SUR 5 APPAREILS

Les mesures ont été effectuées au Co.S.T.I.C. par R. Micard en 1965, en chambre réverbérante. Les débits d'air soufflés à vitesse de rotation maximale sont respectivement de : 425, 500, 360, 1100 et 360 [m³/h]. Il s'agit d'appareils à placer en allège.

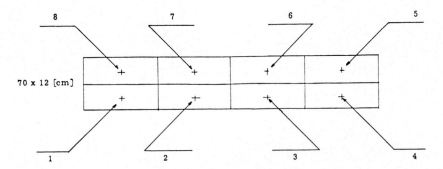

Point	Vitesse en [m/s]	
	900 [tr/mn]	600 [tr/mn]
1	0,7	0,23
2	1,3	0,8
3	2,2	0,9
4	0,8	0,3
5	0,8	0,29
6	0,55	0,17
7	1	0,39
8	0,8	0,32

Fig. 1 — Vitesse d'air au-dessus de la grille de soufflage d'un ventiloconvecteur (fabricant *a*).

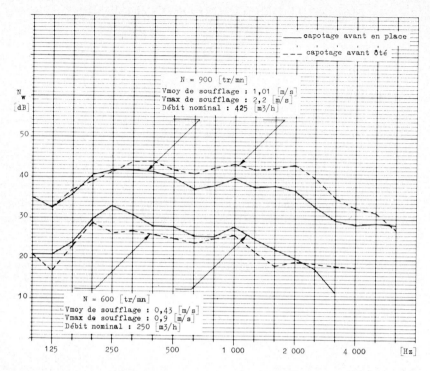

Fig. 2 — Puissance acoustique d'un ventiloconvecteur à 2 vitesses (fabricant *a*), par tiers d'octave (re : 10^{-12} watt).

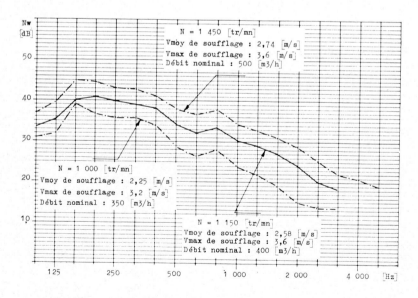

Fig. 3 — Puissance acoustique d'un ventiloconvecteur à 3 vitesses (fabricant *b*), par tiers d'octave (re : 10^{-12} watt).

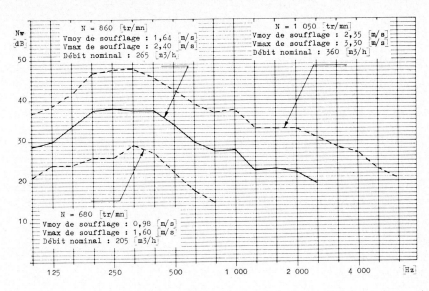

Fig. 4 — Puissance acoustique d'un ventiloconvecteur à 3 vitesses (fabricant *c*), par tiers d'octave (re : 10^{-12} watt).

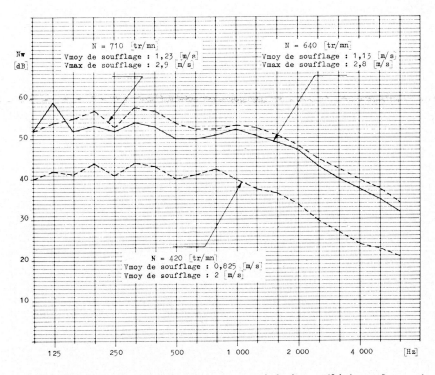

Fig. 5 — Puissance acoustique d'un ventiloconvecteur à 3 vitesses (fabricant *d*), par tiers d'octave (re : 10^{-12} watt). La façade avant de l'appareil est en place.

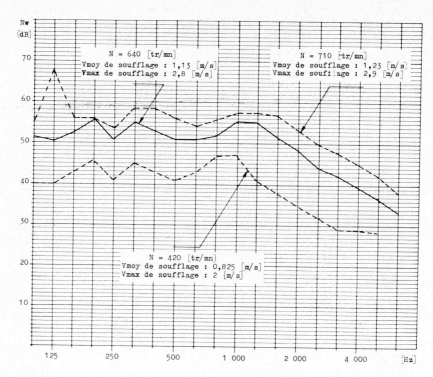

Fig. 6 — Puissance acoustique d'un ventiloconvecteur à 3 vitesses (fabricant *d*), par tiers d'octave (re : 10^{-12} watt). La façade avant de l'appareil est ôtée.

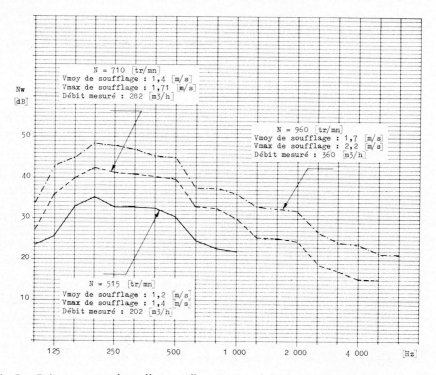

Fig. 7 — Puissance acoustique d'un ventiloconvecteur à 3 vitesses (fabricant *e*), par tiers d'octave (re : 10^{-12} watt).

f [Hz]	Grande Vitesse	Moyenne Vitesse	Petite Vitesse
	Niveau [dB] N_p		
31,5	47	46	46
63	48	43	39
125	61	56	53
250	56	52	49
500	50	46	43
1000	45	41	38
2000	41	37,5	35
4000	38,5	33,5	29
8000	28	22	18
16000	16	15	14,5
dB A	53	49	46
dB B	61	55,5	53
dB C	64	58	56
dB LIN	64	59	57,5
Indice ISO	45/50	40/45	40
	————	············	————

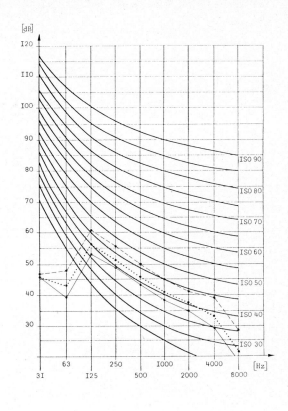

Fig. 8 — Niveaux sonores mesurés à 1,60 [m] du ventiloconvecteur f dans un bureau meublé de 45 [m³].

a) *Mesures aérauliques* : les vitesses d'air à 2 [cm] des grilles de soufflage ont été relevées (voir, à titre d'exemple, la figure 1).

b) *Mesures acoustiques :* Les puissances acoustiques par tiers d'octave sont portées aux figures 2 à 7. Chaque figure est relative à un appareil. Chaque courbe comporte les indications suivantes :

– la vitesse de rotation N du moteur en [tr/mn].
– la vitesse moyenne de soufflage V_{moy} en [m/s]. Cette vitesse est la moyenne des relevés de vitesses d'air effectués au-dessus des grilles. Cette vitesse n'est donc pas le quotient du débit d'air par l'aire de la grille de soufflage.
– la vitesse maximale de soufflage V_{max} mesurée en [m/s].
– le débit en [m³/h]. Il s'agit du débit nominal indiqué dans les catalogues des constructeurs et non du débit mesuré en laboratoire, excepté pour l'appareil *e*.

2.1 *Influence de la vitesse de rotation* (ou du débit d'air soufflé pour un appareil donné).

Installés dans un local meublé, les appareils provoquent des niveaux sonores qui, d'après les critères ISO, sont les plus gênants à 500 ou à 1000 [Hz] selon les appareils.

501

f [Hz]	Grande Vitesse	Moyenne Vitesse	Petite Vitesse
	Niveau [dB] N_P		
31,5	46/48	44/48	46/48
63	38	37	35
125	50	47/48	47
250	50	45	40
500	45	39/40	34
1000	36	30	25
2000	29	23	17
4000	22	18	13
8000	16	13	13
16000	15	14,5	14
dB A	45,5	40	35,5
dB B	52	48	44
dB C	54	50	48
dB LIN	56	53	51
Indice ISO	40	35	35
	− − − −	··········	———

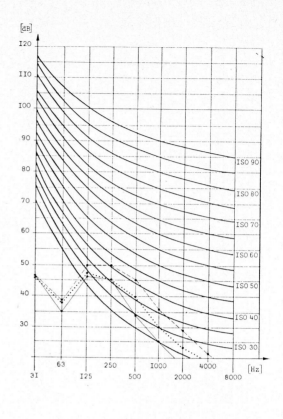

Fig. 9 — Niveaux sonores mesurés à 1,60 [m] du ventiloconvecteur *g* dans un bureau meublé de 45 [m³].

Afin d'uniformiser les comparaisons, on peut constater quelle est l'augmentation de la puissance acoustique à 1000 [Hz] lorsque la vitesse du ventilateur augmente.

a) Appareil *a*:

+ 12 [dB] lorsque Q passe de 250 à 425 [m³/h] soit:

$$12 = x \lg \frac{425}{250}$$

d'où:

$$x = 52$$

D'après cet essai, la loi de variation du niveau sonore à 1000 [Hz] avec le débit d'air est donc:

$$52 \lg \frac{Q_2}{Q_1} \simeq 50 \lg \frac{Q_2}{Q_1}$$

502

f [Hz]	Grande Vitesse	Moyenne Vitesse	Petite Vitesse
	Niveau [dB] N_P		
31,5	46/48		45
63	34		33
125	44		35
250	42		35
500	48		37
1000	43		28
2000	34		20
4000	26		18
8000	17		14
16000	15		14
dB A	48		35
dB B	50		40
dB C	51		44
dB LIN	56		50
Indice ISO	45		35
	————	··········	————

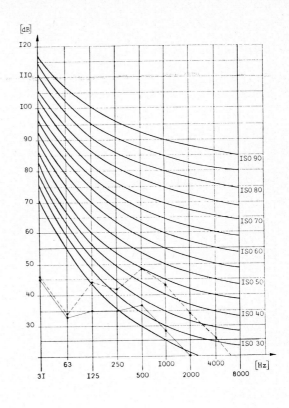

Fig. 10 — Niveaux sonores mesurés à 1,60 [m] du ventiloconvecteur h dans un bureau meublé de 45 [m³].

On a, en suivant un calcul identique, les résultats suivants pour les autres appareils :

b) Appareil b:

$$10 = x \lg \frac{500}{350}$$

d'où :

$$\boxed{x = 65}$$

c) Appareil c:

$$10 = x \lg \frac{360}{265}$$

d'où :

$$\boxed{x = 76}$$

d) Appareil d:

$$13 = x \lg \frac{710}{420}$$

f [Hz]	Grande Vitesse	Moyenne Vitesse	Petite Vitesse
	Niveau [dB] N_p		
31,5	45/57		46/48
63	38		34
125	41		38
250	40		36
500	45		39
1000	40		32
2000	31		24
4000	26		23
8000	19		18
16000			
dB A	44		38
dB B	47		43
dB C	49		47
dB LIN	52		51
Indice ISO	40		35

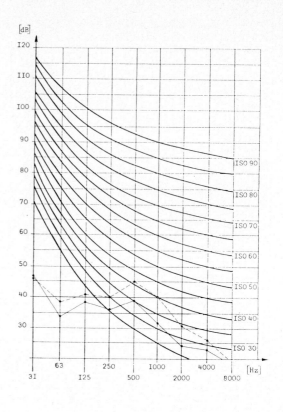

Fig. 11 — Niveaux sonores mesurés à 1,60 [m] du ventiloconvecteur *i* dans un bureau meublé de 45 [m³].

d'où:

$$x = 57$$

D'autre part, cet appareil a été placé dans un bureau meublé. A 1 [m] du sol et à 1,20 [m] de la grille, les niveaux suivants ont été mesurés:

Vitesse de rotation [tr/mn]	710	640	420
dB (A)	53	51.5	43.5
dB (B)	57	55	48
dB (C)	60	59	52
Lin.	61	60	53

Les écarts entre les vitesses de rotation extrêmes sont de 9,5 dB (A) et 8 dB (C) et 13 dB à 1000 [Hz].

f [Hz]	Grande Vitesse	Moyenne Vitesse	Petite Vitesse
	Niveau [dB] N_P		
31,5	46/50		46
63	37		34
125	39		37
250	45		40
500	48		43
1000	44		37
2000	36		28
4000	26		18
8000	17		15
16000			
dB A	48		42
dB B	51		45
dB C	52		48
dB LIN	56		54
Indice ISO	40		35
		– – – – –	·············	————

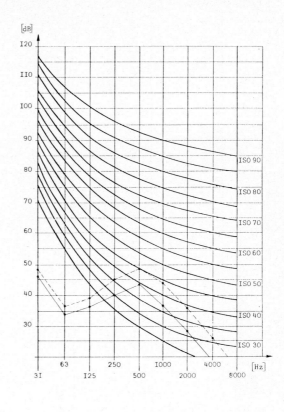

Fig. 12 — Niveaux sonores mesurés à 1,60 [m] du ventiloconvecteur *j* dans un bureau meublé de 45 [m³]. (Erratum : Indice ISO, grande vitesse = 45, petite vitesse = 40).

e) Appareil *e*:

$$14 = x \lg \frac{360}{202} \text{ à } 1000 \, [\mathrm{H}z]$$

d'où:

$$\boxed{x = 56}$$

D'autre part, cet appareil a été placé dans un bureau meublé de 3,50 × 4. Il crée les niveaux sonores suivants à 1 [m] du sol et à 1,60 [m] du centre de la grille de soufflage:

Vitesse de rotation [tr/mn]	960	710	515
dB (A)	43	37	29
dB (B)	48.5	43.5	37
dB (B)	51	47.5	45
Lin.	56	52	50

Tableau 1

CARACTÉRISTIQUES DES VENTILOCONVECTEURS *f, g, h, i*, ET *j* (MESURES DES NIVEAUX SONORES DANS UN BUREAU)

	Puissance calorifique nominale [kcal/h]	Puissance frigorifique nominale [fg/h]	Débit d'air [m³/h]	Vitesse rotation [tr/mn]	Dimensions de la grille de soufflage [mm]	Dimensions L × h × l [mm]	Puissance moteur [W]	Remarque
f	3600 3050 θ eau = 90 [°C] θ air = 20 [°C] q = 450 [1/h]	1950 1700	500 350 250	3 vitesses	460 × 100	762 × 615 × 289	140	1 batterie froide 1 batterie chaude
g	5700 4600 4000 θ eau = 80 [°C] θ air = 20 [°C] q = 350 [1/h]	2030 1700 1450	500	3 vitesses	560 × 100	995 × 635 × 234		
h	4140 2610	fonctionnement non prévu en froid	390 250	2 vitesses	1020 × 115	1040 × 385 × 120	43	
i	2900 1830	»	260 170	»	700 × 115	725 × 385 × 120	31	
j	1870 1550	»	205 145	»	480 × 115	500 × 385 × 120	28	

N.B. Les puissances calorifiques pour les appareils *h, i, j*, sont données pour les conditions suivantes:

θ air   : 20 [°C]
θ eau à l'entrée : 90 [°C]
θ eau à la sortie : 70 [°C]

Les écarts aux vitesses extrêmes sont donc:

14 dB à 1000 [Hz]
14 dB (A)
11,5 dB (B)
 6 dB (C)
 6 (Lin.)

Pour ce spectre et à cette distance de l'appareil, les écarts en dB (A) et à 1000 [Hz] sont donc égaux. De même que pour l'appareil *e*, les écarts en dB (A) sont supérieurs aux écarts en dB (C).

CONCLUSION

La puissance acoustique à 1000 [Hz] varie selon le débit suivant des valeurs comprises entre:

$$52 \lg \frac{Q_2}{Q_1} \quad \text{et} \quad 76 \lg \frac{Q_2}{Q_1}$$

La variation de la gêne en dB (A) d'après les résultats *d* et *e* est:

$$42 \lg \frac{Q_2}{Q_1} \qquad \text{pour l'appareil } d$$

et

$$52 \lg \frac{Q_2}{Q_1} \qquad \text{pour l'appareil } e$$

Des essais plus systématiques permettraient peut-être de dégager des lois plus précises, telles que celles donnant les bruits de bouche et de ventilateur. Néanmoins, un ventiloconvecteur groupe un ventilateur et une bouche de soufflage. Les variations de niveaux sonores en fonction du débit, dues à ces deux catégories d'éléments, étant différentes, il sera sans doute difficile de dégager des lois générales relativement précises d'essais plus complets. C'est en tout cas ce que l'on peut en penser au vu des résultats ci-dessus.

2.2 *Influence de la conception des appareils*

En comparant les puissances acoustiques développées à débit égal ou très voisin, l'on constate que celles-ci dépendent beaucoup de l'appareil.

Exemple:

    appareil capoté *b* 400 [m^3/h] 1150 [tr/mn] (fig. 3)
et   appareil capoté *a* 425 [m^3/h] 900 [tr/mn] (fig. 2).

L'appareil *a* produit une puissance acoustique supérieure à l'appareil *b*:

    de 10 [dB] à 1000 [Hz]
et   de 13 [dB] à 2000 [Hz]

bien qu'il tourne moins vite.

Ceci met en évidence l'importance de l'isolement soigné du compartiment ventilateur et du guidage de l'air entre le ventilateur et la grille de soufflage.

Autre exemple :

appareil *b* 350 [m³/h] 1000 [tr/mn] (fig. 3)
appareil *c* 360 [m³/h] 1050 [tr/mn] (fig. 4).

L'appareil *(c)* est plus bruyant de :

13,5 [dB] à 1000 [Hz]
et de 17,5 [dB] à 2000 [Hz].

2.3 *Influence du capotage*

L'influence du capotage est visible sur les figures 2 (appareil *a*) et 5 et 6 (appareil *d*).
Les différences notées ci-dessous sont précédées d'un signe + lorsque l'appareil sans capot produit un niveau sonore supérieur à l'appareil avec capot.

	Appareil *a*		Appareil *d*		
	425 [m³/h]	250 [m³/h]	710 [tr/mn]	640 [tr/mn]	420 [tr/mn]
1000 [Hz]	+3,5	−2	+4	+3,5	+7
2000 [Hz]	+6	−1	+5	+1	+1

L'influence est, en général, plus marquée aux fortes vitesses. Elle est même curieusement inversée à basse vitesse sur l'appareil *a*.

3. *Niveaux sonores mesurés dans un bureau meublé* (volume 46 m³).

Le sol est revêtu d'un linoléum. Ces mesures effectuées en 1969 sont relatives à des appareils disponibles sur le marché.
Les 5 appareils essayés sont de 3 marques différentes. Leurs caractéristiques essentielles sont portées au tableau 1.
Les appareils *h*, *i* et *j* sont ceux de la même série d'un constructeur.
Les niveaux sonores mesurés à 1 [m] de hauteur et à 1,60 [m] du centre de la bouche de soufflage sont portés aux figures 8 à 12.

RÉFÉRENCE

[1] M. MICARD, Bruit des ventilo et éjectoconvecteurs. Thèse expérimentale effectuée au Co.S.T.I.C. (1965).

NOISE CONTROL
OF AIR CONDITIONING COOLING TOWERS

L. SCHAUDINISCHKY and A. SCHWARTZ

Technion, Israel Institute of Technology, Building Research Station
Haifa (Israel)

Réglage du bruit causé par les tours de refroidissement des installations de conditionnement d'air

RÉSUMÉ : *Ce rapport décrit les problèmes liés au bruit de l'eau dans les tours de refroidissement. On a effectué les expériences sur un type de tour de refroidissement d'une puissance nominale de 3000 à 9000 fg/h. On a prêté une attention particulière à la forme et aux dimensions des orifices dans le bassin supérieur de répartition d'eau et à la profondeur de l'eau dans le bassin inférieur.*

Le rapport montre quatre points importants :

a) *L'importance d'effectuer les mesures acoustiques à des intervalles de 1/3 d'octave;*
b) *L'influence de l'amortissement déterminée par la profondeur de l'eau dans le bassin inférieur et la transformation de l'énergie phonique en énergie à plus haute fréquence;*
c) *La possibilité de déterminer un diamètre optimal des orifices dans le bassin de répartition d'eau pour un rapport donné de l'évacuation d'eau à l'énergie phonique;*
d) *La possibilité de déterminer le rendement de la transformation de l'énergie potentielle de l'eau en énergie phonique et l'existence de moyens pour réduire la valeur de ce rendement.*

INTRODUCTION

As a result of the growth of town population on the one hand, and of the ever increasing standard of living on the other hand, the use of air conditioning appliances is developing continuously.

Together with this development, the nuisance originating from these plants is also increasing. This is especially true for the noise of cooling towers and of air-cooled condensers, because they are generally located on the roofs, or in other places outside the buildings. As far as cooling towers are concerned, the noise is not as high as with air-cooled condensers, and noise control is therefore easier.

The Departement of Applied Acoustics of the Building Research Station, at the Technion-Israel Institute of Technology, carried out a number of investigations and surveys on the following subjects:

1. Special acoustic aspects of air-conditioning cooling towers [1, 2];
2. The legal aspect of acoustic treatment and noise reduction of cooling towers [2, 3];
3. The noise levels of cooling towers and the analysis of existing empiric formulas [4, 5, 6].

In this paper the problems connected with the noise caused by water in cooling towers are especially considered. The paper describes research work (partially sponsored by the Israel Ministry of Commerce and Industry) based on experiments carried out on a cooling tower model, corresponding to a nominal cooling capacity of 1-3 ton refrigeration.

EXPERIMENTAL

The experiments were carried out in a semi-reverberant room having a volume of 100 m³, wherein the cooling tower model was located. Figure 1 shows a part of this

509

room, with the upper basin (water distribution basin), the lower basin and the micro-phones arrangement.

Fig. 1 — The room of the experiments.

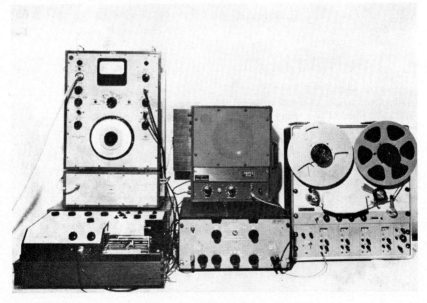

Fig. 2 — The main instruments.

The measurements and their recording were performed in another room. Figure 2 shows the main instruments, namely:

1. Brüel and Kjaer, Type 2112, Audio-Frequency Spectrometer, together with Type 1620 Extension Filter Set;
2. Brüel and Kjaer, Type 2305, High Speed Level Recorder;
3. Ampex, 4 Tracks, Tape Recorder;
4. High and Low Pass Filter;
5. Monitoring Loudspeaker.

In this section of the research, special attention was given to the shape and size of the holes at the bottom of the distribution basin, to the depth of water in the lower basin and to the material of which this basin was made.

The flow through individual interchangeable orifices of cylindrical or conical shape and of different diameters was investigated. Further experimentation was carried out with interchangeable perforated bottom plates. Figure 3 shows some of the orifices and the perforated plates.

Fig. 3 — Some of the orifices and perforated plates.

The orifices had diameters of 5, 10, 15, and 20 mm, they were cylindrical or conical with an angle of 45° or 135°. The perforated plates had cylindrical holes only, of 5, 10, 15 and 20 mm. They had for each size of holes, 25, 49 or 100 holes.

The measurement was based on the comparison of the sound levels to a standard noise source, according to the ASHRAE procedure.

MEASUREMENTS RESULTS AND THEIR ANALYSIS

In this chapter we shall concentrate on the experiments performed on individual orifices. The sound pressure levels, measured in a near field, when water was flowing through the orifices of 5, 10, and 20 mm diam. are represented in figure 4, as a function of the central frequencies of the octave bands (curves marked "a") and of 1/3 octave bands (curves marked "b"). In both cases the absolute physical values dB(Lin) are also represented, as well as the weighed values: dB(A), dB(B), dB(C).

511

Contrary to the usual method, i.e., to perform the spectral analysis of noises on octave bands, it is possible to learn from the curves which were analyzed by 1/3 octave steps that the picture is much clearer and certain phenomena appear which cannot be seen on the octave curves (a). We shall therefore consider curves "b" only.

On curve 1b we can see a continuous increase of the sound level whose rate is 8 dB per octave, with a slight peak at 300 Hz. After this increase, there is a strong decrease of approximately 20 dB between 300 Hz and 400 Hz. From 500 Hz on, the decrease is of 3 dB per octave up to 5,000 Hz and for still higher frequencies the decrease is of 10 dB per octave.

It is interesting to note that the curve corresponding to an orifice of 10 mm diam. (curve 2b) is almost identical to the curve of a 5 mm orifice (1b), but there is a pronounced peak approximately 10 dB higher than in its vicinity, against 2 dB only in the preceding case. In curve 3b the relatively small peak appearing at 80 Hz on curve 2b, becomes the main peak at 90 Hz. Between 100 Hz and 3,000 Hz the curve is practically a straight line (if the sudden decrease mentioned above is not taken into consideration).

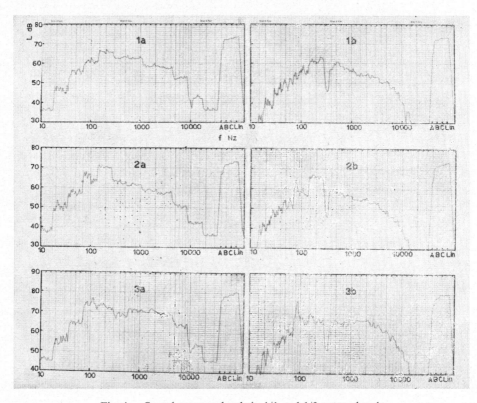

Fig. 4 — Sound pressure levels in 1/1 and 1/3 octave bands.

We can see a pronounced increase of the overall noise level, equal approximately to an upward displacement by 10 dB of the previous curves.

The influence of the water level in the lower basin can be seen in figure 5, in the case of water flowing through an orifice of 20 mm diam. Curve No. 1 represents noise pressure levels as a function of the central frequencies of the octave bands for a water depth of 250 mm. Curve No. 2 corresponds to an almost empty lower basin with the

512

water level at approx. 20 mm. In this case, there is a pronounced maximum between 100 Hz and 400 Hz with its peak at 200 Hz. From a central frequency of 200 Hz to 3,150 Hz the decrease is, in the average, 3 dB per octave. From 3,150 Hz it is 5 dB per octave.

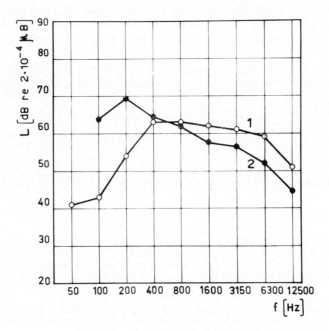

Fig. 5 — The influence of water level on sound pressure levels.

From curve No. 1 we can see that the influence of the strong damping due to water is felt especially at low frequencies. The difference between a full basin and an almost empty one, at a central frequency of 100 Hz, is approx. 20 dB; at 200 Hz it is approx. 16 dB; and between 400 Hz and 800 Hz it is practically equal to zero. From 800 Hz on, the noise of the falling water (when the basin is full) is, on an average, by 5 dB higher than when it is almost empty.

Investigating the efficiency of energy transformation into sound, the ratio between resulting sound energy and the potential energy of the falling water was examined. In general this efficiency was found to be $3 \cdot 10^{-5}$. In the case of a full lower basin, it is possible to consider also a central frequency of 1,000 Hz and then the efficiency is $5 \cdot 10^{-6}$. Experiments were carried out in order to reduce the above efficiency rate, using additional damping means.

CONCLUSION

From this short study of the noise caused by falling water in cooling towers it is possible to draw the following conclusions:

1. The spectral analysis of sound pressure or sound energy, in 1/3 octave steps, gives a cleaner and more relevant picture than the analysis by octave steps performed, according to the normal recommendations of ASHRAE;

2. In addition to the damping effect of a lower basin full of water, the favourable effect of the transformation of sound energy from low to high frequency should be noted.

In spite of a comparatively slight increase of the sound levels at high frequencies, this transformation makes possible the utilization of practical noise control rather easily, because of the great efficiency of most methods (absorption, insulation, barriers) at those high frequencies;

3. The measurement of sound pressure levels at several water discharges (through different orifices as shown by curves 1, 2, 3 in figure 4) makes it possible to foresee that there is an optimum hole diameter which enables the flow of certain quantities of water at relatively minimal sound pressure levels;

4. The efficiency of the transformation into sound energy of falling water potential energy is of 10^{-5} approximately. This value can be reduced by the use of special damping means.

REMARK

This paper will constitute a part of the doctorate work of Mr. A. Schwartz on the subject: "Development of a theoretical way of cooling tower noise level calculation."

REFERENCES

[1] L. SCHAUDINISCHKY and A. SCHWARTZ, "Noise of Cooling Towers in Air Conditioning Plants". *Journal of the Association of Engineers and Architects in Israel*, 1, XXVI (January/February 1968), in Hebrew, pp. 16-18.
[2] A. SCHWARTZ, "Noise Control of Cooling Towers". *Paper presented at the 2nd Convention of Mechanical Engineers in Israel* (April 1968), in Hebrew.
[3] A. SCHWARTZ, "Acoustic Problems of Air Conditioning Cooling Towers". Paper presented at the *IIIᵉ Conférence sur la lutte contre le bruit et les vibrations*, Bucarest (May 1969).
[4] L. SCHAUDINISCHKY and A. SCHWARTZ, "Noise Levels of Cooling Towers". *Building Research Station*, Haifa (BR120) (October 1968), in Hebrew.
[5] L. SCHAUDINISCHKY and A. SCHWARTZ, "Über die Akustischen Probleme der Kühltürme", *Kältetechnik-Klimatisierung*, 21, 1 (January 1969), pp. 2-6.
[6] L. SCHAUDINISCHKY and A. SCHWARTZ, Ditto, *Journal of Applied Acoustics*, 1,4 (October 1968), pp. 309-322.

SECTION 6

**PROBLÈMES D'ACTUALITÉ ET APPLICATIONS SPÉCIALES
DU CONDITIONNEMENT D'AIR**

*TOPICAL PROBLEMS AND SPECIAL APPLICATIONS
OF AIR CONDITIONING*

NEW TENDENCIES FOR
DESIGNING COMFORT-AIR-CONDITIONING PLANTS

H. SCHUCHARDT

Frenger International Corporation, Berne (Switzerland)

Nouvelles tendances de la conception des installations de conditionnement d'air assurant le confort

RÉSUMÉ : *On décrit ici la combinaison d'un plafond métallique rayonnant pour le chauffage et le refroidissement, dans une installation de conditionnement d'air.*

La quantité d'air fourni (air frais) est réglée de façon à obtenir des conditions d'hygiène parfaites dans les locaux climatisés; de plus elle doit absorber toute la chaleur latente. Le plafond lui-même fait face aux besoins de chaleur sensible et de refroidissement.

Les résultats expérimentaux montrent les conditions climatiques obtenues dans les locaux traités par ce système, à l'aide de courbes représentant les températures d'air, la vitesse de l'air, les températures ambiantes et les températures superficielles.

On indique spécialement les avantages de la répartition de l'air frais dans les fentes du plafond rayonnant, grâce auxquelles le vide au-dessus de ce plafond forme chambre de répartition.

Un autre point important est l'évacuation de l'air vicié sous les fenêtres. Ce rapport montre les avantages qui en résultent du point de vue du confort comme des frais d'exploitation. De plus, ces arguments sont confirmés par des résultats expérimentaux.

D'autres essais traitent de la facilité de réglage de l'installation :

a) *du côté du plafond. Comme le plafond rayonnant est très léger (8 kg/m²) et qu'on utilise de l'aluminium pour les panneaux, ceux-ci réagissent immédiatement aux variations des températures de l'eau. En raison de la grande vitesse de réaction on recommande de relier ce plafond à un système à 4 tuyaux pour ne pas avoir de retard du côté de l'eau.*

b) *du côté de l'air. L'air fourni est transporté* toute *l'année à la même* température *à* tous *les locaux ou zones; le point de rosée est réglé avantageusement d'après la température extérieure.*

I suppose you know out of practice all the advantages but also the problems of cooling and heating with metallic radiant underceilings and "ALL AIR"-systems.

The subject of this article is to show that the combination of the advantages of the two systems, the panel-air-system, is in a position to satisfy high demands not only regarding to comfort but also regarding to economy.

First I would like to give you a short description of the tested panel-air-system. Then I will explain the most important details.

The ceiling, as figure 1 shows, consists of aluminium panels "A", which are directly fixed to the tubes "B" by means of clips "C". On the back-side of the panels is an insulating layer "D" as thermal insulation and sound absorbing material. A thin sheet of aluminium "E" glued on the insulating blanket prevents the air from entering the conditioned room out of the plenum through the normally perforated panels.

The conditioned supply air flows through the duct at a constant temperature and a dewpoint temperature, depending preferably on the outside temperature, into the plenum. From the plenum the supply air streams at a maximum velocity of 7 m/sec and a minimum velocity of 3 m/sec into the room to be conditioned (figs. 2 and 3). Thus you will get a perfect uniform distribution of air in the room.

Exhaust air must absolutely be taken under the window, i.e., the "COLD WALL", for that is the only possibility to catch the cold down draught in winter (fig. 3) which also has good effects to the indoor climate near the window.

Now I would like to enter into the details of the tested system. For this I must premise that all measurings had taken place in a test-room of 3.22 m width, 5.38 m length and 3.00 m height.

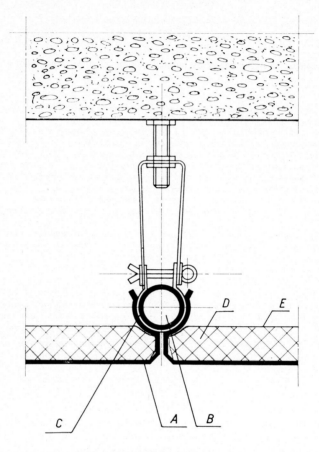

Fig. 1 — Ceiling-construction.

The "outside wall" of this test-room borders on an outside-climate-room, wherein practically all outside conditions can be simulated with one exception, the sun radiation cannot be simulated. I cannot give you more details about construction, possibilities and the system of measuring, because this is another question.

1. Heat transfer from the water in the tube as carrier of energy to the panels.
In this case I would like to refer to the report by Mr. Dr. H.J. Hamaker read on the occasion of the Expo-congress. Its title was "The Theoretic Calculation of the FRENGER-Ceiling".

2. The effect of the panel-air-system increases when the quantity of supply air augments. You may note this in the two diagrams, figures 4 and 5.
Reason: The flow of the supply air through the slots between the panels causes higher air velocity directly below the panels and parallel to them. Thereby the convective heat transfer is increased.

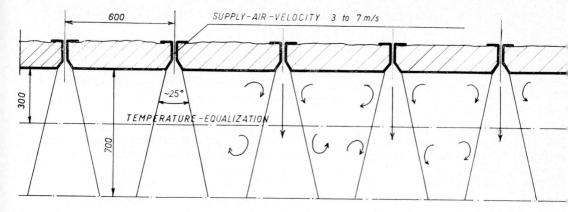

Fig. 2 — Supply-air-distribution.

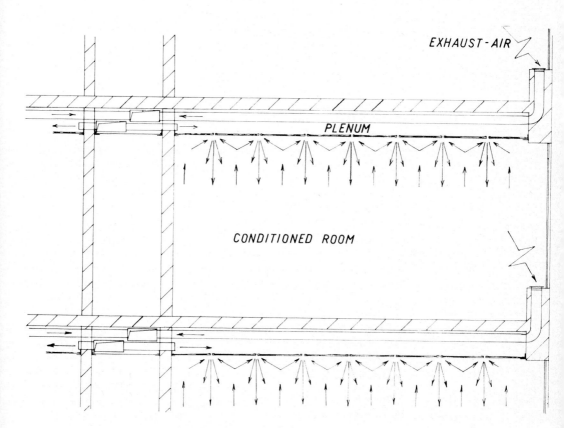

Fig. 3 — Scheme.

3. The cooling and heating capacity of the ceiling covers sensible cooling and heating loads.

4. Supply air—in most cases equal to the needed fresh air quantity—is conditioned in the central unit and flows during the whole year at a constant temperature of about 16°C and a dewpoint preferably regulated according to the outside temperature through the supply air ducts into the plenum above the ceiling.

5. The minimum velocity is 3 m per sec, the maximum velocity 7 m per sec. A smaller velocities than 3 m per sec the induction effect of the supply air stream will be too weak, so that a draughtless air distribution is not guaranteed anymore.

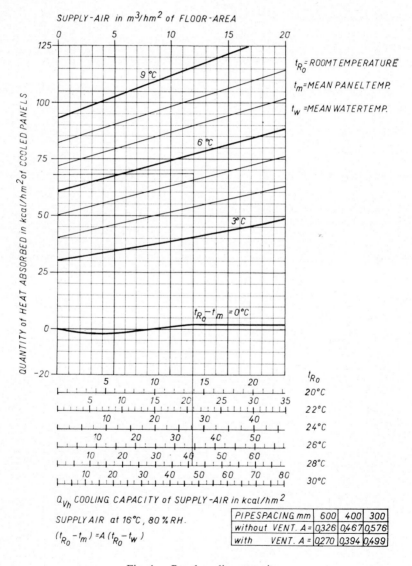

Fig. 4 — Panel cooling capacity.

6. Because of its induction effect the supply air stream being 1.5 mm wide is decomposed on a relative short distance. The stream's range is practically independent of the supply air velocity between 0.5 and 0.7 m (fig. 6).

7. The supply air quantity per square meter ceiling surface is 27 m^3/hr. m^2 at a velocity of 3 m/sec and 63 m^3/hr. m^2 at 7 m/sec. This would correspond to 9 to 21 air changes per hour in a 3 m high room (fig. 7).

But in comfort systems, conditioned according to the panel-air system smaller air changes are usual, for example 3 to 6 times per hour which would correspond to 9 to 18 m^3/hr. m^2 ceiling surface. One part of the slots—carefully selected—will be sealed by means of self-adhering foam rubber strips to keep supply air velocity at the minimum value, 3 m/sec. Later changes of the slot arrangement will be easily possible.

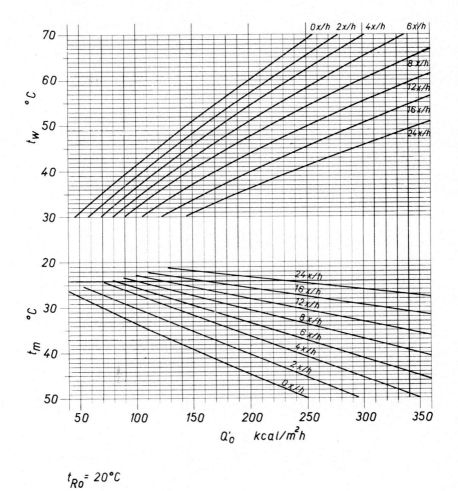

$t_{Ro} = 20°C$

$t_{ZL} = 16°C$

$l = 300 mm$

Fig. 5 — Panel heating capacity.

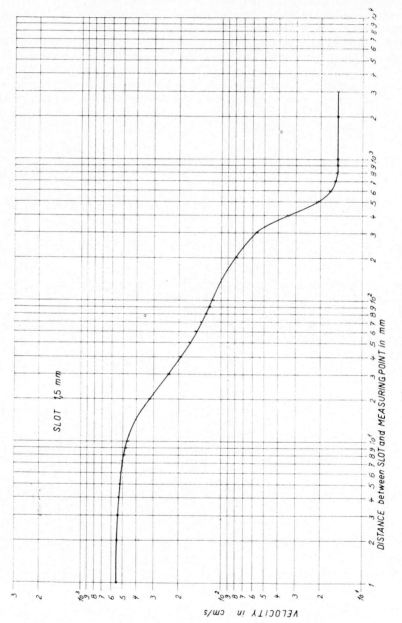

Fig. 6 — Range of supply-air-stream.

SLOTSPACING 600 mm $\cdot 2500 \, mm^2/m^2$

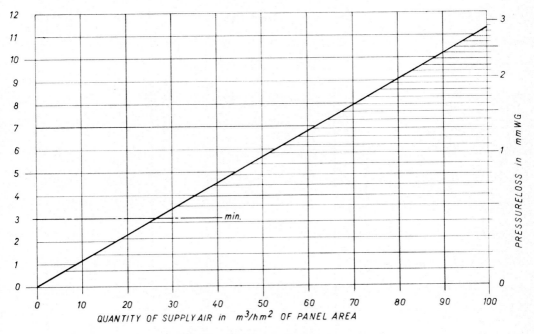

Fig. 7 — Supply-air diagram.

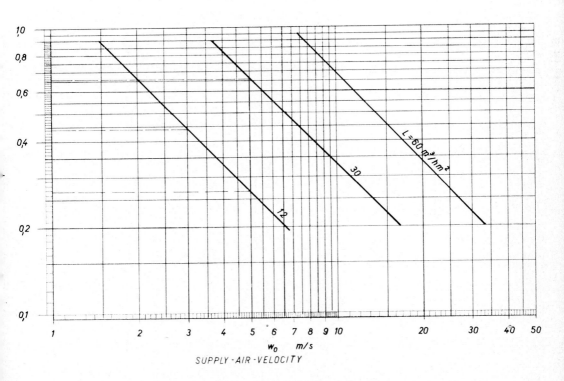

Fig. 8 — According to Prof. J. Rydberg's formula.

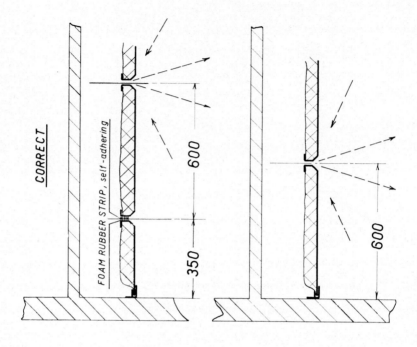

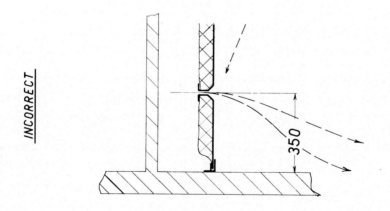

Fig. 9 — Slot-position.

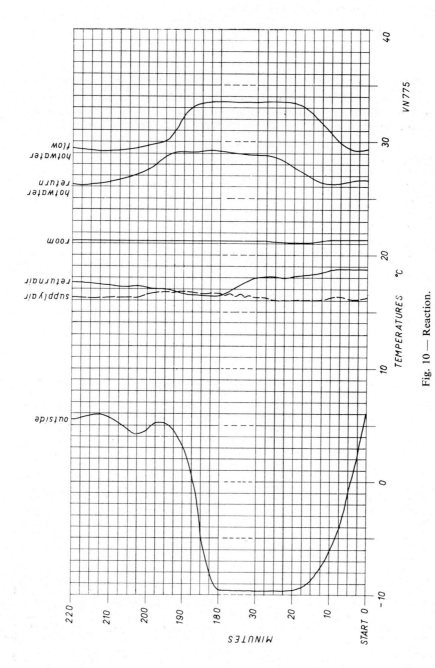

Fig. 10 — Reaction.

525

8. The air quantity effectively circulating in the room is of course bigger because of the induction effect of the supply air stream, namely by the factor 4 to 5, so that the uniform distribution of fresh air in the room is guaranteed.

9. If you apply Prof. J. Rydberg's formula:

$$L = \frac{3600 \times w_0 \times s}{b} \times \frac{A_s}{A_t} \quad [m^3/hr.\ m^2]$$

to the supply air distribution of this system you get the diagram shown in figure 8.

You may read out of this diagram the most favourable data for the supply air velocity and for the ratio of the total ceiling surface to the surface with slots by a given quantity of air.

That is for the examples market in:

$w_0 =$	3	5	5	m/sec
$L\ =$	12	12	30	$m^3/hr.\ m^2$
$s\ =$	0.0015	0.0015	0.0015	m
$b\ =$	0.6	0.6	0.6	m
$t_r - t_0 =$	4	4	4	°C
$\vartheta\ =$	1.41	2.09	2.21	°C

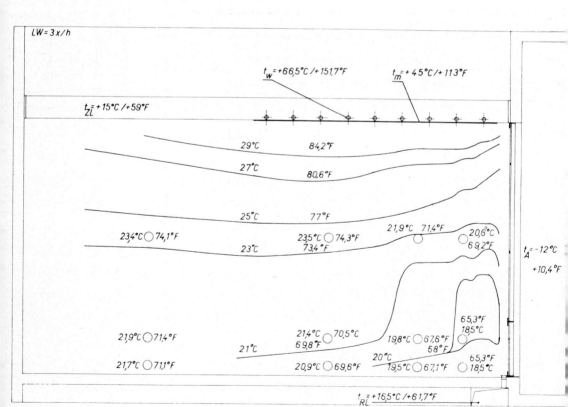

Fig. 11 — Heating and ventilation S-160 VN 111.

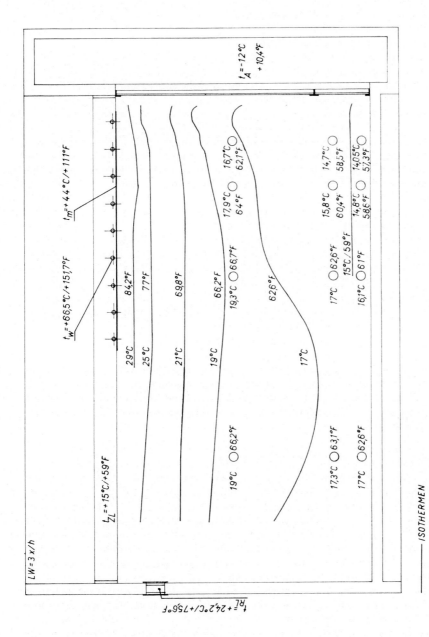

Fig. 12 — Heating and ventilation S-160 VN 115.

——— ISOTHERMEN

○ KUGELTEMPERATUREN GLOBE

527

The values for ϑ are within the limits asked by Prof. Rydberg for comfort air conditioning plants.

The signs used in the formula and in the above-mentioned list mean:

L = quantity of air per hour and per square meter of floor area
w_0 = supply air velocity in the slots m/sec
s = width of the slots 0.0015 m
b = distance between the slots 0.6 m
A_s = ceiling surface with slots
A_t = total ceiling surface
ϑ = effective undertemperature °C at a point in the supply air stream
t_r = air temperature in the room °C
t_0 = supply air temperature °C

10.   Closing my comment on distribution of supply air I would like to mention that a minimum distance between a side-wall and the nearest parallel slot of supply air of 500 mm must be followed. Otherwise the air-stream will go to this wall, which would unfavourably influence air circulation in the room (fig. 9).

11.   The exhaust air opening has to be arranged directly below the window. This is the only way to eliminate the "cold down draught" at the window in winter. The velocity of exhaust air in the exhaust grid has to be about 10 times higher than the velocity of cold down draught near the window, that is in most cases 4 or 5 m per sec. As the temperature of exhausted air is normally below the mean air temperature of the room, this fact will also take good effect in economic regard. The air near the window goes upward in summer. It is impossible to have also an exhaust opening above the window for reason of costs. Especially since in the ceiling zone more cooling load is taken up because of the higher temperature difference. This zone comes into contact with more or less tempered air according to the shading device.

12.   The weight of the ceiling is, according to its construction, small, about 8 kg per square meter, including panels, insulation, tubes with water contents and fixing. It could be expected that this ceiling system has a good basis for fast and exact reaction particularly because 0.75 mm thick aluminium is used. Tests in this respect confirmed these expectations.

I would like to show you as an example one test out of many similar ones:
Test-condition: constant room temperature when altering outside temperature.
We chose a pneumatic hot water control being dependent on outside temperature changes. As shown by figure 10, room temperature decreases by 0.2 deg C, 10 minutes after outside temperature already had decreased from $+5$ °C to -6.2 °C. Then already the effect of the hot water temperature increase which began to work 5 minutes after the start can be noticed. The room temperature remounts to its value of 21.4 °C and remains invariable, although outside temperature still decreases to about -9.5 °C.
Supply air temperature remains nearly constant while exhaust air adapted its temperature corresponding to the load changed.
May I ask you to draw your attention to the fact that exhaust air temperature nearly reached the supply air temperature and this already at an outside temperature of -9.5 °C. Here I must mention that because of the high reaction velocity the ceiling registers should be connected to a pipe-system which is in a position to take advantages of it, for example a four-pipe-system with pneumatic control.

13.   From the economic point of view, it is more favourable when exhaust temperature is lower than room temperature.

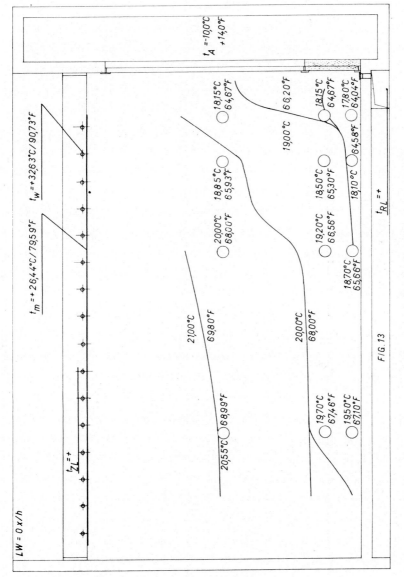

Fig. 13 — Heating and ventilation Frengair S-160 VN 596.

LW = 0 x/h

$t_{ZL} = +$

$t_m = +26{,}4°C / 79{,}9°F$

$t_W = +32{,}63°C / 90{,}73°F$

$t_A = -10{,}0°C$ $+14{,}0°F$

$18{,}15°C$ $64{,}67°F$

$19{,}00°C$ $66{,}20°F$

$18{,}15°C$ $64{,}67°F$

$17{,}80°C$ $64{,}04°F$

$18{,}85°C$ $65{,}93°F$

$20{,}00°C$ $68{,}00°F$

$18{,}50°C$ $65{,}30°F$

$19{,}20°C$ $66{,}56°F$

$18{,}10°C$ $64{,}58°F$

$18{,}70°C$ $65{,}66°F$

$21{,}00°C$ $69{,}80°F$

$20{,}00°C$ $68{,}00°F$

$19{,}70°C$ $67{,}46°F$

$19{,}50°C$ $67{,}10°F$

$20{,}55°C$ $68{,}99°F$

$t_{RL} = +$

FIG. 13

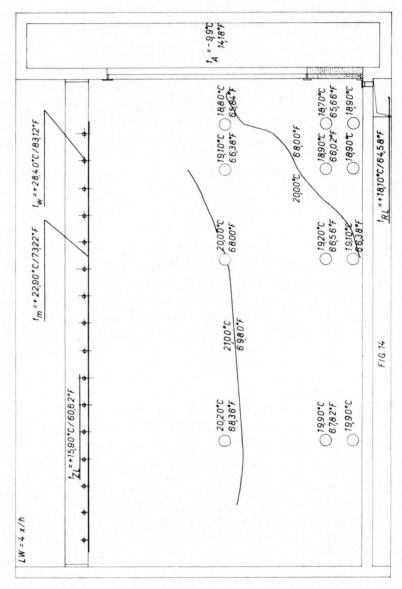

Fig. 14 — Heating and ventilation Frengair S-160 VN 572.

In the following I would like to compare tests with the exhaust opening below the window and others with the exhaust opening above in the rear-wall.

The test results are representative of a series of tests wherein we examined also other climate systems. We got practically the same results with all systems having the same or a similar manner of supplying and exhausting air as the panel air system.

Both figures 11(T-148) and 12 (T-150) show evidently the difference. The test-conditions were the following:

1. Structural:
   a) outerwall single glazed window 100%;
   b) heating with the FRENGER-ceiling;
   c) air distribution through slots;
   d) exhaust opening variable;
       d_1) below the window;
       d_2) above, in the rear-wall.

2. Concerning temperature:
   a) outside temperature constant;
   b) ambient temperature constant;
   c) mean water temperature constant;
   d) supply air temperature constant;
   e) three air changes per hour.

List of conventional signs:

t_{ZL}	supply air temperature
t_{RL}	exhaust air temperature
t_w	mean water temperature
t_m	mean panel temperature
t_A	outside temperature
O	globe temperature

As we had expected the room temperature rose by 4.2 deg C when exhausting air below the window.

Otherwise said: At the same room temperature we have to supply 13% less heat energy into the room where the exhaust opening is situated below the window. Moreover the distribution of air is here more favourable, if we may really speak of favourable, having an outerwall of 100% single glazed window.

14. I still would like to show you with another example how to improve the climate of a room with a panel heating.

Figure 13 (T-335) and figure 14 (T-395) show the same room as before, but with a double glazed window, which makes up 43 per cent of the outside wall.

Test-condition: A constant room temperature at a constant outside temperature but variable air changes (LW).

As we immediately notice, air distribution in the room is at 0 air change rather good, it would absolutely meet normal demands. But near the window we notice directly above the ground an air movement which is about 20 cm/sec but against which people could protest.

At 4 air changes air distribution is essentially better and the cold air stream above the ground is completely eliminated.

DISCUSSION

K. H. SCHEUNEMANN (East Germany) — In figure 6 is shown the influence of distance between slot and measuring point on air-velocity in case of 1.5 mm slot-width.

How do these values change, if there is a slot of 3 or 4 mm width? Are they quite parallel above the function you have shown in the paper?

H. SCHUCHARDT — We did not run tests with slots of 3 or 4 mm width but we can say that function will be quite congruent. The range of the stream will increase with an increasing slot width, for instance, a double slot width will double the range.

J. LEBRUN (Belgique) — Je voudrais simplement souligner l'importance du champ convectif, au ras du sol, induit par une paroi froide non « compensée »; à ce propos, la reprise d'air au bas de la paroi est sans doute très efficace.

Il subsiste cependant une autre source d'inconfort éventuel : c'est l'*anisotropie radiative*.

La température moyenne de rayonnement relative à un élément de surface orienté vers la paroi peut être assez basse avec ce type d'installation.

L'importance de ces phénomènes *locaux* s'est révélé à l'occasion de nos recherches sur le confort en ambiances anisotropes.

H. SCHUCHARDT — Our experience in this respect is the following: There is no discomfort if people, even sitting, observe a distance of 0.7 to 1.0 m between window surfaces and the body; this fact also depends on the window construction, e.g. many installations in hospitals and schools.

OPTIMISATION DU DÉBIT D'AIR EXTÉRIEUR DANS UN CONDITIONNEMENT À EXTRACTION D'AIR PAR LES LUMINAIRES

H. MARCQ

Marcq & Roba, Ingénieurs-Conseils, Bruxelles (Belgique)

Optimising the fresh air flow in an air conditioning system using a ventilated luminaire

SUMMARY: *During periods of room cooling, the air conditioning of non-industrial spaces is usually carried out by using the minimum amount of fresh air required for hygienic and comfort conditions. In cases where the air is being extracted by ventilated luminaires—this technique becoming more frequent due to the increase in artificial lighting—in order to reduce the heat load in rooms, the A. demonstrates that it may be of interest (in economising the refrigerating capacity) to increase the fresh air flow to a figure above the predetermined minimum.*

The report explains a semi-analytic, semi-graphic method for determining the thermally optimal fresh air flow; economic optimisation will, in addition, take into account those expenses besides those involved in supplying refrigeration, expenses which must be estimated for each individual case.

1. INTRODUCTION

L'accroissement considérable dans les dernières années du niveau d'éclairement artificiel réalisé dans les immeubles de bureaux modernes conduit à une puissance absorbée par les luminaires dépassant fréquemment 50 watts par m^2 de surface éclairée. Toute cette puissance se transforme en chaleur; pour éviter qu'elle ne se transmette totalement au local où elle constitue une charge calorifique à combattre par le conditionnement d'air, on combine de plus en plus les luminaires avec l'extraction d'air (luminaires intégrés) afin de faire emporter par l'air extrait une partie de cette chaleur. Cet air extrait peut être en partie évacué et en partie recyclé; l'économie que l'intégration permettra de réaliser sur la puissance frigorifique totale de l'installation dépendra évidemment du débit d'air évacué et par conséquent du débit d'air extérieur traité par l'installation de conditionnement pour compenser cette évacuation.

La présente étude expose une méthode d'optimisation du débit d'air extérieur, et par voie de conséquence, du débit d'air à évacuer par les luminaires.

2. FACTEUR DE REFROIDISSEMENT D'UN LUMINAIRE À EXTRACTION D'AIR

Considérons un luminaire absorbant une puissance électrique totale P et traversé par un débit Q d'air extrait (exprimé en poids d'air sec par unité de temps) entrant dans le luminaire à la température du local t_i°C (ou plus exactement à la température de l'air au voisinage du plafond du local).

Appelons facteur de refroidissement r le rapport à la puissance P de la partie de cette puissance emportée *en fin de compte* par l'air extrait, c'est-à-dire sous déduction de la chaleur restituée aux locaux par le plenum entre faux plafond et plancher si l'extraction se fait par ce plenum, ou indirectement par les conduits si l'extraction se fait par conduits calorifugés ou non. Le facteur de refroidissement r dépendra non seulement de ces conditions d'extraction mais aussi du type de luminaire (conception — nombre et type de lampes — nombre et type de stabilisateurs), du débit d'air Q et de la température t_i.

On peut dire notamment que :

— r est nettement plus élevé si l'air traverse le compartiment des lampes et stabilisateurs (refroidissement direct) que s'il passe autour du luminaire dans un capot adapté sur celui-ci (refroidissement indirect); par exemple, Siemens [1] a obtenu dans des essais sur un luminaire à 2 tubes de 65 W avec grille de défilement, les valeurs suivantes de r mesuré à la sortie du luminaire en fonction du débit d'air volumique Q :

	10 m^3/h	20 m^3/h	30 m^3/h
Refroidissement direct	0,37	0,57	0,68
Refroidissement indirect	0,30	0,45	0,52

— r est plus élevé si la face inférieure du luminaire, sauf à l'endroit de pénétration de l'air, est un écran transparent ou diffusant que si elle est constituée d'une grille de défilement, en raison de l'absorption des rayons infra-rouges à grande longueur d'onde par l'écran [2];
— r croît lorsque le débit d'air Q croît, en tendant asymptotiquement vers un maximum qu'il n'atteindrait que si l'émission d'énergie du luminaire vers le local se limitait au rayonnement lumineux et calorifique sans transmission par convection;
— r doit varier avec la température t_i de l'air entrant dans le luminaire; à notre connaissance, cette variation n'a pas encore été étudiée mais n'a probablement que peu d'importance, d'autant plus qu'en pratique t_i ne variera qu'entre 20 et 25 °C.

3. Position du problème

Il résulte de la définition de r que la puissance calorifique W emportée en fin de compte par l'air extrait au travers du luminaire vaut :

$$W = r P \tag{1}$$

Cette évacuation de chaleur par l'air extrait réduit d'autant les apports calorifiques au local, d'où *réduction correspondante du débit d'air conditionné à pulser dans le local en régime maximum d'été et/ou augmentation possible de la température de cet air en conditionnement tout air ou réduction de la puissance de l'unité terminale en conditionnement air-eau.*

Mais *la chaleur évacuée par l'air extrait ne réduira la puissance frigorifique totale à installer et la consommation de frigories que si cet air* n'est pas recirculé *mais bien évacué à l'extérieur en tout ou en partie.*

Or, en général, l'air extérieur (air de ventilation) préconditionné introduit dans les locaux et qui constitue, selon les systèmes et les cas, tout ou partie de l'air qui y est pulsé, sert partiellement à mettre les locaux en pression et en sort :
— en partie par l'inétanchéité des parois extérieures;
— en partie par l'inétanchéité, éventuellement favorisée par l'établissement de bouches de passage, des parois vers les dégagements, d'où il rejoint le plus souvent des toilettes et des locaux secondaires mis en dépression par une simple extraction mécanique;
— le solde par un réseau d'extraction d'air dont, dans la présente étude, les bouches sont les luminaires.

Il en résulte que *la partie évacuée à l'extérieur du débit d'air extrait au travers des luminaires doit nécessairement rester inférieure au débit d'air extérieur introduit dans les locaux*, la différence devant rester égale à la somme des débits de mise en pression par rapport à l'extérieur et de compensation de l'extraction des toilettes et locaux secondaires.

D'autre part, *le débit d'air extérieur introduit dans les locaux est généralement fixé par le minimum hygiénique et/ou de confort correspondant au nombre d'occupants prévus;* ce minimum est en tout cas maintenu pour limiter les puissances calorifique et frigorifique et les frais d'exploitation :

— tant que les locaux demandent du chauffage;

— dans les périodes où les locaux demandent du refroidissement, tant que la température ou l'enthalpie de l'air extérieur est supérieure à celle de l'air intérieur.

Ce n'est qu'en demi-saison, si la température ou l'enthalpie de l'air extérieur est inférieure à celle de l'air intérieur avec demande de refroidissement dans les locaux que, dans certaines installations, on passe manuellement ou automatiquement à un débit d'air plus élevé que le minimum, afin de réduire les frais d'exploitation.

Avec extraction au travers des luminaires, du fait de la croissance de leur facteur de refroidissement *r* avec le débit extrait, *on doit se demander s'il n'y a pas intérêt à augmenter en période de refroidissement le débit d'air extérieur introduit dans les locaux au-delà du minimum hygiénique et/ou de confort* même lorsque la température ou l'enthalpie de l'air extérieur est supérieure à celle de l'air intérieur.

Pour répondre à cette question, nous allons chercher à déterminer le débit d'air extrait optimal du point de vue thermique, d'une part pour les conditions extérieures extrêmes d'été en vue de rendre minimale la puissance frigorifique installée, d'autre part, dans des conditions extérieures quelconques en vue de rendre minimale la consommation de froid.

4. CAS D'UN CONDITIONNEMENT « AIR-EAU »

Nous considérons d'abord le cas d'un système de conditionnement « air-eau » (zone périphérique, par exemple) à *débit d'air pulsé constant* quelles que soient les charges thermiques positives ou négatives des locaux. *Le minimum de ce débit est égal au minimum hygiénique ou de confort du débit d'air extérieur;* mais on peut choisir lors de la conception de l'installation un débit pulsé supérieur à ce débit d'air extérieur minimum, quitte à recirculer une partie de l'air extrait en période de chauffage et, éventuellement, pour certaines conditions extérieures de la période de refroidissement.

Le débit d'air extrait au travers des luminaires doit nécessairement rester également constant en exploitation; il est égal au débit pulsé moins la somme des débits de mise en pression par rapport à l'extérieur et de compensation de l'extraction des toilettes et locaux secondaires. Une augmentation du débit pulsé à la conception de l'installation entraîne donc la même augmentation du débit extrait.

Cherchons à rendre minimale la puissance frigorifique nécessaire dans les conditions extérieures extrêmes d'été. Nous devons considérer à cet effet deux postes du bilan thermique rapporté à un luminaire dans ces conditions extrêmes :

— la puissance frigorifique W_e nécessaire pour ramener le débit d'air extérieur introduit dans les locaux en supplément du minimum hygiénique ou de confort à l'enthalpie correspondant à la température et à l'humidité relative désirées dans les locaux; si i_e et i_i en kcal/kg sont les enthalpies extérieures et intérieures, Q_m en kg/h le débit d'air à extraire par le luminaire si on se contentait du minimum d'air extérieur et Q en kg/h le débit à extraire avec le débit d'air extérieur majoré, cette puissance supplémentaire vaut :

$$W_e = (Q - Q_m)(i_e - i_i) \qquad (2)$$

— la puissance calorifique W_l provenant du luminaire et évacuée à l'extérieur en supplément de celle qui serait évacuée si on se contentait du minimum de débit d'air extérieur; si r et r_m sont les facteurs de refroidissement du luminaire respectivement pour les débits Q et Q_m, on a, en vertu de (1) :

$$W_l = (r - r_m)\, P \tag{3}$$

La puissance frigorifique totale nécessaire sera diminuée de W_l et majorée de W_e; la diminution nette ΔW vaudra donc :

$$\Delta W = (r - r_m)\, P - (Q - Q_m)(i_e - i_i) \tag{4}$$

et il s'agit de trouver la valeur de Q, que nous appellerons Q_0, qui rend ΔW maximum, compte tenu de ce que r est une fonction expérimentale de Q.

Si cette fonction pouvait se traduire par une expression analytique, la valeur de Q_0 pourrait être calculée. Il est plus simple d'appliquer la méthode graphique suivante (fig. 1).

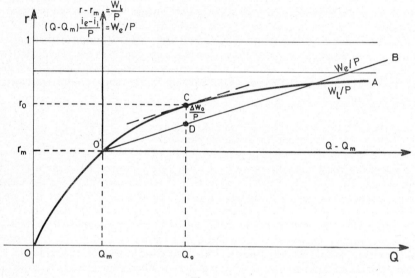

Fig. 1

Dans un diagramme cartésien rectangulaire, on trace la courbe OA, caractéristique du luminaire considéré et du système d'extraction adopté, figurant r en ordonnée en fonction de Q en abscisse; cette courbe part de l'origine O des axes et tend asymptotiquement vers une horizontale d'ordonnée inférieure à 1 (voir § 2).

Sur cette courbe, on repère le point O′ d'abscisse Q_m et dont l'ordonnée vaut donc r_m.

Par le point O′ pris comme origine, on trace un second système d'axes rectangulaires parallèles au premier; la courbe OA représente dans ce 2ᵉ système $r - r_m = W_l/P$, en ordonnée en fonction de $Q - Q_m$ en abscisse.

Dans ce même système, figurons la variation de

$$\frac{W_e}{P} = (Q - Q_m) \cdot \frac{i_e - i_i}{P}$$

en ordonnée en fonction de $(Q-Q_m)$ en abscisse; c'est une droite O'B passant par l'origine O' et de coefficient angulaire

$$\frac{i_e-i_i}{P}.$$

Les différences d'ordonnées entre la courbe O'A et la droite OB représentent, dès lors, la variation de

$$\frac{\Delta W}{P} = (r-r_m) - (Q-Q_m) \cdot \frac{i_e-i_i}{P},$$

et comme P est pratiquement constant*, le maximum ΔW_0 de ΔW correspondra au maximum de ces différences d'ordonnées. Il suffira donc de tracer la tangente à la courbe O'A parallèle à la droite O'B; les coordonnées du point de tangence C dans le 1er système d'axes seront respectivement :

— le débit d'air thermiquement optimal Q_0 au travers du luminaire;

— le facteur de refroidissement r_0 correspondant.

L'économie de puissance frigorifique que l'extraction au travers du luminaire permettra de réaliser comportera dès lors :

— un terme ΔW_0 qui peut être calculé en multipliant la longueur CD par l'échelle des ordonnées r ou $(r-r_m)$ et par la puissance P;

— un second terme correspondant à la chaleur emportée par le débit minimum Q_m et qui vaut $r_m P$.

L'importance relative de ces 2 termes dépendra de l'allure de la courbe OA et de la position de O' sur cette courbe ou du débit Q_m.

Il faut remarquer que le débit Q_0 est le *débit thermiquement optimal* (puissance frigorifique minimale) mais n'est pas le plus économique pour l'ensemble de l'installation car l'augmentation du débit par rapport à Q_m augmente le coût des groupes et des réseaux de pulsion et d'extraction et la consommation d'énergie électrique de ces groupes. La solution économiquement optimale correspondra donc à une valeur de Q comprise entre Q_m et Q_0 et qui devra être déterminée par un calcul économique; l'intérêt de la présente méthode est de délimiter la zone de débits dans laquelle se trouvera le *débit économiquement optimal*.

REMARQUES

a) Si la droite O'B se situe au-dessus de la courbe O'A, cela signifie que le débit thermiquement optimal est inférieur au débit minimal Q_m; il y a lieu alors d'adopter ce dernier débit.

b) Pour des enthalpies de l'air extérieur inférieures à celle des conditions d'été les plus défavorables, on aura intérêt au point de vue économie d'exploitation à maintenir le débit d'air extérieur résultant des considérations précédentes tant que l'enthalpie de l'air extérieur restera supérieure à celle de l'air intérieur. Lorsqu'elle devient inférieure, tout dépend de l'importance et du signe des charges thermiques des locaux.

(*) En fait, la puissance électrique absorbée par un luminaire à tubes fluorescents servant à l'extraction d'air croît en même temps que son flux lumineux lorsque le débit d'air qui le traverse augmente, mais la variation est faible dans le domaine des débits à envisager pratiquement.

Considérons maintenant le cas d'un *système de conditionnement « tout air »* (par exemple, zone intérieure en basse pression avec ou sans rechauffage ou système à double conduit) *à débit d'air pulsé constant* quelles que soient les charges thermiques positives ou négatives des locaux.

Dans ce cas, le débit d'air pulsé est généralement beaucoup plus important que le débit d'air extérieur minimal et il est fixé à la conception par les charges thermiques positives maximales des locaux et par la différence de température maximale admissible entre l'air pulsé et les locaux.

Il en résulte que, quel que soit le débit d'air extérieur utilisé, *le débit d'air extrait au travers des luminaires reste constant; le facteur de refroidissement est donc aussi constant.* Désignons le débit constant d'un luminaire et le facteur r constant par Q_a et r_a.

La puissance calorifique provenant du luminaire et emportée par l'air extrait vaut donc :

$$r_a\, P$$

Désignons par Q_{em} le débit d'air extérieur minimum du point de vue hygiène et confort et par Q_e le débit d'air extérieur adopté supérieur ou égal à Q_{em}; ces débits sont rapportés à un luminaire.

La puissance frigorifique W_e rapportée à un luminaire nécessaire pour ramener le débit d'air extérieur introduit en supplément du minimum à l'enthalpie correspondant aux conditions intérieures désirées vaut :

$$W_e = (Q_e - Q_{em})(i_e - i_i) \tag{5}$$

La puissance calorifique W_l provenant du luminaire et évacuée à l'extérieur en supplément de celle qui serait évacuée si on se contentait du minimum d'air extérieur vaut :

$$W_l = \frac{Q_e - Q_{em}}{Q_a} \cdot r_a\, P \tag{6}$$

La diminution nette de puissance frigorifique due à l'augmentation du débit d'air extérieur au-delà du minimum vaut dès lors :

$$\Delta W = \frac{Q_e - Q_{em}}{Q_a} \cdot r_a\, P - (Q_e - Q_{em})(i_e - i_i)$$

ou

$$\Delta W = (Q_e - Q_{em})\left[\frac{r_a\, P}{Q_a} - (i_e - i_i) \right] \tag{7}$$

Si le crochet est positif, la puissance frigorifique diminue réellement lorsque le débit d'air extérieur adopté augmente au-delà du minimum et la diminution est proportionnelle au supplément de débit d'air extérieur par rapport au minimum. Remarquons qu'en appelant Q_m le débit d'air extérieur minimum Q_{em} moins la somme du débit de mise en pression s'il est nécessaire et du débit de compensation de l'extraction des toilettes et locaux secondaires, on doit bien entendu avoir :

$$Q_e \leqslant Q_a + (Q_{em} - Q_m) \tag{8}$$

Si le crochet est négatif, la puissance frigorifique augmente lorsque le débit d'air extérieur adopté augmente et il n'y a donc aucun intérêt à augmenter ce débit. Il faut remarquer que même s'il en est ainsi pour les conditions les plus défavorables d'été, il n'en sera plus de même lorsque i_e se sera suffisamment rapproché de i_i et a fortiori pour $i_e < i_i$; dès que $i_e - i_i$ sera inférieur à $r_a P/Q_a$, on aura intérêt en exploitation à porter le débit d'air extérieur au maximum possible.

Il faut noter que, dans le cas d'un conditionnement « tout air », l'augmentation du débit d'air extérieur ne donne lieu qu'à une légère augmentation du prix du groupe de pulsion (batterie de refroidissement), mais ne change pas le prix du groupe d'extraction ni des réseaux de pulsion et d'extraction.

Si $i_e - i_i$ est inférieur à $r_a P/Q_a$ et si on réalise alors le débit d'air extérieur maximal possible, c'est-a-dire $Q_e = Q_a + (Q_{em} - Q_m)$, l'économie totale de puissance frigorifique que l'extraction au travers du luminaire permet de réaliser vaudra :

$$r_a P - W_e = r_a P - (Q_e - Q_{em})(i_e - i_i)$$

$$= r_a P - (Q_a - Q_m)(i_e - i_i)$$

$$= r_a P \cdot \frac{Q_m}{Q_a} + (Q_a - Q_m) \left[\frac{r_a P}{Q_a} - (i_e - i_i) \right] \qquad (9)$$

et tout l'air extrait au travers du luminaire sera évacué à l'extérieur.

Si $i_e - i_i$ est supérieur à $r_a P/Q_a$, il faut s'en tenir au minimum d'air extérieur et l'économie de puissance frigorifique se limite à :

$$r_a P \cdot \frac{Q_m}{Q_a} \qquad (10)$$

puisque seule la fraction Q_m/Q_a du débit Q_a d'air extrait au travers du luminaire pourra être évacuée à l'extérieur.

6. REMARQUES GÉNÉRALES

1) Toutes les considérations précédentes ne sont plus valables telles quelles si on réalise un échange de chaleur ou même d'enthalpie entre l'air évacué et l'air extérieur à préconditionner.

2) Il est bien entendu que les débits d'air extérieur optimaux déterminés dans cette étude ne valent que tant que l'air extérieur ne doit pas être chauffé.

RÉFÉRENCES

[1] H.J. HENTSCHEL et A. WALD, Installations intégrées de climatisation dans les grands bâtiments administratifs, *Revue Siemens* (1967), n° 10.
[2] EDISON ELECTRIC INSTITUTE, Electrical Space Conditioning Concepts, New York (1965).
[3] K. MEAKAWA, Evacuation de chaleur des appareils d'éclairage incorporés combinés avec les bouches d'extraction, *Industries Thermiques et Aérauliques* (novembre 1968).

DISCUSSION

J. HANNAY (Belgique) — Dans le cas des installations avec extraction d'air par les luminaires devant la croissance des puissances d'éclairage installées et devant la croissance du nombre d'heures d'utilisation de la lumière artificielle, je voudrais demander à M. MARCQ s'il n'est pas possible d'envisager, dès maintenant, le chauffage des locaux en mi-saison et en hiver par une récupération efficace de la puissance installée pour l'éclairage, par exemple en inversant le circuit d'air.

En cas d'insuffisance des puissances installées il serait certainement aisément possible d'ajoindre un appareil électrique.

Le coût plus élevé de l'utilisation ne serait-il pas compensé par l'économie réalisée sur la simplification de l'installation de conditionnement qui ne fonctionnerait plus qu'en rafraîchissement?

H. MARCQ — Sans aucun doute, il y a intérêt pour des bâtiments à haut niveau d'éclairement artificiel, à étudier dans chaque cas les possibilités techniques et économiques de la récupération de la chaleur dégagée par les luminaires *durant les périodes où on a besoin de chaleur*; ceci n'était pas mon propos puisque je n'ai examiné la question des luminaires intégrés que du point de vue frigorifique (voir § 1 et remarque 2 du § 6).

Mais le problème de la récupération de chaleur est loin d'être simple et ne peut se résoudre par une simple inversion du circuit d'air. Il serait trop long d'exposer au cours de cette discussion tous les éléments de ce problème et les solutions possibles et déjà réalisées.

Il est certain aussi qu'une telle récupération peut dans des cas concrets rendre rentable la solution « tout électrique ». Il ne faut cependant pas s'imaginer que l'installation de conditionnement ne fonctionnerait plus qu'en rafraîchissement; il restera toujours au moins à chauffer l'air de ventilation, à combattre l'effet des parois froides et à assurer un chauffage au ralenti durant les périodes de non occupation.

H. MARCQ — M. F. CLAIN m'a fait remarquer que l'équation (7) peut se mettre sous la forme :

$$\Delta W = (Q_e - Q_m)\left[\left(i_i + \frac{r_a P}{Q_a}\right) - i_e\right] \tag{7'}$$

or $i_i + r_a P/Q_a$ est l'enthalpie de l'air extrait à l'aval du plénum ou des conduits d'extraction. On peut donc dire que la puissance frigorifique nécessaire diminue, lorsqu'on augmente le débit d'air extérieur, dès que l'enthalpie de l'air extrait à l'aval du plénum ou des conduits d'extraction est supérieure à l'enthalpie de l'air extérieur, ce qui est assez intuitif.

MAXIMUM EXPLOITATION OF INSTALLED CAPACITY IN AIR CONDITIONING OF HOSPITALS

J. ROTHSCHILD

Government of Israel, Public Works Dept., Tel-Aviv (Israel)

Exploitation maximale de la puissance installée dans le conditionnement d'air des hôpitaux

RÉSUMÉ : *Dans les pays méditerranéens et des zones subtropicales, l'un des facteurs limitatifs les plus sérieux est le manque de fonds pour le conditionnement d'air total des hôpitaux, aussi bien pour les parties du bâtiment où ce conditionnement d'air est essentiel, que pour celles où il est seulement souhaitable. Les salles de malades, habituellement considérées comme non essentielles, ne bénéficient pas du conditionnement d'air. Dans certains cas, l'air chaud et humide retarde aussi la convalescence des malades.*

Les zones où le conditionnement d'air est indispensable sont les salles d'opération et les salles de rayons X, les salles de dispensaire et les laboratoires. On sait que la pleine puissance pour laquelle le système de conditionnement d'air de ces zones a été conçu n'est normalement nécessaire qu'entre 8 h du matin et 4 h de l'après-midi. Il apparaît donc qu'en dehors des heures de pointe, une partie du moins de la puissance peut être détournée vers les zones considérées comme non essentielles. Cela permettrait de conditionner la plus grande partie des zones de l'hôpital pendant la plus grande partie du jour et pendant la plus grande partie du temps au cours de la saison chaude.

On a étudié le cas d'un hôpital général de 800 lits, dans la région de Tel-Aviv. Les besoins pour la partie de l'hôpital où le conditionnement d'air est essentiel sont d'environ 1 500 000 fg/h. Les besoins pour la zone où il n'est pas essentiel sont d'environ 2 100 000 fg/h.

D'après une étude des conditions atmosphériques, on est arrivé à la conclusion que les heures de pointe sont comprises entre 10 h du matin et 3 h de l'après-midi, pendant deux mois seulement sur les huit mois de la saison chaude.

Ainsi en dehors des heures de pointe, en particulier pendant les heures humides du soir, on peut utiliser le conditionnement d'air pour réduire l'inconfort des malades alités.

Une analyse du coût montrerait que l'augmentation de dépense, pour réaliser le conditionnement d'air total, est surtout due au remplacement des radiateurs ou convecteurs par des groupes serpentins-ventilateurs, à l'augmentation des dimensions de tuyaux et des différentes isolations. Il faut avoir à l'esprit que le chauffage de toutes les zones de l'hôpital est considéré comme un besoin essentiel.

INTRODUCTION

Hospitals today in general require full air conditioning. In the most developed countries all areas of hospitals are air conditioned. This has been found to be essential not only for operating rooms and laboratories, but also for wards and other areas occupied by patients and professional staff. According to medical opinion, air conditioning for patient-occupied areas facilitates recovery and convalescence for many diseases.

In Mediterranean and sub-tropical countries, the above-mentioned problems become even more acute. However, in such countries the financing of modern hospitals presents grave difficulties. Due to lack of funds, air conditioning is installed only in the most essential areas, such as operating rooms, laboratories, X-rays rooms and outpatient clinics.

This analysis intends to show how, in spite of limited funds, air conditioning may be installed in all areas both where it is essential and where it is merely desirable. This can be done with only a slight increase in expenditure above what is required for the essential areas only. This would bring comfortable conditions over most of the day during the season when cooling is required.

The cooling season in the Mediterranean areas extends over a period of eight months. Moreover it will become apparent that during the cooling season, the evening and early morning hours are as uncomfortable as the peak load hours.

In hospitals, the areas considered essential for air conditioning are generally in use during normal working hours. Therefore, during off hours the existing plant capacity is available for cooling elsewhere.

GENERAL ANALYSIS

The following graphical representation of temperature variation, and heat flow through walls and roofs during the various hours of the day and month, will enhance the argument for desirability of providing full treatment of air supply in all departments and parts of hospitals. The data have been compiled from the Carrier design manual. Climatic data have been compiled from information supplied by the meteorological service of the State of Israel.

תנאים אקלימיים הקריה בתל-אביב

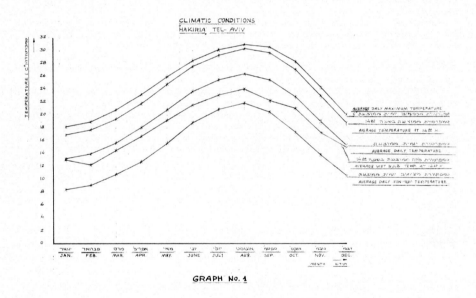

GRAPH No. 1

Graph No. 1 shows the average daily maximum temperature to be above 23 °C from April through November.

The daily average minimum temperatures are above 20°C from June through September.

It is therefore considered that the cooling season lasts for 8 months of the year, of which there are 4 months when air conditioning is needed even during the coolest hours of the day.

Graph No. 2 shows the relation between dry bulb and wet bulb temperatures over the full range of the day during the peak cooling season, namely from June through September.

542

תנאים אקלימיים , הקריה בתל-אביב

CLIMATIC CONDITIONS HAKIRIA TEL-AVIV

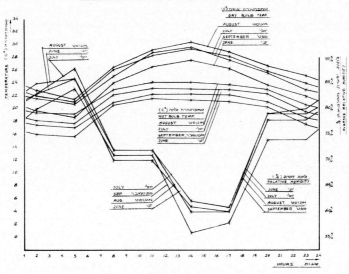

GRAPH No: 2

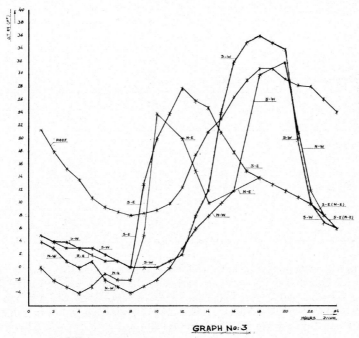

GRAPH No: 3

NOTES:

FROM: HANDBOOK OF AIR CONDITIONING SYSTEM DESIGN
CARRIER AIR CONDITIONING COMPANY.
TABLES No: 19-20
JULY - 30° N.LAT
DAILY RANGE 20°F
ΔT = 15°F
WALLS - DARK COLOR - 60 Btu/ft²hr
ROOF - MEDIUM COLOR 80 °.

543

It is noticed that the relative humidity during all four months is above 70% at dry bulb temperature over 22°C up to 8 a.m. and again from 6 p.m. throughout the night.

Such conditions present physical hardships for patients. Air conditioning is therefore warranted during this period as much as during peak load hours.

Graph No. 3 gives a pictorial representation of heat flow through masonry, walls and roofs based on a 60 lb/sq.ft. wall and 80 lb/sq.ft. roof, at 30°N latitude. Daily temperature range 20°F, and 15°F differential between outside and inside design temperature. The southwest and northwest exposures show a transmission heat gain particularly in the late afternoon. A peak transmission load is noticed at 6 p.m. and it falls off only after 8 p.m. The northeast and southeast exposures show transmission gains from the late morning to the early afternoon.

The case in question is an 800 bed general hospital to be built near Tel Aviv. The cooling load for the essential areas is 500 tons refrigeration. The cooling load for the non essential is 700 tons refrigeration

The refrigeration load distribution and operating hours for air conditioning is as follows:

| | | | Operating hours | |
| | | | | |
Essential Parts	Area aircond. m²	Operating T.R.	24 hours daily	Between 8 a.m to 4 p.m.
			T.R.	T.R.
Operation rooms	1232	189	40	149
X-rays	1560	58	12	46
Laboratories	128	28	—	28
Outpatient	3249	165	—	165
Receiving	542	39	39	—
Library, lecture hall	432	37	—	37
Central supply	1780	32	—	32
	8323	548		
Diversity		500	91	457
Non-essential parts				
Wards	18446	599	599	—
Laboratories	4134	180	—	180
Dining hall	590	40	—	40
	31493	1367		
Diversity		1260	690	677

The four graphs, appearing in the following pages represent the calculated load relation between essential and non-essential areas for the months June through September.

The maximum load for essential areas never rises above 500 tons. For non-essential areas the load from 4 p.m. to 8 a.m. is about 600 tons, except for the month of August between 4 and 7 p.m. when it rises to about 660 tons.

With a 600 ton plant it is possible to provide cooling for the non-essential areas during the off-peak hours and thereby relieve the uncomfortable conditions in the wards and patients' rooms for most of the day and particularly during the hours of greatest humidity

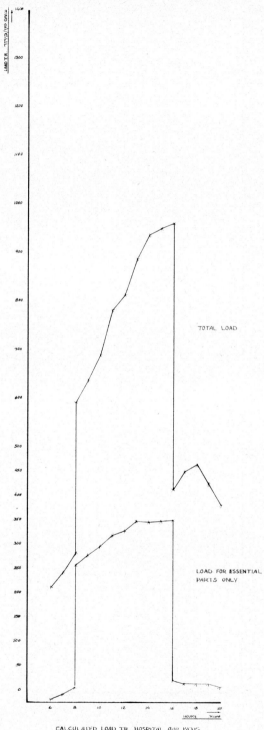

CALCULATED LOAD FR HOSPITAL 800 BEDS
(JUNE)

545

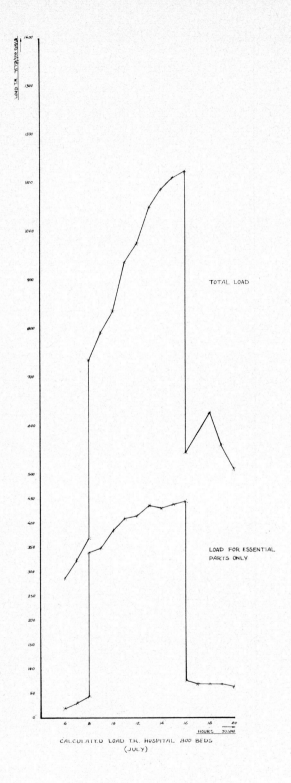

CALCULATED LOAD T.R. HOSPITAL 300 BEDS
(JULY)

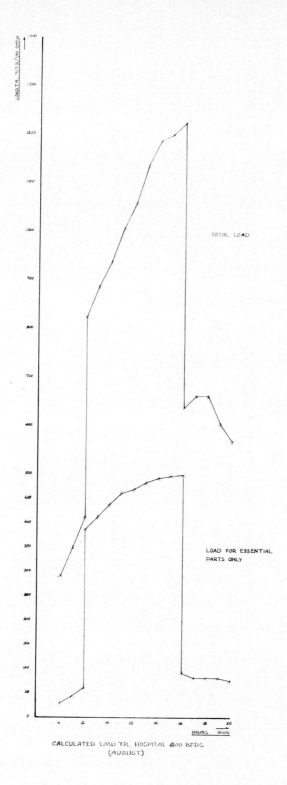

LOAD T.R. 10 בקו"ח/שע

1400

1300

1200

1100

1000

900

800

700

600

500

450

400

300

340

280

200

150

100

50

0

6 8 10 12 14 16 18 20

HOURS   שעות

TOTAL LOAD

LOAD FOR ESSENTIAL
PARTS ONLY

CALCULATED LOAD T.R. HOSPITAL 800 BEDS
(AUGUST)

547

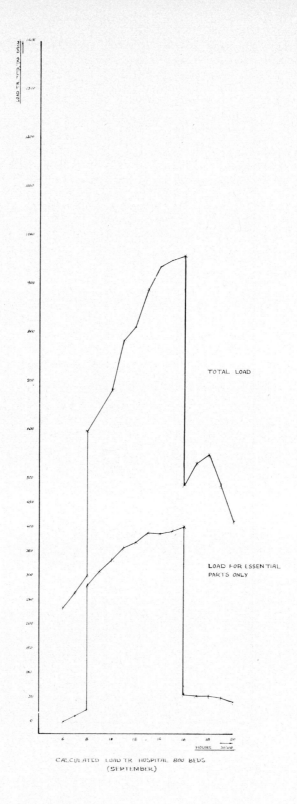

CALCULATED LOAD FR HOSPITAL 800 BEDS
(SEPTEMBER)

548

A further check will show that during the remainder of the cooling season normally from April through May and from September through November, cooling can be provided for the non-essential areas even for the whole day.

SUGGESTIONS FOR A SOLUTION

Cooling the essential areas during normal operating hours, and non-essential areas during the remainder of the day (for 4 peak months, as described above), will require a 600 ton refrigeration plant. This plant is only 20% (100 tons) larger than the 500 tons required for the essential areas.

Heating for the winter season is mandatory. Therefore, at least radiators and a hot water piping system must be installed. It has been found practical, in all hospital construction, to install fan coil units in lieu of radiators and increased diameter piping thus allowing in the future for the use of hot/chilled water system. This will prepare for future air conditioning in all space without major alterations in construction.

COST ANALYSIS

Cost/m^2 (without equiment room) with radiators   IL.   40.–
Cost/m^2 (without equipment room) with fan and coil units IL. 102.–

The extra cost in respect of fan and coil units is a general expense for projected air conditioning, not particular to the problem under discussion, and therefore, cannot be considered as an additional expense for the purposes of this cost analysis.

Cost for equipment per ton refrig.   IL. 1250.–

Differential cost between 500 and 600 ton plant (equipment only) should be considered as an additional expenditure above that required for cooling essential areas only.

The net gains of such a system are as follows:

Air conditioning is provided for an area normally served by a 1225 ton plant, with a 600 ton plant only. This smaller plant means lower initial cost of investment in machinery, installation, electrical supply, as well as a saving in space required for machines and cooling tower.

The savings in terms of cash are the following:

Refrigeration 625 tons × IL. 1250/ton	IL. 781 250.–
Elect. sup. 800 H.P. × IL. 90.–/HP	IL. 72 000.–
Construction 120 m^2 × IL. 450.–/m^2	IL. 54 000.–
Total saving	IL. 907 250.–

In addition, in Israel where purchase tax is levied on refrigeration plants for air conditioning according to capacity, a net saving of IL. 500/T.R. is also realized. This amounts to IL. 312 500 net saving to bring the total saving to IL. 1 219 750.–.

The cost of a 1225 ton plant providing air conditioning for the entire hospital during the entire day and night is as follows:

Cost of refrig. plant 1225 ton × IL.3800.–/ton	IL. 4 655 000.–
Construction cost for machine room and cooling tower:	
550 m^2 × IL. 450.–/m^2	IL. 247 500.–
Electrical supply	IL. 140 000.–
Total	IL. 5 042 000.–
Purchase tax	IL. 630 000.–
	IL. 5 672 000.–

Percentage saving by using the 600 ton plant instead of the 1225 ton plant serving the entire hospital is 22%.

METHOD OF SUPPLY AND CONTROL OF COOLING CAPACITY TO THE ESSENTIAL AND NON-ESSENTIAL PARTS OF THE HOSPITAL

The basic chilled water supply system will consist of primary and secondary pumping loops. The problem arises that during certain periods a 1225 ton water flow will be served by a 600 ton plant. This can be overcome by sizing the primary pumps at half the temperature differential range or twice the normal water flow. The secondary pump will be sized normally. This will assure a constant flow through all pumps and full capacity delivery of chilling plant.

The individual secondary zone pumps can be controlled from a centralized control board.

Within each zone individual air handling systems can be controlled from the centralized control board and their water supply by-passed when supply fan is switched off.

Actual load determinations can be ascertained by a simple method of measuring the return water temperature at individual zones and at the chiller. This will enable cooled water to be supplied to the non-essential areas as the load on the essential area falls off.

The above is only one suggested method of a variety of alternative ways of control applications. It is of secondary importance at this stage to deal with control applications since the above-mentioned system is presented merely as a tentative proposal.

CONCLUSION

The ideas outlined above take into account the limited financial resources available to many hospitals now under construction, and offer a solution as near as possible to the ideal, by alleviating the worst discomfort likely to be suffered by patients, particularly during the humid morning and evening hours.

Furthermore, two days a week a major part of the essential area does not operate. During this period full air conditioning can be provided for the wards.

When air conditioning is supplied to non-essential areas during the night and early morning hours, the storage effect of the structure will keep conditions in the wards semi-comfortable long after the system has been switched off.

REMARQUE

G.F. BERTOLINI (Italie) — Je me rapporte aux communications de MM. MARCQ et ROTHSCHILD et je remercie spécialement M. ROTSCHILD pour avoir affirmé la nécessité de conditionner l'air dans les salles de malades dans lesquelles d'habitude l'air ambiant chaud et humide retarde la convalescence des malades, qui se poursuit dans une ambiance très inconfortable.

La solution proposée par M. ROTSCHILD fait contraste avec la tendance actuelle à employer les salles d'opération pendant toute la journée.

Mais il y a un autre moyen pour résoudre dans les hôpitaux ce problème : c'est-à-dire l'emploi de récupérateur de chaleur air-air.

Dans le cas d'un hôpital à Milan le conditionnement d'air des salles des malades a été prévu et je vous donne à ce sujet quelques données que je pense être intéressantes.

Dans cet hôpital les chambres des malades sont climatisées, et les dépenses pour le chauffage et le refroidissement sont réduites par la récupération.

Le système de climatisation est constitué par la combinaison d'une installation de chauffage par panneaux légers (Stramax) et d'une installation de conditionnement d'air : l'installation de chauffage fournit la chaleur nécessaire pour compenser les déperditions thermiques des locaux; l'installation de conditionnement fournit l'air de ventilation dans la mesure de $2 \div 2,5$ volume/heure d'air frais, entièrement repris et expulsé au-dessus du toit.

Le bâtiment est, donc, traversé à partir du souterrain jusqu'au toit par un flux d'air frais pris à l'extérieur qui est traité pendant toute l'année, filtré, humidifié, chauffé ou refroidi et déshumidifié.

A l'orifice d'expulsion de l'air repris, on peut placer un échangeur de chaleur traversé à contre-courant par l'air expulsé et l'air extérieur qui est chauffé en hiver et refroidi en été, avant d'être envoyé comme air frais aux appareils de climatisation.

Le profit est considérable si nous considérons le grand volume des chambres de malades, mais il est encore plus grand pour les salles d'opération qui nécessitent $12 \div 20$ vol/h d'air frais.

Pour l'hôpital de Milan les données techniques sont les suivantes :
Conditions extérieures :

— Hiver $- 5°C$ et humidité relative 80%
— Été $+32°C$ et humidité relative 50%
Les conditions moyennes intérieures sont :

a) *salles de malades*

— Hiver $+22°C$ et humidité relative 50%
— Été $+27°C$ et humidité relative 50%

b) *salles d'opération*

24-26°C et humidité relative 50% toute l'année.
Le débit d'air expulsé est de 180 000 m³/h.

Les conditions obtenues dans l'air frais à la sortie du récupérateur sont en moyenne :

		Été	Hiver
Température au bulbe sec	°C	28,22	16,61
Température au bulbe humide	°C	21,94	12,22
Vapeur d'eau	g/kg	13,8	7,15
Enthalpie	kcal/kg	14,8	8,2

Les économies qui en résultent se chiffrent par
Été : 498 000 fg/h sur 1 637 000 nécessaires sans récupérateur; soit 30%
Hiver : 1 730 000 kcal/h sur 2 349 000 nécessaires sans récupérateur, soit 72%

Les dépenses supplémentaires pour l'installation des récupérateurs sont amorties après une année de fonctionnement.

THE TOTAL ENERGY SYSTEM AS AN ENERGY PRODUCER IN AIR CONDITIONING INSTALLATIONS

P. LEIJENDECKERS

Technisch Adviesbureau "van Heugten n.v."
Nijmegen (The Netherlands)

Le système à énergie totale en tant que producteur d'énergie dans les installations de conditionnement d'air

RÉSUMÉ : *Le rapport étudie une forme particulière de systèmes produisant simultanément chaleur et énergie. Une turbine ou un moteur à gaz fournit chaleur et électricité qui sont utilisées complètement d'une part dans une machine à absorption et d'autre part dans une machine à compression, en vue de fournir le médium de refroidissement à une installation de conditionnement d'air. Il est possible aussi de n'utiliser que l'électricité ou la chaleur.*
   Le rapport analyse ces trois possibilités au point de vue économique.
   Il semble que le moteur à gaz soit préférable à la turbine et que le système à énergie totale ne devienne intéressant que si le nombre d'heures de fonctionnement dépasse 1000 par an et que si le coût du gaz est supérieur à 0,07 F/m³.

1. INTRODUCTION

The basic principle to be discussed consists of a special form of an already well known system for combined heat and power production. The report does not deal with back-pressure steam turbines.

The operation of such steam installations is characterised by high investment costs and small flexibility in heat power production whereby the immediate production of electrical energy is determined by the demand for heat at any given moment. This could be disadvantageous to the refrigeration plant for an air conditioning system because of the relatively small number of working and full load hours.

The starting point for the various calculations and investigations, the results of which are presented in an installation called "Gas Total Energy System"; it has attracted a lot of attention with the coming of natural gas as a cheap and clean fuel.

The heart of such a total energy system is a prime mover (gas motor or gas turbine) which, as a driving machine, produces electrical energy by means of a generator. Because of the technical limitations inherent in this process, the heat content of the fuel can only partially be used for the production of electrical energy. The thermal efficiency, depending on varying factors, can have values between 15 and 35%.

The remainder of the heat content is freed as exhaust gases and as heat from the motor cooling system (gas motor) and is used in the "total energy" installation to cover certain heat demands.

As a cooling technique it was interesting to investigate as to what (after confining the application of the total energy principle to the cooling part of the air conditioning installation) economical advantages could be provided by a total energy system in general.

Technologically, three principles can be the basis of such an approach. Firstly a complete synthesis of the heat power installation with the cooling system. These ideas are worked out in figure 1.

Here, both the heat and the electrical energy produced by the gas turbine (or gas motor) are completely used up in an absorption- and compression-type refrigerating machine; solely to produce the cooling medium.

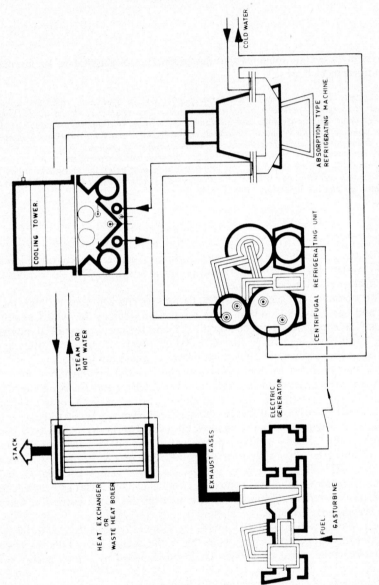

Fig. 1 — Basic total energy — Cooling system combination.

The next possibility is if there is a large enough outlet for the waste heat produced, to use only the electrical energy generated in the refrigerating system, viz., to drive the compression refrigerating machine (fig. 2).

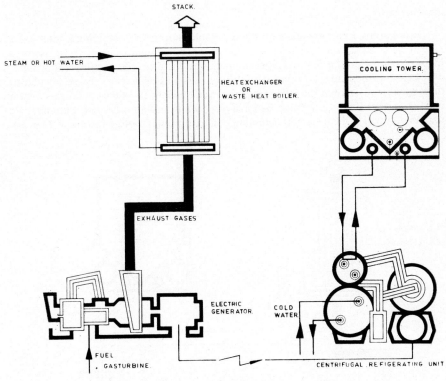

Fig. 2 — Total energy — Cooling system combination.

Finally, the waste heat coming from a total energy installation can be used solely in the production of a cooling medium in an absorption refrigerating machine, while the electrical energy produced can be used for other purposes (fig. 3).

In the following we shall investigate, for each of the three installations, the cost of gas consumption per unit cooling capacity produced.

This shall then be supplemented with the costs necessary for the extra investments to extend the total energy system so that, depending on a number of influencing factors, the total prices per cooling unit in the three different installations types will be available. From a comparison with simular costs per cooling unit produced in a conventional absorption or compression cooling system, an impression of the value of the total energy principle in connection with the cooling technical part of air conditioning system can be obtained.

2. THE ECONOMY OF A "TOTAL ENERGY SYSTEM" USED IN THE ENERGY PRODUCTION OF
COOLING INSTALLATIONS

This analysis is based on the calculation of the costs per unit cooling capacity (here 1,000 kcal/hr) for the different installations. These costs can be put into two categories:

— the costs of fuel and electrical energy consumption;

— interest rate and depreciation over the extra investment for the total energy installations.

2.1 *Fuel costs*

Before going into the details on the economic aspects of combining a total energy system with an absorption or compression refrigerating machine, it is useful to investigate how (in the last two types of machines) the price per cooling unit produced develops itself depending on the energy prices (gas and electricity). These calculations do not take into account the influence of the investment costs of the cooling installation itself. A further limitation which we have put on this and all other calculations is that for the removal of heat from the cooling machines, only cooling towers will be used

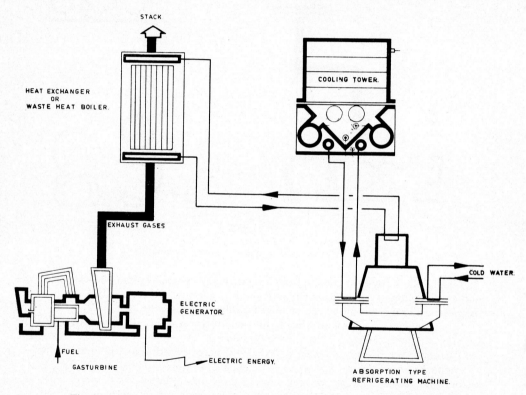

Fig. 3 — Absorption refrigerating machine using heat from total energy system.

In such prices analyses the energy price per 1,000 kcal/hr is plotted against the price per m³ water. Various lines are valid for given gas and kWhr-prices. The change in price is calculated from the following values for the heat rejection factor: absorption machine 2.49; compression machine 1.18. To calculate the heat used by the absorption cooling machine, Dutch natural gas with a heat content of 7,560 kcal/m³ was used as the starting point.

The results thus obtained can be used in the various comparisons of costs of the cooling units produced in a total energy combination; this will be enlarged upon further.

In the introduction, three combinations of the total energy system with a cooling installation were already mentioned which, in principle, could be considered as technical possibilities.

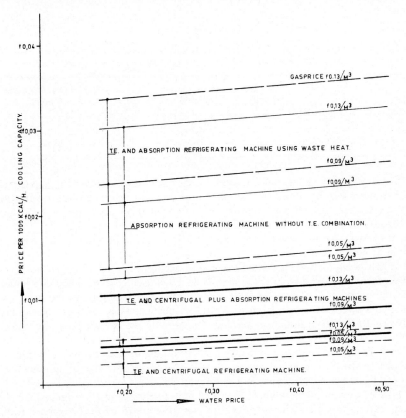

Fig. 4 — Costs for gas and water expenditure per 1,000 kcal/hr cooling capacity for various gas motor T.E. cooling system combinations.

We start our investigations with the system shown in figure 1. As the energy use of both the absorption as well as the compression refrigerating machine through the application of the total energy installation as an energy producer is reduced to the consumption of natural gas alone, the calculation of the energy costs is reduced to an analysis of the cost of gas consumption of the combination. This is determined from values of the heat rejection factor and the efficiency with which the total energy installation converts natural gas into heat and electrical energy.
These efficiencies have roughly the following values:

Table 1

EFFICIENCY OF THE TOTAL ENERGY INSTALLATIONS

	Efficiency of electricity production	Efficiency of usable heat production	Losses
Gas motor	28-33%	50%	20%
Gas turbine	16-20%	55-65%	20%

557

The efficiency figures are determined by the ratio of the production capacity of both types of refrigerating machines. They indicate the ratio of heat and electrical energy produced in the total energy installation.

Roughly speaking the cooling capacity produced can be divided in the following manner:

Table 2

DIVISION OF COOLING CAPACITY OF ABSORPTION AND COMPRESSION
REFRIGERATING MACHINE IN A TOTAL ENERGY INSTALLATION

	Compression refrigerating machine	Absorption refrigerating machine
Gas motor T.E. installation	80%	20%
Gas turbine T.E. installation	70%	30%

Taking all these factors into account when calculating the cooling capacity, the price estimates can be obtained. Here again, depending on the gas and water prices in the cost price for the consumption of both media per cooling unit produced, a graph could be plotted.

In a similar way, the systems from figures 2 and 3 could be analysed from the price per unit of cooling capacity.

The following factors have been taken into account.

The waste heat freed from the system 2, which is used elsewhere and consequently not in the production of cooling units, can be revalued according to the price of an equivalent amount of gas and consequently be subtracted from the gas consumption of the total energy installation needed for driving the compression refrigerating machine. For the system in figure 3, the evaluation of the electrical energy to be used elsewhere according to the price of its equivalent reveals distinct consequencies since due to the poor internal efficiency of the absorption refrigerating machine the price per 1,000 kcal/hr becomes very unfavourable.

This is shown in figure 4 where the calculated prices per unit cooling capacity are plotted and compared with the prices for the absorption cooling machine without total energy combination. Consequently in the final comparison of the last installation, the application of compression or absorption refrigerating machines without a total energy combination for the electrical energy liberated by total energy, the current kWhr price must be used.

2.2 *Investment costs*

The installation of a total energy plant to produce the basic energy for driving the refrigerating machines is, in the first place, attractive because of the low costs of gas consumption as shown in figure 4.

Where the effect of the extra investment on the price per cooling unit is unknown, the insight into the economy of the total energy application is incomplete.

A basic number of price factors has been taken as starting point; viz. the price per unit of power of the total energy system to be installed is, for gas motors, f 400 per kW, and roughly f 550 for gas turbines.

Furthermore we will assume an amortisation of 11% along with an estimated life span of 20 years and an interest rate of 9%. With these data we can calculate the costs for rent and depreciation on the total energy addition per 1,000 kcal of cooling. To relate these sums to the 1,000 kcal/hr unit, they have to be divided by the number of working hours for 1,000 kcal, i.e., full load hours.

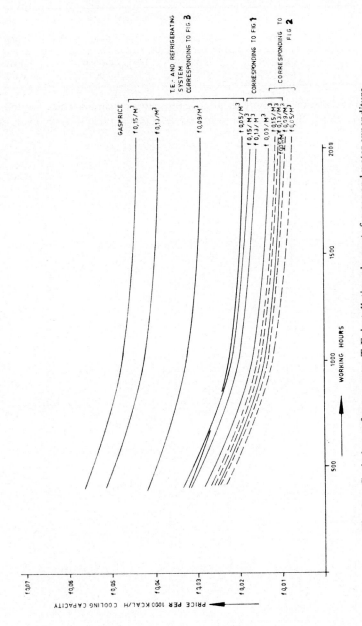

Fig. 5 — Capital costs for gas motor T.E. installation plus costs for gas and water expenditure per 1,000 kcal/hr cooling capacity for various T.E. refrigerating plant systems.

559

For the installation shown in figure 1, the following table was produced:

Table 3

COST OF INTEREST AND DEPRECIATION PER 1,000 kcal/hr BY THE ADDITION
OF A TOTAL ENERGY SYSTEM AS SHOWN IN FIGURE 1

| Type of prime mover | Number of working hours | | | |
	500	1000	1500	2000
Gas motor	ƒ0.0178	ƒ0.0089	ƒ0.0059	ƒ0.0045
Gas turbine	ƒ0.0216	ƒ0.0108	ƒ0.0072	ƒ0.0054

For the installation in figure 2, after subtraction of certain costs for the waste heat installation which is not used in the cooling plant, table 4 gives some idea of cost prices.

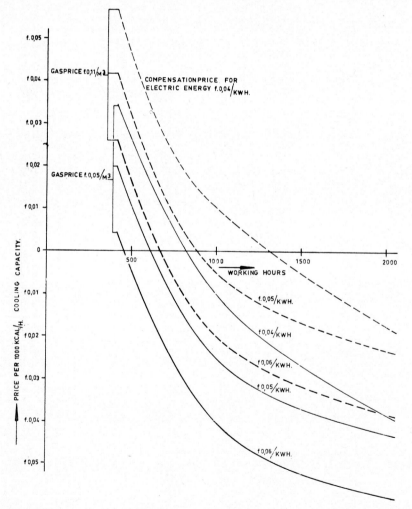

Fig. 6 — Capital costs for gas motor T.E. installation plus costs for gas and water expenditure per 1,000 kcal/hr cooling capacity for T.E. refrigerating system according to figure 4.

Table 4

COST OF INTEREST AND DEPRECIATION PER 1,000 kcal/hr
BY THE ADDITION OF A TOTAL ENERGY SYSTEM AS IN FIGURE 2

Type of prime mover	Number of working hours			
	500	1000	1500	2000
Gas motor	ƒ 0.0180	ƒ 0.0090	ƒ 0.0060	ƒ 0.0045
Gas turbine	ƒ 0.0252	ƒ 0.0126	ƒ 0.0084	ƒ 0.0063

For the type as shown in figure 3 the costs per 1,000 kcal/hr when the installation of a waste heat plant is taken into account are as follows:

Table 5

COST OF INTEREST AND DEPRECIATION PER 1,000 kcal/hr
BY THE ADDITION OF A TOTAL ENERGY SYSTEM AS IN FIGURE 3

Type of prime mover	Number of working hours			
	500	1000	1500	2000
Gas motor	ƒ 0.0143	ƒ 0.0072	ƒ 0.0048	ƒ 0.0036
Gas turbine	ƒ 0.0184	ƒ 0.0092	ƒ 0.0061	ƒ 0.0046

2.3 The total price per unit cooling capacity

The price curves in part 2.1 have already shown that the influence of the cost of water on the price per cooling unit is insignificant. This is partially due to the rather low functional costs of cooling towers.

It is therefore permissible to neglect the water costs as a variable, thus making price calculations simpler. The remaining factors influencing the price per unit of cooling capacity are then the gas costs and the number of working hours.

The price of water was fixed, for simplicity, at ƒ 0.40/m³.

By combining results of various calculations and cost estimates figure 5 is produced.

From the configuration shown in figure 3, the price per 1000 kcal/hr were calculated and are shown in figure 6 depending on the repayments for the electrical energy produced in the total energy installation.

3. CONCLUSION

From the various calculations, the value of the total energy principle in connection with cooling installations is becoming apparent. It seems that there is no special reason to expect more profit from the combination of total energy and cooling installation than from any other combination of total energy with power and heat consuming systems.

From the results of a combination of an absorption refrigerating machine (fig. 3) with a total energy system it is very clear that the importance of a total energy system principally exists in the production of cheap electrical energy and that the available waste heat is in fact a by-product which in itself does not justify a total energy system.

When, however, this by-product is used in combination with the electrical energy produced, there are major financial advantages in the use of a total energy system as shown in figure 6. Therefore the installation in figure 3 cannot be considered as a specific total energy application in cooling techniques but as a supplement to the general total energy principle, namely the use of electrical energy for any purpose which in fact makes the total energy application complete.

Only the systems shown in figures 1 and 2 can be considered as real applications of the total energy system in cooling installations.

For a complete evaluation of the possible advantages we would like to limit ourselves to gas motor installations since it has been shown that gas turbine installations demand too high investment costs for the capacity that can be expected in the framework of this study.

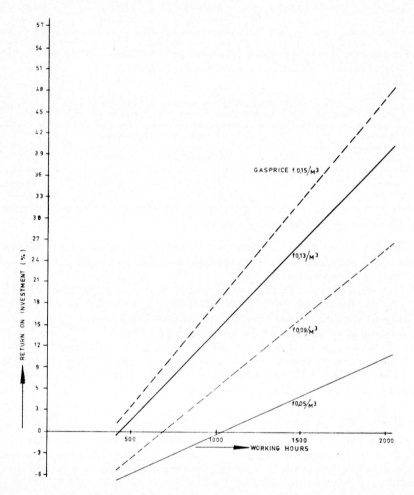

Fig. 7 — Return on investment for T.E. addition to refrigerating system according to figure 2 compared with simple absorption refrigerating plant.

The advantages of a gas turbine, namely low maintenance and operation costs, only begin to be of importance when the purchase cost per unit of power becomes lower. This is only at capacities of 1,000 kW and higher.

The calculated savings on production costs of the cooling capacity which are produced via a combination of a total energy system and a compression cooling machine can only be estimated at their correct value when it is investigated as to what these savings have provided from having made extra investments in the total energy addition to the cooling system.

562

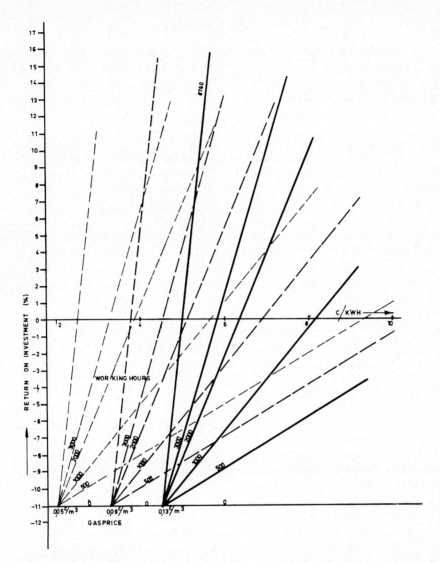

Fig. 8 — Return on investment for T.E. addition to refrigerating system according to figure 2 compared with simple centrifugal chiller plant.

The investment yield is defined as:

$$\frac{\text{saving in cash - annuity over the extra investment}}{\text{extra investment}}$$

Compared with the exploitation costs for the absorption and compression refrigerating machine the yield is calculated for various gas prices and annual working hours for the installations shown in figures 1 and 2.

The results of this calculation are given in figures 7 and 8 which show that the total energy addition only becomes attractive when the number of working hours is greater than 1,000 and the gas price higher than $f\,0.05/m^3$, when comparing with the absorption refrigerating machine and the electricity price at least higher than $f\,0.06/kWhr$ when comparing with the centrifugal refrigerating machine.

563

DISCUSSION

P. VETZEL (France) — 1. Dans le système 1, pour quelles raisons, les condenseurs à eau des machines à compression et à absorption sont-ils en série? Est-ce pour réduire le risque de cristallisation dans la machine à absorption? Mais alors le coefficient de performance :

$$\frac{\text{Production frigorifique}}{\text{Energie fournie en kcal}} \; \text{diminue}$$

2. Les schémas 1, 2 et 3 semblent correspondre à des turbines. Dans le cas où ils concerneraient aussi les moteurs y a-t'il récupération de chaleur sur les chemises d'eau?

3. Au sujet du graphique 4, il semble correspondre à des moteurs à gaz et non à des turbines. À ce sujet les auteurs peuvent-ils indiquer quel est le prix du kWh recueilli à l'alternateur permettant de déduire le prix des 1 000 frigories lorsque l'énergie électrique produite n'est pas utilisée pour la production de froid?

4. Au sujet du tableau 2, si l'on prend comme base : un rendement mécanique de 30% pour les moteurs et 18% pour les turbines, des pertes d'énergie égales à 20% dans les deux cas, le partage de production entre compression et absorption devrait être :

	compression	absorption
moteur	83 %	17 %
turbine	69 %	31 %

Il faut remarquer à ce sujet :

1° le taux de récupération de chaleur sur les moteurs nous semble élevé 1,4 th/kWh;

2° le rendement des turbines est un excellent rendement qui n'est valable qu'à pleine puissance.

5. D'une manière générale, l'auteur semble négliger la solution qui consiste à entraîner les turbocompresseurs directement par des moteurs et des turbines à gaz, le gain étant de $(1)/(\lambda A \; \lambda m)$ %

λA rendement de l'alternateur;

λm rendement du moteur.

Cette observation est d'ailleurs confirmée par les données de la communication de R. T. TAMBLYN, « Evolution of central plant toward a heat reclaim loop » (p. 587).

P. LEIJENDECKERS — 1.1. With condenser temperatures of 26-36°C maximal and evaporator temperatures of 6-12°C, you do not have any crystallization in the absorption machine when the circulating water quantities are constant.

1.2. Whether the efficiency should decrease is not clear. In the event where the evaporator and condenser were not connected in series, the absorption machine should have a much lower efficiency because with parallel connection, the outlet temperature of the cooling water, which is about 10.5°C, should be much lower i.e. 6°C.

For the centrifugal machine which, incidentally, takes care of the greatest part of the load the difference between the evaporator and condenser temperature is smaller and this also slighty improves the situation.

2. Yes.

3. Here we do not discuss the introduction of an alternative price for the generated kWh, but these you can find in figure 6.

4. In the text the figures have been rounded off.

The figures you have specified indeed represent the exact values.

4.1. The given value of 1.4 th/kWh is certainly not too high. Caterpillar states a full load of about 1.5 th/kWh, whilst this figure increases with partial load.

5. It is difficult, in our opinion, to couple the refrigerating machine to the gas engine since breakdown and maintenance of the gas engine will put the refrigerating installation out of action thus making the air conditioning installation vulnerable.

In smaller refrigerating installations it is simpler and more economic to drive these machines directly with electro-motors and to absorb the necessary electrical capacity in the total capacity of the T. E.-installation of a building.

The number of installed gas engines can then be restricted.

CONDITIONS DE RENTABILITÉ DE L'UTILISATION DES MOTEURS À GAZ POUR L'ENTRAÎNEMENT DES COMPRESSEURS FRIGORIFIQUES DANS LES INSTALLATIONS DE CONDITIONNEMENT D'AIR

P. VETZEL

Direction des Services Économiques et Commerciaux,
Gaz de France, Paris (France)

Economic conditions for the use of gas-engine-driven refrigerating compressors in air-conditioning plant

SUMMARY: *Refrigeration for air-conditioning may be supplied by gas-engine-driven refrigerating units. This technology consists in replacing the electric motor and its controls by a gas-engine which not only provides mechanical power but also a high level heat source (hot water at 80°C, steam under 1 bar). This heat may be used both for central heating and for sanitary hot water or any other heat requirements (heating of swimming-pools, central station for producing conditioned air, absorption machines).*

The report gives economic conditions for the use of gas-engines in relation to the different factors involved: investment, annual period of utilization, amortization time of the plant, power costs, heat recovery. The A. discusses the effect on the over-all cost of producing 1,000 fg/hr.

INTRODUCTION

Après avoir examiné la consistante d'une installation de conditionnement d'air utilisant le gaz comme source d'énergie primaire, on définit des coûts unitaires de production des frigories pour en déduire les conditions d'un développement du gaz pour cette application nouvelle en France.

1. ÉLÉMENTS CONSTITUTIFS D'UNE INSTALLATION DE CLIMATISATION ÉTÉ-HIVER

Une installation de ce type comporte les éléments suivants :
— un appareil producteur d'eau chaude;
— un appareil producteur d'eau glacée;
— une tour de refroidissement pour économiser l'eau;
— un réseau de canalisations pour faire circuler l'eau entre la tour de refroidissement et le condenseur de la machine frigorifique;
— un réseau de transport et de distribution des calories et des frigories produites, réseau qui est commun à toutes les solutions, quelle que soit l'énergie primaire utilisée pour la production du froid.

1.1 *Appareil producteur d'eau chaude*

La production des calories nécessaires au chauffage des locaux en hiver est assurée soit par une chaudière à eau chaude, soit par une chaudière à vapeur accouplée à un échangeur, soit par un appareil mixte (chauffage — réfrigération) donnant, suivant la saison, de l'eau chaude ou de l'eau glacée. L'expérience montre que le rendement de ces appareils est identique.

1.2 *Appareil producteur d'eau glacée*

1.21 *Si l'électricité sert d'énergie primaire* pour la production de froid, l'appareil comprend :

— un compresseur frigorifique entraîné par un moteur électrique;
— un condenseur à circulation d'eau;
— un évaporateur donnant l'eau glacée nécessaire au refroidissement des locaux.

Le moteur du compresseur frigorifique a une consommation spécifique de 0,4 kWh par 1000 frigories produites. C'est là une valeur moyenne compte tenu des températures d'évaporation et de condensation habituellement pratiquées dans les installations de conditionnement d'air, et du nombre d'heures de fonctionnement à puissance réduite.

Le condenseur à circulation d'eau permet d'éliminer de la chaleur qui provient des locaux à rafraîchir et de la dégradation de l'énergie dans le compresseur. La quantité de chaleur à évacuer au condenseur est égale à 1,25 fois la quantité de frigories produites. Elle se présente sous forme d'eau tiède à 30-35 °C. La quantité d'eau qui doit transiter dans le condenseur est de 0,2 m^3 par thermie à évacuer pour une différence de température de 5 °C. Il faut noter que dans certaines installations, on remplace le condenseur à eau par un condenseur à air qui est refroidi par un courant d'air. Cette technique

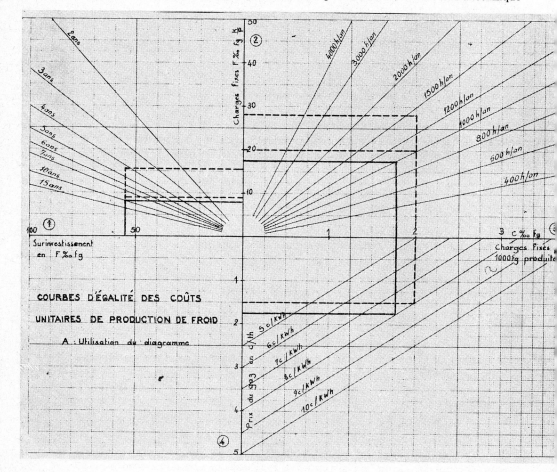

Fig. 1

conduit à un abaissement des performances du compresseur, à une dépense d'énergie pour le ventilateur. On ne tiendra pas compte de ce système dans ce qui suivra.

L'évaporateur est un échangeur où circule de l'eau qui se refroidit par échange de chaleur avec le fluide frigorigène qui s'y vaporise.

1.22 *Si le gaz sert d'énergie primaire* : on remplace le moteur électrique par un moteur à gaz, le reste du groupe frigorifique étant le même que dans le cas de l'électricité.

Dans une machine à compression de ce type, la consommation spécifique est de 0,6 à 1,25 thermie par 1000 frigories produites suivant le rendement mécanique du moteur à gaz. En outre ce moteur devant être refroidi comme tout moteur à explosion, la quantité de chaleur disponible sous forme d'eau à 80-90°C ou de vapeur basse pression est de l'ordre de 0,4 thermie par 1000 frigories produites. Une récupération complémentaire peut être effectuée au moyen d'un échangeur placé sur l'échappement du moteur.

1.3 *Tour de refroidissement*

On a vu que le refroidissement des machines frigorifiques demandait de grandes quantités d'eau; on est donc conduit à chercher à l'économiser en la recyclant après

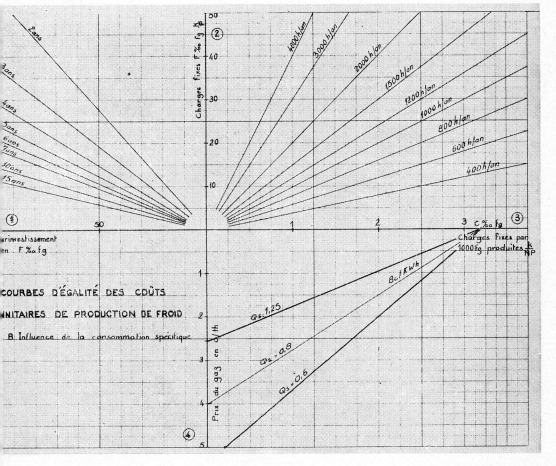

Fig. 2

l'avoir refroidie dans une tour de refroidissement. Ce refroidissement de l'eau est provoqué par une évaporation partielle de celle-ci dans un courant d'air. La consommation d'eau théorique d'une tour de refroidissement est de 1,75 litre par thermie transférée. Le remplacement de l'eau évaporée volume pour volume, conduit à une concentration des sels dissous dans l'eau du circuit de refroidissement d'où risque d'entartrage. Pour éviter cet inconvénient, les fabricants de machines frigorifiques préconisent de remplacer l'eau évaporée par un volume double d'eau *traitée*. Ainsi la consommation d'eau est portée à 3,5 litres par thermie transférée.

Le courant d'air est entretenu dans la tour au moyen d'un ventilateur, dont la consommation d'électricité est de 0,04 kWh par thermie transférée.

1.4 *Réseau d'eau de refroidissement*

La tour de refroidissement est reliée au condenseur de la machine frigorifique par un réseau de canalisations où l'eau circule par l'action d'une pompe. La consommation d'énergie de cette pompe est fonction :

— du débit à transiter;
— des pertes de charges dans les canalisations;

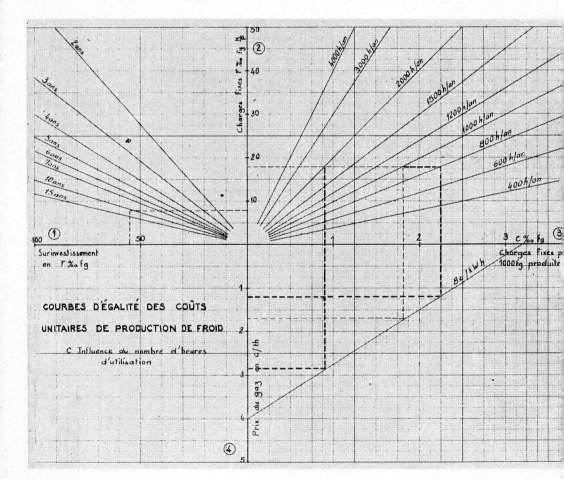

Fig. 3

— de la hauteur manométrique entre le condenseur et la tour.

Pour fixer un ordre de grandeur, la consommation d'électricité a été estimée à 0,02 kWh par thermie transférée.

1.5 Réseau de transport des calories et des frigories

Les calories et les frigories produites en centrale sont transportées sur les points d'utilisation par différents moyens :
— réseau de gaines avec bouches de soufflage et de reprise, l'air étant traité dans un caisson central;
— réseau de tuyauteries avec ventilo-convecteurs alimentés partiellement en air neuf;
— réseau mixte de canalisations et gaines :
  – caisson de traitement par zone d'occupation;
  – éjecto-convecteurs alimentés en air neuf et traités sous pression.

La conception et la réalisation de ce réseau de transport sont indépendantes du mode de production des frigories.

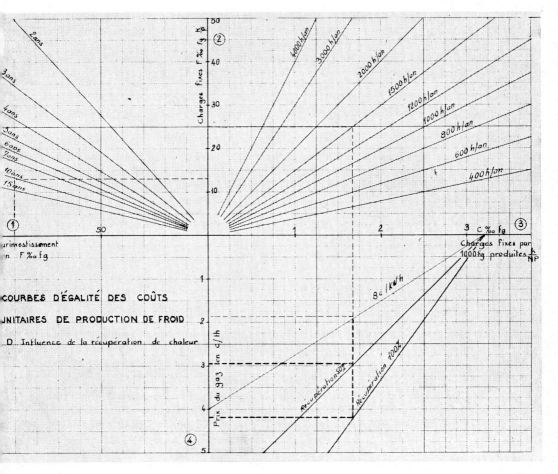

Fig. 4

2. Investissements et amortissements

Les investissements relatifs à une installation de climatisation été-hiver peuvent se ranger en deux catégories :
— les investissements communs, où le mode de production des frigories n'intervient pas, comme le réseau de transport des calories et des frigories;
— les investissements propres au mode de production de froid choisi et qui comportent :
  – le générateur d'eau chaude;
  – le générateur d'eau glacée;
  – le réseau d'eau de refroidissement;
  – la tour de refroidissement;
  – la canalisation d'amenée de l'énergie primaire et ses accessoires.

La suite de l'étude devant conduire à l'établissement de coûts comparatifs de diverses solutions, on ne retiendra pour l'établissement de ces coûts que la différence entre les investissements propres aux modes de production du froid divers, différence que l'on désignera désormais par surinvestissement. Pour juger de la valeur de la solution « gaz » le surinvestissement, s'il existe, doit être compensé par des économies sur le coût de production des frigories.

La durée de l'amortissement peut être définie par la loi dans le cas d'installation à caractère industriel ou commercial, par la durée de vie des matériels dans le cas d'installation particulière non amortissable fiscalement. En prenant un taux d'intérêt de l'argent de 8 %, l'annuité d'amortissement d'un surinvestissement de 1 F est donnée dans le tableau suivant :

Durée en années	Annuité en F
2	0,560
3	0,388
4	0,302
5	0,250
6	0,216
7	0,192
10	0,149
15	0,117

3. Coût global annuel

Le coût global annuel de la climatisation été-hiver a pour expression :

$$\sum (A, D, E, C, F)$$

A — amortissement de l'installation;
D — coût de la distribution de chaleur et de froid : il est indépendant des modes de production pour une installation donnée;
E — coût de l'entretien de l'installation : il comporte trois postes d'importance relative variable en fonction de la puissance frigorifique de l'installation, du mode de distribution de la chaleur et du froid, de la technique adoptée pour la régulation :
  – réseau de distribution;
  – appareils de production de chaleur;
  – appareils de production de froid.

Pour le réseau de distribution d'une installation il est indépendant des modes de production.

Pour les appareils de production de chaleur ce coût est identique puisque les appareils et la source d'énergie sont identiques.

Pour les appareils de production de froid, le coût d'entretien du groupe frigorifique doit être majoré de celui du moteur à gaz.

C — coût de production de la chaleur : il est identique puisque la source d'énergie est la même et que les appareils ont un rendement identique;

F — coût de la production de froid.

Si l'on compare deux solutions on a les coûts suivants :

$$\sum (A_1, D_1, E_1, C_1, F_1)$$

$$\sum (A_2, D_2, E_2, C_2, F_2).$$

Or nous avons vu dans ce qui précède que :

$A_2 = A_1 + A'$ A' étant l'amortissement du surinvestissement éventuel de la solution 2 par rapport à la solution 1.

$D_1 = D_2$

$E_2 = E_1 + E'$ E' étant le coût de l'entretien du moteur à gaz.

$C_1 = C_2$

$F_1 \neq F_2$

Il s'ensuit que les grandeurs comparables sont F_1 et $F_2 + A' + E'$.

4. Coût unitaire de production

En vue de faciliter les comparaisons, on peut définir le coût unitaire de production des frigories comme le quotient des dépenses par le nombre de milliers de frigories produites dans l'année. On a ainsi les relations :

$$F_1 = N \cdot P \cdot f_1 \qquad \text{et} \qquad F_2 = N \cdot P \cdot f_2 + K$$

avec :

F_1, F_2 dépenses globales pour la production de froid.

N nombre d'heures d'utilisation de la puissance installée en froid dans l'année.

P puissance de l'installation frigorifique en milliers de frigories.

f_1, f_2 coûts élémentaires de production des 1000 frigories donnés dans le tableau suivant.

K frais fixes correspondant à l'amortissement du surinvestissement de la solution « gaz » et au coût d'entretien du moteur à gaz.

Pour produire 1000 frigories, il faut suivant les données des § 1-2 :

	Solution gaz	Solution électricité
Eau en m³	0,0044	0,0044
Électricité en kWh	0,075	0,475
Gaz en th	Qs	—

Ainsi les coûts unitaires définis au début du paragraphe sont :

$$f'_1 = f_1^{\tilde{z}} \qquad \text{et} \qquad f'_2 = f_2 + \frac{K}{NP}$$

en posant :

x est le prix de la thermie gaz;
y le prix du kWh d'électricité;
z le prix du m^3 d'eau traitée;

$$f_1 = 0,475\,y + 0,0044\,z$$

$$f_2 = Qs\,x + 0,075\,y + 0,0044\,z\,.$$

Ces coûts sont indépendants de la puissance de l'installation car la production frigorifique ne dépend que des conditions thermodynamiques du cycle et du fluide frigorigène choisi.

5. Conditions d'égalité des coûts unitaires

5.1 *Incidence tarifaire*

Pour que les coûts unitaires f'_1 et f'_2 soient égaux, il faut et il suffit que la relation

$$0,475\,y + 0,0044\,z = Qs\,x + 0,075\,y + 0,0044\,z + \frac{K}{NP}$$

soit vérifiée.

Elle peut d'ailleurs se simplifier et se mettre sous la forme

$$0,4\,y = Qs\,x + \frac{K}{NP} \qquad\qquad (1)$$

en exprimant y, x et K en centimes.

Cette relation montre que le prix de l'eau traitée n'a aucune incidence sur le coût unitaire de production.

5.2 *Courbes d'égalité des coûts*

L'équation (1) établie ci-dessus permet de tracer les courbes d'égalité de coût en fonction des paramètres suivants :
— surinvestissement de la solution gaz;
— durée d'amortissement;
— coût d'entretien annuel du moteur à gaz;
— nombre annuel d'heures d'utilisation;
— prix du courant électrique;
— prix du gaz.

Par contre on s'est fixé des constantes :
— le taux du loyer de l'argent a été fixé à 8 %;
— la consommation spécifique du moteur à gaz a été prise égale à 0,8 th/1000 fg, ce qui est une valeur très acceptable en conditionnement d'air car elle correspond à un rendement mécanique du moteur de 30 % environ, ce qui est considéré comme une valeur moyenne par les fabricants de moteurs.

Ces courbes permettent de déterminer graphiquement, en fonction des paramètres définis ci-dessus :

a) les charges fixes annuelles par 1000 fg installées;

b) l'incidence de ces charges sur le coût de production de 1000 fg, rapporté aux 1000 fg produites;

c) le prix d'équivalence du gaz par rapport à l'électricité;

d) la rentabilité de la solution gaz;

e) l'amortissement maximum possible compte tenu du prix des énergies;

f) l'économie annuelle réalisée pendant la période d'amortissement;

g) l'économie annuelle réalisée après la période d'amortissement;

h) la durée minimale d'amortissement;

i) la durée minimale de fonctionnement de l'installation.

5.3 *Exemple d'application*

Soit une installation de conditionnement d'air dont les caractéristiques technico-économiques sont les suivantes :

	Solution électricité	*Solution gaz*
Puissance installée	290 000 fg/hr	290 000 fg/hr
Vitesse de rotation du compresseur	1450 t/m	1750 t/m
Prix du groupe frigorifique sans moteur	96 000 F	80 000 F
Prix du moteur et armoire de contrôle	15 000 F	46 890 F
	111 000 F	126 890 F

Amortissement de l'installation en 15 ans;

Nombre d'heures d'utilisation 1000 h/an;

Prix du courant électrique 8 c/kWh;

Contrat d'entretien garantie totale au moteur 3400 F;

On calcule d'abord le surinvestissement rapporté aux milliers de frigories installées $(126\,890 - 111\,000)/290 = 55$ F.

On porte cette valeur sur l'axe ① du diagramme A.

On trace la verticale passant par ce point : elle coupe l'oblique correspondant à la durée d'amortissement fixée (15 ans); au point d'intersection on trace une horizontale qui coupe l'axe ② en donnant l'amortissement du surinvestissement pour 1000 fg/h installées (6,435 F/an).

A cette valeur, on ajoute l'incidence du coût d'entretien $3400/290 = 11,725$ F, ce qui donne comme charges fixes, par 1000 fg/h installées, une somme de 18,16 (*a*) que l'on porte sur l'axe ②.

A partir de ce point, on trace une horizontale qui coupe l'oblique correspondant au nombre d'heures de fonctionnement à pleine puissance (1000 h), du point d'intersection on trace une verticale qui coupe :

— l'axe ③ en donnant l'incidence des charges fixes par 1000 fg produites soit 1,816 c‰ fg (*b*).

— l'oblique correspondant au prix du courant électrique 8 c/kWh.

De cette dernière intersection, on trace l'horizontale qui coupe l'axe ④ des prix du gaz donnant ainsi le prix d'équivalence du gaz soit 1,73 c/th (*c*).

Comme ce prix est supérieur au prix de vente du gaz, la solution « gaz » est rentable.

A partir du prix de vente du gaz 1,5 c/th, on trace une horizontale jusqu'au point de rencontre correspondant au prix de vente de l'électricité, puis de là, on trace une verticale qui donne sur l'axe ③ des charges fixes correspondant à l'égalité des coûts, soit 2 c‰ fg (e).

A partir de ces valeurs (b) et (e), on tire :

— l'économie annuelle réalisée pendant la période d'amortissement, en multipliant la différence entre ces deux valeurs par le nombre de milliers de frigories produites annuellement :

$$(2-1,816) \times 290 \times 1000 = 53\,360 \text{ c ou } 533,60 \text{ F.}$$

— l'économie annuelle réalisée après la période d'amortissement en multipliant la différence entre la valeur maximale de charges fixes admissibles (e) et l'incidence du coût d'entretien du moteur à gaz $(3400 \times 100)/(290 \times 1000)$ soit 1,1725 c par le nombre de milliers de frigories produit annuellement :

$$(2-1,1725) \times 290 \times 1000 = 233\,975 \text{ c} = 2\,339,75 \text{ F } (g)$$

On peut à partir du même diagramme rechercher quelle est la durée minimale d'amortissement du surinvestissement : connaissant la différence entre la valeur maximale des charges fixes admissibles (e) et l'incidence du coût d'entretien : 1,1725 c, c'est-à-dire la valeur des amortissements maximaux, on porte cette valeur sur l'axe 3, on élève la verticale correspondante jusqu'au point d'intersection avec l'oblique du nombre d'heures de fonctionnement, d'où l'on tire une horizontale qui coupe la verticale correspondant au surinvestissement en un point représentatif de la durée minimale d'amortissement, soit ici une durée voisine de 10 ans.

De même, on peut rechercher en fonction d'une durée d'amortissement fixée de manière arbitraire le minimum d'heures de fonctionnement par an, nécessaire pour établir l'égalité des coûts.

Pour cela, on trace à partir de l'intersection de la verticale des investissements pour 1000 fg/h installées avec l'oblique des durées d'amortissement (4 ans) une horizontale jusqu'à son intersection avec l'axe ②. On lit sur celui-ci la charge fixe relative à l'amortissement du surinvestissement de la solution gaz. On majore cette valeur de l'incidence de l'entretien du moteur (11,725 F). De ce nouveau point de l'axe ② on trace une horizontale jusqu'à son intersection avec la droite des charges fixes maximales par 1000 frigories/heure produites. Cette durée minimale est de :

$$1416 \text{ h } \quad (i) \, \rceil$$

5.4 *Influence de la consommation spécifique*

La consommation spécifique est une fonction inverse du rendement mécanique du moteur. Les droites d'égal coût représentant la fonction $0,4\,y - Qs\,x - (K/NP) = 0$ voient leurs pentes varier quand le rendement du moteur, c'est-à-dire la consommation spécifique, varie.

La famille de droites établie, pour un même coût de l'électricité, a la même origine sur la droite des charges fixes par 1000 fg produites. Pour un prix du courant électrique égal à 8 c/kWh, on a tracé sur le diagramme les droites limites qui correspondent aux rendements minimaux et maximaux des moteurs. Pour les autres prix, les droites se déplacent parallèlement les unes aux autres pour un même rendement de moteur.

5.5 *Influence du nombre d'heures d'utilisation dans l'année*

La fonction représentative des courbes d'égal coût $0,4\,y - Qs\,x - (K/NP) = 0$ comporte un terme K/NP qui varie de manière inversement proportionnelle au nombre d'heures d'utilisation.

Si l'on reprend le diagramme avec les données suivantes :

— surinvestissement 55 F‰ fg/h;

— durée d'amortissement 15 ans;

— prix du courant électrique 8 c/kWh ;

le prix d'équivalence du gaz qui dans ces conditions était de 1,75 c/th pour 1000 h d'utilisation monte à 2,85 c/th pour 2000 h et tombe à 1,20 c/th pour 800 h.

5.6 *Influence de la récupération de chaleur*

Les moteurs thermiques doivent être refroidis:

— au niveau des chemises d'eau des moteurs au moyen d'un échangeur branché au secondaire sur le réseau général de chauffage, un deuxième échangeur dit de sécurité évite de faire revenir au moteur une eau trop chaude, cet échangeur étant branché au secondaire sur le réseau des tours de refroidissement;

— au niveau des refroidisseurs d'air des turbines de suralimentation et des refroidisseurs d'huile dans le cas de moteur suralimenté au moyen d'une circulation d'eau raccordée sur la tour de refroidissement;

— en outre, le moteur peut être raccordé sur une chaudière de récupération pouvant fournir soit de l'eau surchauffée à 120 °C soit de la vapeur basse pression.

Dans le cas défini ci-dessus, les quantités de chaleur récupérables sont de 61,6 th/h sur les chemises d'eau et de 40 th/h sur l'échappement, soit au total 101,6 th/h.

Si on rapporte cette dernière valeur au millier de frigories, la récupération possible est de $101,6/290 = 0,35$ th, le rendement d'une chaudière susceptible de fournir cette chaleur est de 0,8, la consommation de gaz correspondante est de $0,35/0,8 = 0,44$ th.

Mais l'installation de récupération de chaleur nécessite un investissement supplémentaire de 10 000 F pour le groupe frigorifique considéré. En reprenant les données de 5.3., on a :

Puissance frigorifique installée 290 000 fg/h.

Surinvestissement de la solution gaz 25 890 F.

Amortissement de l'installation en 10 ans.

Nombre d'heures d'utilisation 1500 heures.

Prix du courant électrique 8 c/kWh.

Prix du gaz 1,5 c/th.

Surinvestissement pour 1000 fg 90 F.

La fonction représentative des courbes d'égal coût qui s'exprime sous la forme

$$0,4 \, y - 0,8 \, x + 0,44 - (K/NP) = 0$$

devient $\quad 0,4 \, y - 0,36 \, x - (K'/NP) = 0$ si la récupération est totale,

ou $\qquad 0,4 \, y - 0,58 \, x - (K'/NP) = 0$ si elle n'est que de 50 %.

La famille de droites établie pour un même coût de l'électricité a la même origine sur la droite des amortissements.

On a tracé en traits pleins sur le diagramme pour un prix du courant électrique égal à 9 c/kWh les droites correspondant à différents taux de récupération.

Suivant les données ci-dessus énoncées, le prix du gaz équivalent est de 2,90 c/th dans le cas d'une récupération à 50 % et 4,15 c/th dans le cas d'une récupération à 100 %. On voit de suite l'intérêt d'un tel système car les bénéfices réalisés passent à 1343 F pour une récupération nulle, à 4214 F pour une récupération totale. De même, dans le cas d'une installation de récupération de chaleur coûtant autant et pour un fonctionnement de 1000 h/an, le taux de récupération minimum est de 50 %, il s'avère donc que dans le cas où le taux de récupération escompté est inférieur à 50 %, il y a lieu de simplifier l'installation en ne récupérant que sur les chemises d'eau afin de rester rentable.

6. Conclusion

On peut dire en conclusion que l'utilisation du moteur à gaz est d'autant plus intéressante :

— que la durée d'utilisation est plus longue, car dans le coût annuel la part des charges fixes supplémentaires (amortissement et entretien) est prépondérante;

— que la récupération de chaleur est plus importante, car elle permet de majorer rapidement les bénéfices d'exploitation.

LES CENTRALES AUTONOMES AUTOMATIQUES DE PRODUCTION D'ÉNERGIE TOTALE, AVEC MOTEURS À GAZ À REFROIDISSEMENT PAR ÉBULLITION ET MACHINES À ABSORPTION

L. LEFAURE

Éts. Bergerat Monnoyeur et Cie, Paris (France)

Individual automatic central stations for the production of total energy with boiling-cooling gas engines and absorption machines

SUMMARY: *Created in the U.S.A. during the last ten years, individual automatic central stations for the production of total energy are based on a simple principle:*
High performance boiling-cooling natural gas (or fuel oil) heat engines drive electric generators and produce the electricity required for supplying the centre.
The cooling water from heat engines, recovered in the form of steam at 125 °C, is used for:
— *supplying a heating circuit via a steam-water exchanger;*
— *supplying the air-conditioning circuit via absorption machine.*

The cooling of engines is controlled by means of a balance exchanger.
The production of ice water is controlled by means of a stand-by steam-generator or a stand-by compressor unit.
Total energy is therefore, at the same time, electric power, heating and air conditioning, obtained automatically and economically in an on-site individual central station supplied with natural gas or fuel oil.

GÉNÉRALITÉS

Énergie Totale, c'est d'abord la traduction en français de Total Energy, expression américaine qui désigne l'ensemble des techniques de récupération et d'utilisation de la chaleur perdue des moteurs thermiques.

L'idée de récupérer utilement la chaleur perdue des moteurs thermiques n'est pas récente : de nombreuses installations industrielles, anciennes de plusieurs dizaines d'années, utilisent :
— l'eau chaude de refroidissement des moteurs pour le chauffage;
— la chaleur des gaz chauds d'échappement pour produire de la vapeur basse pression.

Toutefois, l'industrialisation de cette idée remonte seulement à quelques années et coïncide avec l'apparition sur le marché américain de nouvelles gammes de moteurs à gaz, suralimentés, à haut rendement, à refroidissement par ébullition et capables d'assurer un service continu de plusieurs dizaines de milliers d'heures sans surveillance permanente et sans défaillance.

De tels moteurs utilisés pour entraîner, par exemple, des alternateurs et associés à des machines frigorifiques à absorption et à des échangeurs, sont susceptibles de réaliser de véritables centrales automatiques de production d'énergie électrique, de chauffage et de conditionnement d'air. En fait, les deux concepts : conditionnement d'air et Énergie Totale, sont étroitement liés, et il est démontré qu'au cours des dix dernières années le développement rapide du conditionnement d'air a largement facilité le premier essor de l'Énergie Totale. De même, il est aujourd'hui facile de prévoir que dans les prochaines années l'Énergie Totale permettra à son tour un nouvel essor du conditionnement d'air.

Ces quelques remarques nous permettent de donner deux définitions de l'Énergie Totale :

— Théoriquement, dans un sens large, c'est une Centrale thermique autonome, fonctionnant au gaz naturel ou au fuel, capable d'alimenter totalement en énergie un Centre limité;
— Pratiquement, dans le cas le plus usuel, l'Énergie Totale c'est l'énergie électrique, le chauffage et le conditionnement d'air d'un ensemble limité, produits par une Centrale thermique autonome et automatique alimentée de préférence au gaz naturel.

Pour qu'une telle solution ait une chance de s'imposer sur le marché face à la solution traditionnelle, elle doit posséder au moins trois qualités essentielles :
— La FIABILITÉ qui doit être au moins égale à celle de la solution traditionnelle constituée par le réseau électrique général et un groupe électrogène de secours. Cette condition est remplie :
  – par l'emploi de matériel de très haute qualité;
  – par la réalisation d'un programme de maintenance rigoureux basé sur les interventions préventives et seul capable de permettre une marche ininterrompue et sûre pendant 12 ans;
  – par l'installation d'un groupe électrogène supplémentaire maintenu en permanence en réserve et qui permet d'assurer la maintenance dans de bonnes conditions.
— La RENTABILITÉ qui doit être au moins égale à celle de la solution traditionnelle. Cette condition est remplie :
  – par l'emploi de matériel sélectionné de très haut rendement;
  – par l'optimisation, dans chaque cas particulier, des conditions d'exploitation.
— Les PERFORMANCES de l'installation qui doivent être au moins égales à celles de la solution traditionnelle. Cette condition est remplie, comme dans le cas de la solution traditionnelle, par l'emploi de systèmes de régulation et d'insonorisation appropriés.

En résumé, pour espérer atteindre la dimension industrielle, l'Énergie Totale doit apporter à l'utilisateur tout ce que lui apporte déjà la solution traditionnelle avec en plus « quelque chose ».
Ce « quelque chose », nous allons maintenant le montrer : c'est un PROFIT supplémentaire.

2. Principe de fonctionnement et réalisation

Dans le cas le plus général, une Centrale de production d'Énergie Totale comprend cinq parties principales :

1. Un ensemble de groupes électrogènes à commande automatique et à refroidissement par ébullition, produisent simultanément l'énergie électrique au réseau électrique d'utilisation et la vapeur basse pression à 115-125 °C au réseau d'utilisation de vapeur.

2. Un échangeur de chauffage alimenté au primaire par le circuit de vapeur fournit au secondaire l'eau chaude du circuit de chauffage.

3. Une machine frigorifique à absorption alimentée au primaire par le circuit de vapeur assure la production d'eau glacée nécessaire à l'alimentation du circuit de conditionnement d'air.

4. Un générateur de vapeur d'appoint assure la régulation de la production de vapeur aux faibles charges des groupes électrogènes.

5. Un échangeur d'équilibre associé à un réfrigérant atmosphérique évacue à l'atmosphère la chaleur excédentaire du circuit de vapeur.

Dans le cas général, cet ensemble d'équipements sera commandé automatiquement par une armoire centrale et contrôlé en permanence par un système complet d'alarme-préalerte; de plus l'installation sera réalisée dans un ou plusieurs locaux spécialement aménagés, ventilés et insonorisés.

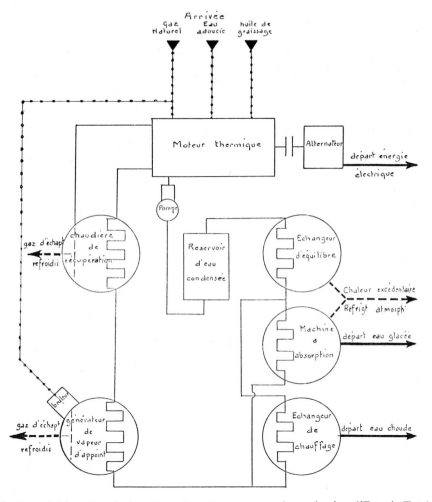

Fig. 1 — Schéma de principe d'une Centrale autonome de production d'Energie Totale.

L'installation traditionnelle équivalente comprendrait elle aussi cinq parties principales :

1. Un ensemble de transformateurs alimentant en énergie électrique le circuit d'utilisation.

2. Un ensemble de chaudières alimentant en eau chaude le circuit de chauffage.

3. Un ensemble de groupes électrocompresseurs frigorifiques alimentant en eau glacée, le circuit de conditionnement d'air.

581

4. Un groupe électrogène de secours chargé d'alimenter le circuit électrique prioritaire en cas de défaillance du réseau général.

5. Un réfrigérant atmosphérique évacuant à l'atmosphère la chaleur excédentaire de l'installation.

Comme précédemment, cet ensemble d'équipements est installé dans un ou plusieurs locaux convenablement aménagés, insonorisés et ventilés.

Comment fonctionne une Centrale de production d'énergie totale…? Le principe en est simple (voir fig. 1) :

— Les groupes électrogènes, alimentés de préférence au gaz naturel et commandés par une horloge électrique et des relais de mesure wattmétriques, fournissent continuellement, automatiquement, et avec le rendement optimum, l'énergie électrique au réseau d'utilisation du Centre.

— La chaleur de refroidissement des moteurs thermiques est récupérée dans des chaudières de récupération spéciales, sous forme de vapeur basse pression à 115-125°C. Cette vapeur basse pression alimente, suivant les conditions climatiques :

    – soit le circuit de chauffage par l'intermédiaire d'un échangeur vapeur-eau;
    – soit le circuit d'eau glacée, par l'intermédiaire d'une machine frigorifique à absorption.

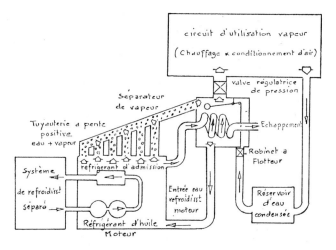

Fig. 2 — Schéma de principe d'un moteur à piston à refroidissement par ébullition (Système Caterpillar).

Lorsque la quantité de vapeur produite par les moteurs thermiques est supérieure à la consommation du réseau d'utilisation, la chaleur excédentaire est évacuée à l'atmosphère par l'intermédiaire d'un échangeur d'équilibre.

Lorsque la quantité de vapeur produite par les moteurs thermiques est inférieure à la consommation du réseau d'utilisation, la chaleur d'appoint est fournie par un (ou plusieurs) générateur d'appoint branché en parallèle avec les chaudières de récupération des moteurs thermiques.

Ainsi, quels que soient les besoins en énergie du Centre, la centrale de production d'énergie totale fournit une énergie régulée aux trois circuits : électrique, chauffage, et conditionnement d'air.

A l'exception des moteurs thermiques et de leur chaudière de récupération, l'ensemble du matériel utilisé dans les centrales de production d'énergie totale est du matériel standard.

L'originalité principale des moteurs thermiques réside dans leur système de refroidissement qui est du type par ébullition et dans leur très grande durée de vie.

Pour un moteur à gaz naturel de ce type, la consommation moyenne à charge normale est de 2,8 Th (pci)/kW.h, la chaleur récupérée d'environ 1,25 Th/kW.h et le rendement moyen d'une installation complète d'environ 77%.

3. ÉCONOMIE

Pour tenter de définir le champ d'applications et les avantages des Centrales Autonomes Automatiques de Production d'Énergie Totale, dans le cadre du marché français facilement extrapolables au cadre du marché européen, nous avons choisi d'analyser deux cas concrets, extraits des projets réels étudiés au cours des derniers mois.

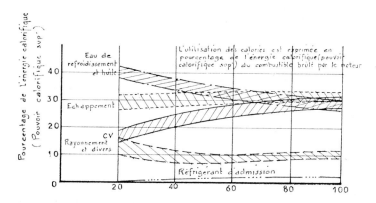

Fig. 3 — Pourcentage de la puissance du moteur en service continu.

Les deux exemples choisis sont deux supermarchés ayant respectivement des superficies de vente de $2\,000$ m^2 (supermarché A) et $20\,000$ m^2 (supermarché B).

Pour mettre nettement en évidence l'incidence du conditionnement d'air sur la rentabilité de la centrale de production d'énergie totale, nous avons étudié l'installation dans le cadre des 4 hypothèses suivantes :

— avec et sans conditionnement d'air;

— alimentation par solution traditionnelle et par système à Energie Totale;

— pour plus de clarté nous avons également supposé que les conditions d'aménagement, d'insonorisation et de ventilation du local technique étaient les mêmes dans tous les cas;

— l'analyse du tableau ci-dessous appelle les commentaires suivants :

   – dans tous les cas le bénéfice global supplémentaire obtenu avec le système Énergie Totale sur une période de 12 ans est positif;

   – pour calculer le coût d'exploitation annuel, nous avons supposé constants sur toute la période de 12 ans :

      i) d'une part la consommation d'énergie du supermarché;

      ii) d'autre part le coût du gaz naturel et de l'énergie électrique.

Tableau Comparatif

Système à énergie totale — Solution traditionnelle

		Supermarché A (2000 m²)				Supermarché B (20 000 m²)			
		Solution traditionnelle		Système énergie totale		Solution traditionnelle		Système énergie totale	
		Avec Cdt d'air	Sans Cdt d'air	Avec Cdt d'air	Sans Cdt d'air	Avec Cdt d'air	Sans Cdt d'air	Avec Cdt d'air	Sans Cdt d'air
I — Puissances installées :									
Électrique	kW	300	220	300	300	3 000	2 200	2 500	2 500
Chauffage	Th/h	600	600	600	600	6 000	6 000	6 000	6 000
Conditionnement d'air	$10^3 \times$ Fg/h	240	—	240	—	2 400	—	2 400	—
II — Consommations annuelles d'énergie									
Électrique	$10^3 \times$ kW.h	600	600	600	600	6 000	6 000	6 000	6 000
Chauffage	$10^3 \times$ Th	900	900	900	900	9 000	9 000	9 000	9 000
Conditionnement d'air	$10^3 \times$ Fg	240	—	240	—	2 400	—	2 400	—
III — Coût annuel d'exploitation									
Énergie électrique	$10^3 \times$ FF	72	72	25,62	25,62	720	720	256,20	256,20
Chauffage	$10^3 \times$ FF	15,25	15,25	4,57	4,57	152,25	152,25	45,70	45,70
Conditionnement d'air	$10^3 \times$ FF	9,60	—	1,35	—	96	—	13,50	—
Maintenance	$10^3 \times$ FF	30	19	63	50	120	62	260	198
Frais financiers	$10^3 \times$ FF	21,40	15,60	30,80	22,90	101,75	48,95	178,20	139,15
Coût total	$10^3 \times$ FF	148,30	121,82	126,34	103,09	1 190,00	983,20	753,60	639,05
IV — Investissements et bénéfices									
— Investissement initial	$10^3 \times$ FF	390	284	560	417	1 850	890	3 240	2 530
— Investissement suppl.	$10^3 \times$ FF	—	—	170	133	—	—	1 390	1 640
— Coût d'exploitation annuel supplémentaire	$10^3 \times$ FF	21,960	18,730	—	—	436,40	344,15	—	—
— Période d'amortissement de l'invest¹. suppl.	ANS	—	—	7,8	7,1	—	—	3,2	4,7
— Bénéfice global suppl. sur la période d'amort¹. de 12 ans	$10^3 \times$ FF	—	—	93,52	91,76	—	—	384,68	2 489,80

Nota : Prix du gaz naturel : 1,525 10^{-2} F/Th (p. c. i.)   Durée d'exploitation envisagée : 12 ans
Prix de l'énergie électrique (réseau E.D.F.) −0,12 F/kW. h   Frais financiers sur investissements : 10 %.

Pour tenir compte des variations imprévisibles de ces grandeurs et garantir à coup sûr à l'utilisateur un bénéfice global positif, nous appliquons la règle empirique suivante :

— une installation d'Énergie Totale est « faisable » lorsque le bénéfice supplémentaire global calculé sur la période d'amortissement donnée est au moins égal à l'investissement initial.

En rapprochant cette règle empirique des deux tableaux ci-dessus, nous arrivons aux conclusions suivantes :

— avec ou sans conditionnement d'air, il n'est pas rentable d'équiper le supermarché de 2 000 m^2 d'une centrale autonome de production d'énergie totale;

— avec conditionnement d'air, il est nettement profitable d'équiper le supermarché de 20 000 m^2 d'une centrale autonome de production d'énergie totale;

— sans conditionnement 'd'air, il n'est pas certain que l'équipement en Énergie Totale du supermarché de 20 000 m^2 s'avère être une opération profitable.

Cet exemple montre clairement le rôle essentiel joué par le conditionnement d'air dans le développement des centrales autonomes de puissance moyenne (1000 à 10 000 kW) :

— Le conditionnement d'air est l'élément qui, en assurant une charge permanente du circuit de production de vapeur, rend définitivement rentable, dans la plupart des cas, l'emploi d'une centrale automatique autonome de production d'énergie totale, pour l'alimentation d'un centre.

CONCLUSION

L'Énergie Totale, ce n'est ni la panacée universelle, ni une idée théorique au succès éphémère mais une immense possibilité de profit supplémentaire offerte, grâce à l'emploi des moteurs thermiques modernes, à un grand nombre d'entreprises des secteurs industriel et commercial de toutes tailles ayant des besoins suffisants en énergies mécanique et thermique.

Pour réussir dans cette activité, le candidat installateur doit posséder au moins trois atouts essentiels :

— Un bureau d'études compétent;

— Une gamme d'équipements de hautes performances et très grande fiabilité;

— Un service après vente organisé.

Le domaine d'application de l'Énergie Totale ce n'est pas seulement les supermarchés et les centres commerciaux, mais c'est également les immeubles d'appartements, les immeubles de bureaux, les hôpitaux, les écoles, les hôtels et motels, les usines et ensembles industriels, les patinoires et les piscines, etc. A priori rien ne permet d'affirmer que telle ou telle installation peut ou non être équipée d'une centrale autonome de production d'énergie totale, seul un calcul comparatif détaillé de l'investissement et du coût d'exploitation permet d'éclairer l'utilisateur éventuel.

DISCUSSION

M. A. ENGELS, (The Netherlands) — Is it possible to get saturated steam at 125°C from a gasmotor-cooling system, in regard to the construction of cooling jacket that does not permit high pressures?

L. LEFAURE — Qui. Ces moteurs à gaz (Caterpillar) sont conçus de façon à pouvoir supporter en service continu la pression de 1 bar correspondant à la vapeur saturée à 125°C.

EVOLUTION OF CENTRAL PLANT
TOWARD A HEAT RECLAIM LOOP

R.T. TAMBLYN

R. T. Tamblyn and Partners Ltd
Toronto (Canada)

Récupération de chaleur à partir d'un circuit centralisé de distribution

RÉSUMÉ : *Les installations centralisées de distribution de chaleur et d'eau froide connaissent une réelle vogue pour les ensembles de bâtiments, parce qu'elles permettent de réduire la pollution de l'air et qu'elles sont plus économiques que les équipements individuels.*

Ce rapport présente les avantages économiques à attendre des distributions centralisées, lorsque de la chaleur perdue par certains circuits de conditionnement d'air peut être réutilisée pour d'autres; pour des bâtiments commerciaux modernes, la production globale de chaleur peut être réduite de plus de 50%

Paramètres de conception :

1. La distribution d'eau chaude doit permettre d'utiliser la chaleur perdue aux condenseurs des installations frigorifiques; l'eau chaude en fin de circuit doit toujours être maintenue en dessous de 38°C (100°F); 95% du chauffage des bâtiments commerciaux peut être ainsi produit.

2. La condensation de l'eau, à des températures comprises entre 32°C (90°F) et 60°C (140°F), exige des tuyauteries plus importantes pour le chauffage que lorsque l'on utilise la vapeur. Mais on peut employer des tuyaux non métalliques, sans joints de dilatation et sans protection cathodique.

Avantages : 1. Diminution de la consommation totale du fuel; et par conséquent; 2. Réduction de la pollution de l'air; 3. Diminution et quelquefois élimination des traversées de parois par des conduites.

1. PURPOSE OF PAPER

1. To examine and compare the operating cost of several types of prime movers for water chilling equipment.
2. To demonstrate the savings in annual fuel cost where a Central Plant uses a heat reclaim technique to supplement the use of generated heat.
3. To consider the best method of distributing heat from a Central Plant where some or all of the heat is reclaimed from its customers.

2. BACKGROUND

Prior to 1950 our buildings were erected as single units with a simple steam heating plant in each one. Occasionally a governmental type Owner phased the construction of a series of buildings in one location and arranged a central steam heating plant among the buildings. This provided the economy of one single plant and one set of operators. Now and then such plants found it worthwhile to expand so that they could deliver steam to neighbouring buildings at a profit.

Today the parameters for Central Plant have undergone changes and the rationale is much more convincing. Look at the new factors which affect commercial buildings.

1) Construction density in the downtown core of cities has more than doubled in the past 40 years.
2) Buildings are now being erected in *groups* for commercial complexes, governmental complexes and college campuses.
3) Mechanical refrigeration has now become a necessity for all buildings as ceilings have been lowered and internal heat gains have risen from 10 BTU/sq.ft. to, in some cases, 30 BTU/sq.ft.

Col. 1	Col. 2	Col. 3	Col. 4	Col. 5
Type of Prime Mover	Type of Fuel	Unit Cost of Fuel	Gross BTU Input to Plant to Produce One Daily Ton of Refrigeration (0.95 BHP/Ton)	Fuel Cost per BTU ¢
Electric Motor	Electricity	2.5 ¢/kWh	65 000	.00073
Steam Absorption @ 13 psi Inlet	Bunker Oil	8.4 ¢/Imp. Gal.	600 000	.000058
Condensing Steam Turbine @ 275 psi Inlet	Bunker Oil	8.4 ¢/Imp. Gal.	545 000	.000058
Duplex System with Back Pressure Steam Turbine and Absorption	Bunker Oil	8.4 ¢/Imp. Gal.	525 000	.000058
Non Regenerative Natural Gas Turbine	Natural Gas Diesel Oil Standby	65 ¢/MCF 14 ¢/Imp. Gal.	293 000	.000083
Duplex System Gas Turbine w/Steam Turbine driven by Waste Heat	Natural Gas Diesel Oil Standby	65 ¢/MCF 14 ¢/Imp. Gal.	212 000	.000083
Natural Gas Engine	Natural Gas Diesel Oil Standby	65 ¢/MCF 14 ¢/Imp. Gal.	165 000	.000083
Duplex System Gas Engine w/Absorption Chiller driven by Waste Heat	Natural Gas Diesel Oil Standby	65 ¢/MCF 14 ¢/Imp. Gal.	145 000	.000083

4) Heating has become less important as the need for winter humidification has forced better insulation in walls, as floor plans have grown larger, and as internal heat gains have risen to match ventilation and fabric losses. Maximum heating for most buildings is down from 40 to 25 BTU/sq.ft. in locations with 0 °F design temperature.

5) Air Pollution Authorities are stressing the need for fewer and more expensive chimneys built high enough to dissipate pollutants above the living zone.

6) Cooling diversity among buildings permits central water chillers to supply maximum instantaneous cooling requirement with 70% of the capacity required for on-site units.

7) Larger water chillers which can be used in central plant are significantly cheaper in first cost and space requirement than on-site machinery.

8) Large water chillers permit the selection of more sophisticated prime movers which produce the same chilling effect as electric on-site drives for a small fraction of the energy cost.

9) The cost of labour for maintenance and operation is reduced.

and most importantly :

10) The waste heat available in certain types of buildings can be made available to other types of buildings on the same loop which are deficient in heat. Such

Col. 6	Col. 7	Col. 8	Col. 9	Col. 10
Fuel Cost per Daily Ton ¢	Fuel Cost/Yr/Ton based on 1000 EFL hrs or 41.7 Daily Tons/Ton $ (Col. 6 × 41.7)	Installed Cost of Prime Mover per Ton of Capacity $	Annual Amortization of Prime Cost per Ton (Col. 8 × 0.10) $	Annual Owning Cost/Ton (Col. 7 + Col. 9) $
47.5	19.80	25	2.50	22.30
34.8	14.50	27	2.70	17.20
31.6	13.20	45	4.50	17.70
30.5	12.75	33	3.30	16.05
24.3	10.15	165	16.50	26.65
17.6	7.35	150	15.00	22.35
13.7	5.72	95	9.50	15.22
12.0	5.00	85	8.50	13.50

reclaimed heat can be made available more cheaply than generated heat. Thus, economies can be enjoyed by both the Plant and its customers.

3. CHOICE OF PRIME MOVER FOR WATER CHILLERS

For on-site installations it is seldom feasible to extend the economic analysis beyond the simple choice of electric driven centrifugal and steam or hot water driven absorption assemblies. The unit cost of other prime movers is usually too high to be considered for small plants.

For central water chilling installations, the smallest capacity module can be in excess of 1,000 tons and the largest module in the range of 5,000 to 10,000. At this level the unit cost of a wider selection of prime movers has narrowed and a careful analysis is necessary to determine the best single drive or mix of drives to be used.

Table 1 outlines the main elements of selection which might be considered if approaching this analysis. The figures of prime and operating cost are pertinent for Southern Ontario at this time in the range of 1,000 to 5,000 HP.

The owning costs indicated in col. 10 do not include other statistics such as maintenance and equipment space, which might affect the choice in some situations.

589

Since the natural gas engine drive shows up well thus far, it will be used as a basis for further discussion of a heat reclaim cycle.

Figure 1 illustrates the distribution of heat from a typical duplex assembly delivering one daily ton of refrigeration from a gas engine driven centrifugal and an absorption chiller.

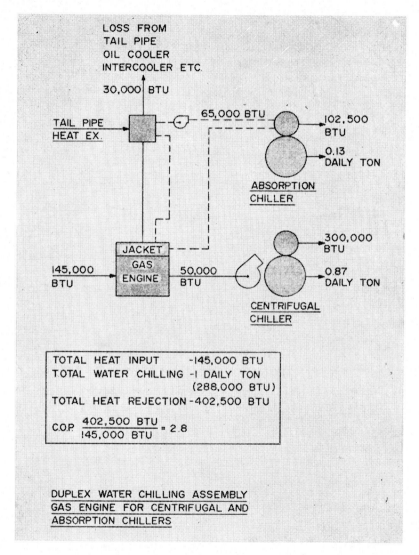

Fig. 1

4. POTENTIAL FOR HEAT RECLAIM ON CENTRAL PLANT

Condenser heat obtained from the process of mechanical cooling is normally wasted. Where this low temperature heat can be reused, it may be provided more cheaply than if the same heat were generated.

Central cooling plant will have an abundance of free condenser heat available at temperatures above 45°F when mechanical cooling is required. Further mechanical cooling below this temperature is normally required to the level of 5% of the maximum summer demand in order to service computer areas and other systems not able to use free cooling with outdoor air.

Unfortunately, the greatest availability of free condenser heat does not coincide with the greatest requirement for heating. But even at maximum heating design there will be interior air handling systems which require cooling beyond their ventilation needs. It may be shown that the substitution of mechanical cooling for free cooling in these systems yields condenser heat more cheaply than the same heat generated in a boiler.

i) From figure 1 – 145,000 BTU input reclaims 402,500 BTU;

ii) Cost of 145,000 BTU @ 65¢/1,000,000 BTU cost of natural gas
= $(145,000/1,000,000) \times 65$ or 9.43 ¢;

iii) Thus 402,500 BTU can be reclaimed for fuel cost of 9.43 ¢;

iv) Cost/million BTU for reclaimed heat is $(1,000,000/402,500) \times 9.43 = 23.4$ ¢;

v) Using natural gas @ 80% efficiency, cost/million BTU generated heat is $(1,000,000/800,000) \times 65$ or 81.3 ¢;

vi) Using bunker oil @ 80% efficiency @ 8.4 ¢/Imp. Gal., cost/million BTU generated heat is $(1,000,000/182,000 \times .8) \times 8.4$ or 57.8 ¢;

vii) Summary of cost/million BTU

Reclaimed Heat	23.4 ¢
Oil Generated Heat	57.8 ¢
Gas Generated Heat	81.3 ¢

Thus it becomes feasible to minimize the use of "free" cooling in buildings and substitute mechanical cooling to the extent that condenser heat can be used by the system.

Table 2

OCCUPIED PERIOD (360 HRS/MONTH)

Building type	Gross area	Loss/gain	BTU/hr/sq.ft. @ −10°F Ambient			
			Perimeter	Interior	Parking or other	Average
Office Buildings (Avge 350 000 sq.ft.)	4 000 000	Loss	57	17	49	33
		Gain	22	19	3	17
Shopping Centres	1 000 000	Loss	30	20	30	16
		Gain	21	21	—	12
Dept. Stores (Avge 25 000 sq.ft.)	750 000	Loss	39	16	28	21
		Gain	21	21	15	19
Schools, Laboratories	750 000	Loss	69	13	23	28
		Gain	15	15	15	15
Hotels, Convention	500 000	Loss	30	—	—	30
		Gain	5	—	—	5

When the water chillers are operated to reclaim waste heat from the system, it is worth examination to see if waste heat from the prime mover can be reused along with condenser heat. The hardware for this purpose includes double condensers, waste heat boilers and other forms of heat exchanger.

Any analysis of waste heat which may be recaptured from building air handling systems involves the determination of two curves:

1) *Net Heating Requirement*. The heating required after all normal heat gain and loss equations have taken place in the building, such as lights vs. ventilation loss and perimeter lights vs. building fabric loss.

2) *Maximum Heat Availability*. The greatest amount of heat which can be reclaimed after internal heat gain is balanced against essential heat losses.

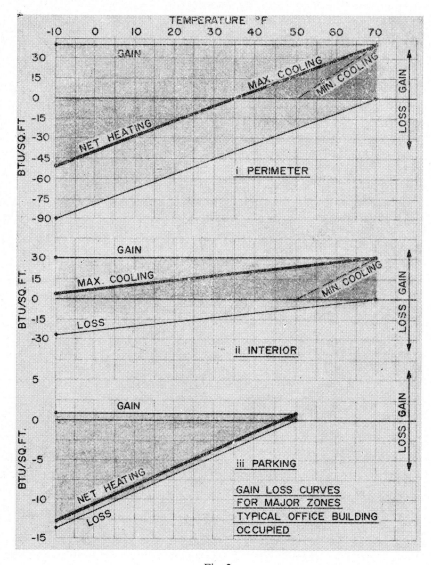

Fig. 2

These curves should be derived for each building on the system for both occupied and unoccupied periods. Care must be taken to analyze building systems separately if they have different balance points. This would be true of perimeter systems as opposed to interior systems or garage ventilation/heating systems.

Table 2 illustrates a mix of new buildings of typical construction which could be contemplated for a modern central plant. If this complex is analyzed for heat reclaim possibilities, the following steps should be taken:

1) Develop T line graphs for each building on the system. Graphs should be done separately for individual systems having different balance points. See figure 2. Determination should also be made for unoccupied periods when most lights are off, ventilation off and some air handling systems shut down.

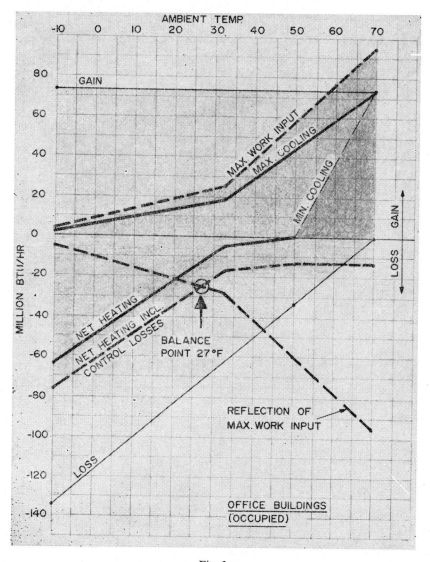

Fig. 3

Table 3

SUMMARY OF MAXIMUM AVAILABLE HEAT (MAH) AND NET HEATING REQUIREMENT (NHR)
FOR EACH BUILDING (OCCUPIED PERIOD)
MILLIONS BTU/HRS

Outdoor Temp. °F	-10		0		10		20		30		40		50		60		70		80	
	MAH	NHR	MAH	NHR	MAH	NHR	MAH	NHR	MAH	NHR	MAH	NHR	MAH	NHR	MAH	NHR	MAH	NHR	MAH	NHR
Office	7	77	12	63	16	48	20	34	25	20	40	15	57	14	75	14	93	14		4
Shopping Centres	6	9	7	8	8	6	9	5	10	4	11	3	12	2	13	—	15	—		
Dept. Stores	2	4	3	3	4	2	5	1	8	—	11	—	13	—	15	—	17	—		
Hotels	—	12	—	11	—	9	—	7	—	5	—	3	—	1	1	—	3	—		
Schools	1	10	1	9	2	7	2	6	4	4	6	2	8	—	10	—	14	—		
SYSTEM TOTAL	16	112	23	94	30	72	36	53	47	33	68	23	90	17	114	14	142	14		4

594

2) Combine graphs for each building to determine the maximum cooling or heat availability vs. net heating requirement. See figure 3. Repeat for unoccupied period.
3) Tabulate values for maximum cooling and net heat requirement for each building, for each 10°F band and for occupied and unoccupied periods. See Table 3. In each case the heat of compression and other waste heat available from the prime mover should be added to the maximum cooling values in order to find maximum heat reclaim.
4) Plot maximum heat reclaim curve against net heating requirement to find the balance point of the whole system. See figure 4. For this typical system, the balance points are at 25°F for occupied periods and 37°F for unoccupied periods.
5) Tabulate heating requirement for each 10°F band for reclaimed and for generated heat. Multiply this requirement by the average hours in each degree band as

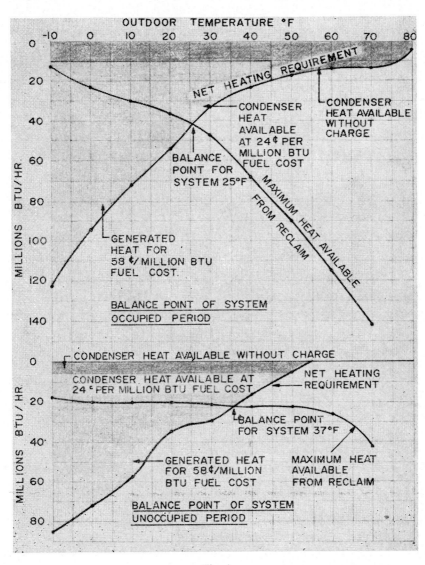

Fig. 4

determined from local weather information and find the total annual reclaimed and total generated heat required by the system. Repeat for unoccupied periods. See table 4.

6) Find occupied requirement, unoccupied requirement and total annual requirement for reclaimed and generated heat.

HEAT REQUIRED

Period	By Reclaim Million BTU	% Total	By Generation Million BTU	% Total	Total
Occupied	84 400	71.4	33 800	28.6	118 200
Unoccupied	50 500	51.0	48 600	49.0	99 100
Total	134 900	62.0	82 400	38.0	217 300

Table 4

SUMMARY OF ANNUAL HEATING REQUIREMENT FOR SYSTEM OCCUPIED PERIOD

Degree band °F	hrs/Yr	Heating million BTU/hr			Annual requirement million BTU		
		By reclaim	By generation	Total	By reclaim	By generation	Total
90	68						
80	370	4	—	4	1 500	—	1 500
70	730	14	—	14	10 200	—	10 200
60	600	14	—	14	8 400	—	8 400
50	500	17	—	17	8 500	—	8 500
40	570	23	—	23	13 100	—	13 100
30	525	33	—	33	17 300	—	17 300
20	375	36	18	54	13 500	6 800	20 300
10	250	30	42	72	7 500	10 500	18 000
0	150	23	71	94	3 500	10 600	14 100
−10	55	16	107	123	900	5 900	6 800
Total					84 400	33 800	118 200

Note that 70% of necessary heat can be reclaimed from waste heat sources during occupied periods and 50% during unoccupied periods for an average reclaim of 60% of annual requirement.

7) Determine annual saving of reclaimed heat over generated heat as shown below.

Type Plant	By Reclaim @ 23.4 ¢/million BTU $	By Generation @ 57.8 ¢/million BTU $	Total $
Reclaim Plant	16 000	47 600	63 600
Conventional Plant	—	125 800	125 800

8) Determine the additional cost of system to furnish reclaim hardware, e.g.:

   i) Oversizing of heating coils in buildings so that 110°F water will suffice for heating requirements at 25°F, and/or 150°F water will suffice for heating requirement at −10°F.

ii) Double condensers and heat exchangers in central plant necessary to salvage all useful heat which can be reused economically.

iii) Additional control necessary to link central plant with each building so that ventilation schedules in the buildings may be remotely controlled from the plant to schedule mechanical cooling to the extent that condenser heat can be used.

9) Find in most cases that the additional cost of this investment may be returned in 5 years with savings in the cost of fuel.

5. CHOICE OF CIRCULATION MEDIUM FOR DISTRIBUTION OF HEAT

The conventional medium for the distribution of heat in commercial buildings today is water in the range of 180°F for central heating coils and water in the range of 90-120°F for perimeter coils such as those found in induction units.

In single buildings which use a heat reclamation technique, heating coils are oversized to perform normal heating duty with water not exceeding 110°F. This permits the acceptance of condenser heat within the temperature range of conventional water chillers.

Where heat reclamation is to be used on a larger scale, as *among* buildings, the medium of distribution must be water, and the portion of the water range between 90 and 110°F must be served with enough water flow that maximum useful reclaimed heat can be redistributed. For ultimate heating requirement, the range can be extended, say up to 150°F, with boiler generated heat providing that there is enough flow to make use of the condenser heat at the lower end of the range.

Flow rate for the heating complex illustrated would need to be 4,000 U.S. GPM in order to absorb the maximum useful reclaimed heat of 40,000,000 BTU/hr between 90°F and 110°F. This flow rate would be maintained below the balance point of 25°F and the range would be extended to 150°F to permit the ultimate heating requirement of 120,000,000 BTU/hr to be achieved at $-10°F$.

Pipe sizing for 5,000 feet of main might be 14″ diameter. On this sizing the pumping energy for an average flow rate of 3,000 GPM against friction head and 15 psi differential would be about 110 HP. These requirements could be compared with 12″ pipe sizing for steam and 6″ for condensate together with an average condensate pumping requirement of 40 HP for half load at 175 psi generation.

While pumping requirements for steam are less than for water at a 60°F range, the heat losses of the system are decidedly in favour of water. Feedwater heating and main losses for steam may expect to total 10% of the load while losses for warm water can be limited to 5%.

The overall consideration of steam vs. hot water should show an economic advantage for water.

A further consideration for water, in systems of less than 200 psi working pressure, is that it may be circulated in buried insulated asbestos pipe with the following advantages over steel pipe which must be used for steam:

1) Freedom from corrosion on both inside and outside of piping.
2) Elimination of tunnelling cost often required to guarantee the inspection of steel piping for corrosion failure.
3) Elimination of expansion joints and steam traps.

VI. FINALLY

There is an interesting rationale for the combination of reclaimed and generated heat in the type of central system described above. Features which will make the plant attractive to future users include:

1) The possibility of selling heat at lower cost through reclamation of 60% of the total requirement from waste heat sources and reducing the annual cost of fuel by 50%.
2) Elimination of steel piping with consequent cost of tunnel and/or corrosion problems.
3) Elimination of more complex steam auxiliaries, water treatment, trap maintenance, high heat losses, etc.
4) Elimination of municipal insistence upon registered plant operators and constant equipment surveillance as required for steam in Ontario.

DISCUSSION

R. LANDSBERG (Israel) — 1. The term "Daily Ton" in figure 1 does not seem to be well defined. 1 Ton = 288,000 Btu *per 24 hours*, this disagrees with "Daily".

2. Figure 2 shows the ordinate "Btu/sq. ft." which should probably be Btu/(hr). (sq. ft.).

R.T. TAMBLYN — 1. A "Daily ton" is 288,000 Btu of cooling.
This differs from a "ton of refrigeration" which is a *rate of cooling* equal to 288,000 Btu/day.

2. Yes, correct!

A.W. BOEKE (The Netherlands) — How do you take into account the difference between clear days and overcast days when establishing the balance gain-loss curves?

R.T. TAMBLYN — An amount for average solar gain is added to the internal heat gain curve. This usually is shown as a smaller value at 0°F than 70°F to represent the lesser hours of sunlight in winter.

INSTITUT INTERNATIONAL DU FROID
INTERNATIONAL INSTITUTE OF REFRIGERATION
177, boulevard Malesherbes, PARIS (17e)

PUBLICATIONS

1. BULLETIN

6 fascicules par an (en français et en anglais). Chaque numéro (environ 250 pages) comporte une revue analytique, des études originales, des informations sur les recherches dans le domaine du froid, des renseignements sur les congrès et réunions diverses, une bibliographie.

6 issues per year (in French and English). Each issue (about 250 pages) includes: analytical reviews, original studies, information on current research in refrigeration, information on congresses and various meetings, book reviews.

Abonnement — *Subscription rate :* 80 F - $ 16.00.

2. OUVRAGES DE BASE — BASIC PUBLICATIONS

— 1er Guide Bibliographique du Froid 1953-1960 — *1st Bibliographic Guide to Refrigeration 1953-1960.* 10.000 ref. (100 F - $ 20.00).

— 2e Guide Bibliographique du Froid 1961-1964 — *2nd Bibliographic Guide to Refrigeration 1961-1964.* 6.000 ref. (100 F - $ 20.00).

— 3e Guide Bibliographique du Froid 1965-1968 — *3rd Bibliographic Guide to Refrigeration 1965-1968.* 7.000 ref. (150 F - $ 33.00).

— Dictionnaire International du Froid (anglais, français, allemand, russe, espagnol, italien). 1600 expressions, dont 400 avec définition — *International Dictionary of Refrigeration (English, French, German, Russian, Spanish, Italian). 1,600 expressions, 400 of which are defined.* 300 p. (60 F - $ 12.00).

— Conditions recommandées pour l'entreposage frigorifique des produits périssables — *Recommended conditions for cold storage of perishable produce.* 1967. 100 p. (12 F - $ 2.50).

— Conditions recommandées pour le transport terrestre des denrées périssables — *Recommended conditions for land transport of perishable foodstuffs.* 1963. 11 p. (5 F - $ 1.00).

— Recommandations de l'I.I.F. pour la préparation et la distribution des aliments congelés — *I.I.R. recommendations for the processing and handling of frozen foods.* 1964. 123 p. (15 F - $ 3.00).

— Projet de code de pratiques pour le poisson congelé — *Draft code of practice for frozen fish.* 1969. 74 p. (7 F - $ 1.80).

— Guide pratique de l'entreposage frigorifique — *Practical guide to refrigerated storage.* 1965. 240 p. (40 F - $ 8.00).

— Essais d'isolants thermiques — *Thermal insulating measurements.* 1968. 33 p. (8 F - $ 1.60).

— Les techniques frigorifiques dans les pays chauds en voie de développement — *Refrigeration techniques in developing countries.* 1964. 116 p. (12 F - $ 2.50).

— Règles pour machines frigorifiques. 1960. Traduction en français de la 5e édition de « Kältemaschinen Regeln ». 152 p., diag. h.t. (35 F - $ 7.00).

3. CONGRÈS INTERNATIONAUX — INTERNATIONAL CONGRESSES

— Comptes rendus du 10e Congrès, Copenhague, 1959. Publication en 3 volumes des travaux des différentes Commissions — *Proceedings of the 10th Congress, Copenhagen, 1959. Three volume publication on the work of Commissions.* 1680 p. (1 vol. 100 F - $ 20.00; 3 vol. 275 F - $ 55.00).

— Comptes rendus du 11ᵉ Congrès, Munich, 1963. Publication en 3 volumes des travaux des différentes Commissions — *Proceedings of the 11th Congress, Munich, 1963. Three volume publication on the work of Commissions.* 2000 p. (3 vol. 260 F - $ 52.00).

— Comptes rendus du 12ᵉ Congrès, Madrid, 1967. Publication en 4 volumes des travaux des différentes Commissions — *Proceedings of the 12th Congress, Madrid, 1967. Four volume publication on the work of Commissions.* 3700 p. (1 vol. 90 F - $ 18.00; 4 vol. 280 F - $ 56.00).

4. COMPTES RENDUS DES RÉUNIONS DES COMMISSIONS SCIENTI-FIQUES ET TECHNIQUES — PROCEEDINGS OF THE MEETINGS OF SCIENTIFIC AND TECHNICAL COMMISSIONS

Commission I. Cryophysique et cryogénie — *Cryophysics and cryoengineering.*

— (1960-1). Méthodes de réfrigération, de liquéfaction et de séparation aux tempé-ratures basses et très basses. Propriétés thermodynamiques des fluides, des mélanges et des solides aux basses températures. Applications des basses températures. — *Refrigeration, liquefaction and separation processes at low and very low temperatures. Thermodynamical properties of fluids, mixtures and solids at low temperatures. Applications of low temperatures.* Eindhoven (Pays-Bas - The Netherlands). 1960. 298 p. (30 F - $ 6.00).

— (1961-5). Méthodes de production à grande échelle de gaz liquéfiés, conservation, transport. Méthodes pour la production et la mesure des températures inférieures à 1 K. — *Methods for the large scale production of liquefied gases, their storage and transport. Methods for the production and measurement of temperatures below 1 K.* Londres - London (Royaume-Uni - United Kingdom). 1961. 196 p. (25 F - $ 5.00).

— (1965-2). Transfert de chaleur au-dessous de 100 K et ses applications techniques — *Heat flow below 100 K and its technological applications.* Grenoble (France). 1965. 365 p. (40 F - $ 8.00).

— Hydrogène liquide; propriétés; production et applications. — *Liquid hydrogen: properties; production and applications.* Grenoble (France). 1965. 395 p. (40 F - $ 8.00).

— (1966-5). Technologie de l'hélium liquide. — *Liquid helium technology.* Boulder, (U.S.A.). 1966. 555 p. (60 F - $ 12.00).

— (1969-1). Cryoélectrotechnique. — *Low temperatures and electric power.* Londres - London (Royaume-Uni - United Kingdom). 1969. 450 p. (60 F - $ 11.00).

— (1969-2). Gaz naturels liquéfiés. — *Liquefied natural gas.* Londres - London (Royaume-Uni - United Kingdom). 1969. 720 p. (42 F - $ 8.00).

Commission II. Transfert de chaleur et de masse. — *Heat and mass transfer.*

— (1962-1). Conductivité thermique. Refroidissement thermoélectrique. Compres-seurs. Conditionnement d'air. Aliments congelés. — *Heat conductivity. Ther-moelectric cooling. Compressors. Air conditioning. Frozen foods.* Washington, (U.S.A.). 1962. 685 p. (50 F - $ 10.00).

— (1964-2). Conductivité et diffusivité thermiques. Transfert de chaleur et de masse. Échangeurs. Fluides frigorigènes. — *Thermal conductivity and diffusivity. Heat and mass transfer. Heat exchangers. Refrigerants.* Turin (Italie - Italy). 1964. 190 p. (20 F - $ 4.50).

— (1966-2). Études sur le transfert de chaleur dans le domaine du froid (isolants; échangeurs de chaleur). — *Studies on heat transfer in refrigeration (insulants; heat exchangers).* Trondheim (Norvège - Norway). 1966. 320 p. (50 F - $ 10.00).

Commission III. Machines frigorifiques. — *Refrigerating machinery.*

— (1960-3). Évaporateurs. Machines à absorption et à éjecteurs de vapeur. Préem-ballage des aliments. Qualité des produits congelés. Réfrigération des fleurs et bulbes. Entrepôts frigorifiques. Portes isothermes et rideaux d'air. — *Evaporators.*

Absorption machines and steam jet systems. Prepackaging of foodstuffs. Quality of frozen foods. Cooling of flowers and bulbs. Cold stores. Insulated doors and air-curtains. Marseille (France). 1960. 555 p. (45 F - $ 9.00).

— (1962-1). Voir Commission II. — See Commission II.

— (1965-4). Matériel frigorifique dans le conditionnement d'air et l'industrie chimique; le froid en médecine et biologie. — *Refrigerating equipment in air conditioning and the chemical industry; refrigeration in medicine and biology.* Tchécoslovaquie - Czechoslovakia. 1965. 700 p. (50 F - $ 10.00).

Commission IV. Le froid appliqué aux produits périssables. — *Refrigeration of perishable produce.*

— (1960-3). Voir Commission III. — See Commission III.

— (1961-1). Entreposage frigorifique des fruits et légumes. — *Refrigerated storage of fruit and vegetables.* Wageningen (Pays-Bas - The Netherlands). 1961. 375 p. (35 F - $ 7.00).

— (1962-1) Voir Commission II. — See Commission II.

— (1962-3). Physiologie, qualité et transport des bananes. — *Physiology, quality and transportation of bananas.* Groupe de Travail - Working Party. 1962. 120 p. (15 F - $ 3.00).

— (1965-1). Viande préemballée. Œufs. Volaille. — *Prepacked meat. Eggs. Poultry.* Karlsruhe (Allemagne - Germany). 1965. 262 p. (25 F - $ 5.00).

— (1966-1). Conservation des fruits et légumes (notamment en atmosphère contrôlée). Entrepôts frigorifiques. — *Storage of fruit and vegetables (including in controlled atmosphere). Cold stores.* Bologne. (Italie - Italy) 1966. 670 p. (50 F - $ 10.00).

— (1969-6). Aliments congelés. Qualité des produits. Techniques de congélation (notamment cryogéniques). — *Frozen foods. Product quality. Freezing techniques (especially cryogenic methods).* Budapest (Hongrie - Hungary). 1969. 328 p. (60 F - $ 11).

Commission V. Entreposage frigorifique. — *Cold storage facilities.*

— (1960-3). Voir Commission III. — See Commission III.

— (1962-2). Entrepôts et transports frigorifiques. — *Cold stores and refrigerated transport.* Santiago de Compostela (Espagne - Spain). 1962. 210 p. (20 F - $ 4.00).

— (1964-1). Entrepôts frigorifiques. — *Refrigerated warehouses.* Dublin (Irlande - Ireland). 1964. 262 p. (25 F - $ 5.00).

— (1966-1). Voir Commission IV. — See Commission IV.

— (1968-1). Chambres froides à atmosphère contrôlée. Entreposage des produits surgelés. — *Controlled atmosphere cold rooms. Storage of quick frozen products.* Avignon (France). 1968. 144 p. (25 F - $ 5.00).

— (1969-6). Voir Commission IV. — See Commission IV.

Commission VI. Conditionnement d'air. — *Air conditioning.*

— (1962-1). Voir Commission II. — See Commission II.

— (1965-4). Voir Commission III. — See Commission III.

Commission VII. Transports frigorifiques terrestres. — *Refrigerated land transport.*

— (1962-2). Voir Commission V. — See Commission V.

— (1965-3). Véhicules frigorifiques : construction, isolation, moyens de refroidissement. — *Refrigerated vehicles: construction, insulation and refrigerating equipement.* Cracovie - Cracow (Pologne - Poland). 1965. 210 p. (20 F - $ 4.00).

— (1966-4). Véhicules frigorifiques : entrée de chaleur dans les véhicules isolés; économie des moyens de refroidissement. — *Refrigerated vehicles: evaluating the heat loss from insulated vehicles; economics of refrigerating systems.* Londres - London (Royaume-Uni - United Kingdom). 1966. 270 p. (30 F - $ 6.00).

— (1969-3). Véhicules et conteneurs frigorifiques. — *Refrigerated vehicles and containers.* Vienne - Vienna (Autriche - Austria). 1969. 213 p. (35 F - $ 6.50).

Commission VIII. Transports frigorifiques maritimes. — *Refrigerated sea transport.*

— (1962-3). Voir Commission IV. — See Commission IV.

— (1965-5). Installations frigorifiques marines. — *Marine refrigeration.* Suède - Sweden. 1965. 380 p. (35 F - $ 7.00).

— (1969-5). Transports frigorifiques maritimes. Questions d'actualité. — *Refrigerated sea transport. Topical problems.* Le Havre (France). 1969. 203 p. (35 F - $ 6.50).

Commission IX. Applications du froid au génie chimique, civil et industriel. — *Applications of refrigeration to chemical, civil and industrial engineering.*

— (1965-4). Voir Commission III. — See Commission III.

— (1966-3). Concentration et purification par congélation. — *Concentration and purification by freezing.* Delft (Pays-Bas - The Netherlands). 1966. 200 p. (25 F - $ 5.00).

Commission X. Cryobiologie et lyophilisation. — *Cryobiology and freeze-drying.*

— (1961-4). Progrès récents en cryobiologie. — *Advancement in cryobiology.* Belgrade (Yougoslavie - Yugoslavia). 1961. 102 p. (10 F - $ 2.00).

— (1969-4). Développements récents en lyophilisation. Tome I. Aspects thermodynamiques (transfert de chaleur et de masse) — *Recent developments in freeze-drying. Tome I. Thermodynamic aspects (heat and mass transfer).* Lausanne (Suisse - Switzerland). 1969. 209 p. (35 F - $ 6.50).

5. COLLOQUES RÉGIONAUX — REGIONAL SYMPOSIUMS

— (1964-3). Les applications du froid en pays tropical. — *Refrigeration applications in tropical countries.* Abidjan (Côte d'Ivoire - Ivory Coast). 1964. 455 p. (40 F - $ 8.00).

— (1966-6). Le froid dans les pays méditerranéens (applications agricoles et alimentaires). — *Refrigeration in Mediterranean countries (applications to foodstuffs).* Athènes - Athens (Grèce - Greece). 1966. 390 p. (40 F - $ 8.00).

(B891)   Imprimé en Belgique par Ceuterick s.a.
         Brusselse straat 153   3000-Louvain
         Adm.-dir. L. Pitsi   Bertemse baan 25   3008-Veltem-Beisem